Geotechnical Engineering Calculations and Rules of Thumb

Geotechnical Engineering Calculations and Rules of Thumb

SECOND EDITION

Ruwan Rajapakse, PE, CCM, CCE, AVS

AMSTERDAM • BOSTON • HEIDELBERG • LONDON
NEW YORK • OXFORD • PARIS • SAN DIEGO
SAN FRANCISCO • SINGAPORE • SYDNEY • TOKYO
Butterworth-Heinemann is an imprint of Elsevier

Butterworth-Heinemann is an imprint of Elsevier
The Boulevard, Langford Lane, Kidlington, Oxford OX5 1GB, UK
225 Wyman Street, Waltham, MA 02451, USA

Library of Congress Cataloging-in-Publication Data
A catalog record for this book is available from the Library of Congress

British Library Cataloguing-in-Publication Data
A catalogue record for this book is available from the British Library

ISBN: 978-0-12-804698-2

For information on all Butterworth-Heinemann publications
visit our website at http://store.elsevier.com/

Publisher: Joe Hayton
Acquisition Editor: Andre Gerhard Wolff
Editorial Project Manager: Mariana Kühl Leme
Editorial Project Manager Intern: Ana Claudia A. Garcia
Production Project Manager: Sruthi Satheesh
Marketing Manager: Louise Springthorpe
Cover Designer: Harris Greg

Contents

Part 1

Geotechnical Engineering Fundamentals

Geology and geotechnical engineering

1

1.1 Introduction

Geologists are interested in the classification of soils and rocks and their history. Geotechnical engineers are primarily interested in the strength of soils and rocks. Geotechnical engineers would like to know what is the load that can be placed on a given soil or rock stratum.

1.2 Strength of soils

Let us first look at strength of soils. Strength of soils comes from two parameters: friction and cohesion.

1.2.1 Friction

Friction is a physical process. Friction of soil occurs due to friction between soil particles. Let us look at Figure 1.1. On the left side, there is a pyramid of spheres made of glass. On the right side, there is a pyramid built of oranges.

It is obvious that the pyramid built with oranges would be much more stable than the pyramid built with glass spheres. The friction between two orange surfaces is much higher than the friction between two glass surfaces. The same can be said of soils. When the friction between individual soil particles is high, the load that can be placed on soil will be greater. The issue is how to measure the friction in soil. A number of methods have been developed to measure the friction in soils. We would discuss these methods later in this book.

Figure 1.2 shows a failed building. The building load was too much for the soil. The friction in soil was not enough to accommodate the building load.

1.2.2 Cohesion

As mentioned earlier, friction is a physical process. In addition to friction, there is another process – cohesion between particles. Cohesion is an electrochemical process. Cohesion mainly occurs in clay soils. Cohesion between particles occurs due to the different chemical properties in particles. Different concentrations of ions will give rise to negative and positive charges.

Unlike sand particles, clay particles are not round but platy (see Figure 1.3). In addition, compared to sand particles, clay particles are extremely small and cannot be seen with the naked eye.

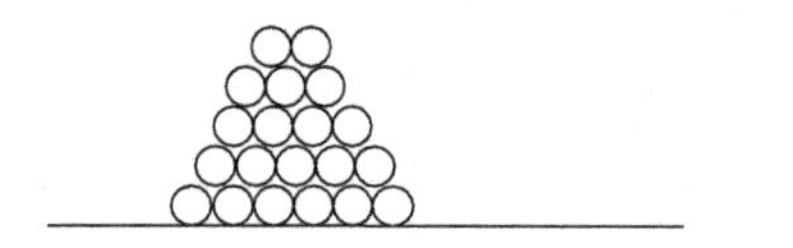

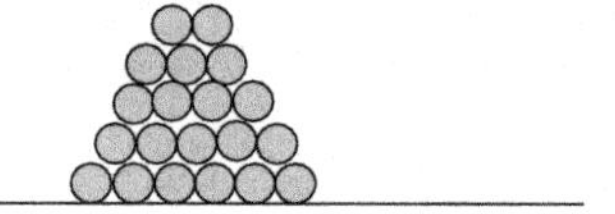

Figure 1.1 Friction in soil particles.

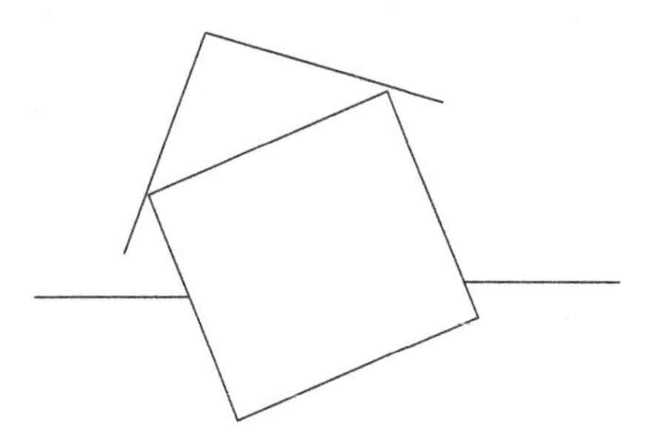

Figure 1.2 Bearing failure.

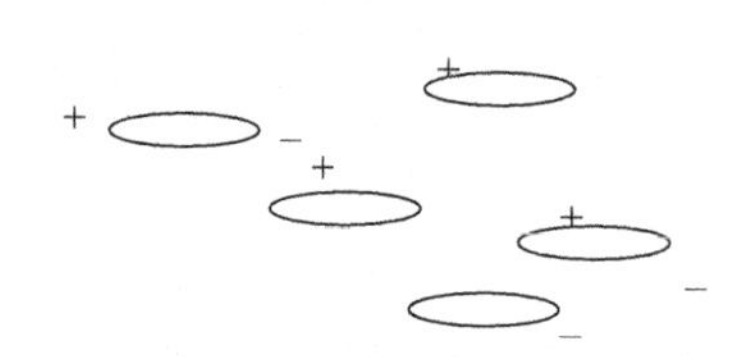

Figure 1.3 Clay particles.

1.2.2.1 Gravel, sand, silt, and clay

Gravel particles are larger than sand particles. Sand particles are larger than silt particles. Generally, silt particles are larger than clay particles. Silt and clay particles cannot be seen with the naked eye.

1.2.2.2 Pure silt has no cohesion

One important fact is that pure silt has no cohesion. Silt is considered to be frictional soil. However, in real life, in many cases silt particles are mixed with clay particles. Silty clays will have both cohesive properties and frictional properties. Silts with cohesion are known as plastic silts.

Quiz

1. What processes give strength to soils?
2. Silty soils mainly obtain strength from what process?

Answers

1. Friction and cohesion
2. Friction

1.3 Origin of rocks and sand

Which came first? Rocks or sand? The answer is rocks. There were rocks on earth before sands. Let us see how this happened. The universe was full of dust clouds. Dust particles due to gravity became small-size objects. They were known as planetesimals. These planetesimals smashed on to each other and larger planets such as earth was formed.

Dust cloud → Small planets (planetesimals) → Collisions → Planets

Many dust clouds can be seen in the space with telescopes. Horsehead Nebula is the most famous dust cloud. Dust clouds gave rise to millions of smaller planets that bashed on to each other. The early earth was a very hot place due to these collisions. The earth was too hot to have water on the surface. All water evaporated and existed in the atmosphere.

Due to extreme heat, the whole earth was covered with a lava ocean.

1.3.1 Earth cools down

Many millions of years later, the earth cooled down. Water vapor that was in the atmosphere started to fall on the earth. Rain started to fall on earth for the first time. This rain could have lasted for millions of years. Oceans were formed. The molten lava ocean became a huge rock. Unfortunately, we have not found a single piece of rock that formed from the very first molten lava ocean.

1.3.2 Rock weathering

Due to the flow of water, change in temperatures, volcanic actions, chemical actions, earthquakes, and falling meteors rocks broke into much smaller pieces. Many millions of years later, rocks broke down to pieces so small that we differentiate them from rocks by calling them sand. Hence, for our first question, which came first? Rock or sand? The rocks came first.

Dust cloud → Small planets (planetesimals) → Planets → Lava ocean
→ Lava ocean cools down to form rocks → First rains → Oceans
→ Weathering of rocks → Formation of sand and clay particles

1.4 Rock types

Brief overview of rocks – All rocks are basically divided into the following three categories:

- Igneous rocks
- Sedimentary rocks
- Metamorphic rocks

1.4.1 Igneous rocks

Igneous rocks were formed from solidified lava. Earth diameter is measured to be approximately 8000 miles and the bedrock is estimated to be only 10 miles. The earth has a solid core with a diameter of 3000 miles and the rest is all lava or known as the mantle (Figure 1.4). Occasionally, lava comes out of the earth during volcanic eruptions, cools down, and becomes rock. Such rocks are known as igneous rocks.

Some of the common igneous rocks are as follows:

- Granite
- Diabase
- Basalt
- Diorite

Igneous rocks have the following two main divisions:

Extrusive igneous rocks: We have seen lava flowing on the surface of earth on television. When lava is cooled on the surface of the earth, extrusive igneous rocks are formed. Typical examples for this type of rocks are andesite, basalt, obsidian, pumice, and rhyolite.

Intrusive igneous rocks: This type of rocks is formed when lava flow occurs inside the earth. Typical examples are diorite, gabbro, and granite (Figure 1.5).

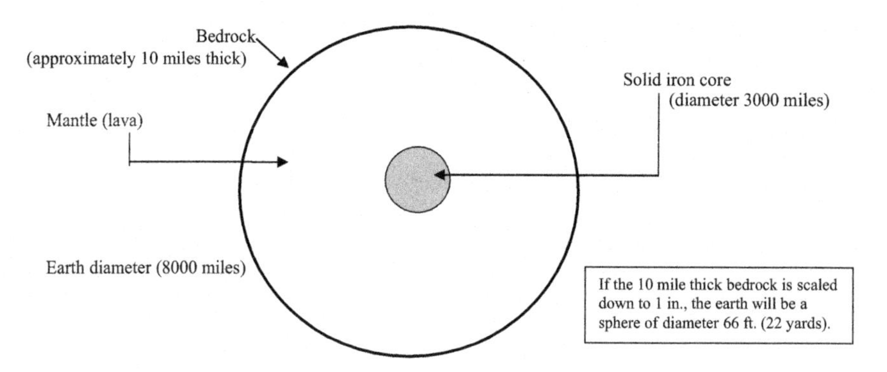

Figure 1.4 Earth.

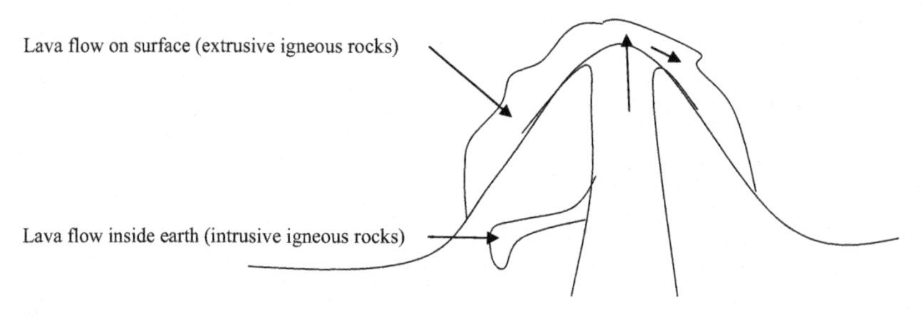

Figure 1.5 Intrusive and extrusive rocks.

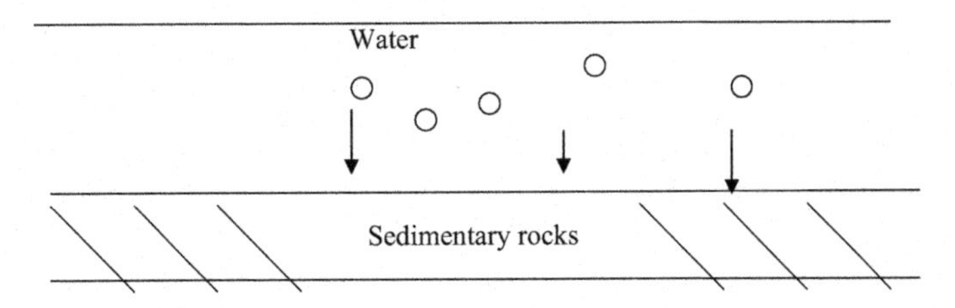

Figure 1.6 Sedimentary process.

The structures of intrusive igneous rocks and extrusive igneous rocks are different.

1.4.2 Sedimentary rocks

Volcanic eruptions are not the only process of rock origin. Soil particles constantly deposit in lakebeds and ocean floor. Over millions of years, these depositions solidify and convert into rocks. Such rocks are known as sedimentary rocks (Figure 1.6).

Some of the common sedimentary rocks are as follows:

- Sandstones
- Shale
- Mudstone
- Limestone
- Chert

Imagine a bowl of sugar left all alone for few months. Sugar particles would bind together due to chemical forces acting between them and solidify. Similarly, sandstone is formed when sand is left under pressure for thousands of years. Limestone, siltstone, mudstone, and conglomerate are formed in a similar fashion.

Lime → Limestone

Mud → Mudstone (also known as shale)

Silt → Siltstone

Mixture of sand, stone, and mud → Conglomerate

1.4.3 Metamorphic rocks

Other than these two rock types, geologists have discovered another rock type. The third rock type is known as metamorphic rocks. Imagine a butterfly metamorphosing from a caterpillar. The caterpillar transforms into a completely different creature.

Early geologists had no problem in identifying sedimentary rocks and igneous rocks. However, they were not sure what to do with this third type of rocks. These rocks did not look like solidified lava or sedimentary rocks. Metamorphic rocks were formed from previously existing rocks due to heavy pressure and temperature. Volcanoes, meteors, earthquakes, and plate-tectonic movements could generate huge pressures and very high temperatures (Figure 1.7).

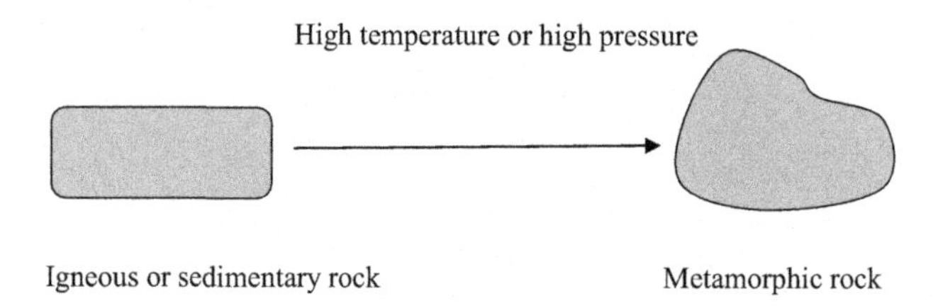

Figure 1.7 Metamorphic rocks.

1.4.3.1 Formation of metamorphic rocks

Both sedimentary rocks and igneous rocks can transform into metamorphic rocks. For example, when shale or mudstone is subjected to very high temperature, it would become schist. When limestone is subjected to high temperature, it would transform into marble. Some metamorphic rocks and their parent rocks are shown here:

Metamorphic rock gneiss → Parent rock could be shale or granite
Metamorphic rock schist → Parent rock is shale
Metamorphic rock marble → Parent rock is limestone

Shale produces the greatest diversity of metamorphic rocks. Metamorphic rocks such as slate, phyllite, schist, and gneiss all come from shale.

1.5 Soil strata types

Soils strata are formed due to forces in nature. The following main forces can be identified for the formation of soil strata:

- Water (soil deposits in river beds, ocean floor, lake beds, and river deltas)
- Wind (deposits of soil due to wind)
- Glacial (deposits of soil due to glacier movement)
- Gravitational forces (deposits due to landslides)
- Organic (deposits due to organic matter such as trees, animals, and plants)
- Weathered *in situ* (rocks or soil can weather at the same location)

1.5.1 Water

1.5.1.1 Alluvial deposits (river beds)

The soils deposited in riverbeds are known as alluvial deposits. Some textbooks use the term, “Fluvial deposits” for the same thing. The size of particles deposited in riverbeds depends on the speed of flow. If the river flow is strong, only large cobble-type material can deposit.

1.5.1.2 Marine deposits

Soil deposits on ocean beds are known as marine deposits or marine soils. Though oceans can be very violent, the seabeds are very calm for the most part. Hence, very small particles would deposit on seabeds. The texture and composition depends on the proximity to land and biological matter.

1.5.1.3 Lacustrine deposits (lakebeds)

Typically very small particles deposit on lakebeds due to the tranquility of water. Lakes deposits are mostly clays and silts.

1.5.2 Wind deposits (eolian deposits)

Wind can carry particles and create deposits. These wind borne deposits are known as eolian deposits.

- Loess: Loess soil is formed due to wind effects. The main characteristic of loess is that it does not have stratifications. Instead, the whole deposit is one clump of soil.
- Eolian sand (sand dunes): Sand dunes are known as eolian sands. Sand dunes are formed in areas that have desert climate.
- Volcanic ash: Volcanic ash is carried away by wind and deposited.

1.5.3 Glacial deposits

Ice ages come and go. The last ice age came 10,000 years ago. During ice ages, glaciers march down to the southern part of the world from the north. During the last ice age, New York City was less than 1000 ft. of ice. Glaciers bring in material. When the glaciers are melted away, the material that was brought in by glaciers is left.

1.5.3.1 Characteristics of glacial till (moraine)

Glacial till contains particles of all sizes. Boulders to clay particles can be seen in glacial till. When the glacier is moving it scoops up any material on the way.

1.5.3.2 Glacio-fluvial deposits

During summer months, the glaciers would melt. The meltwater would flow away with materials and deposits along the way. Glacio-fluvial deposits are different from glacial till.

1.5.3.3 Glacio-lacustrine deposits

Deposits made by glacial meltwater in lakes. Due to low energy in lakes, the glacio-lacustrine deposits are mostly silts and clays.

1.5.3.4 Glacio-marine deposits

As the name indicates, glacial deposits in ocean are known as glacio-marine deposits.

1.5.4 Colluvial deposits

Colluvial deposits occur due to landslides. The deposits due to landslides contain particles of all sizes. Large and heavy particles would be at the bottom of the mountain.

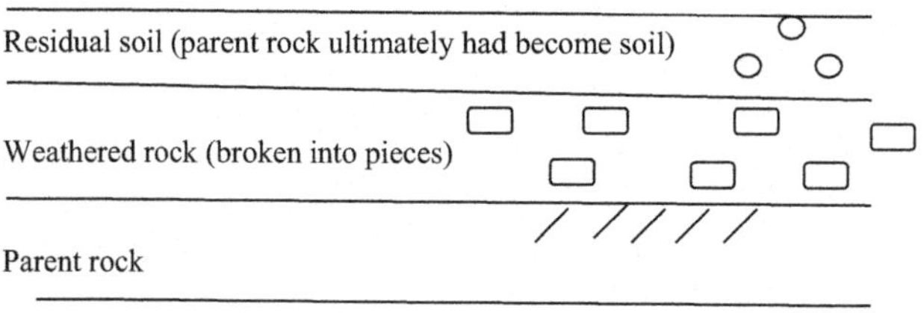

Residual soil (parent rock ultimately had become soil)

Weathered rock (broken into pieces)

Parent rock

Figure 1.8 Residual soil.

1.5.5 Residual soil (weathered in situ *soil)*

Residual soils are formed when soils or rocks weather at the same location due to chemicals, water, and other environmental elements, without being transported. Another name for residual soils is laterite soils. The main process for weathering in residual soils is chemicals (Figure 1.8).

Quiz

1. Name three sedimentary rock types.
2. Are eolian deposits made by rivers or wind?
3. Are fluvial deposits made by rivers or lakes?
4. Residual soils are formed due to what process?

Answers

1. Limestone, conglomerate, and sandstone
2. Wind
3. Rivers
4. Weathered *in situ* or weathering of soil at the same place of origin

Site investigation

2

Soils are of interest to many professionals. Soil chemists are interested in the chemical properties of soil. Geologists are interested in the origin and history of soil strata formation. Geotechnical engineers are interested in the strength characteristics of soil.

Soil strength depends on cohesion and friction between soil particles.

2.1 Cohesion

Cohesion is developed due to the adhesion of clay particles generated by electromagnetic forces. Cohesion is developed in clays and plastic silts while friction is developed in sands and nonplastic silts (Figure 2.1).

2.2 Friction

Sandy particles when brushed against each other will generate friction (Figure 2.2). Friction is a physical process, whereas cohesion is a chemical process. Soil strength generated due to friction is represented by friction angle (φ).

Cohesion and friction are the most important parameters that determine the strength of soils.

2.2.1 *Measurement of friction*

The friction angle of soil is usually obtained from the correlations available with the standard penetration test value known as SPT (N). The Standard penetration test is conducted by dropping a 140 lb hammer from a distance of 30 in. onto a split spoon sample and count the number of blows required to penetrate 1 ft. The SPT value is higher in hard soils and low in soft clays and loose sand.

2.2.2 *Measurement of cohesion*

Cohesion of soil is measured by obtaining a Shelby tube sample and conducting a laboratory unconfined compression test.

2.3 Origin of a project

Civil engineering projects are originated when a company or a person requires a new facility. A company wishing to construct a new facility is known as the owner.

Geotechnical Engineering Calculations and Rules of Thumb

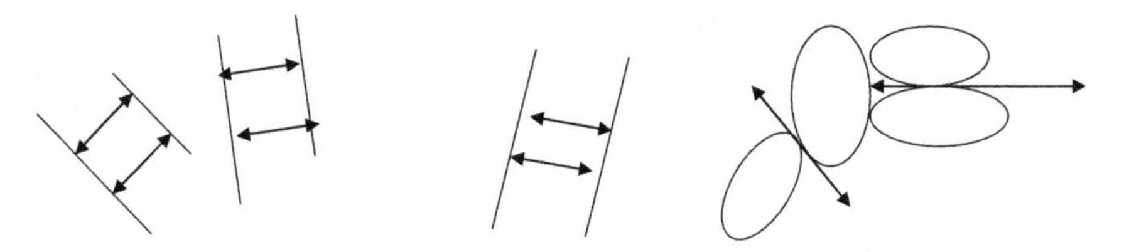

Figure 2.1 Electrochemical bonding of clay particles that gives rise to cohesion.

Figure 2.2 Friction in sand particles.

The owner of the new proposed facility would consult an architectural firm to develop an architectural design. The architectural firm would lay out room locations, conference halls, rest rooms, heating and cooling units, and all other necessary elements of a building, as specified by the owner.

Subsequent to the architectural design, structural engineers and geotechnical engineers would enter the project team. Geotechnical engineers would develop the foundation elements while the structural engineers would design the structure of the building. Usually loading on columns will be provided by structural engineers (Figure 2.3).

2.4 Geotechnical investigation procedures

After receiving information regarding the project, the geotechnical engineer should gather necessary information for the foundation design. Usually, the geotechnical engineers start with a literature survey.

After conducting a literature survey, he or she would make a field visit followed by the subsurface investigation program (Figure 2.4).

2.5 Literature survey

A geotechnical engineer's first step while conducting a geotechnical investigation procedure is to conduct a literature survey. There are many sources available to obtain information regarding topography, subsurface soil conditions, geologic formations, and groundwater conditions. The sources for literature survey are local libraries, Internet,

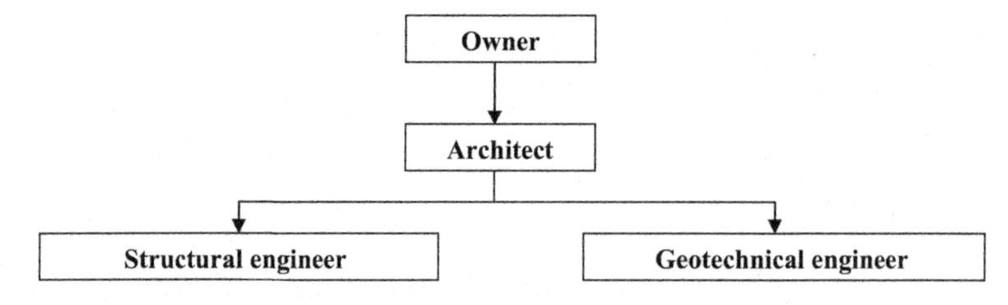

Figure 2.3 Relationship between owner and other professionals.

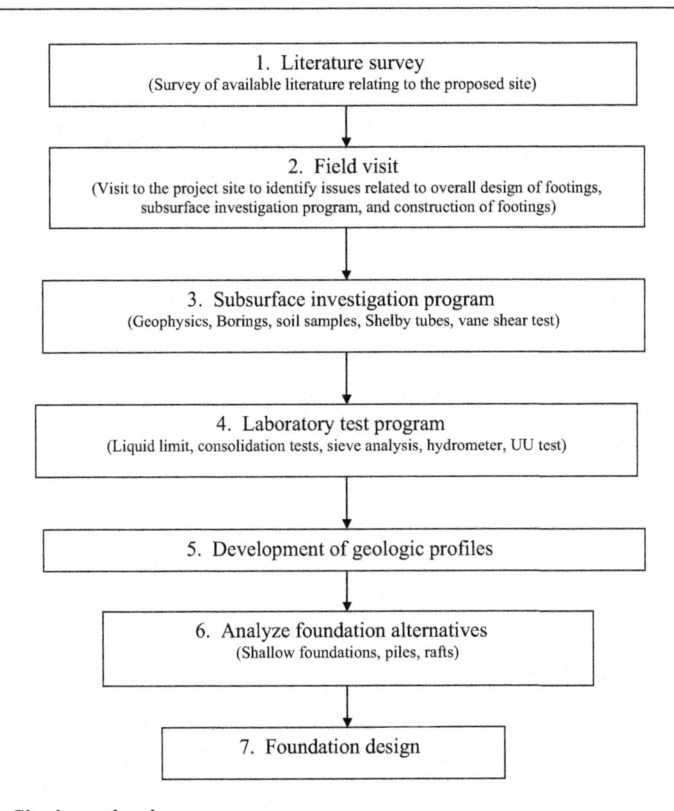

Figure 2.4 Site investigation program.

local universities, and national agencies. National agencies usually conduct geological surveys, hydrogeological surveys, and topographic surveys. These surveys usually provide very important information to the geotechnical engineer. Topographic surveys are useful in identifying depressed regions, streams, marshlands, man-made fill areas, organic soils, and roads. Depressed regions (Figure 2.5) may indicate weak bedrock or settling soil conditions. Construction in marsh areas will be costly. Roads, streams,

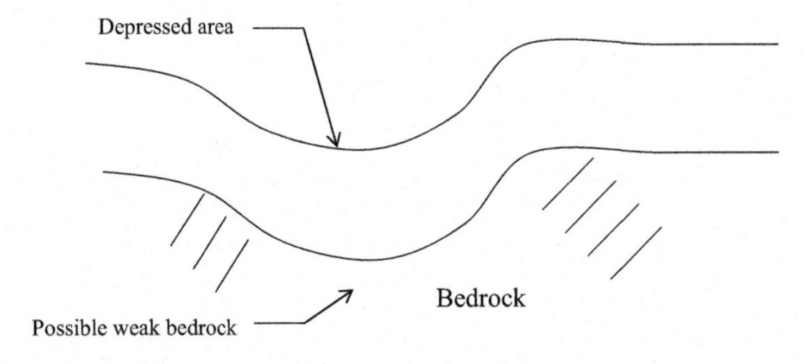

Figure 2.5 Depressed area.

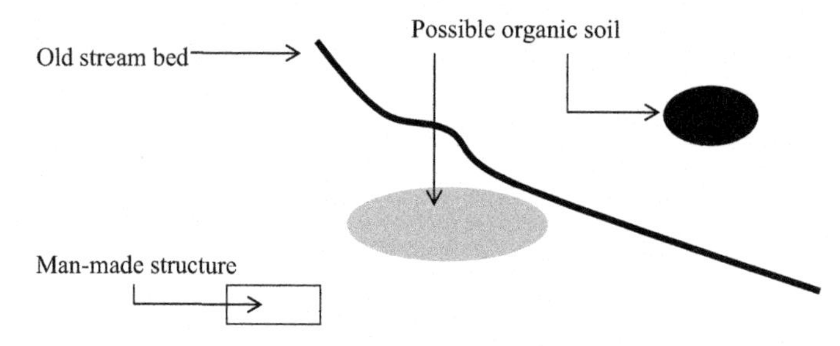

Figure 2.6 Important items in an aerial photograph.

utilities, and man-made fills will interfere with the subsurface investigation program or the construction.

2.5.1 Adjacent property owners

If there are buildings in the vicinity of the proposed building, the geotechnical engineer may be able to obtain site investigation studies conducted in the past.

2.5.2 Aerial surveys

Aerial surveys are done by various organizations for city planning, utility design and construction, traffic management, and disaster-management studies. Geotechnical engineers should contact the relevant authorities to investigate whether they have any aerial maps in the vicinity of the proposed project site.

Aerial maps can be a very good source of preliminary information for the geotechnical engineer. Aerial surveys are expensive to conduct and only large scale projects may have budget for aerial photography (Figure 2.6).

Dark patches may indicate organic soil conditions, different type of soil, contaminated soil, low drainage areas, fill areas, or any other oddity that needs attention of the geotechnical engineer. Darker than usual lines may indicate old stream beds or drainage paths, or fill areas for utilities. Such abnormalities can be easily identified from an aerial survey map.

2.6 Field visit

After conducting a literature survey, the geotechnical engineer should make a field visit. Field visits provide information regarding surface topography, unsuitable areas, slopes, hillocks, nearby streams, soft grounds, fill areas, potentially contaminated locations, existing utilities, and possible obstructions for site investigation activities. The geotechnical engineer should bring a hand auger to the site so that he or she can observe the soil few feet below the surface (Figure 2.7).

Nearby streams could provide excellent information regarding the depth to groundwater.

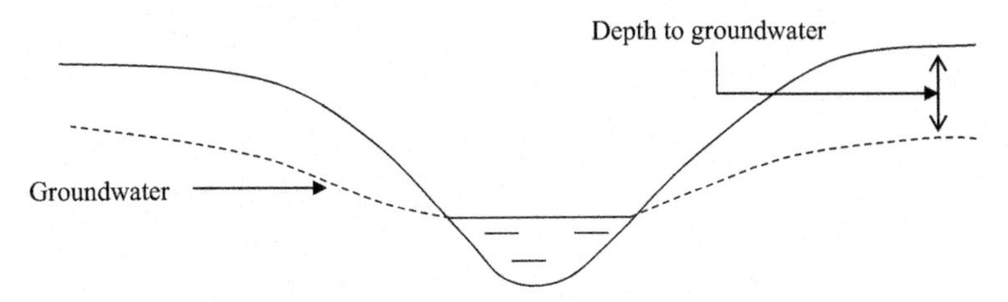

Figure 2.7 Groundwater level near a stream.

2.6.1 Hand auguring

Hand augurs can be used to obtain soil samples to a depth of approximately 6 ft. depending upon the soil conditions (Figure 2.8). Downward pressure (P) and a torque (T) are applied to the hand auger. Due to the torque and the downward pressure, the hand auger would penetrate into the ground. The process stops when the human strength is not capable of generating enough torque or pressure.

2.6.2 Sloping ground

Steep slopes in a site escalate the cost of construction because a compacted fill is required. Such areas need to be noted for further investigation (Figure 2.9).

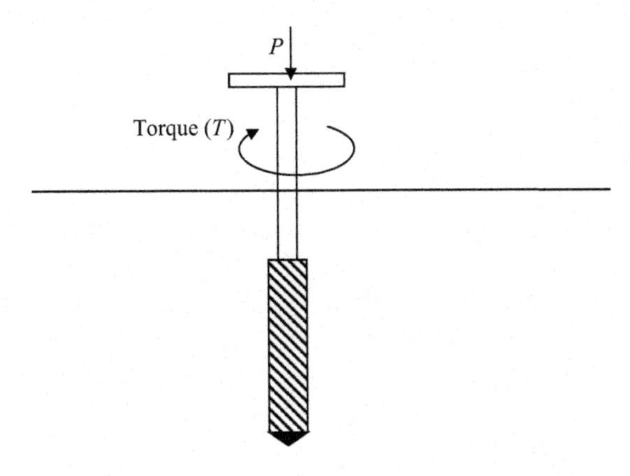

Figure 2.8 Hand auger.

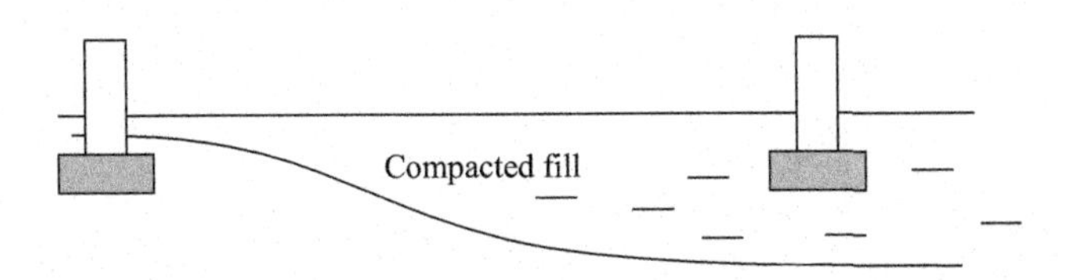

Figure 2.9 Sloping ground.

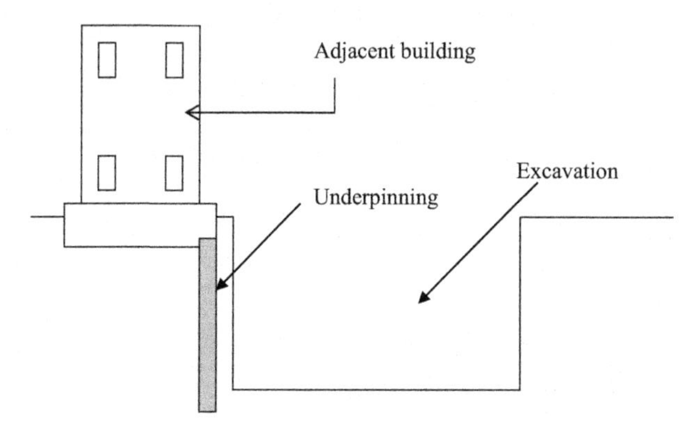

Figure 2.10 Underpinning of nearby building for an excavation.

2.6.3 Nearby structures

Nearby structures could pose many problems for proposed projects. It is always good to identify these issues at the very beginning of a project.

Distance from nearby buildings, schools, hospitals, and apartment complexes should be noted. Pile driving may not be feasible if there is a hospital or school close to the proposed site. In such situations, jacking of piles can be used to avoid noise.

If the proposed building has a basement, underpinning of nearby buildings may be necessary (Figure 2.10).

Other methods such as secant pile walls or heavy bracing can also be used to stabilize the excavation.

Shallow foundations of a new building could induce negative skin friction on the pile foundation of a nearby building. If compressible soil is present at a site, the new building may induce consolidation. The consolidation of a clay layer may generate negative skin friction in piles of nearby structures (Figure 2.11).

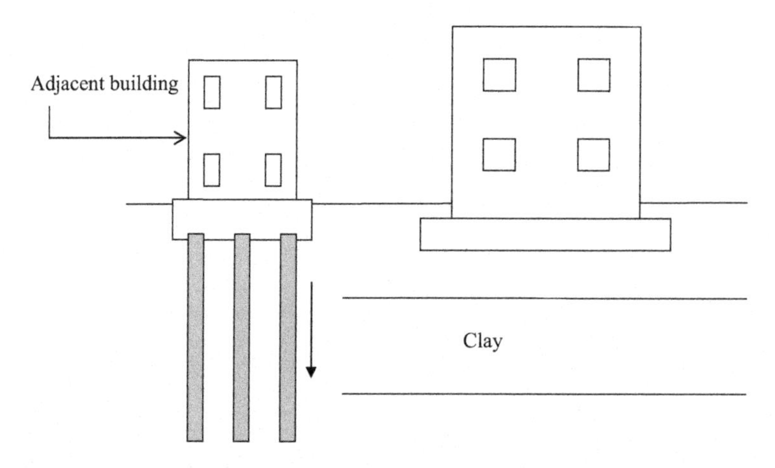

Figure 2.11 Negative skin friction in piles due to new construction.

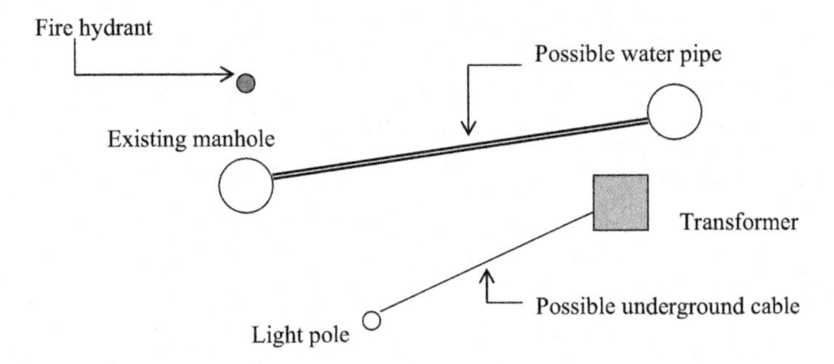

Figure 2.12 Observation of utilities.

2.6.4 Contaminated soils

Soil contamination is a common problem in many urban sites. Contaminated soil can increase the cost of a project or, in some cases, could even kill a project entirely. Identifying contaminated soil areas at early stages of a project is desirable.

2.6.5 Underground utilities

It is a common occurrence for drilling crews to accidentally puncture underground power cables or gas lines. Therefore, early identification of the existing utilities is important. Electrical poles, electrical manhole locations, gas lines, water lines should be noted so that drilling can be done without breaking any utilities.

Existing utilities may have to be relocated or undisturbed during construction.

Existing manholes may indicate drainpipe locations (Figure 2.12).

2.6.6 Overhead power lines

Drill rigs need to keep a safe distance from overhead power lines during the site-investigation phase. Overhead power line locations need to be noted during the field visit.

2.6.7 Man-made fill areas

Most urban sites are affected by human activities. Man-made fills may contain soils, bricks, and various types of debris. It is possible to compact some man-made fills so that they could be used for foundations. This is not feasible when the fill material contains compressible soils, tires, or rubber. The fill areas need to be further investigated during the subsurface investigation phase of the project.

2.6.8 Field-visit checklist

The geotechnical engineer needs to pay attention to issues during the field visit with regard to the following:

1. Overall design of the foundations
2. Obstructions for the boring program (overhead power lines, marsh areas, slopes, or poor access may create obstacles for drill rigs).

3. Issues relating to the construction of foundations (high groundwater table, access, existing utilities).
4. Identification of possible man-made fill areas
5. Nearby structures (hospitals, schools, courthouses etc).

Geotechnical engineer's checklist for the field visit (Table 2.1)

Table 2.1 Field-visit check list

Item	Impact on site investigation	Impact on construction	Cost impact
Sloping ground	May create difficulties for drill rigs	Impact on the maneuverability of construction equipment	Cost impact due to cut and fill activities
Small hills	Same as above	Same as above	Same as above
Nearby streams	Groundwater monitoring wells may be necessary	High groundwater may impact deep excavations (pumping)	Impact on cost due to pumping activities
Overhead power lines	Drill rigs have to stay away from overhead power lines	Construction equipment have to keep a safe distance	Possible impact on cost
Underground utilities (existing)	Drilling near utilities should be done with caution	Impact on construction work	Relocation of utilities will impact the cost
Areas with soft soils	More attention should be paid to these areas during subsurface investigation phase	Possible impact	Possible impact on cost
Contaminated soil	Extent of contamination need to be identified	Severe impact on construction activities	Severe impact on cost
Man-made fill areas	Broken concrete or wood may pose problems during the boring program	Unknown fill material generally not suitable for construction work.	Possible impact on cost
Nearby structures		Due to nearby hospitals and schools some construction methods may not be feasible such as pile driving Excavations for the proposed building could cause problems to existing structures Shallow foundations of new structures may induce negative skin friction in piles in nearby buildings	Possible impact on cost

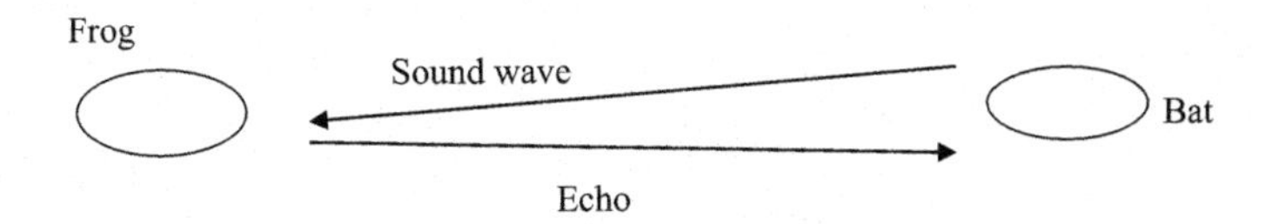

Figure 2.13 Bat and a frog.

2.7 Geophysics

Geophysical methods can be used to obtain soil parameters. Let us look at an example. As you know, bats can locate objects using sound waves. A bat would send a sound wave and wait to hear the reflection or the echo. From the echo, the bat can identify a frog (Figure 2.13).

Geophysical methods are based on the same concept. In case of ground-penetrating radar, the radar signal is sent and wait for the signal to be back. In the case of seismic methods, blows are given to the ground surface with a hammer and the seismic wave back are collected.

2.7.1 GPR methods

In ground penetrating radar (GPR) methods, a radar signal is sent and wait for the signal to come back. Returned signal is analyzed to obtain the soil profile. The radar system is widely used in aircraft identification. Same radar can also be used to identify soil profiles. Radar is a radio wave. Radio waves belong to the family of electromagnetic waves. Light, X-rays, and radar belong to the electromagnetic wave family.

2.7.1.1 General methodology

Radar travels at the speed of light. Light would travel at approximately 1 ft./1 ns. One nanosecond is equal to 10^{-9} s. Let us assume that a radar signal bounces back after 10 ns, we can assume that there is an obstruction 5 ft. below the surface.

In Figure 2.14, the velocity of radar in soil is known. Travel time (t) can be measured. Hence, the depth can be calculated. In Figure 2.14, the depth is multiplied by 2 to account for the wave to reach to a depth of "D" and come back again. The travel distance of the wave form is $2 \times D$.

Let us look at Figure 2.15. As mentioned before, the depth of an object can be computed using the travel time. Figure 2.15 shows two boundaries (three layers) of different soils. It may not be feasible to tell the exact soil type using GPR data. Hence, borings should be conducted. The GPR could be used to properly identify the soil strata boundaries.

Figure 2.15 shows a GPR survey. The GPR lines will curve near a pipe.

2.7.1.1.1 Single borehole GPR

In this method, a borehole is drilled. The radar transmitter and the receiver is set at different depths and data are collected. (Figure 2.16)

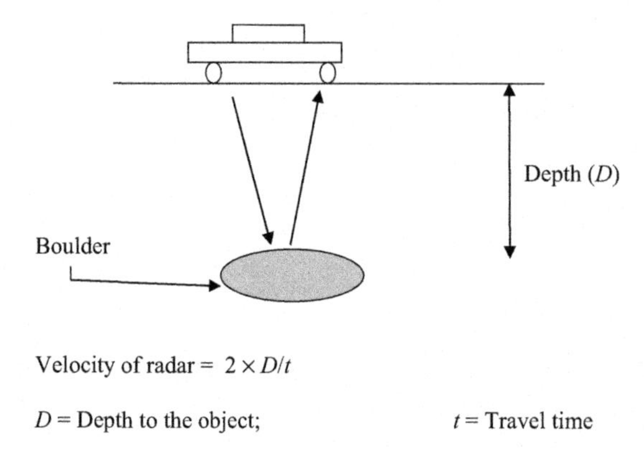

Figure 2.14 Radar.

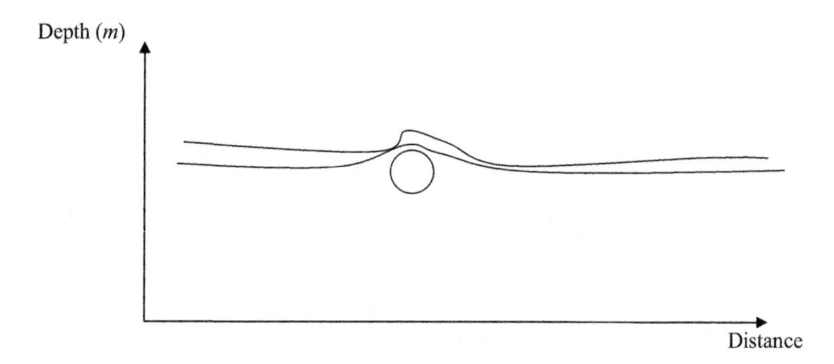

Figure 2.15 GPR example.

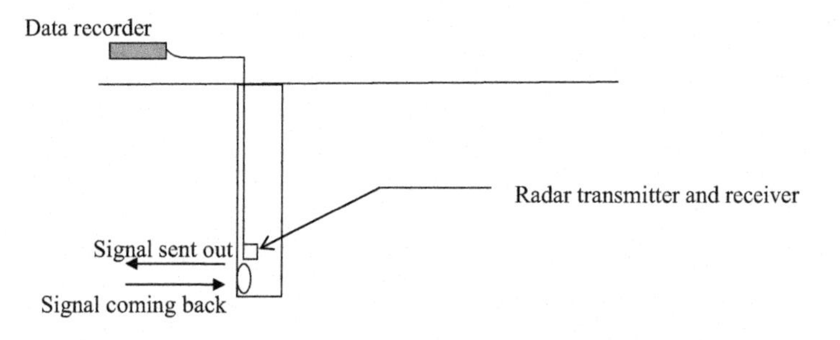

Figure 2.16 GPR in single borehole.

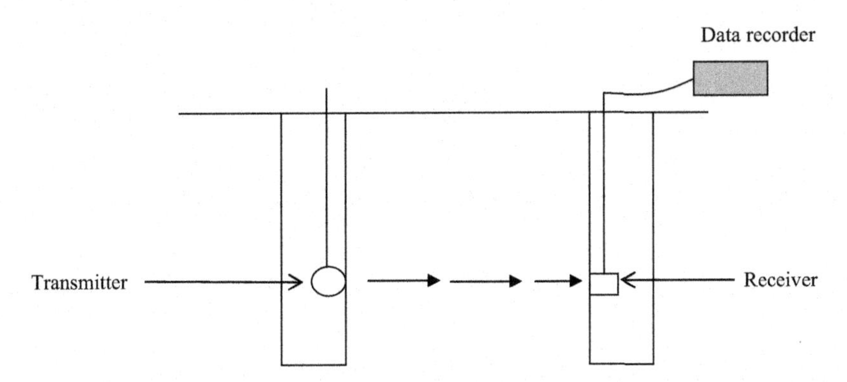

Figure 2.17 Crosshole method in GPR.

Procedure

1. Place the radar transmitter and the receiver at any depth. Record the data.
2. Change the depth and repeat the process.

Many different GPR data sets can be obtained this way. Also, the transmitter and receiver can be placed at different depth.

2.7.1.1.2 Crosshole GPR

In this method, two boreholes are needed. Transmitter is placed at one borehole and the receiver is placed at the other borehole.

In Figure 2.17, a radar transmitter is shown in one borehole. The transmitted waveform is picked up by a receiver placed in a second borehole. From the GPR data, it is possible to ascertain the ground conditions between the two boreholes.

2.7.2 Seismic method

The concept in seismic method is similar to GPR. Instead of a radio signal, a seismic signal is sent out and whatever comes back is recorded. A seismic signal is produced by hammering a piece of wood placed on ground with a hammer. A seismic signal would travel to the seismic sensors. The seismic sensors are attached to a data logger.

Figure 2.18 shows hammering a piece of wood placed on earth. The seismic wave would travel to geophones or seismic sensors. The seismic sensors are attached to a data logger.

- Reflected seismic waves versus refracted seismic waves

  Unlike a radar, the seismic waves could have two path ways: reflected path and refracted path.
- Seismic P-waves and S-waves

  P-waves are known as body waves. They travel faster than S-waves. In a P-wave, the soil particles move in the same direction as the wave.

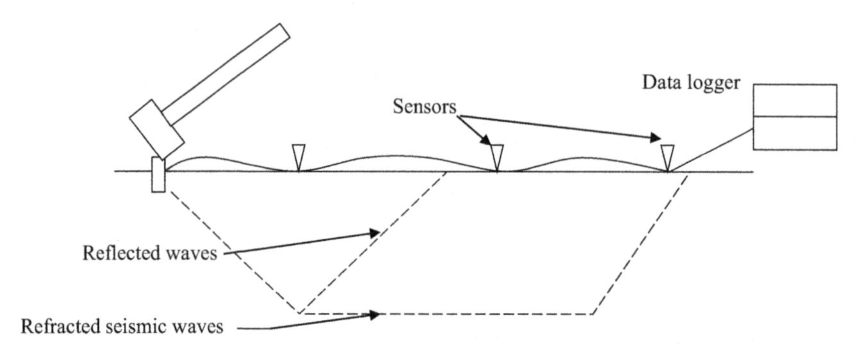

Figure 2.18 Seismic method.

Figure 2.19a shows the soil particles prior to seismic energy. Figure 2.19b shows the soil particles getting compressed due to seismic energy. The arrow shows the particle movement. Compression wave of soil particles would travel creating a P-wave.

- S-waves

  In the case of S-waves, the soil particles move perpendicular to the wave direction. The S-waves are known as shear waves. Shear waves will not occur in air or water (Figure 2.20).

  The arrow in Figure 2.20 shows the particle movement. In S-waves, the soil particles move perpendicular to the wave.
- Surface Waves

  As the name indicates, surface waves travel along the surface. These waves are known as Rayleigh waves.

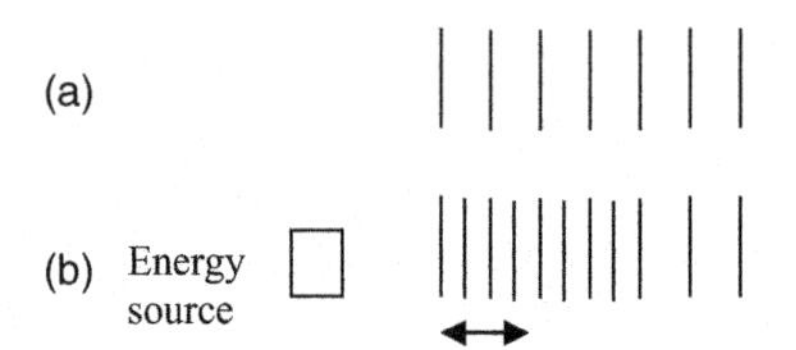

Figure 2.19 Seismic waves. (a) Soil particles prior to seismic energy. (b) Soil particles getting compressed due to seismic energy.

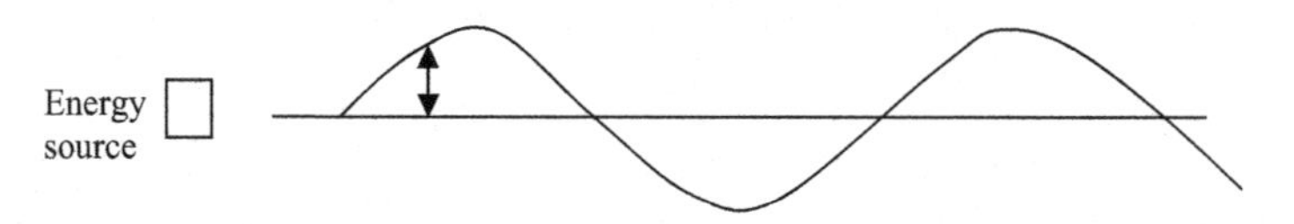

Figure 2.20 Seismic S-waves.

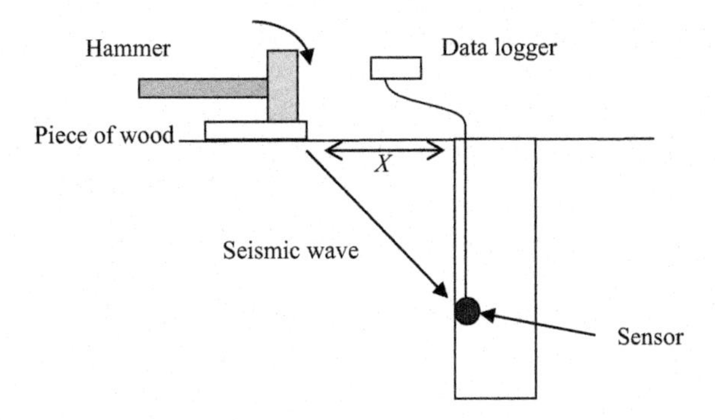

Figure 2.21 Down hole seismic testing.

2.7.2.1 *Down hole seismic testing*

In this method, a borehole is drilled. Seismic energy is applied at a known distance. Seismic sensors are placed inside the borehole (see Figure 2.21). The data will provide soil strata information.

After obtaining data at a given depth, another set of data is obtained by changing the depth of the sensor.

2.7.2.2 *Crosshole seismic testing*

Crosshole seismic testing is similar to GPR crosshole testing (see Figure 2.22).

Seismic energy is applied from one borehole. Generally, an air burst or a shock wave is applied. Seismic sensors (receivers) are placed in the other bore hole. New set of data can be obtained by changing the depths of the sensor and energy source.

Crosshole seismic testing is shown in the Figure 2.23. In Figure 2.23, "E" in the figure indicates seismic energy source and "S" indicates geophones or sensors. In this case, number of geophones are located along the depth of the borehole.

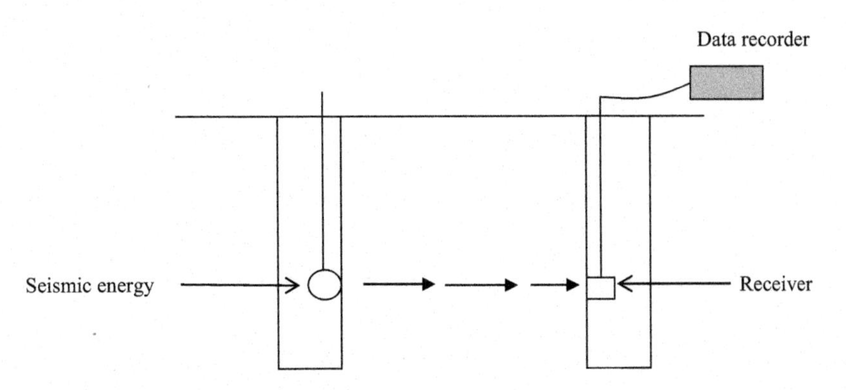

Figure 2.22 Crosshole seismic testing.

2.8 Subsurface investigation phase

Soil strength characteristics of subsurface are obtained through a drilling program. In a nutshell, the geotechnical engineer needs the following information for foundation design work:

- Soil strata identification (sand, clay, silt etc.);
- Depth and thickness of soil strata;
- Cohesion and friction angle (the two parameters responsible for soil strength); and
- Depth to groundwater.

2.8.1 Soil strata identification

Subsurface soil strata information is obtained through drilling.

The most common drilling techniques are

- Augering
- Mud rotary drilling

2.8.1.1 Augering

In the case of augering, the ground is penetrated using augers attached to a rig (Figure 2.24). The rig applies a torque and a downward pressure to the augers. The same principal as in hand augers is used for the penetration into the ground (Plate 2.1).

2.8.1.2 Mud rotary drilling

In the case of mud rotary drilling, a drill bit known as roller bit is used for the penetration. Water is used to keep the roller bit cool so that it does not overheat and stops functioning. Usually the drillers mix Bentonite slurry (also known as drilling mud) to the water to thicken the water. The main purpose of Bentonite slurry is to keep the sidewalls from collapsing (Figure 2.25).

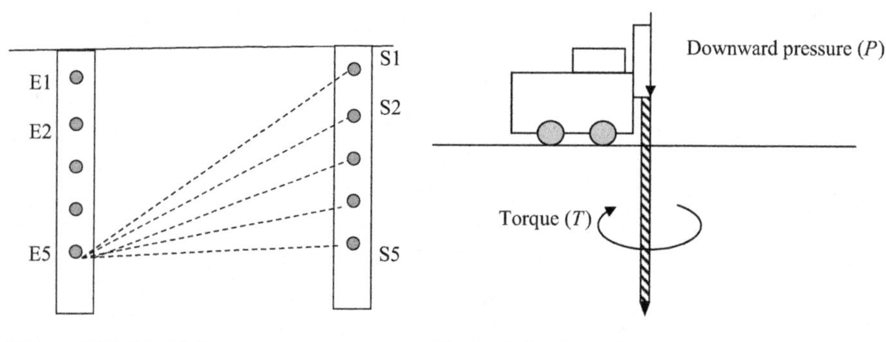

Figure 2.23 Multiple sensors.

Figure 2.24 Augering.

Plate 2.1 Auger drill rig.

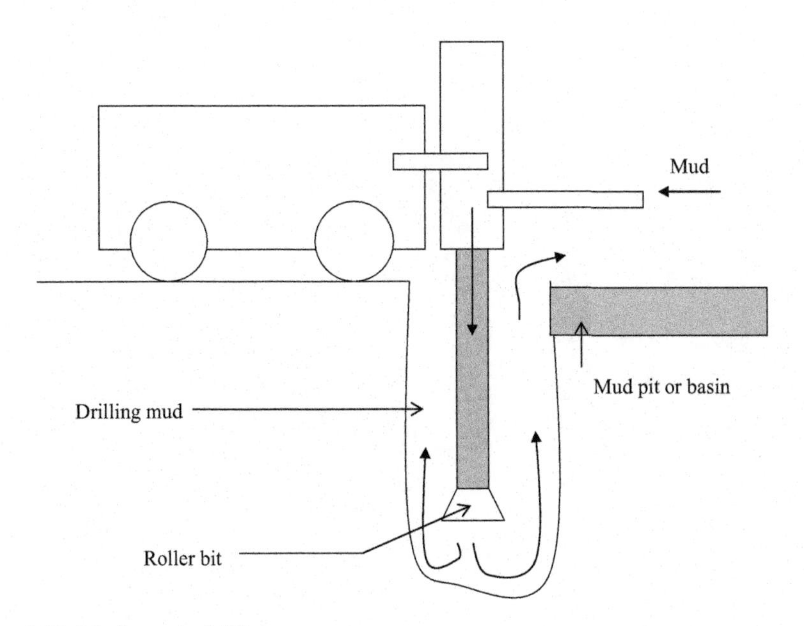

Figure 2.25 Mud rotary drilling.

The drilling mud goes through the rod and the roller bit and comes out from the bottom removing the cuttings. The mud is captured in a basin and is recirculated (Plate 2.2).

2.8.1.2.1 Boring program

The number of borings that need to be constructed may sometimes be regulated by the local codes. For example, the New York City building code requires one boring per each 2500 sq. ft. It is important to conduct borings as close as possible to column locations and strip footing locations. However, this may not be feasible in some cases.

Typically borings are constructed 10 ft. below the bottom level of the foundation.

2.8.1.2.2 Test pits

In some situations, test pits would be more advantageous than borings. Test pits can provide information down to 15 ft. below the surface. Unlike borings, the soil can be visually observed from the sides of the test pit (Plate 2.3).

2.8.1.2.3 Soil sampling

Split spoon samples are obtained during boring construction. Split spoon samples are typically 2 in. diameter and have a length of 2 ft. The soil obtained from split spoon samples are adequate to conduct sieve analysis, soil identification, and Atterberg limit tests. Consolidation tests, triaxial tests, and unconfined compressive strength tests

Plate 2.2 Mud rotary drill rig.

Plate 2.3 Roller bit that cut through soil. A hole is provided to send water to keep the bit cool.

need large quantity of soil and in such situations, Shelby tubes are used. Shelby tubes have a larger diameter than split spoon samples.

2.8.1.2.4 Hand digging prior to drilling

Damage of utilities should be avoided during the boring program. Most utilities are rarely deeper than 6 ft. Hand digging the first 6 ft. prior to drilling the boreholes is found to be an effective way to avoid damaging the utilities. During excavation activities, the backhoe operator is advised to be aware of the utilities. The operator should check for fill materials, since in many instances the utilities are backfilled with select fill material. It is advisable to be cautious since there could be situations where the utilities are buried with the same surrounding soil. In such cases, it is a good idea to have a second person present exclusively to watch the backhoe operation.

2.9 Geotechnical field tests

2.9.1 SPT (N) value

During the construction of borings, SPT (*N*) values are of soils obtained. The SPT (*N*) value provides information regarding the soil strength. SPT (*N*) value in sandy soils indicates the friction angle in sandy soils and in clay soils indicates the stiffness of the clay stratum.

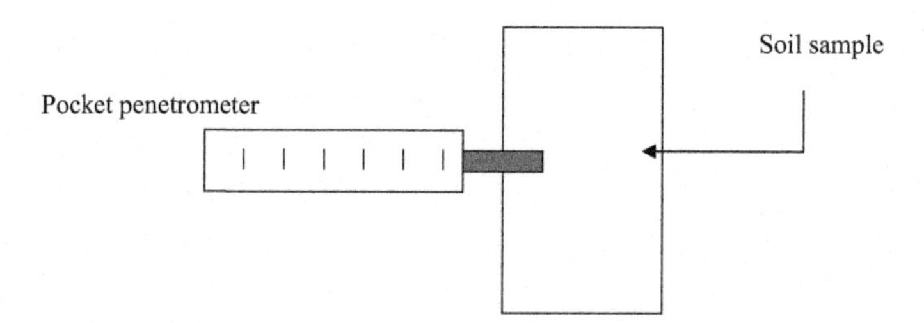

Figure 2.26 Pocket penetrometer.

2.9.2 Pocket penetrometer

Pocket penetrometers can be used to obtain the stiffness of clay samples. The pocket penetrometer is pressed into the soil sample and the reading is recorded. The reading would indicate the cohesion of the clay sample (Figure 2.26).

2.9.3 Vane shear test

Vane shear tests are conducted to obtain the cohesion (C) value of a clay layer. An apparatus consisting of vanes is inserted into the clay layer and rotated. The torques of the vane is measured during the process. Soils with high cohesion values register high torques (Figure 2.27).

2.9.3.1 Vane shear test procedure

- A drill hole is made with a regular drill rig.
- The vane shear apparatus is inserted into the clay.
- The vane is rotated and the torque is measured.
- The torque would gradually increase and reach a maximum. The maximum torque achieved is recorded.
- At failure, the torque would reduce and reach a constant value. This value refers to the remolded shear strength (Figure 2.28).

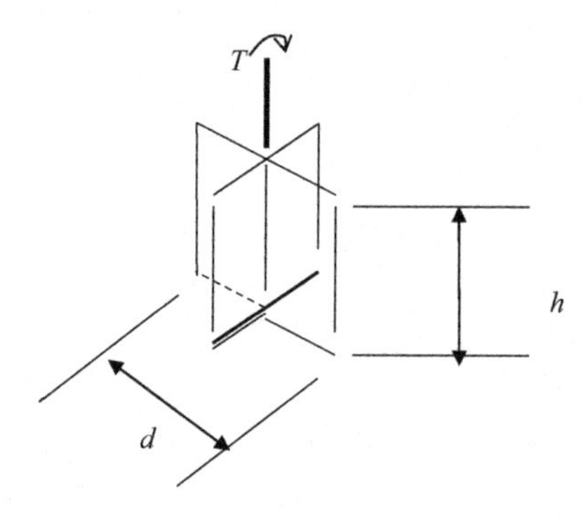

Figure 2.27 Vane shear apparatus.

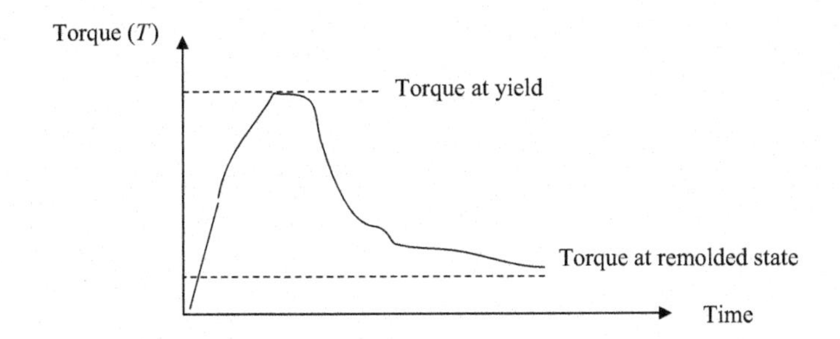

Figure 2.28 Torque versus time curve.

The cohesion of clay is given by

$$T = C\pi(d^2h/2 + d^3/6)$$

T, torque (measured); *C,* cohesion of the clay layer; *d,* width of vanes; *h,* height of vanes.

The cohesion of the clay layer is obtained by using the maximum torque. Remolded cohesion is obtained by using the torque at failure.

2.10 Correlation between friction angle (φ') and SPT (*N*) value

Friction angle (φ') is a very important parameter in geotechnical engineering. The soil strength in sandy soils depends solely on friction.

Correlations have developed between SPT (*N*) value and the friction angle and are given in Table 2.2.

2.10.1 Hatakanda and Uchida Equation

After conducting numerous tests, Hatakanda and Uchida (1996) provided the following equation to compute the friction angle using the SPT (*N*) values:

$$\varphi' = 3.5 \times (N)^{1/2} + 22.3$$

φ', friction angle; *N*, SPT value.

This equation ignores the particle size and most tests are done on medium to coarse sands. For a given "*N*" value, fine sands will have a lower friction angle while coarse sands will have a larger friction angle. Hence, the following modified equations are proposed:

Fine sand – $\varphi' = 3.5 \times (N)^{1/2} + 20$
Medium sand – $\varphi' = 3.5 \times (N)^{1/2} + 21$
Coarse sand – $\varphi' = 3.5 \times (N)^{1/2} + 22$

Table 2.2 Friction angle, SPT "N" values, and relative density, Bowles (2004)

Soil type	SPT (N_{70} value)	Consistency	Friction angle (φ')	Relative density (D_r)
Fine sand	1–2	Very loose	26–28	0–0.15
	3–6	Loose	28–30	0.15–0.35
	7–15	Medium	30–33	0.35–0.65
	16–30	Dense	33–38	0.65–0.85
	<30	Very dense	<38	<0.85
Medium sand	2–3	Very loose	27–30	0–0.15
	4–7	Loose	30–32	0.15–0.35
	8–20	Medium	32–36	0.35–0.65
	21–40	Dense	36–42	0.65–0.85
	<40	Very dense	<42	<0.85
Coarse sand	3–6	Very loose	28–30	0–0.15
	5–9	Loose	30–33	0.15–0.35
	10–25	Medium	33–40	0.35–0.65
	26–45	Dense	40–50	0.65–0.85
	<45	Very dense	<50	<0.85

Design Example 2.1

Find the friction angle of fine sand with an SPT (*N*) value of 15.

Solution

Use the modified Hatakanda and Uchida equation for fine sand.

$$\begin{aligned}\varphi' &= 3.5\times(N)^{1/2}+20\\ &= 3.5\times(15)^{1/2}+20\\ &= 33.5\end{aligned}$$

2.10.2 SPT (N) value versus total density

Correlations between SPT (*N*) value and total density have been developed. These values are given in Table 2.3.

2.11 SPT (*N*) value computation based on drill rig efficiency

Most geotechnical correlations were done based on the SPT (*N*) values obtained during the 1950s. Drill rigs today are much more efficient than the drill rigs of the 1950s.

Table 2.3 **SPT (*N*) value and soil consistency**

Soil type	SPT (N_{70} value)	Consistency	Total density
Fine sand	1–2	Very loose	70–90 pcf (11–14 kN/m^3)
	3–6	Loose	90–110 pcf (14–17 kN/m^3)
	7–15	Medium	110–130 pcf (17–20 kN/m^3)
	16–30	Dense	130–140 pcf (20–22 kN/m^3)
	<30	Very dense	<140 pcf < 22 kN/m^3
Medium sand	2–3	Very loose	70–90 pcf (11–14 kN/m^3)
	4–7	Loose	90–110 pcf (14–17 kN/m^3)
	8–20	Medium	110–130 pcf (17–20 kN/m^3)
	21–40	Dense	130–140 pcf (20–22 kN/m^3)
	<40	Very dense	<140 pcf < 22 kN/m^3
Coarse sand	3–6	Very loose	70–90 pcf (11–14 kN/m^3)
	5–9	Loose	90–110 pcf (14–17 kN/m^3)
	10–25	Medium	110–130 pcf (17–20 kN/m^3)
	26–45	Dense	130–140 pcf (20–22 kN/m^3)
	<45	Very dense	<140 pcf < 22 kN/m^3

If the hammer efficiency is high, then to drive the spoon 1 ft., the hammer would require lesser amount of blows compared to a less efficient hammer.

For example, let's assume that one uses an SPT hammer from the 1950s and he required 20 blows to penetrate 12 in. However, if a modern hammer is used, a 12-in. penetration may be achieved with lesser number of blows. Therefore, a high efficiency hammer requires less number of blows as compared to a low efficiency hammer, which would require more number of blows.

The following equation is used to convert blow count from one hammer to another. It gives the conversion from 50% efficient hammer to a 70% efficient hammer.

$$N_{70}/N_{50} = 50/70$$

Design Example 2.2

An SPT blow count of 20 was obtained using a 40% efficient hammer. What blow count is expected if a hammer with 60% efficiency is used.

Solution

$$N_{60}/N_{40} = 40/60$$

$$N_{60}/20 = 40/60$$

$$N_{60} = 13.3$$

Design Example 2.3

An SPT blow count of 15 was obtained using a 55% hammer. A second hammer was used and the blow count of 25 was obtained. What was the efficiency of the second hammer?

Solution

Let's say that the efficiency of the second hammer is N_x.

$$N_x/N_{55} = 55/X$$

$$N_x = 25 \quad \text{and} \quad N_{55} = 15$$

$$\text{Hence, } 55/X = N_x/N_{55} = 25/15 = 1.67$$

$$X = 55/1.67 = 32.9$$

Efficiency of the second hammer is 32.9%

2.12 Cone penetration testing (CPT)

Cone penetration tests are done to identify the soil type. In this test, a cone is pushed into the ground. In the case of SPT tests, a spoon was driven with a hammer. In CPT testing, the cone is pushed instead of driving.

Figure 2.29 shows the following items:

1. *Vehicle* – Vehicle is required to provide power to the pushing mechanism.
2. *Pushing mechanism* – Hydraulic mechanism is used to push the cone into the ground.
3. *Cone at the tip* – Cone at the tip measures the tip resistance. There is an electronic sensor just above the tip.
4. *Friction sleeve* – Friction sleeve measures the friction in the shaft. There is a sensor that measures the friction in the sleeve.
5. *Pore pressure sensor* – This sensor would measure pore pressure in the soil.

The cone used is 37.5 mm in diameter and has an angle of 60°. The friction sleeve is 133.7 mm. The surface area of the skin friction sleeve is 15,000 mm^2 (Figure 2.30).

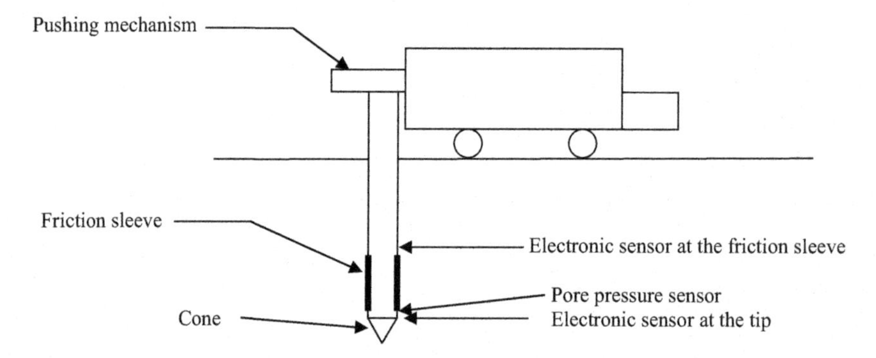

Figure 2.29 CPT testing.

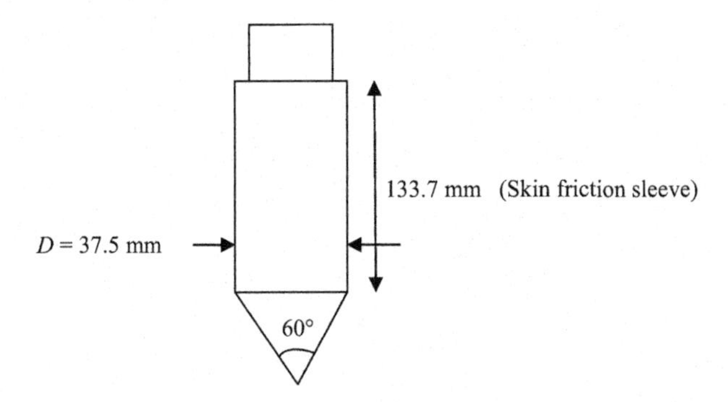

Figure 2.30 CPT apparatus.

Perimeter area of the skin friction sleeve = $\pi D \times$ length = $\pi \times 35.7 \times 133.7 = 15{,}000$ mm^2.

2.12.1 Measurements obtained

- Depth to the cone
- Tip resistance
- Skin friction in the sleeve
- Pore pressure

The resistance to cone penetration comes from the following two processes:

1. Resistance at the tip
2. Resistance in the sleeve due to skin friction

Clay soils have high skin friction and sandy soils have high tip resistance. This is a general statement and values are dependent on individual soils. Very stiff clay may have higher tip resistance than a loose sand layer. Also clay soils will have high excess pore pressure as compared to sandy soils.

2.12.2 Friction ratio

Friction ratio is the ratio between skin friction and tip resistance, expressed as a percentage.

$$\text{Friction ratio } (f_R) = f/Q_u \times 100$$

f, skin friction in the sleeve measured using electronic sensors (tsf); Q_u, tip resistance measured using electronic sensors (tsf).

The following general statements can be said of the CPT data:

1. Gravelly sand – Very low friction ratio and very high tip resistance
2. Sand – Low friction ratio moderate tip resistance
3. Sandy silt or silty sand – Moderate friction ratio and moderate tip resistance
4. Clays – High friction ratio and low tip resistance
5. Peat and organic clays – Very low friction ratio and very low tip resistance

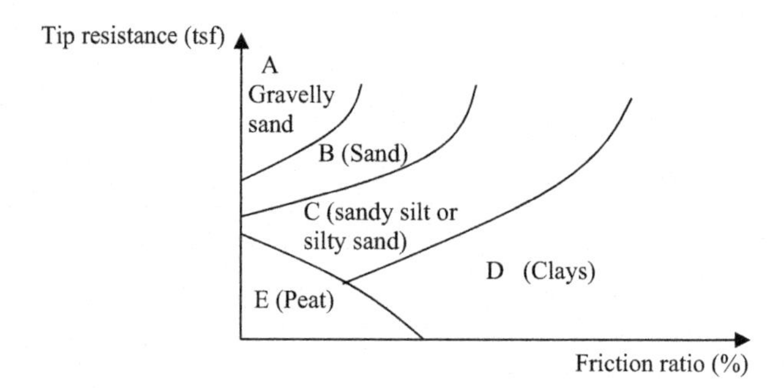

Figure 2.31 Tip resistance versus friction.

The soil types (A–E) can be represented in a graph as shown in Figure 2.31.

Note that the representation in Figure 2.31 is very general in nature and should not be used without supporting data. It is very possible that a stiff clay could be area C or B. Generally, a friction ratio of 1–2% is considered being very low and 10–15% is considered to be very high. Also, the tip resistance of 1 to 1–2 tsf is in the lower category and 8–10 in the high category.

Practice Problem

The following CPT data was obtained from a site (Figure 2.32). Predict the soil types for each layer.

Solution

Layer 1: Tip resistance is very low and friction ratio is very high. This layer could be identified as a clay layer.

Layer 2: Tip resistance is moderate and friction ratio is also moderate. This layer could be identified as a sandy silt or silty sand layer.

Layer 3: Tip resistance is low and friction ratio is high. This layer could be identified as a clay layer.

Layer 4: Tip resistance is moderate and the friction ratio is low. This layer can be identified in the silty sand category.

2.13 Pressuremeter testing

In this test, a hole is predrilled. Then a pressure meter is inserted and inflated. The force required to inflate the pressure meter and the strain is recorded (Figure 2.33).

When the pressure meter is inflated, the stress in the pressure meter and the strain in soil is recoded. The following two methods are typically used:

The Equal Pressure Increment Method – In this method, the pressure in the pressuremeter is increased at equal intervals. If the pressure of the pressuremeter is 10 tsf and the increment is 2 tsf, the following pressure values will be maintained: 10, 12, 14, 16, 18 tsf ...

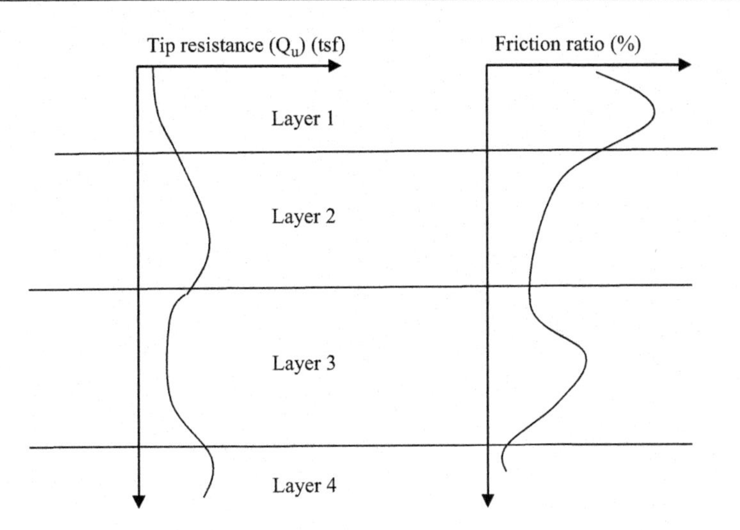

Figure 2.32 CPT data in layered soil.

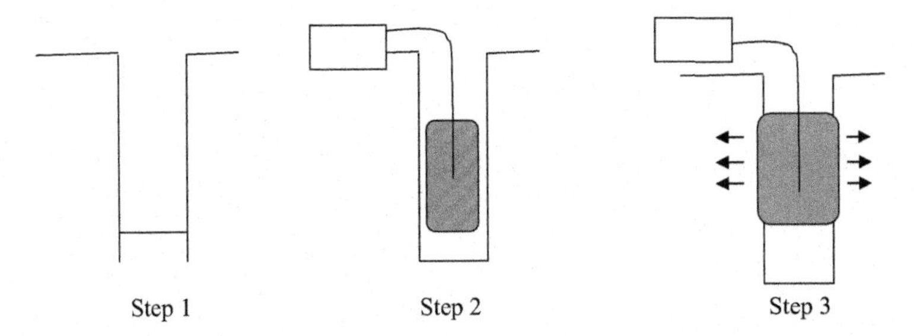

Figure 2.33 Pressure meter procedure. Step 1: Drill a hole. Step 2: Insert the pressure meter. Step 3: Inflate the apparatus.

The Equal Volume Increment Method – In this method, the volume of the pressuremeter is increased at equal intervals. If the volume of the pressuremeter is 10 in.^{3} and the increment is 2 in.^{3}, the following volume values will be maintained: 10, 12, 14, 16, 18 in.^{3}…
A typical graph obtained from a pressuremeter test is shown in Figure 2.34.

When the volume of the pressuremeter is increased, the pressure inside the pressuremeter will go up. Initially, when the pressuremeter starts to expand, there is a slight gap between the soil and the pressuremeter. Also the soil at the wall of the borehole is disturbed. Initial region of the graph is not of much use. After that approximately straight line graph is seen. This region is named nearly elastic or pseudo elastic. Soils are rarely fully elastic. After that the plastic region is reached. Failure load is known as p_L.

The following equation is used to calculate a parameter known as a pressuremeter modulus.

$$E_{pm} = (1+v)2V(\Delta p/\Delta V)$$

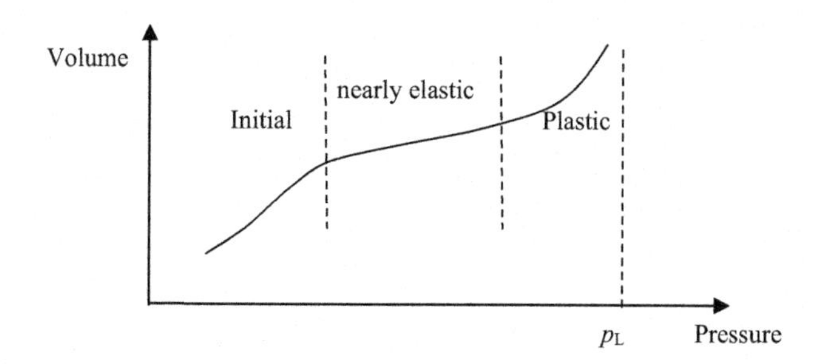

Figure 2.34 Pressure meter graph.

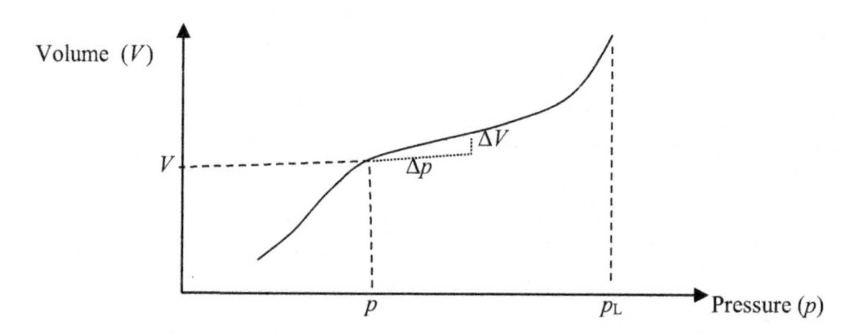

Figure 2.35 Pressure meter graph analysis.

E_{pm}, pressuremeter modulus; ν, Poisson's ratio; V, volume of the pressuremeter at a pressure of p; Δp, presssure increment; ΔV, volume increment; p_L, failure pressure.

See Figure 2.35 for explanation.

In the equation discussed earlier, $(\Delta p/ \Delta V)$ is the gradient of the graph.

Once the graph is obtained, E_{pm} can be computed.

What is the purpose of computing E_{pm}?

There are correlations between E_{pm} and various soil types (Table 2.4). Hence, the soils can be identified using the pressuremeter test data.

Table 2.4 **Pressure meter table**

	E_{pm}/p_L
Clay	>16
Silt	>14
Sand	>12
Sand and gravel	>10

Obtained from Terzaghi, et al. (1996).

Table 2.4 is valid only for overconsolidated soils. Most soils are overconsolidated. Normally, consolidated soils are extremely rare.

Practice Problem

A pressuremeter test was done and the following data were obtained. At a pressure of 4000 psf, the volume of the pressuremeter was measured to be 13 in.^{3}. The pressure is slightly increased by 1000 psf and the volume increase was measured to be 0.25 in.^{3}. Poisson's ratio of soil is known to be 0.3. The failure stress (p_L) is obtained from the graph and found to be 8500 psf. How would you characterize the soil?

Solution

Step 1: Write down the known values.

$p = 4000$ psf; $V = 13$ in^3.
$\Delta p = 1000$ psf; $\Delta V = 0.4$ in^3., $p_L = 8500$ psf
$E_{pm} = (1 + \nu)\, 2V.\, (\Delta p/\, \Delta V)$
$E_{pm} = (1 + 0.3)\, 2 \times 13.\, (1{,}000/0.25) = 135{,}200$
$E_{pm}/p_L = 135{,}200/8{,}500 = 15.9$

The value 15.9 is greater than 14. Hence, the soil is most likely a silt.
Note: E_{pm}/p_L is unit less. Make sure that the same units are maintained for Δp and p_L.

2.14 Dilatometer testing

Dr Silvano Marchetti developed the dilatometer test apparatus in 1975 (Figure 2.36). The dilatometer blade has a cross sectional area of about 14 cm^2 and can be pushed into soil with a rig. A dilatometer blade has a membrane attached to it. Once the dilatometer blade is in place where the test has to be done, the membrane is inflated. When the membrane is inflated, the soil would be stressed. Stronger soils would obviously provide a higher resistance.

Tests can be successfully performed in all penetrable soils, including clay, silt, and sand. Dilatometer is not good to be used in gravelly soil.

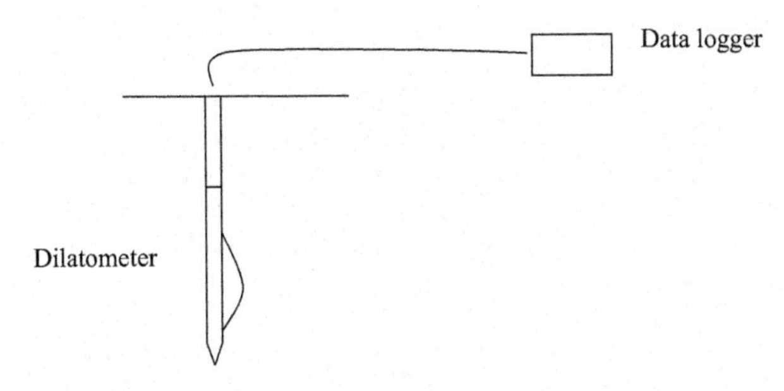

Figure 2.36 Dilatometer.

Table 2.5 SPT–CPT correlations for clays and sands

Soil type	Mean grain size (D_{50}) (measured in mm)	Q_c/N
Clay	0.001	1
Silty clay	0.005	1.7
Clayey silt	0.01	2.1
Sandy silt	0.05	3.0
Silty sand	0.10	4.0
Sand	0.5	5.7
	1.0	7.0

Source: Robertson et al. (1983).

2.15 SPT–CPT correlations

In USA, standard penetration test (SPT) is used extensively. On the other hand, CPT (cone penetration test) is popular in Europe. A standard cone has a base area of 10 cm^2 and an apex angle of 60°.

The following correlation between SPT and CPT shown in Table 2.5 can be used to convert CPT values to SPT number or vice versa.

Q_c, CPT value measured in bars, (1 bar = 100 kPa); N, SPT value; D_{50}, size of the sieve that would pass 50% of the soil.

Design Example 2.4

SPT tests were done on a sandy silt with a D_{50} value of 0.05 mm. An average SPT (N) value for this soil is 12. Find the CPT value.

Solution

From Table 2.5, for sandy silt with an D_{50} value of 0.05 mm,

$Q_c/N = 3.0$

$N = 12$; Hence $Q_c = 3 \times 12 = 36$ bars = 3600 kPa.

References

Bowles, J., 2004. Foundation Analysis and Design. McGraw Hill Book Company, NY.

Hatakanda, M., Uchida, A., 1996. Empirical correlation between penetration resistance and effective friction angle of sandy soil. Soils Found. 36 (4), 1–9.

Robertson, P.K., et al.,1983. SPT-CPT Correlations. ASCE Geotech. Eng. J. 109, 1449–1468.

Terzaghi, K., Peck, R., Mesri, G., 1996. Soil Mechanics in Engineering Practice, third ed. John Wiley and Sons, New York.

Groundwater

3

Water existing in the atmosphere, in clouds and air, is known as meteoric water. Surface water could be fresh water or oceanic water. Groundwater is a part of the water cycle. Groundwater would seep into streams and then get evaporated. The evaporated water falls back on earth and seeps back into the ground (Figure 3.1).

Man-made parking lots and cities prevent the infiltration of water into ground. The water falling on a big city would be transported to drains and then it flows to the ocean or large rivers through storm pipes. This would bypass the groundwater (Figure 3.2).

Magmatic water: when volcanoes erupt, magma would flow. The water in magma gets released to the atmosphere and to the ground. Magmatic water can be identified from its mineral content.

Connate water: in Chapter 1, we discussed the formation of sedimentary rocks. During the formation of sedimentary rocks, water gets trapped inside the rocks. This water has not been in contact with the atmosphere for a long period of time.

Metamorphic water: metamorphic rocks are formed due to high pressure and high temperature. During the formation of metamorphic rocks, water gets trapped inside metamorphic rocks. The chemical composition of metamorphic water is different from connate water.

Juvenile water: magmatic water, connate water, or metamorphic water newly entering into the water cycle is known as juvenile water.

3.1 Vertical distribution of groundwater

The zone above the groundwater level is known as vadose zone. The vadose zone is also known as the zone of aeration. The vadose zone is divided into soil-water zone, intermediate-vadose zone, and capillary zone (Figure 3.3).

Soil-water zone: this zone is exposed to the atmosphere. During rains, this zone gets saturated. Other times, this zone is semisaturated.

Intermediate vadose zone: water in this zone depends on the soil type. Some soils would be able to store a significant amount of water.

Capillary zone: due to capillary action, the groundwater rises up. The amount of water in the capillary zone depends on the soil type.

3.2 Aquifers, aquicludes, aquifuges, and aquitards

Aquifer: Water-bearing soil and rock formations are known as aquifers. The aquifers are capable of absorbing water and also transmitting water. Typically, sandy soils and sedimentary rock formations are considered to be aquifers.

Aquifers are of two types: confined aquifers and unconfined aquifers.

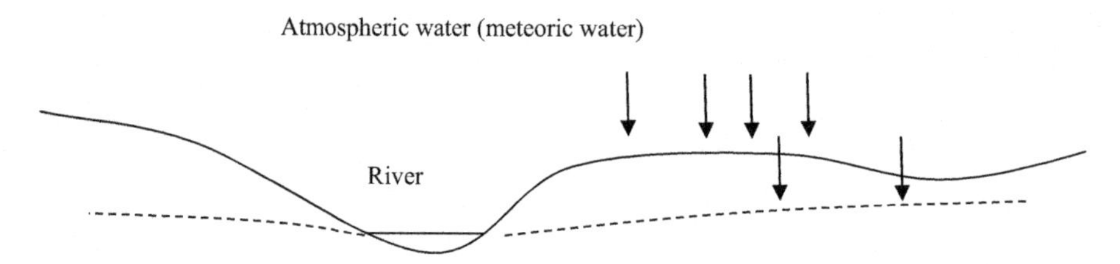

Figure 3.1 Water cycle.

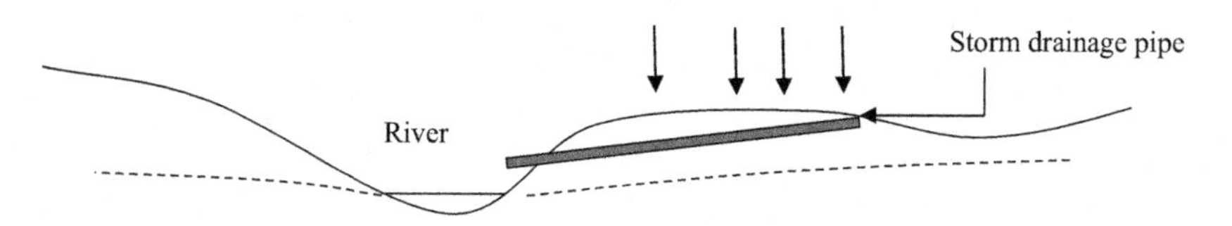

Figure 3.2 Storm drainage.

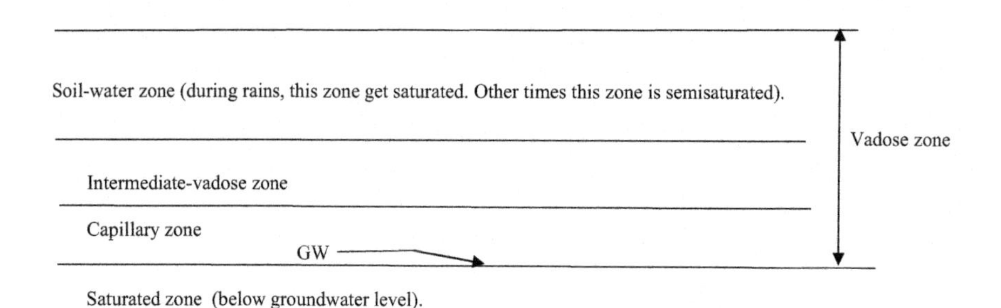

Figure 3.3 Groundwater zones.

Confined aquifers are not open to the atmosphere. Confined aquifers sometimes may be under pressure. When a confined aquifer is under pressure, it is known as an *artesian aquifer*.

Figure 3.4 shows the water flowing to the surface without a pump. This is due to a confined aquifer under pressure, also known as an artesian aquifer.

Aquiclude: Aquicludes can absorb water but would not yield an appreciable quantity of water.

Aquitard: Aquitards, as in the case of aquicludes, can absorb water; however, they would not yield an appreciable quantity of water. But unlike aquicludes, the aquitards can transmit water from adjacent aquifers. Apparently, there is a slight difference between aquicludes and aquitards. Following are the definitions given by US Geological Survey (USGS):

> *Aquiclude – A hydrogeologic unit which, although porous and capable of storing water, does not transmit it at rates sufficient to furnish an appreciable supply for a well or spring. See preferred term confining unit.*

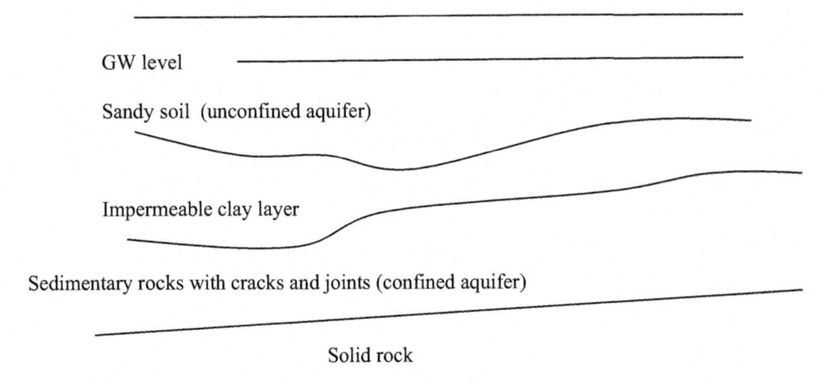

Figure 3.4 Aquifers.

> *Aquitard – A confining bed that retards but does not prevent the flow of water to or from an adjacent aquifer; a leaky confining bed. It does not readily yield water to wells or springs, but may serve as a storage unit for ground water (AGI, 1980). See preferred term confining unit.*
> *Confining unit – (1) A hydrogeologic unit of impermeable or distinctly less permeable material bounding one or more aquifers and is a general term that replaces aquitard, aquifuge, aquiclude (after AGI, 1980).*

Although, aquitards do not have much water in store, they can transmit water from nearby aquifers. However, aquicludes are unable to do so. Also it seems that the USGS prefers to use the term "confining unit" for aquicludes, aquitards, and aquifuges.

Aquifuges: Aquifuges are not capable of absorbing water. Hence, there is no water to transmit.

3.3 Piezometers

Piezometers are used to measure the pore water pressure in soil.

Figure 3.5 shows the main items of a piezometer. Figure 3.5 shows a piezometer tip. The piezometer tip is equipped with an electronic device known as a pressure transducer that can measure the water pressure. The piezometer tip is backfilled with a filter sand. Bentonite seal is placed above the filter sand. Bentonite cement grout is placed on top of the bentonite seal. This is a much cheaper product than bentonite seal.

Piezometer tip: The tip has an electronic device to measure the pore pressure.
Filter sand: Filter sand allows the water to flow to the piezometer tip.
Bentonite seal: The bentonite seal stops the water migrating from top aquifers.
Bentonite cement grout: The bentonite cement grout serves the same purpose as the bentonite seal. This is a much cheaper product.

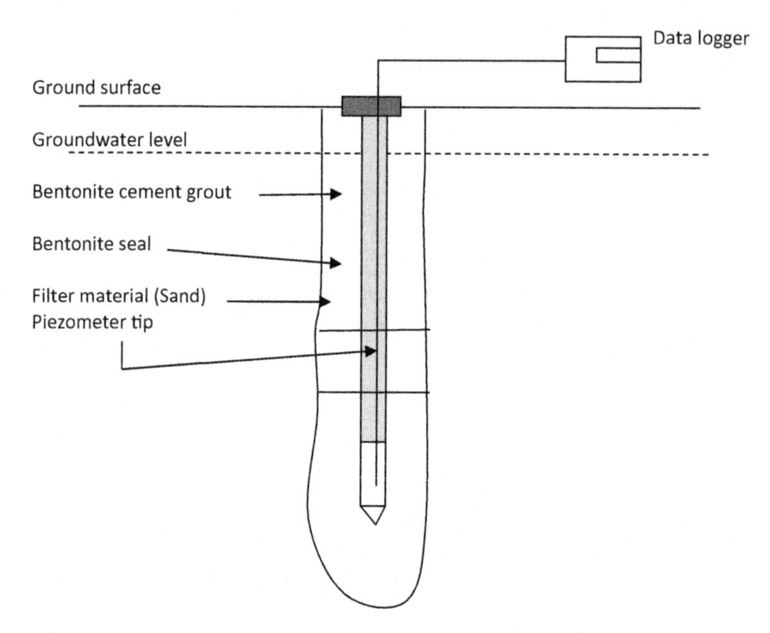

Figure 3.5 Piezometer.

3.3.1 Piezometric surface versus groundwater level

In some situations, piezometric surface is not the same as the groundwater level. This happens in confined aquifers that are under pressure. Figure 3.6 shows an aquifer. The groundwater level and piezometric level is same in aquifers.

3.3.2 Aquitard under pressure

Figure 3.7 shows an aquitard under pressure. In this case, the piezometric level will be higher than the groundwater level.

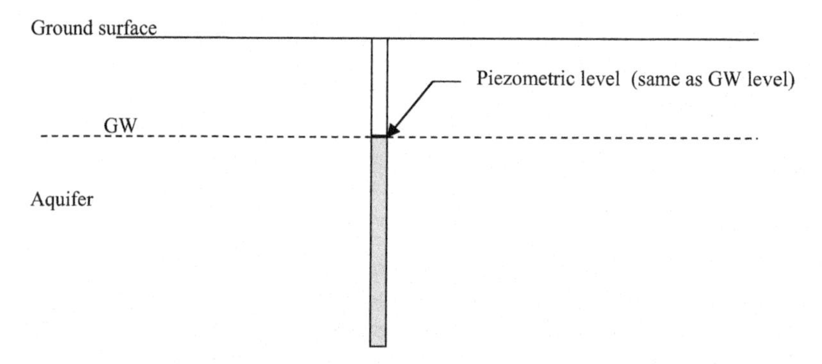

Figure 3.6 Piezometric level.

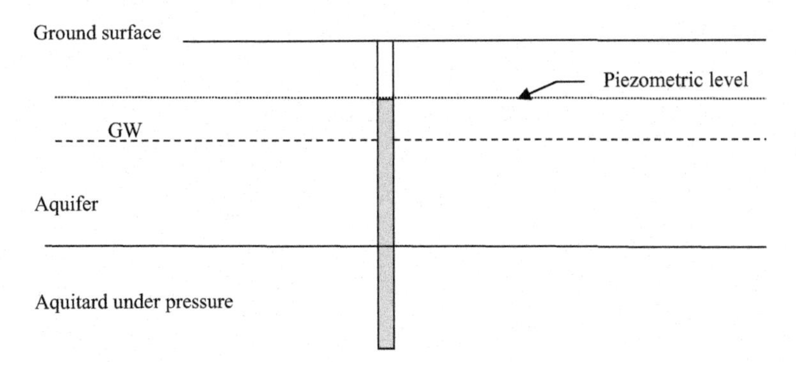

Figure 3.7 Aquitard under pressure.

3.3.3 Vertical upward groundwater flow

Figure 3.8 shows the number of soil layers and their piezometric level.

If you look at Figure 3.8, you would see that the pressure head in soil layer 4 is higher than the pressure head in soil layer 3. Hence, the water will flow from layer 4 to layer 3. Similarly, the pressure head in soil layer 3 is higher than the pressure head in soil layer 2. Hence, water will flow from layer 3 to layer 2. This is called upward vertical flow.

3.3.4 Vertical groundwater flow

Figure 3.9 shows the number of soil layers and their piezometric level.

3.3.5 Monitoring wells

Monitoring wells are installed to obtain groundwater elevation (Figure 3.10). Monitoring wells are typically constructed using PVC pipes.

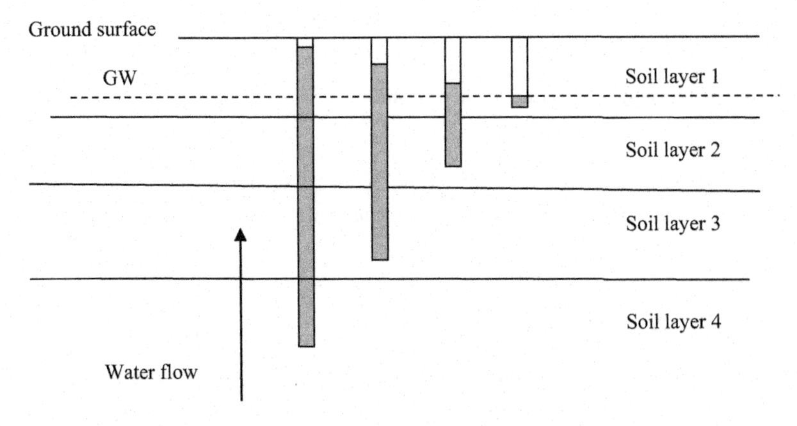

Figure 3.8 Vertical upward flow.

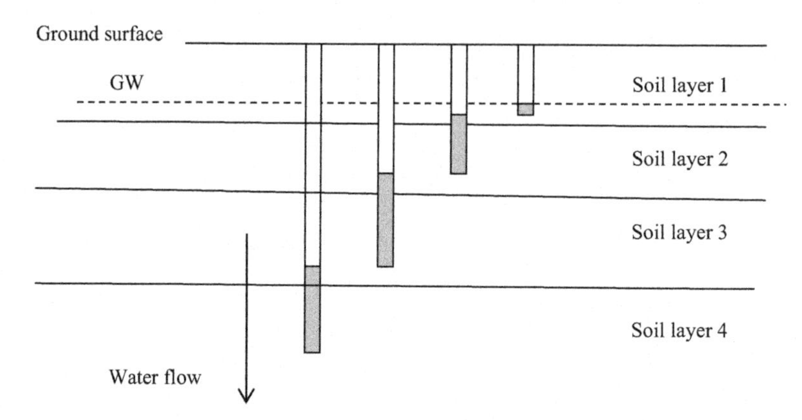

Figure 3.9 Vertical downward flow.

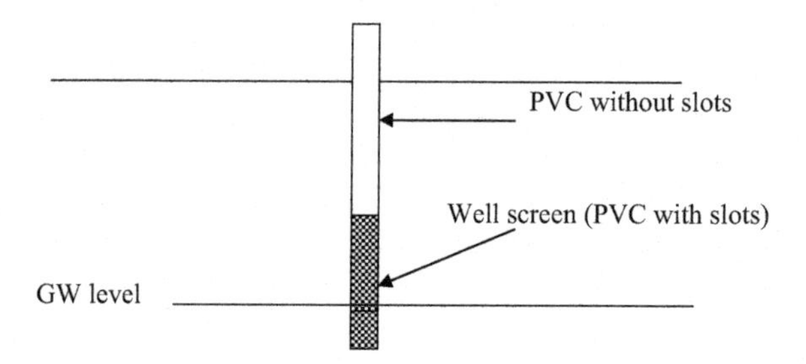

Figure 3.10 Groundwater monitoring well.

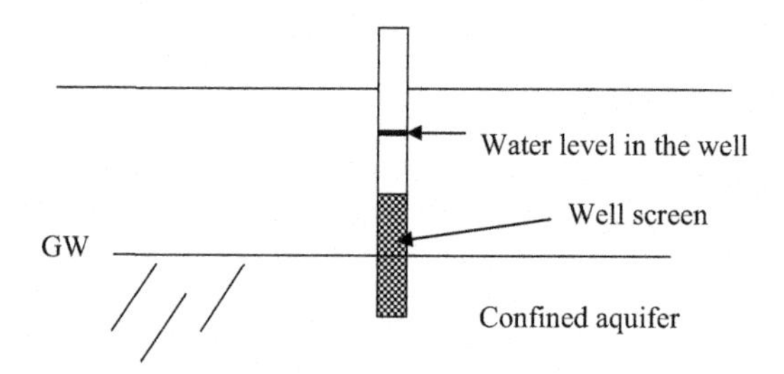

Figure 3.11 Monitoring well in a confined aquifer.

Slotted section of the PVC is known as the well screen and it allows the water to flow into the well. If there is no pressure, the water level in the well indicates the groundwater level.

3.3.6 Aquifers with artesian pressure

Groundwater in some aquifers can be under pressure. The monitoring wells will register a higher water level than the groundwater level. In some cases, water would spill out from the well due to artesian pressure (Figure 3.11).

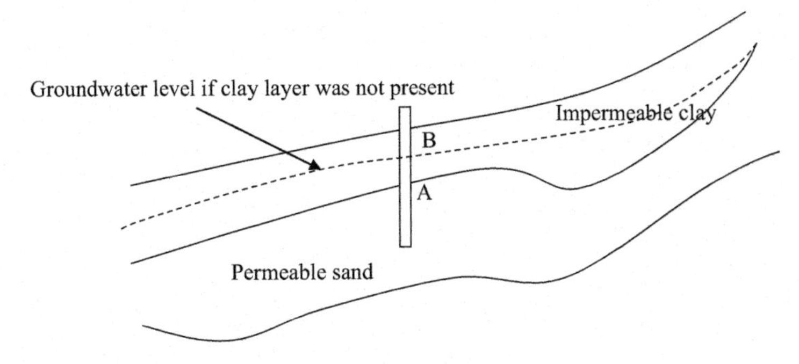

Figure 3.12 Artesian conditions.

Water level in the well is higher than the actual water level in the aquifer since the aquifer is under pressure.

In Figure 3.12, the impermeable clay layer is shown lying above the permeable sand layer. The dotted line shows the groundwater level if the clay layer is absent. Due to the impermeable clay layer, the groundwater can not reach the level shown by the dotted line. Hence, the groundwater level is confined to point A in the monitoring well. When a well is installed, the water level will rise to point B, higher than the initial water level due to artesian pressure.

Soil laboratory testing

4

After completion of the boring program, laboratory tests are conducted on the soil samples. Laboratory test program is dependant upon the project requirements. Some of the laboratory tests done on soil samples are as follows:

1. Sieve analysis
2. Hydrometer
3. Water content
4. Atterberg limit tests (liquid limit and plastic limit)
5. Permeability test
6. UU tests (undrained unconfined tests)
7. Density of soil
8. Consolidation test
9. Triaxial tests
10. Direct shear test

4.1 Sieve analysis

Sieve analysis is conducted to classify soil into sands, silts, and clays. Sieves are used to separate soil particles and group them based on their size. This test is used for the purpose of classification of soil.

Standard sieve sizes are shown in Tables 4.1 and 4.2.

Hypothetical sieve analysis test based on selected group of sieves is given in Figure 4.1 as an example.

If we know the percentage of soil passed from a given sieve, we can find the percentage of soil retained in that sieve.

- Sieve no. 4 (size 4.75 mm): All soil went past Sieve no. 4.
  Percent retained at this sieve is 0%.
  Percent passed = 100%.
- Sieve no. 16 (1.18 mm): 20% of soil was retained in Sieve no. 16.
  Percent retained at Sieve no. 16 = 20%
  Percent passed = 100 – 20 = 80%.
- Sieve no. 50 (0.30 mm): 25% of soil was retained in Sieve no. 50.
  Total retained, so far = 20 + 25 = 45%
  Percent passed = 100 – 45 = 55%
- Sieve no. 80 (0.18 mm): 20% retained in Sieve no. 80.
  Total retained so far = 20 + 25 + 20 = 65%
  Percent passed = 100 – 65 = 35%.
- Sieve no. 200 (0.075 mm): 30% retained in Sieve no. 200.
  Total retained so far = 20 + 25 + 20 + 30 = 95%
  Percent passed through Sieve no. 200 = 100 – 95 = 5%.

Table 4.1 **US sieve number and mesh size**

Sieve number	Mesh size (mm)
4	4.75
6	3.35
8	2.36
10	2.00
12	1.68
16	1.18
20	0.85
30	0.60
40	0.425
50	0.30
60	0.25
80	0.18
100	0.15
200	0.075
270	0.053

Notes: gravel, sizes greater than #4 sieve is considered to be gravel; sands, #4–#200; silts and clays, smaller than #200.

Table 4.2 **British sieve number and mesh size**

British sieve number	Size (mm)
8	2.057
16	1.003
30	0.500
36	0.422
52	0.295
60	0.251
85	0.178
100	0.152
200	0.076
300	0.053

Notes: gravel, sizes greater than #4 sieve is considered to be gravel; sands, #4–#200; silts and clays, smaller than #200.

Now it is possible to draw a graph indicating percent passing at each sieve (Figure 4.2).

4.1.1 D_{60}

D_{60} is defined as the size of the sieve that allows 60% of the soil to pass. This value is used for soil classification purposes and it frequently appears in geotechnical engineering correlations.

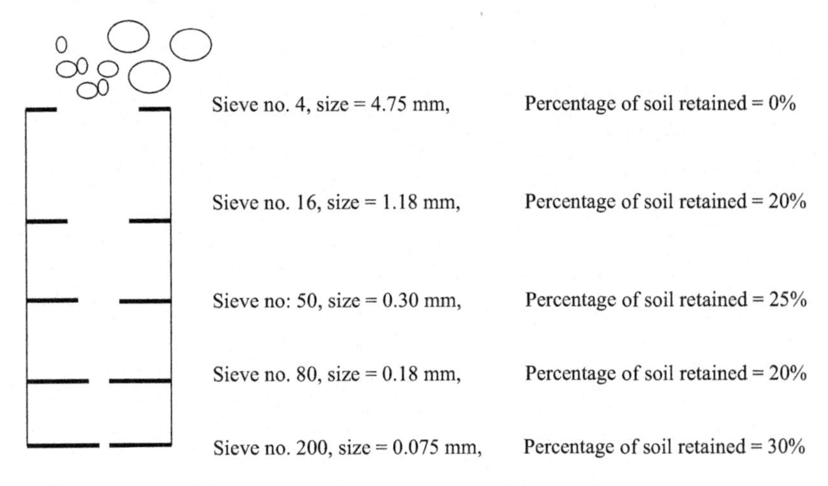

Figure 4.1 Sieve analysis.

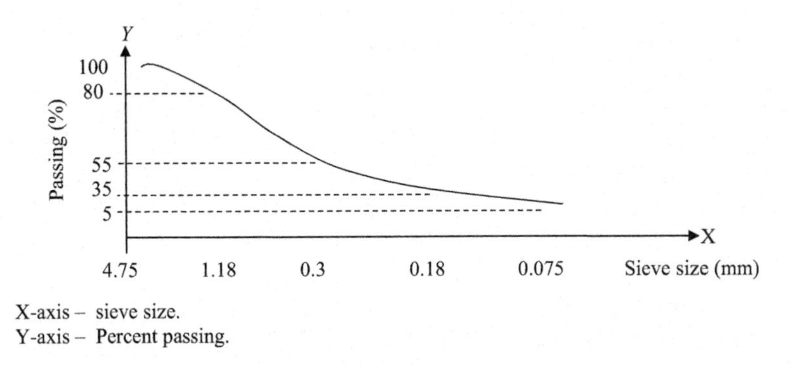

Figure 4.2 Percent passing versus particle size.

If you look at the 0.3 mm sieve, 55% of the soil would pass though this sieve. If we make the sieve bigger, more soil would pass. On the other hand, if we make the sieve smaller, then less than 55% of the soil would pass.

To find the D_{60} value, draw a line at 60% passing point. Then drop it down to obtain the D_{60} value (Figure 4.3).

In this case, D_{60} is closer to 0.5 mm.

4.1.2 Find D_{30}

As earlier, draw a line at 30% passing line. In this case, D_{30} happened to be at approximately 0.1 mm (Tables 4.3 and 4.4).

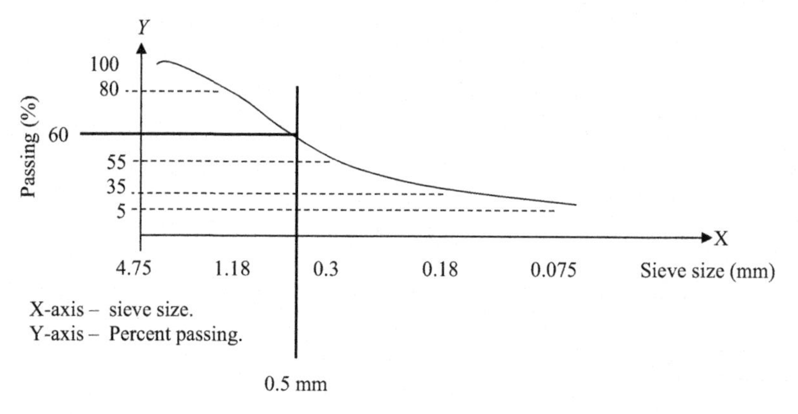

Figure 4.3 Finding D_{60}.

Table 4.3 **Size ranges for soils and gravels**

Soil	Size in (in.)	Size in (mm)	Comments
Boulders	6 or larger	150 or larger	
Cobbles	3–6	75–150	
Gravel	0.187–3	4.76–75	Greater than #4 sieve size
Sand	0.003–0.187	0.074–4.76	Sieve #200 to sieve #4
Silt	0.00024–0.003	0.006–0.074	Smaller than sieve #200
Clay	0.00004–0.00008	0.001–0.002	Smaller than sieve #200
Colloids	Less than 0.00004	Less than 0.001	

Table 4.4 **Specific gravity (G_s)**

Soil	Specific gravity
Gravel	2.65–2.68
Sand	2.65–2.68
Silt (inorganic)	2.62–2.68
Organic clay	2.58–2.65
Inorganic clay	2.68–2.75

4.2 Hydrometer

Hydrometer tests are conducted to classify particles smaller than #200 sieve. (0.075 mm).

When soil particles are mixed with water, larger particles would settle fast. On the other hand, smaller particles tend to float or settle at a low velocity (Figure 4.4).

Since D_2 is greater than D_1, the velocity of sand particles (V_2) will be greater than the velocity of silt particles (V_1). Similarly, V_3 will be greater than V_2.

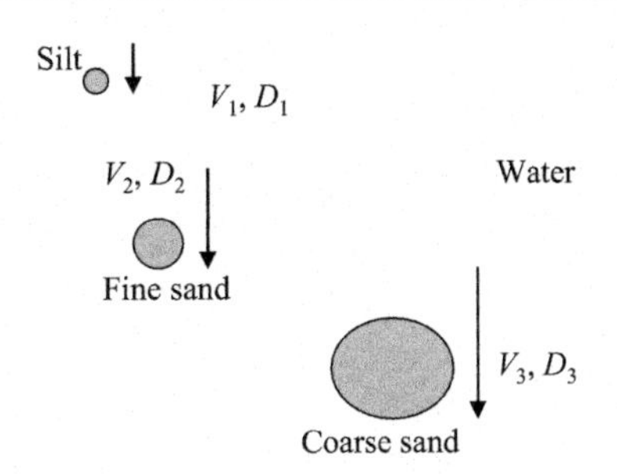

Figure 4.4 Settling soil particles.

Velocity of settling particles is given by the "Stokes Law".

$$V = \frac{980 \times (G - G_w) \times D^2}{30\eta}$$

G, specific gravity of soil; G_w, specific gravity of water; D, particle diameter (mm); V, velocity of particles (mm/s); η, absolute viscosity of water (in Poises).

"V" and "D" are unknown quantities in the given equation.

If the settling velocity "V" can be found by an experiment then the particle diameter "D" can be computed.

4.2.1 Hydrometer test procedure

The reader is referred to the ASTM D422 for a full explanation of the hydrometer test procedure. Overview of the test is provided here (Figure 4.5).

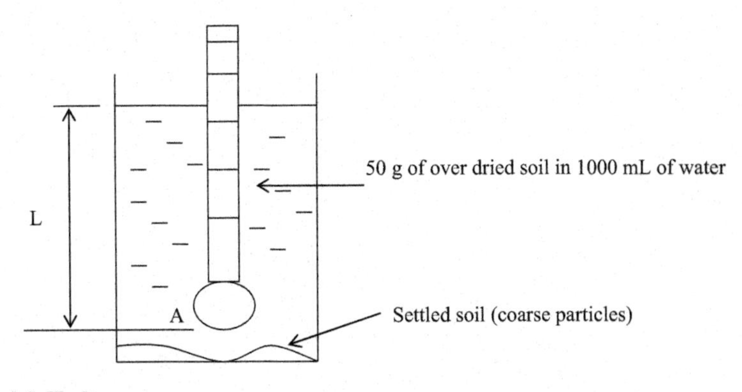

Figure 4.5 Hydrometer.

Hydrometer reading is dependant upon the density of the liquid. When soil is mixed with water very fine particles will suspend in water while the heavy particles will settle at the bottom. If more soil were to suspend in water hydrometer reading, "L" will be smaller. The reading "L" is an indication of the amount of fine particles suspended in water.

4.2.1.1 Procedure

Mix 50 g of oven dried soil and 1000 mL of distilled water.

- Mix soil and water thoroughly.
- Insert the hydrometer and obtain the hydrometer readings, "L".
- The hydrometer reading provides grams of soil in suspension per liter of solution.
- Obtain readings for different time periods. Typically 2, 5, 15, 30, 60, 250, and 1440 min.
- Compute "D" for every reading.
- Plot a graph between grams of soil per liter of water versus "D".
- Soil particles will increase the density of water/soil mixture. Hence, initially "L" would be a lower value. When time passes, the larger diameter particles would settle. The density of the soil/water mixture would go down and the hydrometer would sink. ("L" value would increase).
- Lets say at time T_1, the length was measured to be L_1. Use the Stokes equation to find D_1.

$$V_1 = \frac{L_1}{T_1} = \frac{980(G_1 - G_w)D_1^2}{30\eta}$$

- Since L_1 and T_1 are known, D_1 can be computed.
- L_1 is the hydrometer reading and T_1 is the time passed.
- G_1 is the specific gravity of soil and has to be measured separately.
- According to the Stokes equation, all particles larger than D_1 have settled below the hydrometer level.
- Obtain the hydrometer reading (grams of soil in suspension). The "D" value computed from the Stokes equation gives the largest particles that possibly could be in suspension. In other words, all the particles in the solution are smaller than the "D" value obtained using the Stokes equation.
- Hence, the hydrometer reading is similar to the percent weight passing reading given by a sieve.
- Use the weights passing readings to obtain the percent passing readings by dividing the weight per each size by the total sample.
- Fill the table below.

Sieve analysis

Sieve size number	Percent passing
4	________
10	________
40	________
200	________

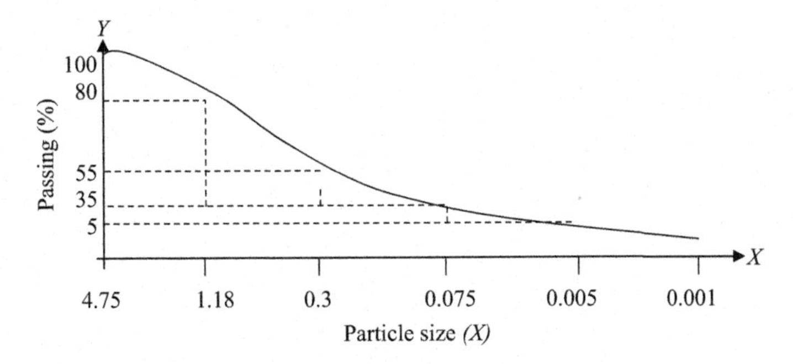

Figure 4.6 Hydrometer readings incorporated to sieve analysis curve.

Hydrometer analysis

0.074 mm	______
0.005 mm	______
0.001 mm	______

Combine the hydrometer readings with sieve analysis readings and obtain one graph (Figure 4.6).

4.3 Liquid limit, plastic limit, and shrinkage limit (Atterberg limit)

4.3.1 *Liquid limit*

Liquid limit is the water content where the soil starts to behave as a liquid.

Liquid limit is measured by placing a clay sample in a standard cup and making a separation (groove) using a spatula. The cup is dropped till the separation vanishes. The water content of the soil is obtained from this sample. The test is performed again by increasing the water content. Soil with low water content would yield more blows and soil with high water content would yield less blows.

A graph is drawn between number of blows and the water content (Figure 4.7).

Liquid limit of a clay (LL) is defined as the water content that corresponds to 25 blows.

What's the significance of liquid limit?

Liquid limits for two soils are shown in Figure 4.8. Soil 1 would reach a liquid-like state at water content of LL1. On the other hand, soil 2 would attain this state at water content LL2.

In Figure 4.8, LL1 is higher than LL2. In other words, soil 2 loses its shear strength and becomes liquid-like at a low water content than soil 1.

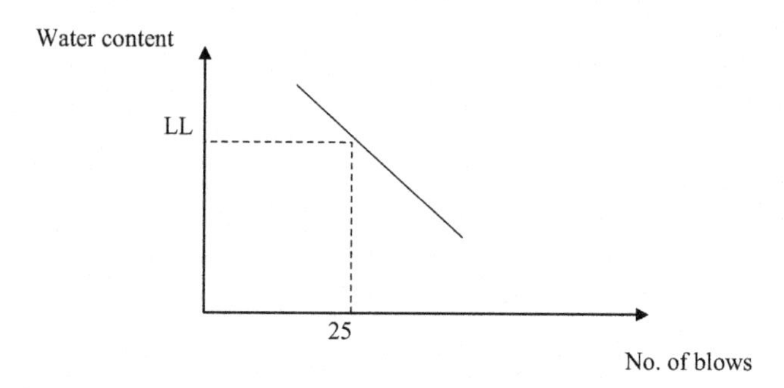

Figure 4.7 Graph for liquid limit test.

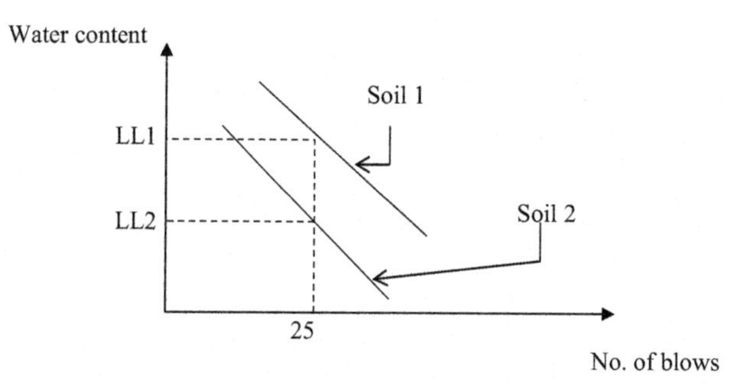

Figure 4.8 Liquid limits for two soils.

4.3.2 Plastic limit

Plastic limit is measured by rolling a clay sample to a 3 mm diameter cylindrical shape. During continuous rolling at this size, the clay sample tends to lose moisture and cracks start to appear. The water content where cracks start to appear is defined to be the plastic limit (Figure 4.9).

4.3.3 Practical considerations of liquid limit and plastic limit

The water content where a soil converts to a liquid-like state is known as the liquid limit. Consider two slopes as shown in Figure 4.10. Assuming all other factors to be equal, which slope would fail first?

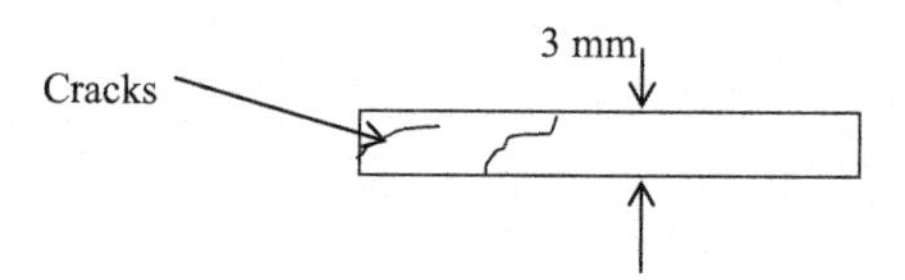

Figure 4.9 Plastic limit test.

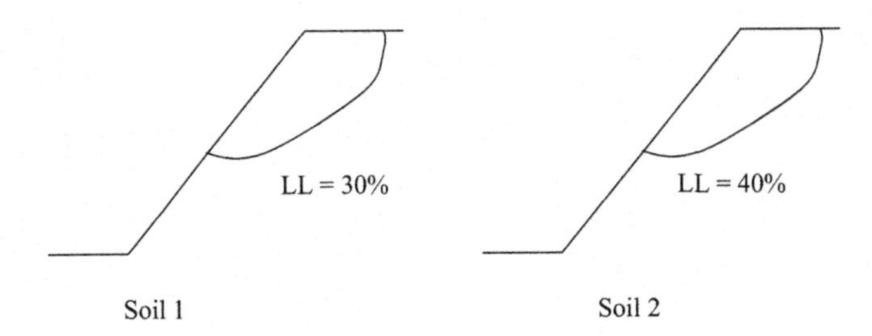

Figure 4.10 Slope stability in different soils.

During a rain event, soil 1 would reach the liquid limit prior to soil 2. Hence, soil 1 would fail before soil 2.

During earthquakes, water tends to rise. If soils with low liquid limit were to be present, those soils would lose their strength and fail.

The plastic limit indicates the limit of plasticity. When the water content goes below the plastic limit of a soil, then cracks would start to appear in that soil. Soils lose their cohesion below the plastic limit.

4.3.4 Shrinkage limit

When the moisture content of a cohesive soil reduces due to evaporation, the soil would shrink or a reduction of volume would occur(Figure 4.11). The volume reduction would come to a stop below the shrinkage limit. In other words, below shrinkage limit, reduction in moisture content will not cause reduction in volume (Figure 4.12).

The liquidity index is calculated as: LI = (NM − PL)/PI

Where, NM, the soil's natural moisture content in percent; PL, plastic limit; PI, plasticity index.

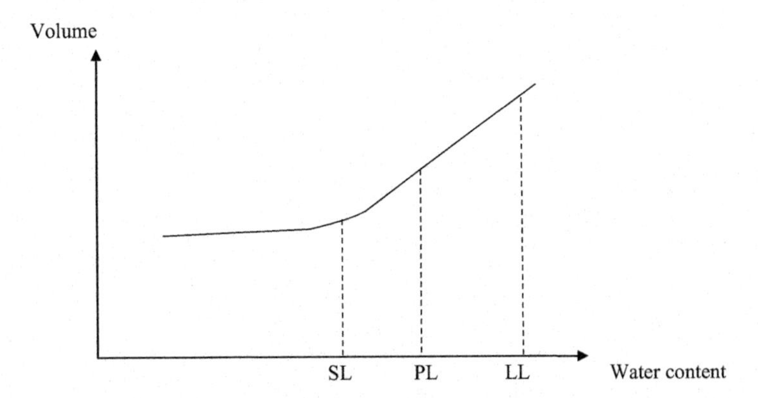

Figure 4.11 Volume versus water content.

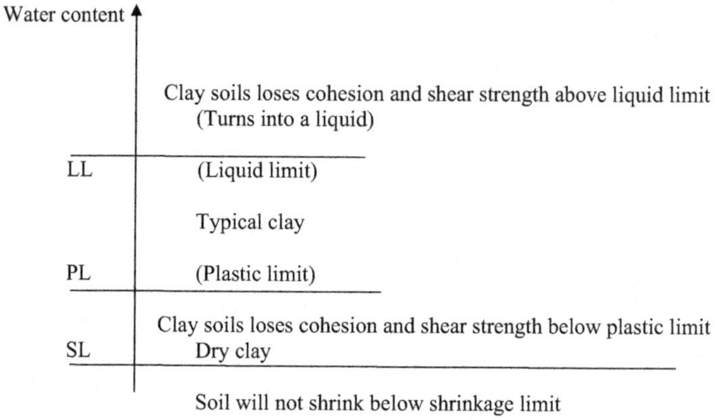

Water content

Clay soils loses cohesion and shear strength above liquid limit
(Turns into a liquid)

LL (Liquid limit)

Typical clay

PL (Plastic limit)

Clay soils loses cohesion and shear strength below plastic limit
SL Dry clay

Soil will not shrink below shrinkage limit

Figure 4.12 Liquid limit and plastic limit.

4.4 Permeability test

Transport of water through soil media depends on the pressure head, velocity head, and the potential head due to elevation. In most cases, the most important parameter is the potential head due to elevation.

In Figure 4.13, water travels from "A" to "B" due to high potential head. The velocity of traveling water is given by the Darcy equation.

$$v = k \times i \qquad \text{(Darcy's equation)}$$

v, velocity; k, coefficient of permeability (cm/s or in./s); i, hydraulic gradient = h/L; L, length of soil.

$$\text{Volume of water flow} = Q = A \times v$$

A, area; v, velocity.

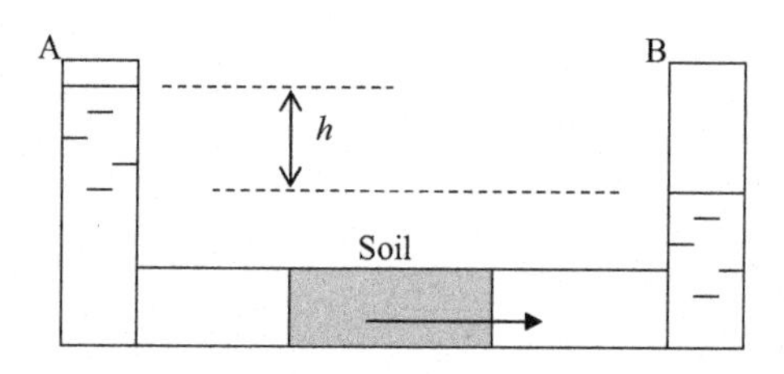

Figure 4.13 Water flowing through soil.

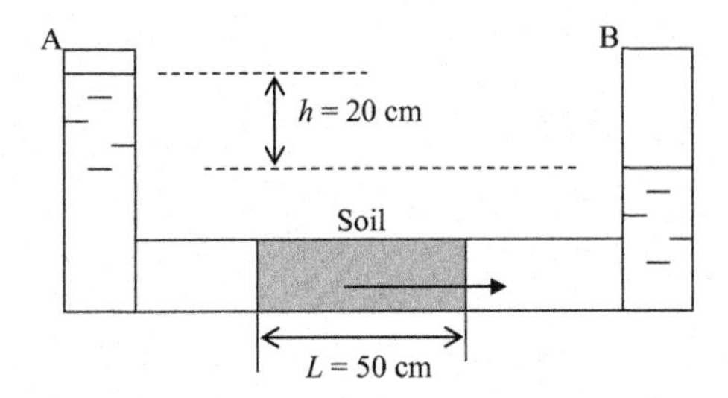

Figure 4.14 Water flow due to 20 cm gravity head.

Design Example 4.1

Find the volume of water flowing in the pipe shown in Figure 4.14. Soil permeability is 10^{-5} cm/s. Area of the pipe is 5 cm^2. Length of soil plug is 50 cm.

Solution

Apply the Darcy equation.

$$v = k \times i \qquad \text{(Darcy's equation)}$$

v, velocity; k, coefficient of permeability (cm/s or in./s); i, hydraulic gradient = h/L; L, length of soil.

$$v = k \times (h/L)$$
$$v = 10^{-5} \times 20/50 = 4 \times 10^{-6} \text{ cm/s}$$

Volume of water flow = $A \times v = 5 \times 4 \times 10^{-6}$ cm^3/s = 2×10^{-5} cm^3/s

---

4.4.1 Seepage rate

Water movement in soil occurs through the voids inside the soil fabric. More voids mean more water can flow through (Figure 4.15).

We could see that the velocity of water seepage through a soil mass is dependant upon the void ratio or porosity of the soil mass (Figure 4.16).

$$\text{Water volume travel through soil} = Q = v \times A$$

A, Area; v, Velocity.

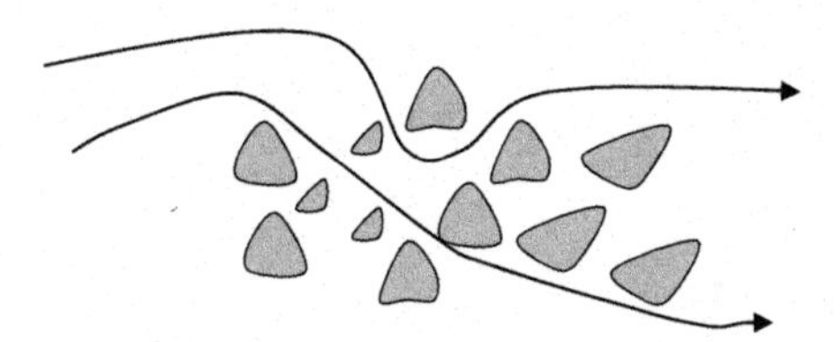

Figure 4.15 Seepage path.

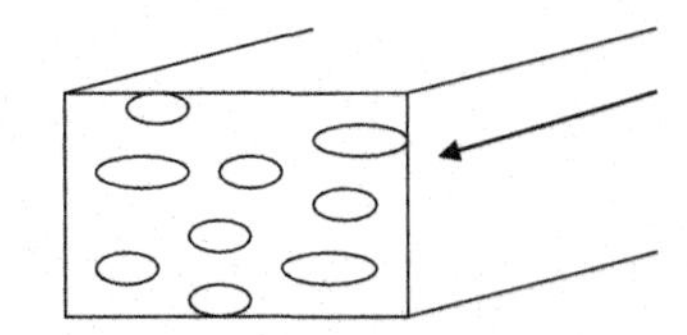

Figure 4.16 Seepage velocity.

Total cross sectional area, A; void area, A_v; porosity (n) is defined as V_v/V.

$$n = V_v/V$$

V_v,volume of voids $= L \times A_v$; A_v, area of voids; V, total volume $= L \times A$; L, length; A, total cross sectional area.

Hence,

$$n = (L \times A_v)/(L \times A) = A_v/A$$
$$A_v = n \times A$$
$$Q = v \times A$$

Velocity of water traveling through voids (v_s) is known as the seepage velocity.

$Q = v_s \times A_v$

$Q = v \times A = v_s \times A_v = v_s \times (n \times A)$

$v \times A = v_s \times (n \times A)$

$$v_s = v/n$$

4.5 Unconfined–undrained compressive strength tests (UU tests)

Unconfined compressive strength test is designed to measure the shear strength of clay soils. This is the easiest and most common test done to measure the shear strength.

Since the test is done with the sample in an unconfined state and the load is applied fast so that there is no possibility of draining, the test is known as unconfined–undrained test (UU test) (Figures 4.17, 4.18, and 4.19).

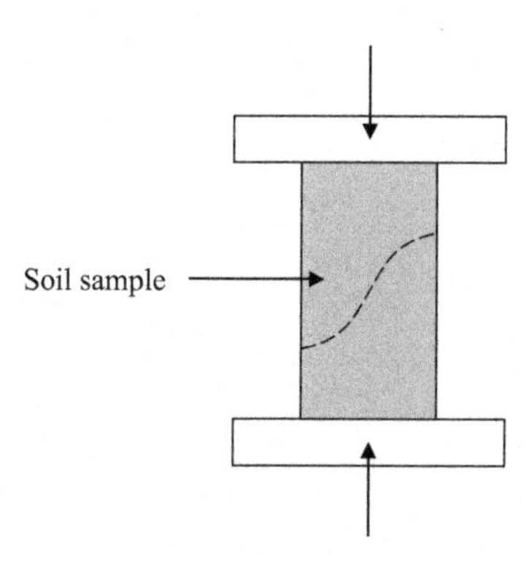

Figure 4.17 UU test apparatus.

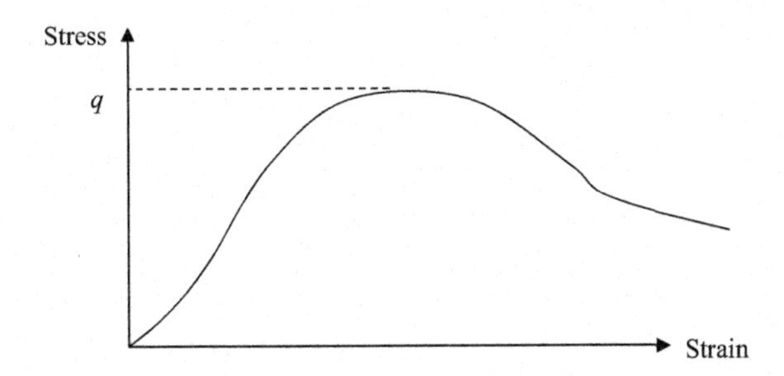

Figure 4.18 UU apparatus and stress–strain curve. Stress at failure, q.

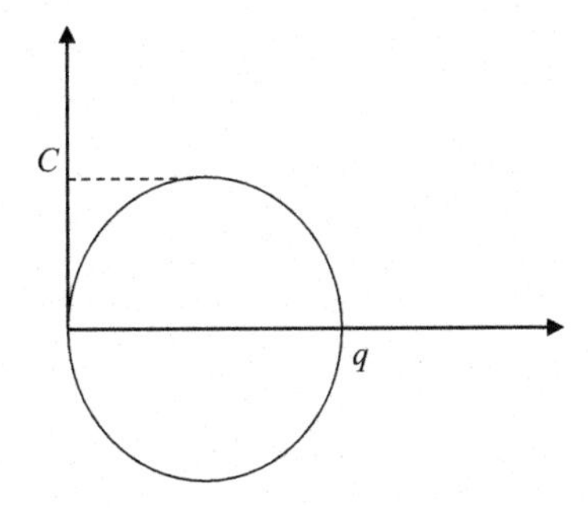

Figure 4.19 Mohr's circle for the UU test. Stress at failure, q; C, cohesion = $q/2$.

A soil sample is placed in a compression machine and compressed till failure. Stresses are recorded during the test and plotted.

4.6 Tensile failure

When a material is subjected to a tensile stress, it would undergo tensile failure.

Figure 4.20 shows material failure under tension. The tensile failure of soil is not as common as shear failure and tensile tests are rarely conducted. On the other hand,

Figure 4.20 Tensile strength test.

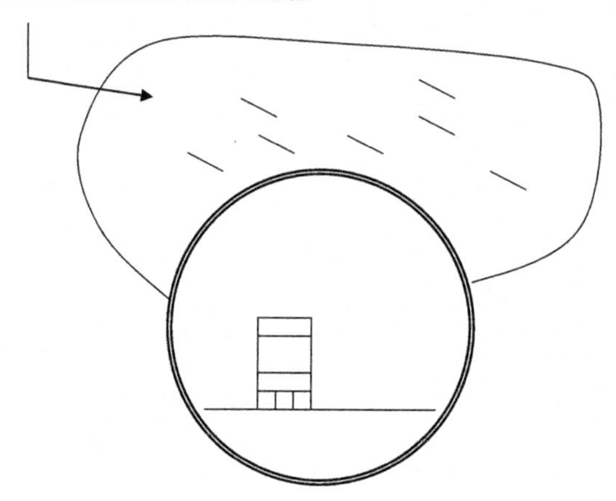

Figure 4.21 Rock under tension.

tensile failure is common in tunnels. Rocks in tunnel roofs are subjected to tensile forces and proper supports need to be provided (Figure 4.21).

Geotechnical engineering theoretical concepts

Geotechnical engineering concepts that are needed for design will be discussed in this chapter.

5.1 Vertical effective stress

Almost all problems in geotechnical engineering require computation of effective stress.

Before venturing into the concept of effective stress, let us look at a solid block sitting on a table (Figure 5.1). The density of the block is given to be "γ" and the area at the bottom of the block is "A".

Volume (V) = Ah

Weight of the block = density × volume = $(\gamma \times Ah)$

Vertical stress at the bottom = weight/area = $(\gamma \times Ah)/A = \gamma h$

What happens when there is water? (see Figure 5.2).

Due to buoyancy, the effective stress is reduced when water is introduced. This fact can be experienced when one steps into a pool. Inside the pool, we feel less weight due to buoyancy.

New vertical stress at the bottom = $(\gamma - \gamma_w) \times h$

γ_w = density of water

What happens when water is partially filled?

What is the pressure at the base, if the water is filled to a height of "y" ft. as shown in Figure 5.3.

New vertical stress at the bottom = $\gamma h - \gamma_w y$

Soils also have a lesser effective stress under the water table. Let's look at an example.

Design Example 5.1

Find the effective stress at point "A" (no water present) as shown in Figure 5.4.

Solution

Effective stress at point "A" = 18.1 × 7 = 126.7 kN/m^3

Since there is no groundwater, total density of soil is obtained to compute the effective stress.

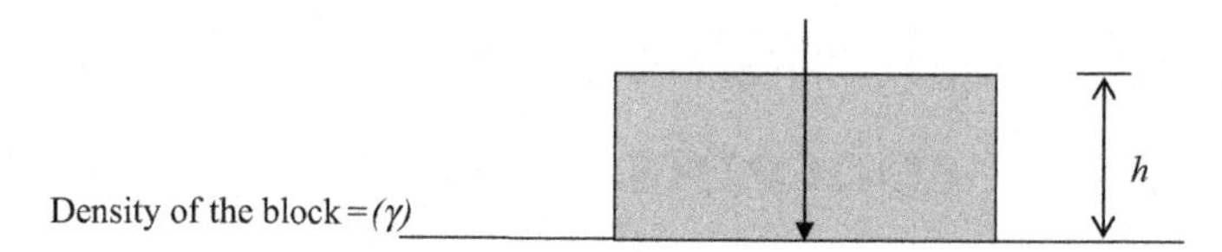

Figure 5.1 Block on top of a plane.

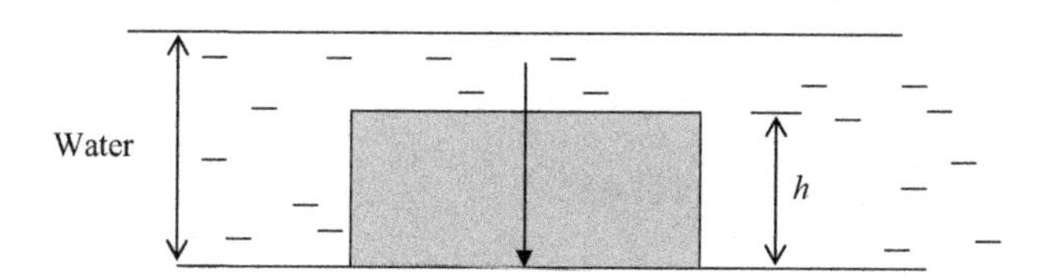

Figure 5.2 Block under buoyant forces.

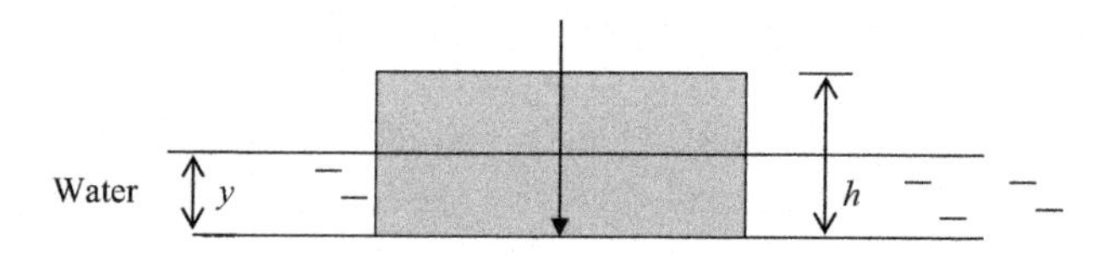

Figure 5.3 Partially submerged block.

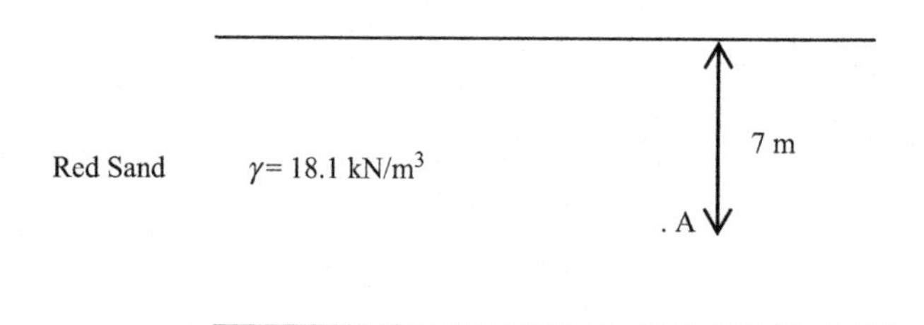

Figure 5.4 Effective stress.

Design Example 5.2

Find the effective stress at point "A." Groundwater is 2 m below the surface (Figure 5.5).

Solution

Effective stress at point "A" = $18.1 \times 2 + (18.1 - \gamma_w) \times 5\ \text{kN/m}^3$

$\gamma_w = 9.81\ \text{kN/m}^3$

Effective stress at point "A" = $18.1 \times 2 + (18.1 - 9.81) \times 5\ \text{kN/m}^3 = 77.7\ \text{kN/m}^3$.

$\gamma_w = 9.81\ \text{kN/m}^3$

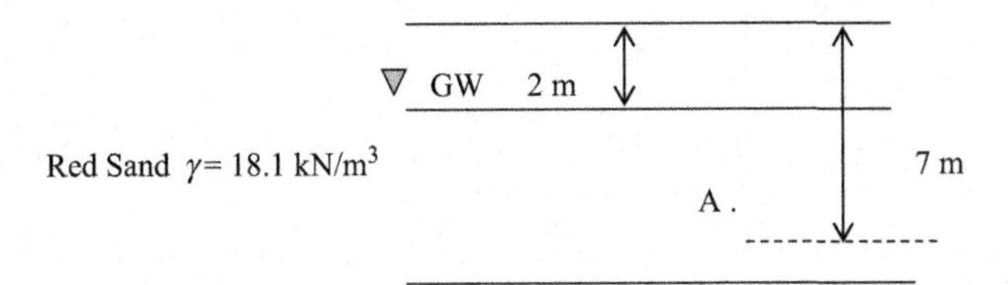

Figure 5.5 Effective stress when groundwater is present.

Note: There is no buoyancy acting on first two feet of the soil. Hence, density was NOT reduced. There is buoyancy acting on the soil below groundwater level. Hence the density of soil was reduced by 9.81 kN/m^3 to account for the buoyancy.

5.2 Lateral earth pressure

Once the vertical effective stress is found, it is a simple matter to calculate the lateral earth pressure.

Let's look at water pressure at point A (Figure 5.6).

Vertical pressure at point "A" = $\gamma_w \times h$

Horizontal pressure at point "A" = $\gamma_w \times h$

In the case of water, vertical pressure and horizontal pressure at a point are the same.

This is not the case with soil. In soil, the horizontal pressure (or stress) is different from the vertical stress (Figure 5.7).

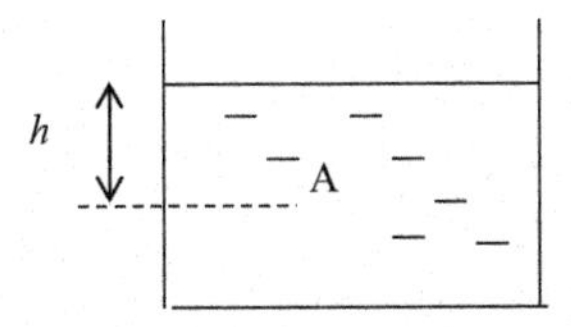

Figure 5.6 Pressure in water.

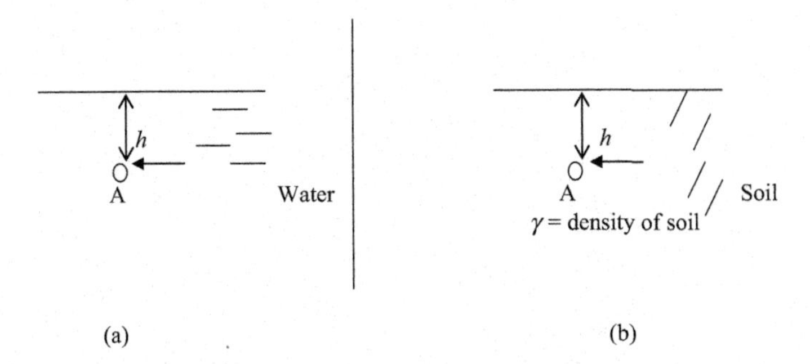

Figure 5.7 Lateral earth pressure in water and soil. (a) Water. (b) Soil.

Horizontal and vertical pressure at point "A" in water = $\gamma_w \times h$

As mentioned earlier, the pressure in water is the same in all directions.

In the case of soil, experimentally it has been found that the horizontal pressure is given by the following equation.

Horizontal pressure at point "A" in soil at rest = $K_0 \times$ (vertical effective stress)

K_0 = lateral earth pressure coefficient at rest.

Horizontal pressure at point "A" in soil at rest = $K_0 \times \gamma \times h$

γ = density of soil

h = height of soil

Jaky in 1948 found the following relationship to find K_0.

$$K_0 = 1 - \sin\varphi' \quad \text{(A)}$$

φ' = internal friction angle of soil

This equation is valid only for normally consolidated soils. For overconsolidated soils, the following equation is proposed:

$$(K_0)_{OC} = (1 - \sin\varphi')\text{OCR}^{\sin\varphi'} \quad \text{(B)}$$

$(K_0)_{OC}$, lateral earth pressure coefficient at rest for overconsolidated soil; OCR, overconsolidated ratio, (p_c'/p_0'); p_c', past preconsolidated pressure; p_0', present pressure.

Undisturbed soil samples are needed to obtain the OCR. It is not practical to obtain undisturbed soil samples for sandy soils. Due to this reason, Equation B is rarely used.

The retaining wall will move slightly to the left due to the earth pressure (Figure 5.8). Due to this slight movement, pressure on one side will be relieved and the other side will be amplified. K_a is known as the active earth pressure coefficient and K_p is known as the passive earth pressure coefficient. Passive earth pressure coefficient is larger than the active earth pressure coefficient.

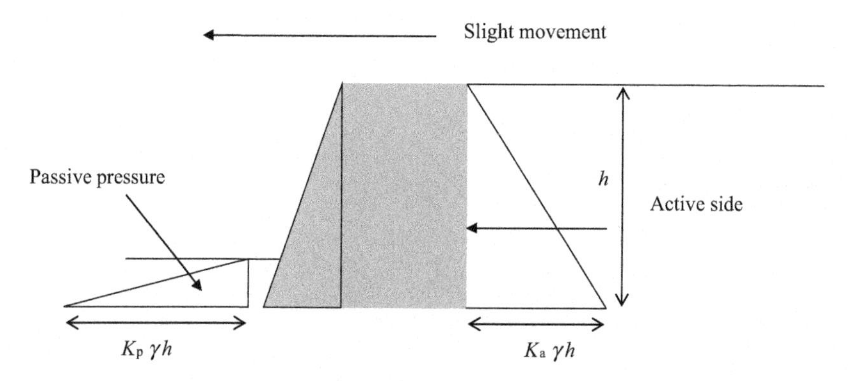

Figure 5.8 Active and passive earth pressure.

$$K_a = \tan^2(45 - \varphi'/2)$$
$$K_p = \tan^2(45 + \varphi'/2)$$
$$K_a < K_0 < K_p$$

5.3 Stress increase due to footings

When a footing is placed on soil, the stress of the soil layer would increase, whereas, the stress of the footing would reduce due to distribution.

Assume a 3 m × 3 m square footing as shown in Figure 5.9.

Column load = 180 kN

The stress on the soil just below the footing is 180/9 = 20 kPa.

Length of one side at plane A with 2:1 stress distribution = AB + BC + CD = 3 + 2 + 2 = 7 m

Stress at plane A = 180/(7 × 7) = 3.67 kN/m^2.

Note: Length AB and CD are equal to 2 m since the stress distribution is assumed to be 2:1. The depth to plane A from the bottom of the footing is 4 m. Hence, length AB and CD are 2 m each.

5.3.1 Loading on strip footings

In Figure 5.10, *D*, depth to the layer of interest measured from the bottom of the footing; pressure at footing level, *q*; length of the wall footing, *L*; width of the wall footing, *B*.

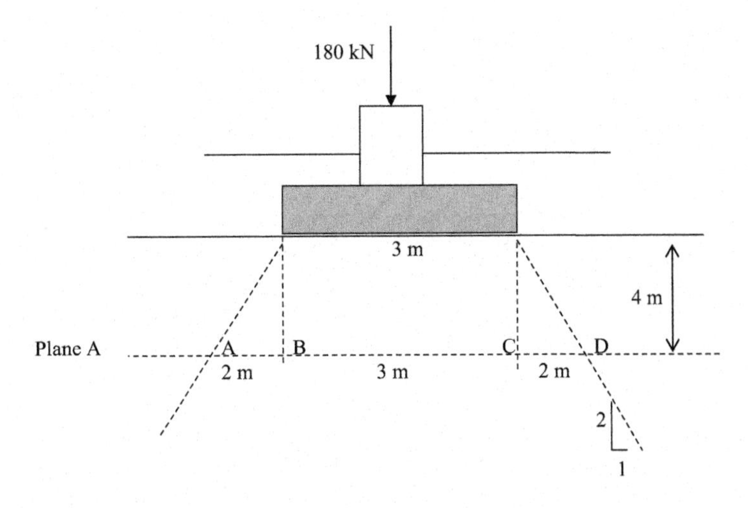

Figure 5.9 Stress distribution in a column footing.

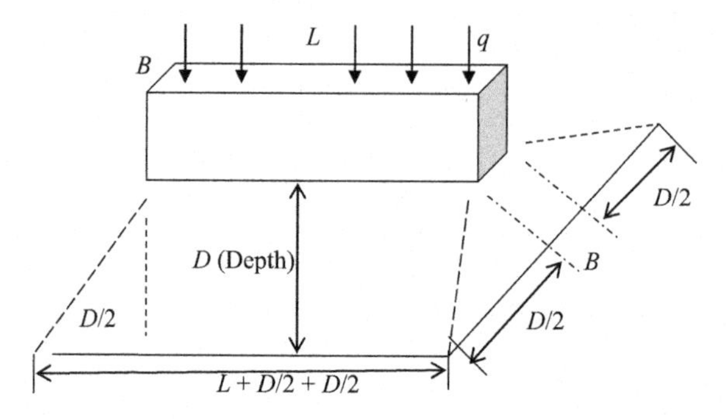

Figure 5.10 Pressure distribution under a wall footing assuming a 2:1 distribution.

Total load at footing level = $q \times (L \times B)$

Pressure at "D" m below the bottom of footing

$$= \frac{q \times (L \times B)}{(L + D/2 + D/2) \times (B + D/2 + D/2)}$$

$$= \frac{q \times (L \times B)}{(L + D) \times (B + D)}$$

Design Example 5.3

Find the stress increase at a point 2 m below the bottom of footing. The strip foundation has a load of 200 kN per 1 m length of footing and has a width of 1.5 m (Figure 5.11). Assume 2:1 stress distribution.

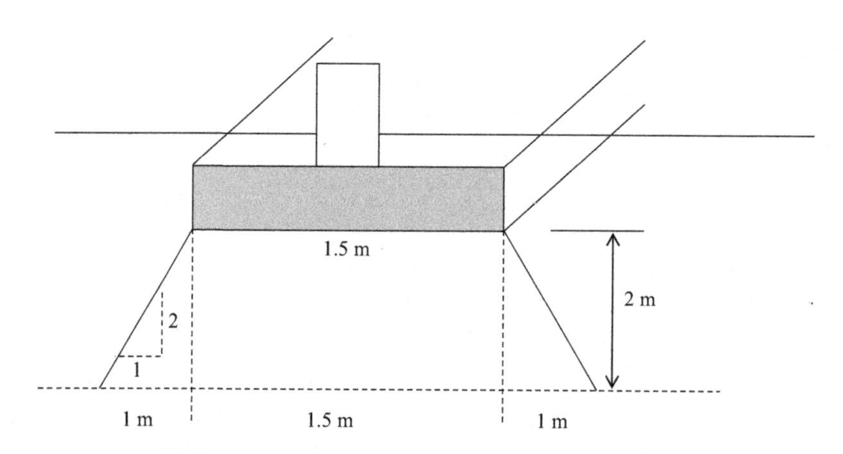

Figure 5.11 Strip footing loading.

Solution

Loading at bottom of footing per 1 m length of the footing = 200 kN

Stress increase at 2 m below the bottom of footing = 200/(3.5 × 1)

= 57.1 kN per 1 m length of footing

Note: In this case, the load distribution at the ends of the strip footing has been neglected.

5.4 Overconsolidation ratio

Overconsolidation ratio is defined as the ratio of past maximum stress and present existing stress. The existing stress in a soil can be computed based on the effective stress method. In the past, the soil most probably would have been subjected to a much higher stress.

p_c', maximum stress that soil ever being subjected to in the past p_c'; p_0', present stress; over consolidation ratio, p_c'/p_0'.

Soils have been subjected to larger stresses in the past due to glaciers, volcanic eruptions, groundwater movement, appearance, and disappearance of oceans and lakes.

5.4.1 Overconsolidation due to glaciers

Ice ages usually come and go approximately every 20,000 years. During an ice age, large percentage of land is covered with glaciers. The glaciers generate huge stresses in underneath soil (Figure 5.12).

p_c' = maximum effective stress encountered by clay (Figure 5.13).

When the load is removed, the clay layer will rebound.

Effective stress after the load is removed = p_0'

Overconsolidation ratio = p_c'/p_0'

When the glacier is melted, the stress on soil is relieved. Hence $p_c' > p_0'$

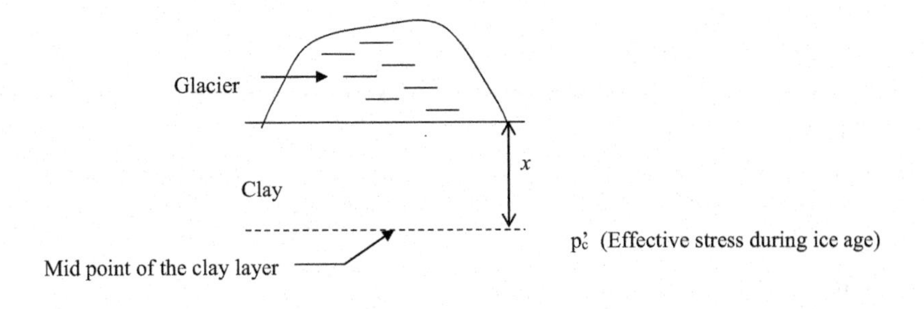

Figure 5.12 High stress levels in soil during ice ages due to glaciers.

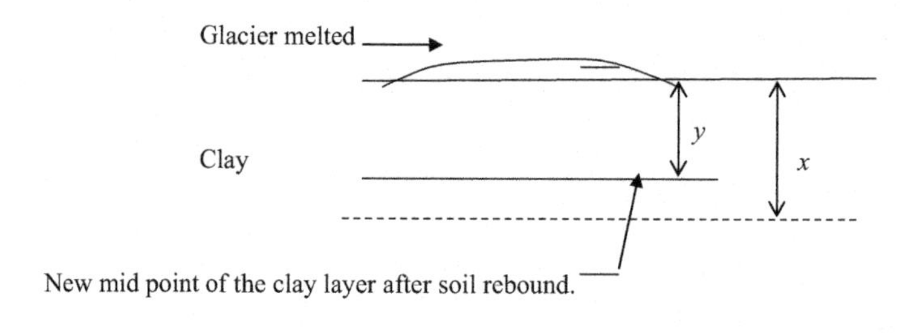

Figure 5.13 Rebound of the clay layer due to melting of the glacier.

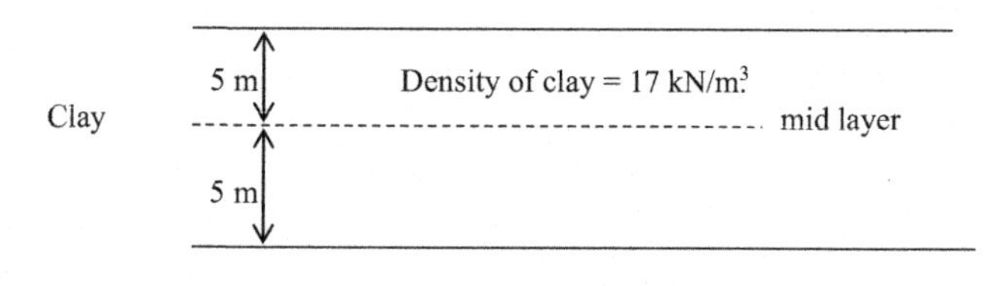

Figure 5.14 Overconsolidation ratio.

Design Example 5.4

Find the overconsolidation ratio of the clay layer at the mid point (Figure 5.14). The maximum stress that the soil was subjected to in the past is 150 kN/m^2.

Solution

Step 1: Find the present stress at the mid point of the clay layer (p_0')

$$p_0' = 17 \times 5 = 85 \text{ kN/m}^2.$$

Step 2: Find the overconsolidation ratio (OCR)
Overconsoidation ratio = past maximum stress/present stress
Past maximum stress is given to be 150 kN/m^2.
OCR = 150/85 = 1.76

Design Example 5.5

Find the overconsolidation ratio of the clay layer at the mid point. The maximum stress that the soil was subjected to in the past is 150 kN/m^2. Groundwater is at 3 m below the surface (Figure 5.15).

Solution

Step 1: Find the present stress at the mid point of the clay layer (p_0')

$$p_0' = 17 \times 3 + (17 - \gamma_w) \times 2$$

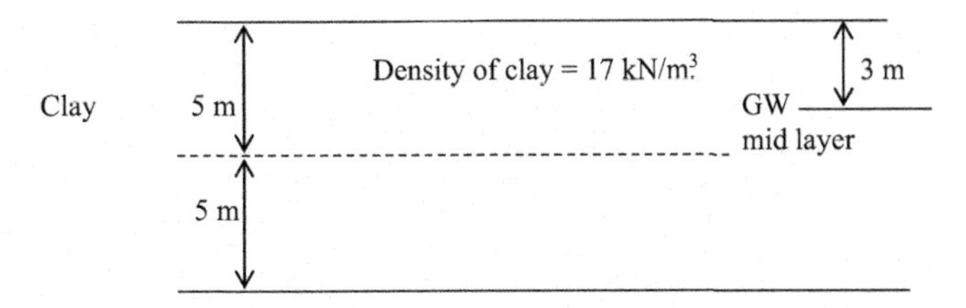

Figure 5.15 Soil profile for clay layer with groundwater at 3 m below the surface.

γ_w = density of water = 9.81 kN/m^3.
p_0' = 65.4
Step 2: Find the overconsolidation ratio (OCR):
Overconsoidation ratio = past maximum stress/present stress
Past maximum stress is given to be 150 kN/m^2.
OCR = 150/65.4 = 2.3

5.4.2 Overconsolidation due to groundwater lowering

When groundwater is lowered, the effective stress at a point will rise. This could be seen from an example.

Design Example 5.6

Find the effective stress at point "A". The density of soil is found to be 17 kN/m^3 and the groundwater is at 2 m below the surface (Figure 5.16).

Solution
Effective stress at point "A" = $17 \times 2 + (17 - \gamma_w) \times 6$
γ_w = 9.81 kN/m^3.
Effective stress at point "A" = 77.1 kN/m^2.
What would happen to the effective stress, if the groundwater is lowered to 4 m below the surface (Figure 5.17)?
New effective stress at point "A" = $17 \times 4 + (17 - \gamma_w) \times 4$ = 96.8 kN/m^2.
When the groundwater was lowered, the effective stress increased.

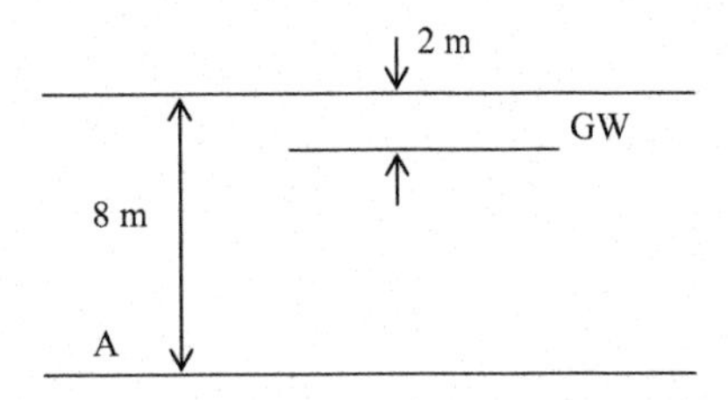

Figure 5.16 Soil profile with groundwater at 2 m below the surface.

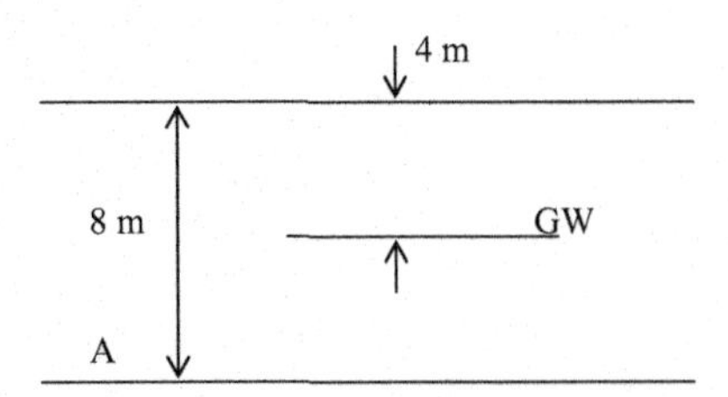

Figure 5.17 Soil profile with groundwater at 4 m below the surface.

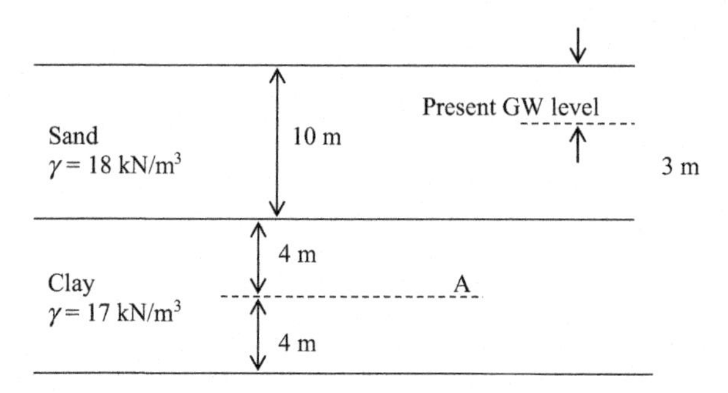

Figure 5.18 Soil profile and present groundwater level.

Design Example 5.7

A 10 m thick sand layer is underlain by a 8 m thick clay layer. Groundwater is found to be at 3 m below the surface, at present time (Figure 5.18). Old well log data shows that the groundwater was as low as 6 m below the surface in the past. What is the overconsolidation ratio (OCR) at the mid point of the clay layer?
Density of sand is 18 kN/m^3 and density of clay is 17 kN/m^3.

Solution

Step 1:

Present: Find the effective stress at the mid point of the clay layer (point A):

$= 3 \times 18 + 7 \times (18 - \gamma_w) + 4 \times (17 - \gamma_w)$

Density of water = $\gamma_w = 9.81$ kN/m^3.

Effective stress at point "A" = $3 \times 18 + 7 \times (18 - 9.81) + 4 \times (17 - 9.81) = 140.1$ kN/m^2.

Step 2: Find the effective stress at mid point of the clay layer (point A) in the past:

In the past, groundwater level was 6 m below the surface (Figure 5.19).

Find the effective stress at the mid point of the clay layer (point A):

$= 6 \times 18 + 4 \times (18 - \gamma_w) + 4 \times (17 - \gamma_w)$

Density of water = $\gamma_w = 9.81$ kN/m^3.

Effective stress at point "A" = $6 \times 18 + 4 \times (18 - 9.81) + 4 \text{ x } (17 - 9.81) = 169.5$ kN/m^2.

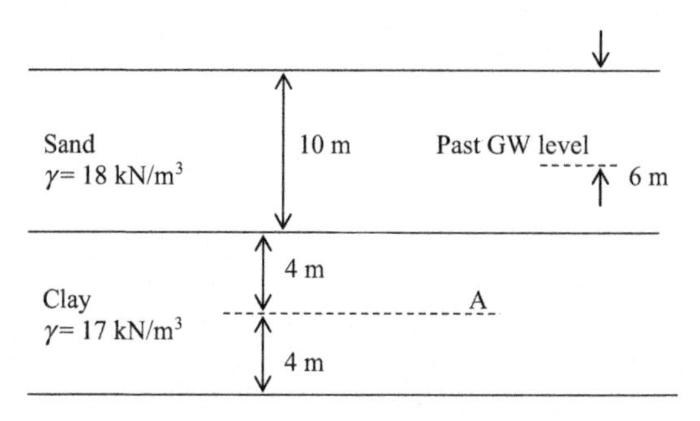

Figure 5.19 Soil profile and past groundwater level.

Step 3: Find the overconsolidation ratio due to past groundwater lowering (OCR):
OCR = Past maximum stress/present stress
OCR = 169.5/140.1 = 1.21.

5.5 Soil compaction

Shallow foundations can be rested on controlled fill, also known as an engineered fill or a structural fill. Typically, such fill material are carefully selected and compacted to 95% of the modified Proctor density.

Modified Proctor test is conducted by placing soil in a standard mould and compacted with a standard ram.

5.5.1 Modified Proctor test procedure

Step 1: The soil that needs to be compacted is placed in a standard mould and compacted (Figure 5.20).

Step 2: Compaction of soil is done by dropping a standard ram 25 times for each layer of soil from a standard distance. Typically, the soil is placed in 5 layers and compacted.

Step 3: After compaction of all the 5 layers, the weight of the soil is obtained. The soil contains solids and water. Solid is basically soil particles.

$$M = M_s + M_w$$

M, total mass of soil including water; M_s, mass of solid portion of soil; M_w, mass of water.

Step 4: Find the moisture content of the soil.
Moisture content is defined as M_w/M_s.
A small sample of soil is taken and placed in the oven and measured.

Step 5: Find the dry density of soil.
Dry density of soil is given by M_s/V
M_s is the dry weight of soil and "V" is the total volume.

Step 6: Repeat the test a few times with different moisture content and plot a graph between dry density and moisture content (Figure 5.21).

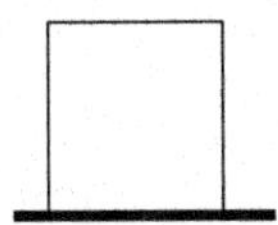

Figure 5.20 Standard mould.

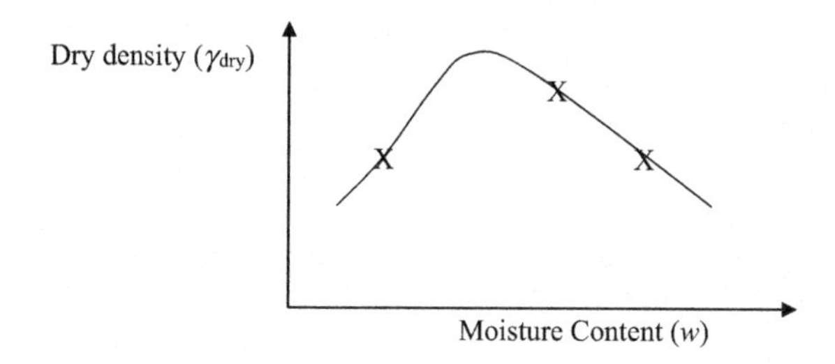

Figure 5.21 Dry density and moisture content.

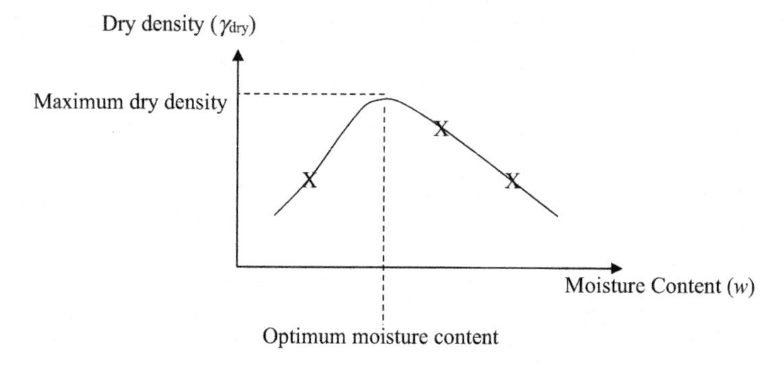

Figure 5.22 Optimum moisture content.

Step 7: Obtain the maximum dry density and the optimum moisture content (Figure 5.22).

For a given soil, there is an optimum moisture content that would provide the maximum dry density.

It is not easy to attain the optimum moisture content in the field. Usually, soil that is too wet is not compacted. If the soil is too dry, water is added to increase the moisture content.

5.5.2 Controlled fill applications

A controlled fill can be used to build shallow foundations, roadways, parking lots, building slabs, and equipment pads (Figure 5.23).

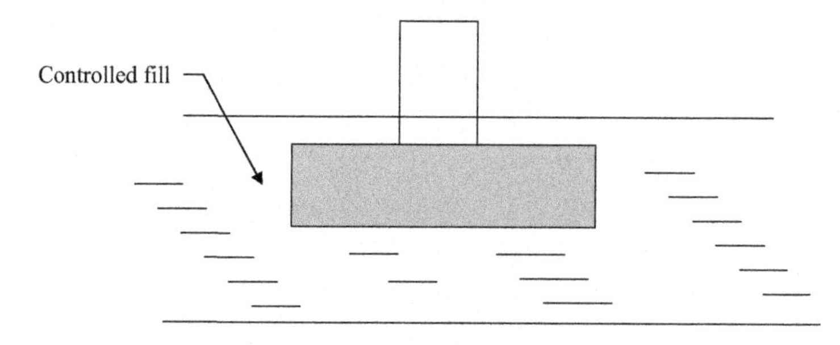

Figure 5.23 Shallow foundation placed on top of controlled fill.

5.6 Borrow pit computations

The fill material for civil engineering work is obtained from borrow pits. The question is how much soil should be removed from the borrow pit for a given project?

Usually, the final product is the controlled fill or the compacted soil. Total density, optimum moisture content, and dry density of the compacted soil will be available. This information can be used to obtain the mass of solids required from the borrow pit. If the soil in the borrow pit is too dry, water can always be added in the site. If the water content is too high, then soil can be dried prior to use. This could take some time in the field since one has to wait for few sunny days to get rid of the water.

Water can be added or removed from the soil.

What cannot be changed is the mass of solids.

5.6.1 Procedure

- Find the mass of solids required for the controlled fill.
- Excavate and transport the same mass of solids from the borrow pit.

Design Example 5.8

A road construction project needs compacted soil to construct a road 10 ft. wide, 500 ft. long. The project needs 2 ft. layer of soil.

Modified Proctor density was found to be 112.1 pcf at an optimum moisture content at 10.5%. (Figure 5.24).

The soil in the borrow pit has following properties:

Total density of the borrow pit soil = 105 pcf.

Moisture content of borrow pit soil = 8.5%.

Find the total volume of soil that need to be hauled from the borrow pit.

Solution

Step 1: Find the mass of solids (M_s) required for the controlled fill:

Volume of compacted soil required = $500 \times 10 \times 2 = 10{,}000$ ft.^{3}

Modified Proctor dry density = 112.1 pcf

Moisture content required = 10.5%

Draw the phase diagram for the controlled fill (Figure 5.25):

M_a, mass of air (usually taken to be zero); V_a, volume of air; M_w, mass of water; V_w, volume of water; M_s, mass solids; V_s, volume of solids; M, total mass of soil = $M_s + M_w$; V, total volume of soil = $V_s + V_w + V_a$.

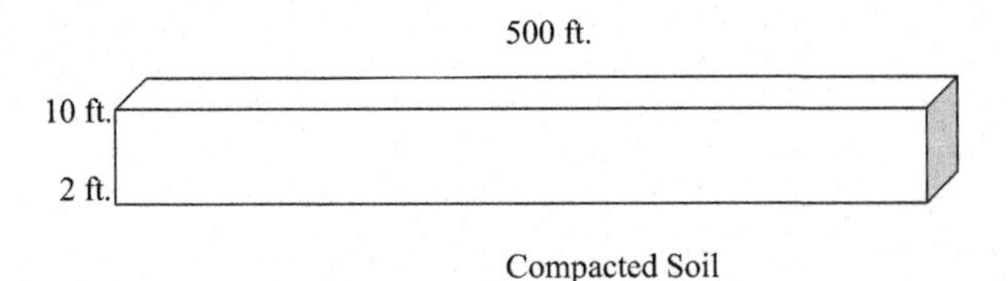

Figure 5.24 Compaction of soil.

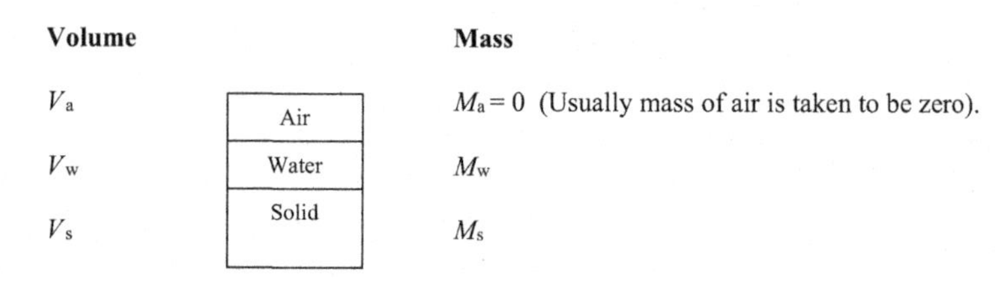

Figure 5.25 Soil-phase diagram.

Soil in the site after compaction has a dry density of 112.1 pcf and moisture content of 10.5%.

Dry density = M_s/V = 112.1 pcf

Moisture content = M_w/M_s = 10.5% = 0.105

The road needed 2 ft. layer of soil at a width of 10 ft. and length of 500 ft.

Hence the total volume of soil = 2 × 10 × 500 = 10,000 ft.^{3}

V = total volume = 10,000 ft.^{3}

Since M_s/V = dry density

M_s/10,000 = 112.1

M_s = 1,121,000 lbs.

M_s is the mass of solids. This mass of solids should be hauled in from the borrow pit.

Step 3: Find the mass of water in the controlled fill:

Moisture content in the controlled fill = M_w/M_s = 10.5% = 0.105

M_w = 0.105 × 1,121,000 lbs = 117,705 lbs

Step 4: Find the total volume of soil that needs to be hauled from the borrow pit:

The contractor needs to obtain 1,121,000 lbs of solids from the borrow pit.

Contractor can add water to the soil in the field if needed.

Mass of solids needed (M_s) = 1,121,000 lbs.

Density and moisture content of borrow pit soil is available.

Density of borrow pit soil = M/V = 105 pcf

Moisture content of borrow pit soil = M_w/M_s = 8.5% = 0.085

Since M_s = 1,121,000 lbs, (M_s is the mass of solids required).

M_w = 0.085 × 1,121,000 lbs = 95,285 lbs.

Solid mass of 1,121,000 lbs of soil in the borrow pit contains 95,285 lbs of water.

Total weight of borrow pit soil = 1,121,000 + 95,285 = 1,216,285 lbs

Total density of borrow pit soil is known to be 105 pcf.

Total density of borrow pit soil = $M/V = (M_w + M_s)/V$ = 105 pcf

Insert known values for M_s and M_w.

$M/V = (M_w + M_s)/V$

(95,285 + 1,121,000)/V = 105 pcf

Hence, V = 11,583.7 ft.^{3}

The contractor needs to extract 11,583.7 ft.^{3} of soil from the borrow pit.

The borrow pit soil comes with 95,285 lbs of water.

Compacted soil should have 117,705 lbs of water (see Step 3).

Hence water needs to be added to the borrow pit soil.

Amount of water needs to be added to the borrow pit soil = 117,705 − 95,285 = 22,420 lbs.

Weight of water is usually converted to gallons. One gallon is equal to 8.34 lbs.
Amount of water that needs to be added = 2,688 gallons.

5.6.2 Summary

Step 1: Obtain all the requirements for compacted soil.

Step 2: Find M_s or the mass of solids in the compacted soil.
This is the mass of solids that needs to be obtained from the borrow pit.

Step 3: Find the information about the borrow pit. Usually the moisture content in the borrow pit and total density of the borrow pit can be easily obtained.

Step 4: The contractor needs to obtain M_s of soil from the borrow pit.

Step 5: Find total volume of soil that needs to be removed in order to obtain M_s mass of solids.

Step 6: Find M_w of the borrow pit. (mass of water that comes along with soil).

Step 7: Find M_w (mass of water in compacted soil).

Step 8: The difference between the above two masses is the amount of water that needs to be added.

5.7 Short course on seismology

5.7.1 Introduction

A general understanding of seismology is needed to design foundations to resist seismic events. The earth is a dynamic system. Under the bedrock (also known as the earth crust) there exists a huge reservoir of lava. Earthquakes and volcanoes occur when the pressure of lava is released. Ancient people found out that earthquakes can be predicted by observing pendulums.

Figure 5.26a shows a pendulum moving back and forth, drawing a straight line prior to an earthquake event. Figure 5.26b shows the same pendulum drawing a wavy line during an earthquake event.

- Seismographs are designed using the above principal.
- Due to the movement of the pile cap, piles would be subjected to additional shear forces and bending moments.
- Earthquakes occur due to the disturbances occurring inside the earth's crust. Earthquakes would produce the following three main types of waves (Figure 5.27):

1. *P-waves (Primary waves)*: P-waves are also known as compression waves or longitudinal waves.
2. *S-waves (Secondary waves)*: S-waves are also known as shear waves or transverse waves.
3. *Surface waves:* Surface waves are the shear waves that travel near the surface.

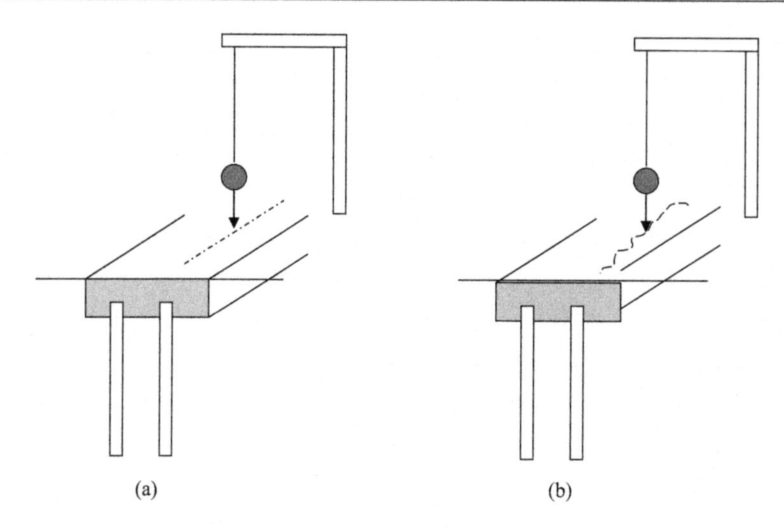

Figure 5.26 Movement of a pendulum during an earthquake. (a) No earthquake. (b) Earthquake event.

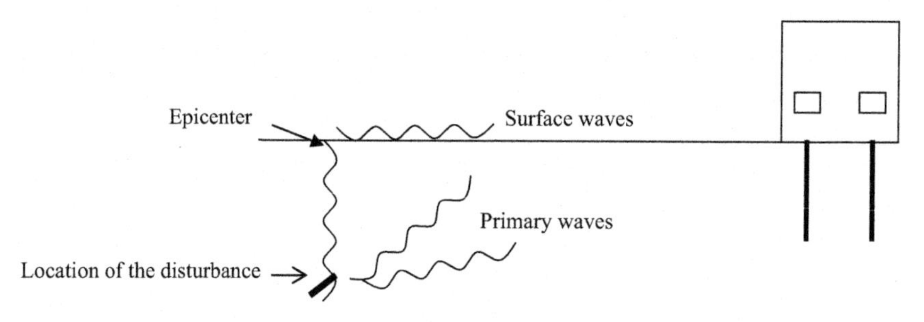

Figure 5.27 Seismic waves.

5.7.2 *Faults*

Faults are a common occurrence in the earth. Fortunately, most faults are inactive and will not cause earthquakes. Fault is a fracture where a block of earth had moved with relative to the other.

5.7.2.1 Horizontal fault

- In a horizontal fault, one earth block had moved horizontally relative to the other (Figure 5.28). The movement is horizontal.

5.7.2.2 Vertical fault (strike slip faults)

In this type of faults, one block moves in downward direction with relative to the other (Figure 5.29).

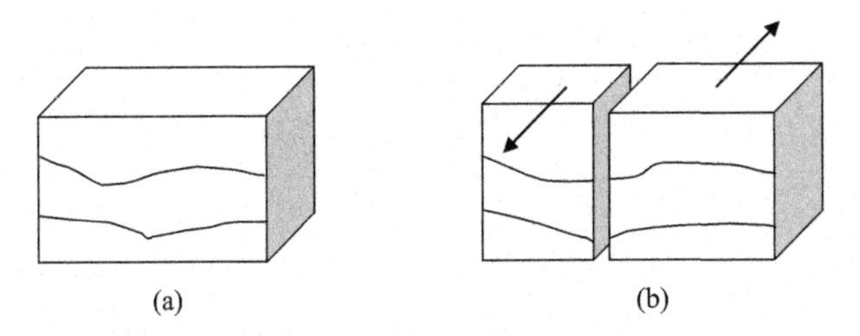

Figure 5.28 Movement in a horizontal fault. (a) Original Ground. (b) One block moves horizontally relative to the other.

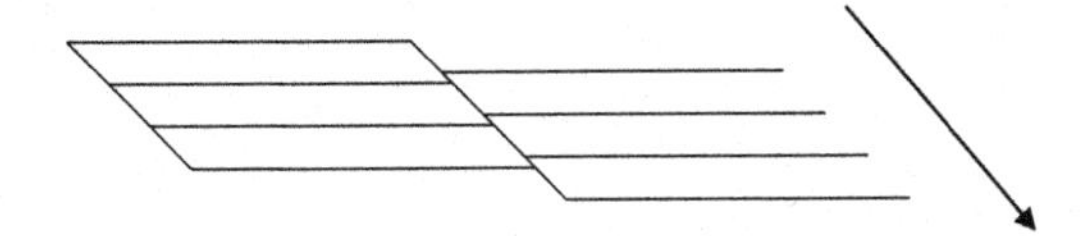

Figure 5.29 Movement in a vertical fault.

5.7.2.3 *Active fault*

An active fault is defined as a fault which has an average historic slip rate of 1 mm/year or more during the last 11,000 years (International Building Code, IBC).

5.7.2.4 *Richter magnitude scale (M)*

$$M = \text{Log}(A) - \text{Log}(A_0) = \text{Log}(A / A_0)$$

M, richter magnitude scale; A, maximum trace amplitude during the earthquake; A_0, standard amplitude (standard value of 0.001 mm is used for comparison. This corresponds to a very small earthquake).

Design Example 5.9

What is the Richter magnitude scale for an earthquake that recorded an amplitude of (a) 0.001 mm, (b) 0.01 mm, (c) 10 mm?

Solution

(a) $A = 0.001$ mm

$M = \text{Log}\ (A/A_0) = \text{Log}\ (0.001/0.001) = \text{Log}\ (1) = 0$

$M = 0$

(b) $A = 0.01$

$M = \text{Log}\ (A/A_0) = \text{Log}\ (0.01/0.001) = \text{Log}\ (10) = 1$

$M = 1$

Table 5.1 Largest earthquakes recorded

Location	Date	Magnitude
Chile	1960	9.5
Prince William, Alaska	1964	9.2
Aleutian Islands	1957	9.1
Kamchatka	1952	9.0
Ecuador	1906	8.8
Rat Islands	1965	8.7
India–China border	1950	8.6
Kamchatka	1923	8.5
Indonesia	1938	8.5
Kuril Islands	1963	8.5

Source: USGS Website (www.usgs.org United States Geological Survey).

(c) $A = 10$ mm

$M = \text{Log}(A/A_0) = \text{Log}(10/0.001) = \text{Log}(10{,}000) = 4$

$M = 4$

Table 5.1 shows the largest earthquakes recorded.

5.7.3 Peak ground acceleration

This is a very important parameter for geotechnical engineers. During an earthquake, soil particles would accelerate. Acceleration of soil particles could be either horizontal or vertical.

5.7.4 Seismic waves

The following partial differential equation represents seismic waves.

$$G(\vartheta^2 u/\vartheta x^2 + \vartheta^2 u/\vartheta z^2) = \rho(\vartheta^2 u/\vartheta t^2 + \vartheta^2 v/\vartheta t^2)$$

G, hear modulus of soil; ϑ, partial differential operator; u, horizontal motion of soil; v, vertical motion of soil; ρ, density of soil.

Fortunately, geotechnical engineers are not called upon to solve this partial differential equation.

A seismic wave form described by the above equation, would create shear forces and bending moments on piles.

5.7.5 Seismic wave velocities

The velocity of seismic waves is dependent upon the soil/rock type (Table 5.2). Seismic waves travel much faster in sound rock than in soils.

Table 5.2 **Seismic wave velocity**

Soil/rock type	Wave velocity (ft./s)
Dry silt, sand, loose gravel, loam, loose rock, and moist fine grained top soil	600–2,500
Compact till, gravel below water table, compact clayey gravel, cemented sand, and sandy clay	2,500–7,500
Weathered rock, partly decomposed rock, and fractured rock	2,000–10,000
Sound shale	2,500–11,000
Sound sandstone	5,000–14,000
Sound limestone and chalk	6,000–20,000
Sound igneous rock (granite, diabase)	12,000–20,000
Sound metamorphic rock	10,000–16,000

Source: Peck et al. (1974).

5.7.6 Liquefaction

5.7.6.1 Theory

Sandy and silty soils tend to lose their strength and turn into a *liquid*-like state during an earthquake (Figure 5.30). This happens due to the increase of pore pressure during an earthquake event in the soil caused by seismic waves.

- The liquefaction of soil was thoroughly studied by Bolten Seed and IM Idris during the 1970s. As one would expect, liquefaction behavior of soil cannot be expressed in one simple equation. Many correlations and semiempirical equations have been introduced by researchers. Due to this reason, Professor Robert W. Whitman convened a workshop in 1985 (IBC) on behalf of the National Research Council (NRC). Experts from many countries participated in this workshop and a procedure was developed to evaluate the liquefaction behavior of soils.
- It should be mentioned here that only sandy and silty soils tend to liquefy. Clay soils do not undergo liquefaction.

5.7.7 Impact due to earthquakes

Imagine a bullet hitting a wall (Figure 5.31).

The extent of the damage to the wall due to the bullet depends on a number of parameters.

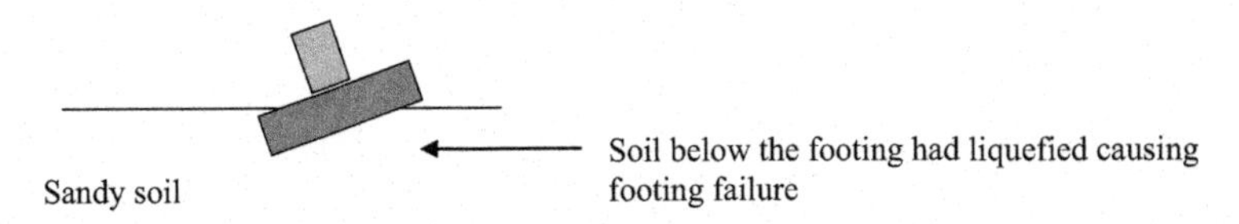

Figure 5.30 Foundation failure due to liquefaction.

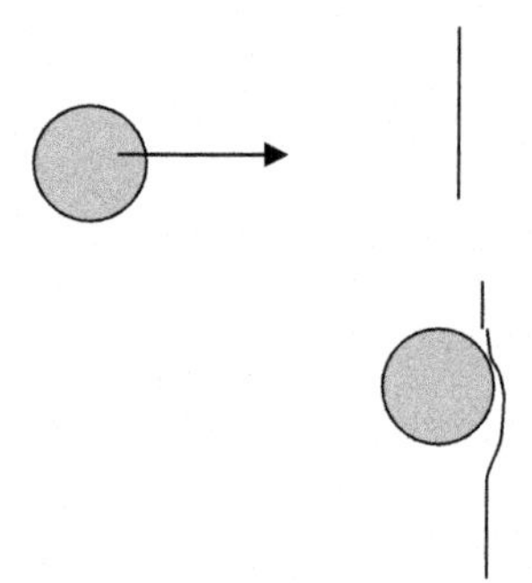

Figure 5.31 Bullet hitting a wall.

Bullet properties

1. Velocity of the bullet.
2. Weight of the bullet.
3. Hardness of the bullet material.

Wall properties

1. Hardness of the wall material.
2. Type of wall material.

The parameters that affect liquefaction are as follows:

Earthquake properties

- Magnitude of the earthquake
- Peak horizontal acceleration at the ground surface (a_{max})

Soil Properties

- Soil strength (measured by Standard Penetration Test (SPT) value)
- Effective stress at the point of liquefaction
- Content of fines (fines are defined as particles that pass through the #200 sieve)
- Earthquake properties that affect the liquefaction of a soil are amalgamated into one parameter known as cyclic stress ratio (CSR).

$$\text{CSR} = 0.65\,(a_{max}/g) \times (\sigma/\sigma') \times r_d \quad (1)$$

- a_{max} = peak horizontal acceleration at the ground surface; σ = total stress at the point of concern; σ' = effective stress at the point of concern; r_d = stress reduction coefficient (this parameter accounts for the flexibility of the soil profile).

$$r_d = 1.0 - 0.00765\,Z \quad \text{for } Z < 9.15\text{ m} \quad (1.1)$$

$$r_d = 1.174 - 0.0267\,Z \quad \text{for } 9.15\text{ m} < Z < 23\text{ m} \quad (1.2)$$

Z = depth to the point of concern in meters.

5.7.8 Soil resistance to liquefaction

- As a rule of thumb, any soil that has an SPT value higher than 30 will not liquefy.
- As mentioned earlier, resistance to liquefaction of a soil depends on its strength measured by the SPT value. Researchers have found that resistance to liquefaction of a soil depends on the content of *fines* as well.

The following equation can be used for a clean sand (clean sand is defined as a sand with less than 5% fines).

$$CRR_{7.5} = \frac{1}{[34-(N_1)_{60}]} + \frac{(N_1)_{60}}{135} + \frac{50}{[10.(N_1)_{60}+45]^2} - \frac{2}{200}. \tag{2}$$

$CRR_{7.5}$ = soil resistance to liquefaction for an earthquake with a magnitude of 7.5 Richter.

Correction factor needs to be applied to any other magnitude. This process is described later in this chapter. Above equation can be used only for sands (content of fines should be less than 5%). Correction factor has to be used for soils with a higher content of fines; this procedure is described later in this chapter.

$(N_1)_{60}$ = corrected to a 60% hammer and an overburden pressure of 100 kPa.

(Note that the above equations were developed in metric units.)

5.7.8.1 How to obtain $(N_1)_{60}$

$$(N_1)_{60} = N_m \times C_N \times C_E \times C_B \times C_R \tag{3}$$

- N_m = SPT value measured in the field
- C_N = overburden correction factor = $(Pa/\sigma')^{0.5}$
  Pa = 100 kPa; σ' = Effective stress of soil at point of measurement.
- C_E = energy correction factor for the SPT hammer
  For donut hammers C_E = 0.5 to 1.0; For trip type donut hammers C_E = 0.8 to 1.3
- C_B = borehole diameter correction
  For boreholes 65 mm to 115 mm use C_B = 1.0
  For borehole diameter of 150 mm use C_B = 1.05
  For borehole diameter of 200 mm use C_B = 1.15
- C_R = rod length correction (rods attached to the SPT spoon would exert their weight on the soil. Longer rods would exert a higher load on soil and in some cases, the spoon would go down due to the weight of rods without any hammer blows. Hence, the correction is made to account for the weight of rods).
  For rod length <3 m use C_R = 0.75;
  For rod length 3 m to 4 m use C_R = 0.8;
  For rod length 4 m to 6 m use C_R = 0.85; For rod length 6 m to 10 m use C_R = 0.95;
  For rod length 10 m to 30 m use C_R = 1.0.

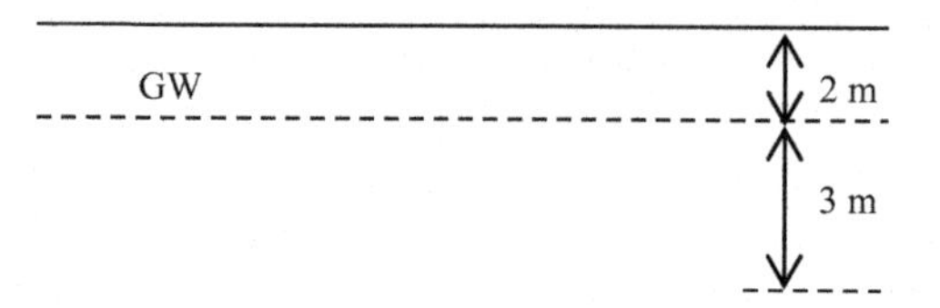

Figure 5.32 Soil profile.

Design Example 5.10

Consider a point at a depth of 5 m in a sandy soil (Fines < 5%) (Figure 5.32). The total density of soil is 1800 kg/m^3. The groundwater is at a depth of 2 m. The corrected $(N_1)_{60}$ value is 15. The peak horizontal acceleration at the ground surface (a_{max}) was found to be 0.15 g for an earthquake of magnitude 7.5. Check to see whether the soil at a depth of 5 m would liquefy under an earthquake of 7.5 magnitude.

Solution

Step 1: Find the cyclic stress ratio (CSR).

$$CSR = 0.65(a_{max}/g)\times(\sigma/\sigma')\times r_d \quad (1)$$

- a_{max} = peak horizontal acceleration at the ground surface = 0.15 g
- σ = total stress at the point of concern; σ' = Effective stress at the point of concern
- r_d = stress reduction coefficient (this parameter accounts for the flexibility of the soil profile).

$$r_d = 1.0 - 0.00765\ Z \quad \text{for } Z < 9.15 \text{ m} \quad (1.1)$$

(Z = depth to the point of concern in meters)

$$r_d = 1.174 - 0.0267Z \text{ for } 9.15\text{m} < Z < 23\text{m} \quad (1.2)$$

$\sigma = 5 \times 1800 = 9000$ kg/m^2; $\sigma' = 2 \times 1800 + 3\ (1800 - 1000) = 6000$ kg/m^2.
(density of water = 1000 kg/m^2)

Since the depth of concern is 5 m (which is less than 9.15 m), use the above first equation to find r_d.

$r_d = 1.0 - 0.00765$ Z for Z < 9.15 m; $r_d = 1.0 - 0.00765 \times 5 = 0.962$

$$\text{Hence, } CSR = 0.65\times(0.15)\times\frac{9000}{6000}\times 0.962 = 0.1407$$

Step 2: Find the soil resistance to liquefaction

$$CRR_{7.5} = \frac{1}{[34-(N_1)_{60}]} + \frac{(N_1)_{60}}{135} + \frac{50}{[10.(N_1)_{60}+45]^2} - \frac{2}{200}. \quad (2)$$

$(N_1)_{60}$ value is given to be 15. Hence $CRR_{7.5} = 0.155$

Since soil resistance to liquefaction (0.155) is larger than the CSR value (0.1407), the soil at 5 m depth will not undergo liquefaction for an earthquake of magnitude 7.5.

5.7.8.2 Correction factor for magnitude

As you are aware equation 2 (for $CRR_{7.5}$) is valid only for earthquakes of magnitude 7.5. The correction factor is proposed to account for magnitudes different than 7.5.

Table 5.3 **Magnitude scaling factors**

Earthquake magnitude	MSF	MSF
5.5	2.2	2.8
6.0	1.76	2.1
6.5	1.44	1.6
7.0	1.19	1.25
7.5	1.00	1.00
8.0	0.84	
8.5	0.72	

Factor of safety (FS) is given by = ($CRR_{7.5}$/CSR)

$CRR_{7.5}$ = resistance to soil liquefaction for a magnitude of 7.5 earthquake and CSR = cyclic stress ratio (which is a measure of the impact due to the earthquake load).

Factor of safety for any other earthquake is given by following equation.

$$\text{Factor of Safety (FS)} = (CRR_{7.5}/CSR) \times MSF \tag{4}$$

MSF = Magnitude scaling factor.

MSF is given by Table 5.3.

The participants of the 1985 NRC (National Research Council) conference gave the freedom to the engineers to select either of the values suggested by Idris or Andrus and Stokoe. As you could see, the Idris values are more conservative and the engineers may be able to use the Andrus and Stokoe values in noncritical buildings such as warehouses.

Design Example 5.11

The $CRR_{7.5}$ value of a soil found to be 0.11. The CSR value for the soil was computed to be 0.16. Will this soil liquefy for an earthquake of 6.5 magnitude?

Solution

Step 1: Find the factor of safety

Factor of safety (FS) = ($CRR_{7.5}$ /CSR) × MSF

Magnitude scaling factor (MSF) for an earthquake of 6.5 = 1.44 (Idris)

Factor of safety = (0.11/0.16) × 1.44 = 0.99 (soil would liquefy).

Use MSF given by Andrus and Stokoe.

Factor of safety = (0.11/0.16) × 1.6 = 1.1 (soil would not liquefy).

5.7.8.3 *Correction factor for content of fines*

The equation 2, was developed for clean sand with fines content less than 5%. Correction factor is suggested for soils with higher fines contents.

$$\text{CRR}_{7.5} = \frac{1}{[34-(N_1)_{60}]} + \frac{(N_1)_{60}}{135} + \frac{50}{[10.(N_1)_{60}+45]^2} - \frac{2}{200}. \tag{2}$$

Corrected $(N_1)_{60}$ value should be used in the above equation for soils with higher fines content.

Following procedure should be followed to find the correction factor.

- Compute $(N_1)_{60}$ as in the previous case.
- Use the following equations to account for the fines content.

$(N_1)_{60\,C} = a + b\,(N_1)_{60}$ $\quad$ $(N_1)_{60\,C}$ = corrected $(N_1)_{60}$ value

$$a = 0 \quad \text{for FC} < 5\% \text{ (FC = Fines Content)} \tag{5.1}$$

$$a = \exp[1.76 - 190/\text{FC}^2] \quad \text{for } 5\% < \text{FC} < 35\% \tag{5.2}$$

$$a = 5.0 \quad \text{for FC} > 35\% \tag{5.3}$$

$$b = 1.0 \quad \text{for FC} < 5\% \tag{5.4}$$

$$b = [0.99 + (\text{FC}^{1.5}/1000)] \quad \text{for } 5\% < \text{FC} < 35\% \tag{5.5}$$

$$b = 1.2 \quad \text{for FC} > 35\% \tag{5.6}$$

Design Example 5.12

The $(N_1)_{60}$ value for soil with 30% fines content was found to be 20. Find the corrected $(N_1)_{60\,C}$ value for that soil.

Solution

Step 1: $(N_1)_{60\,C} = a + b\,(N_1)_{60}$

For FC = 30%;

$$a = \exp\,[1.76 - 190/\text{FC}^2] = \exp\,[1.76 - 190/30^2] = 4.706 \tag{5.2}$$

For FC = 30%

$$b = [0.99 + (\text{FC}^{1.5}/1000)] = 1.154 \tag{5.5}$$

$$(N_1)_{60\,C} = 4.706 + 1.154 \times 20 = 27.78$$

Design Example 5.13

Consider a point at a depth of 5 m in a sandy soil. (Fines = 40%). Total density of soil is 1800 kg/m^3. The groundwater is at a depth of 2 m. (γw = 1000 kg/m^3). Corrected $(N_1)_{60}$ value is 15. (All the correction parameters C_N, C_E, C_B, C_R are applied, except for the fines content). Peak horizontal acceleration at the ground surface (a_{max}) was found to be 0.15 g for an earthquake of magnitude 8.5. Check to see whether the soil at a depth of 5 m would liquefy under this earthquake load.

Solution

Step 1: Find the CSR.

$$CSR = 0.65(a_{max}/g) \times (\sigma/\sigma') \times r_d \quad (1)$$

$$\sigma = 5 \times 1800 = 9000 \text{ Kg/m}^2; \quad \sigma' = 2 \times 1800 + 3(1800 - 1000) = 6000 \text{ Kg/m}^2$$

Since the depth of concern is 5 m (which is less than 9.15 m) use Equation (1.1)

$$r_d = 1.0 - 0.00765\ Z \quad \text{for} \quad Z < 9.15 \text{ m}; \quad r_d = 1.0 - 0.00765 \times 5 = 0.962$$

$$\text{Hence } CSR = 0.65 \times (0.15) \times \frac{9000}{6000} \times 0.962 = 0.1407$$

Step 2: Provide the correction factor for fines content

For soils with 40% fines → $a = 5$ and $b = 1.2$ (Equations (5.3) and (5.6)).

$(N_1)_{60C} = a + b\,(N_1)_{60}$

$(N_1)_{60C} = 5 + 1.2 \times 15 = 23$

Step 3: Find the soil resistance to liquefaction

$$CRR_{7.5} = \frac{1}{[34 - (N_1)_{60}]} + \frac{(N_1)_{60}}{135} + \frac{50}{[10.(N_1)_{60} + 45]^2} - \frac{2}{200}. \quad (2)$$

$(N_1)_{60C}$ value is found to be 23. (see Step 2)

Hence $CRR_{7.5} = 0.25$ (from Equation 2)

$$\text{Factor of Safety (FS)} = (CRR_{7.5}/CSR) \times MSF \quad (4)$$

MSF = Magnitude scaling factor.

Obtain MSF from Table 5.1.

MSF = 0.72 for an earthquake of 8.5 magnitude. (Table 5.3)

CSR = 0.1407 (see Step 1)

Factor of safety (FS) = (0.25/0.1407) × 0.72 = 1.27

The soil would not liquefy.

References

Peck, R.B., Hanson, W.E., Thorburn, T.H., 1974. Foundation Engineering. John Wiley and Sons, New York.

USGS Website, United States Geologic Survey, www.usgs.org

Part 2

Shallow Foundations

Shallow foundation fundamentals

6

6.1 Introduction

Shallow foundations are the very first choice of foundation engineers. They are first and foremost cheap and easy to construct compared to other alternatives such as pile foundations and mat foundations. Shallow foundations need to be designed for bearing capacity and excessive settlement. There are number of methods available to compute the bearing capacity of shallow foundations. Terzaghi's bearing capacity equation, which was developed by Karl Terzaghi still widely used.

6.2 Buildings

Shallow foundations are widely used to support buildings. In some cases doubt is cast upon the ability of a shallow foundation to carry necessary loads due to loose soil underneath. In such situations, engineers use piles to support buildings. Shallow foundations are much more cost effective than piles. Building footings are subjected to wind, earthquake forces, and bending moments (Figure 6.1).

6.2.1 Buildings with basements

Footings in buildings with basements need to be designed for lateral soil pressure as well. In most cases, the building frame has to support forces due to soil (Figure 6.2).

6.3 Bridges

Bridges consist of abutments and piers. Due to large loads many engineers prefer to use piles for bridge abutments and piers. Nevertheless if the site conditions are suitable, shallow foundations can be used for bridges (Figure 6.3).

Shallow footings in bridge abutments are subjected to large lateral loads due to approach fill in addition to vertical loads.

Scour is another problem that needs to be overcome in bridges. Scour is the process of erosion due to water. Shallow foundations could fail if scouring is not noticed and corrected (Figure 6.4).

Shallow foundations are vulnerable for scouring effect and due to this reason many bridge engineers prefer piles for bridge abutments and piers (Figure 6.5).

Scouring is not a problem for bridges over highways and roadways (Figure 6.6).

Geotechnical Engineering Calculations and Rules of Thumb

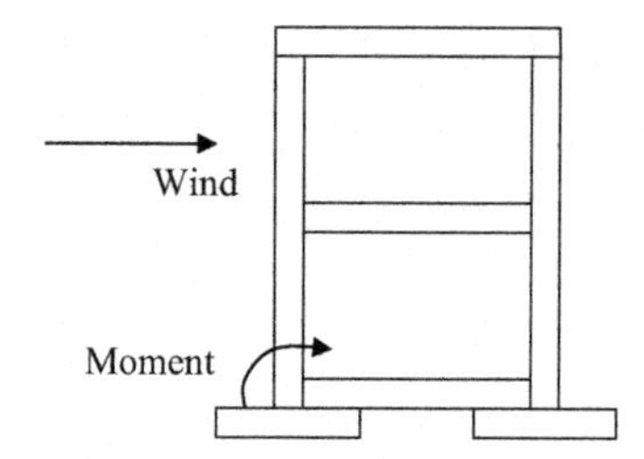

Figure 6.1 Building footings are subjected to both vertical and lateral forces.

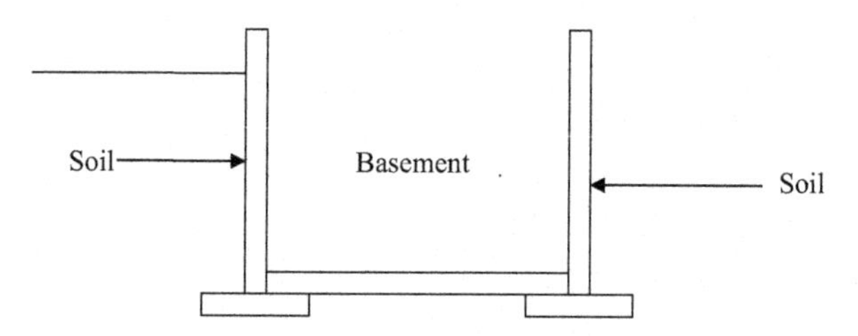

Figure 6.2 Building with basement.

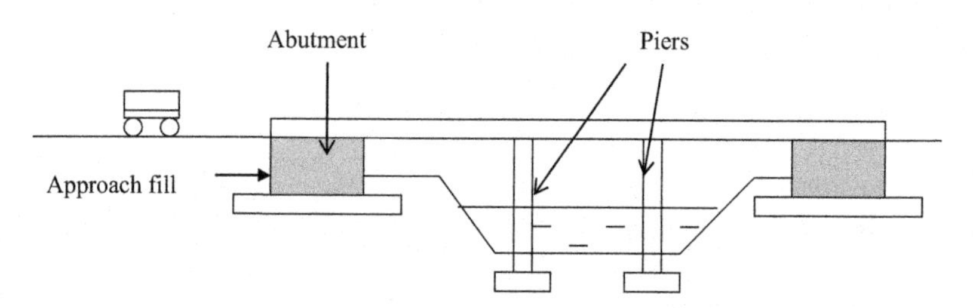

Figure 6.3 Lateral forces in a bridge abutment due to approach fill.

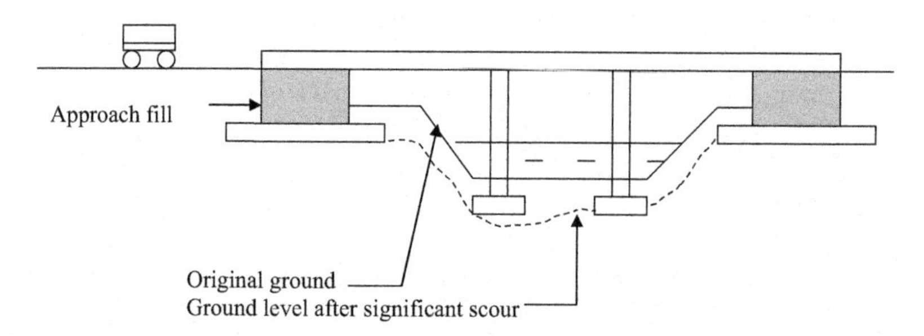

Figure 6.4 Effect of scour (erosion of soil due to water).

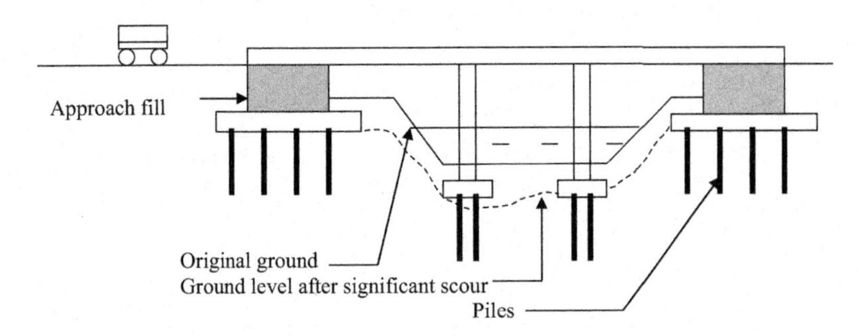

Figure 6.5 Use of piles to counter scour.

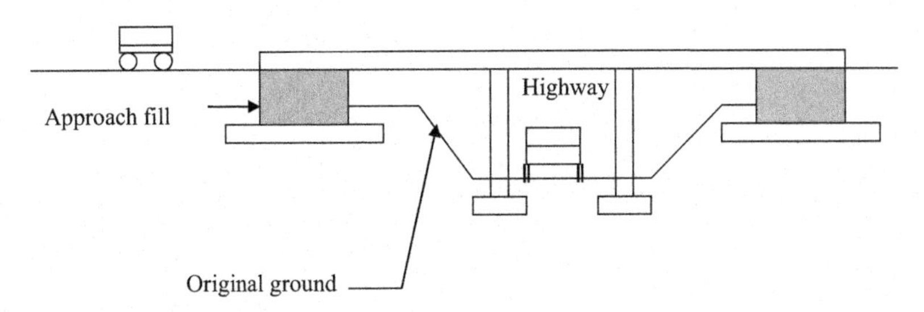

Figure 6.6 Bridge over a highway.

6.4 Frost depth

Shallow foundations need to be placed below frost level. During winter time periods, soil at the surface will be frozen. During the summer frozen soil will melt again. This freezing and thawing of soil generates a change in volume. If the footing is placed below the frost depth, freezing and melting of soil would generate upward forces and eventually cause cracking in concrete (Figures 6.7 and 6.8).

During summer, the water in soil will melt and the footing would settle. Due to this reason all footings should be placed below the frost depth in that region (Figure 6.9).

Frost depth is dependant upon the region. Frost depth in Siberia and some parts of Canada could be as high as 7 feet while in places like New York it is not more than 4 ft. Since there is no frost in tropical countries, frost depth is not an issue.

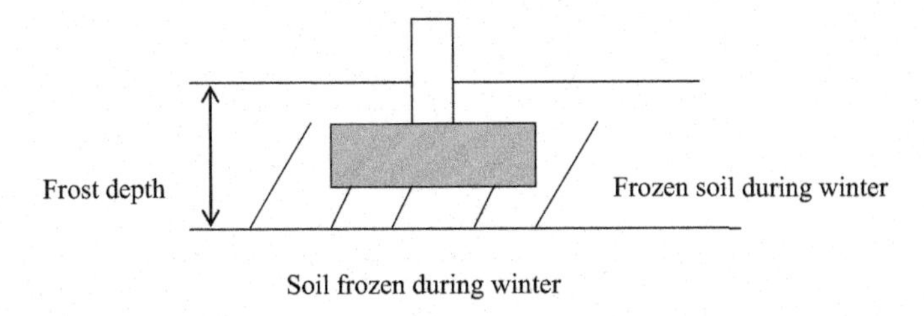

Figure 6.7 Footing on frozen soil.

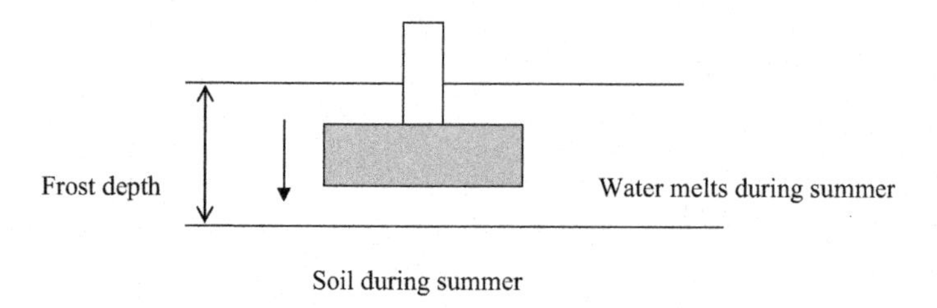

Figure 6.8 Freezing and thawing of soil.

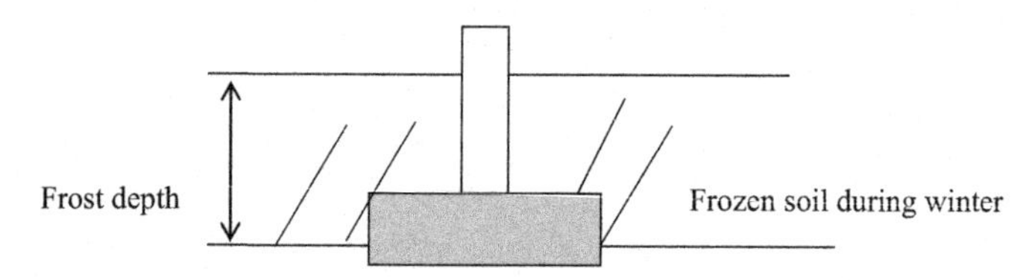

Figure 6.9 Footing placed below the frost depth.

Bearing capacity – rules of thumb 7

7.1 Introduction

Many engineers use bearing capacity equations developed by Terzaghi, Hansen, Meyerhof, and Vesic to find the bearing capacity of foundations. On the other hand, there are rule of thumb methods that have been used for many years by engineers. In this chapter, some of the rule of thumb methods to compute bearing capacity will be presented.

7.2 Bearing capacity in medium to coarse sands (drained analysis)

Permeability of sandy soils is higher than clay soils. Buildings are constructed gradually. First, the footing is constructed, then the columns and walls, then upper floors. Full load on footing does not materialize overnight. Depending upon the speed of contraction, full load on footing may take months or years. Hence, there is plenty of time for the sandy soils to drain excess pore pressure. Hence bearing capacity computations in sandy soils are done based on drained friction angle parameter. As you know cohesion in sandy soils and nonplastic silts is zero.

c = Cohesion = 0 for sandy soils under drained condition.

φ' = Friction angle of sandy soils. This is a nonzero quantity.

Therefore, strength of sandy soils is dependent upon the friction angle.

Bearing capacity in coarse to medium sands can be obtained using the average SPT (N) value.

- (fps) Bearing capacity of coarse to medium sands (allowable) = $0.2\ N_{\text{average}}$ (ksf) (not to exceed 12 ksf)
- (SI) Bearing capacity of coarse to medium sands (allowable) = $9.6\ N_{\text{average}}$ (kPa) (not to exceed 575 kPa)

Procedure to use the above equation:

Step 1: Find the average SPT (N) value below the bottom of footing to a depth equal to width of the footing

Step 2: If the soil within this range is medium to coarse sand, above rule of thumb can be used.

If the average SPT (N) value is less than 10, soil should be compacted.

Design Example 7.1

Shallow foundation is placed on coarse to medium sand. SPT (N) values below the footing are as shown. The footing is 2 m × 2 m. The bottom of the footing is 1.5 m below the ground surface.

Find the allowable load bearing capacity of the footing (Figure 7.1).

Solution

Step 1: Find the average SPT (N) value below the footing to a depth that is equal to the width of the footing.

Average SPT (N) = (10 + 12 + 15 + 9 + 11 + 10)/6

Average SPT (N) = 11

Step 2:

$$\text{Allowable bearing capacity} = 9.6 \times \text{SPT}(N)\ \text{kPa}$$
$$= 9.6 \times 11\ \text{kPa}$$
$$= 105\ \text{kPa}$$

$$\text{Total allowable load on the footing} = \text{Allowable bearing capacity} \times \text{Area of the footing}$$
$$= 105 \times (2 \times 2)\ \text{kN}$$
$$= 420\ \text{kN} \quad (94\ \text{kips})$$

7.3 Bearing capacity in fine sands

Same equation can be used for fine sands, but subjected to a lower maximum value.

$$\text{Bearing capacity (allowable)} = 0.2\,N_{\text{average}} \quad \text{(ksf)} \quad \text{(not to exceed 8 ksf)} \qquad \text{fps units}$$
$$\text{Bearing capacity (allowable)} = 9.6\,N_{\text{average}} \quad \text{(kPa)} \quad \text{(not to exceed 380 kPa)} \qquad \text{SI units}$$

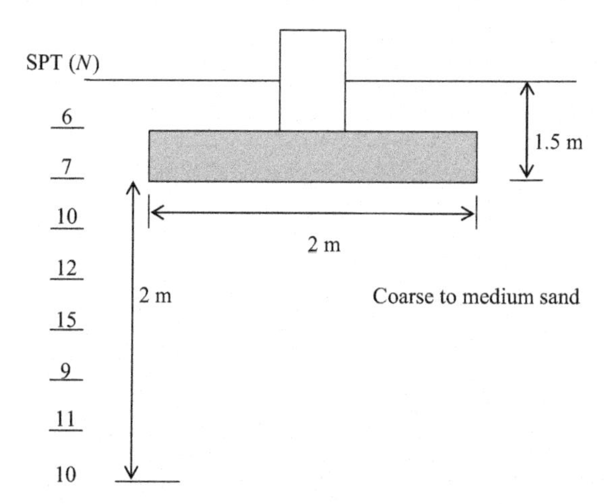

Figure 7.1 Load bearing capacity of footings.

Bearing capacity computation (general equation for cohesive and noncohesive soils)

Terzaghi theorized that the soil underneath the footing would generate a triangle of pressure as shown in Figure 8.1.

Terzaghi considered the angle between the bottom of the footing and the pressure triangle to be $(45 + \varphi/2)$, as shown in Figure 8.1.

The bearing capacity is developed in a footing due to the following three properties of soil:

1. Cohesion
2. Friction
3. Density of soil

8.1 Terms used in Terzaghi's bearing capacity equation

Ultimate bearing capacity (q_{ult}): Ultimate bearing capacity of a foundation is the load that a foundation would fail beyond any usefulness.

Cohesion (c): Cohesion is a chemical process. Clay particles tend to adhere to each other due to the electrical charges present in clay particles.

Friction: Unlike cohesion, friction is a physical process. The higher the friction between particles, the higher is the capacity of the soil to carry footing loads. Usually, sands and silts inherit friction. For all practical purposes, clays are considered friction free. Friction of a soil is represented by the friction angle (φ).

B = width of the footing

$(45 + \varphi/2)$ = The angle of the soil pressure triangle

The general Terzaghi's bearing capacity equation is as follows.

$$q_{ult} = \underbrace{cN_c s_c}_{\text{Cohesion term}} + \underbrace{qN_q}_{\text{Surcharge term}} + \underbrace{0.5BN_\gamma \gamma s_\gamma}_{\text{Density term}}$$

8.2 Description of terms in the Terzaghi's bearing capacity equation

q_{ult} = ultimate bearing capacity of the foundation (tsf or psf).

Cohesion term ($cN_c\, s_c$): This term represents the strength due to cohesion.

Geotechnical Engineering Calculations and Rules of Thumb

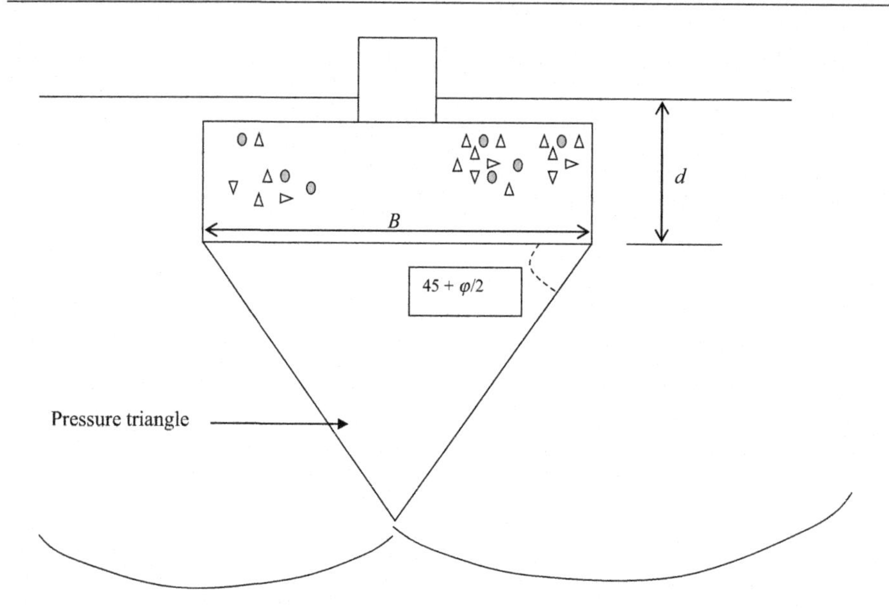

Figure 8.1 Bearing capacity computation.

c, cohesion of the soil; N_c, Terzaghi's bearing capacity factor (obtained from Table 8.1); s_c, shape factor obtained from Table 8.2.

Surcharge term (qN_q): This term represents the bearing capacity strength developed due to surcharge. This term deserves to be explained further.

Surcharge load is the pressure exerted due to soil above the bottom of footing.

"q" is the effective stress at bottom of footing;

$$q = \gamma d$$

Table 8.1 **Terzaghi's bearing capacity factors**

φ (Friction angle)	N_c	N_q	N_γ
0	5.7	1.0	0.0
5	7.3	1.6	0.5
10	9.6	2.7	1.2
15	12.9	4.4	2.5
20	17.7	7.4	5.0
25	25.1	12.7	9.7
30	37.2	22.5	19.7
35	57.8	41.4	42.4
40	95.7	81.3	100.4
45	172.3	173.3	297.5
50	347.5	415.1	415.1

Table 8.2 **Shape factors for Terzaghi's bearing capacity equation**

Shape factors	S_c	S_γ
Square footings	1.3	0.8
Strip footings (wall)	1.0	1.0
Round footings	1.3	0.6

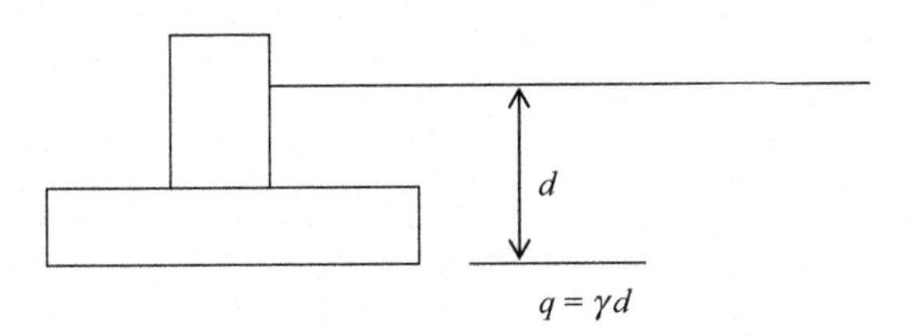

Figure 8.2 Effective stress at bottom of footing.

In Figure 8.2, d is the distance from ground surface to the bottom of the footing and γ is the effective density of soil.

Surcharge term = $q \times N_q$

N_q = Terzaghi bearing capacity factor obtained from Table 8.1. The bearing capacity factor, N_q depends upon the friction angle of soil (φ).

For example, if one places the bottom of the footing deeper, the term "d" in the equation would increase. Hence, the bearing capacity of the footing would also increase. Placing the footing deeper is a good method to increase the bearing capacity of a footing. This may not be a good idea when softer soils are present at deeper elevations.

Density term ($0.5BN_\gamma \gamma s_\gamma$): This term represents the strength due to density of soil. If the soil has a higher density, the bearing capacity of the soil would be higher.

B, width or the shorter dimension of the footing; N_γ, Terzaghi's bearing capacity factor obtained from Table 8.1.

The bearing capacity factors depend on the friction angle of soil (φ).

γ, effective density of soil; s_γ, shape factor obtained from Table 8.2.

Many others have given different tables after Terzaghi which have widely been used.

Additionally, the following equations can be used to find N_c, N_q, and N_γ.

Meyerhof gave the following equations to find N_c, N_q, and N_γ:

$$N_c = (N_q - 1)/\tan\varphi'$$
$$N_q = e^{\pi \tan\varphi'} \tan^2[45 + \varphi'/2]$$
$$N_\gamma = (N_q - 1)\tan(1.4\,\varphi')$$

Note that S_q is 1.0 for all the cases mentioned in Table 8.2.

8.3 Terzaghi's bearing capacity equation (discussion)

It is important to see how each term in the Terzaghi's bearing capacity equation affects the bearing capacity of a footing.

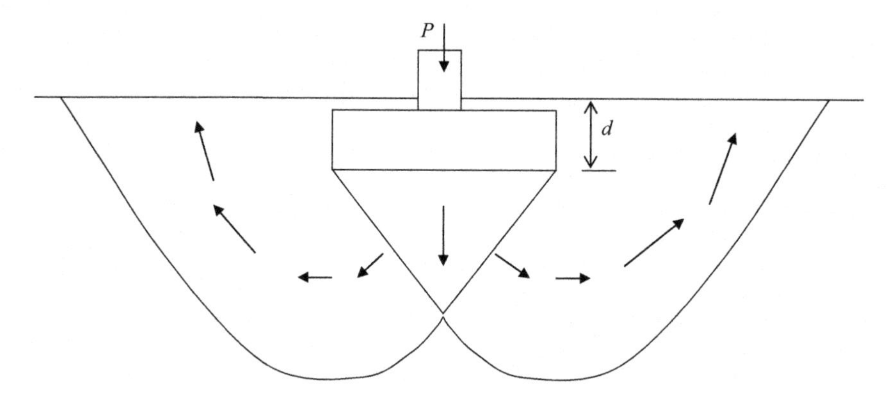

Figure 8.3 Movement of soil below a shallow footing at failure.

In Figure 8.3, when the footing load (P) is increased, the pressure triangle that is below the footing would be pressed down. The soil on sides would get pushed upwards. At failure, soil from the sides of the footing would be heaved. If there is a larger surcharge or the dimension "d" is increased, the soil will be locked in and the bearing capacity would increase. It is possible to increase the bearing capacity of a footing by increasing "d".

The bearing capacity increase due to surcharge (q) is given by the second term (qN_q) as discussed earlier.

Increase in bearing capacity when a layer of thickness "X" is added is shown in Figure 8.4.

In Figure 8.4, an additional soil layer of thickness, "X" is added. The new ultimate bearing capacity will be given by the following equation:

$$q_{\text{ult}} = cN_c s_c + qN_q + 0.5BN_\gamma \gamma s_\gamma$$

The first and the third terms will not undergo any change due to the addition of the new soil layer.

New surcharge "q" in the second term will be $q = \gamma\,(d + X)$.

The foundation engineer can increase the bearing capacity of a footing by adding more surcharge as shown earlier.

Effect of density ($0.5BN_\gamma \gamma s_\gamma$):

The effect due to density of soil is represented by the third term ($0.5BN_\gamma \gamma s_\gamma$).

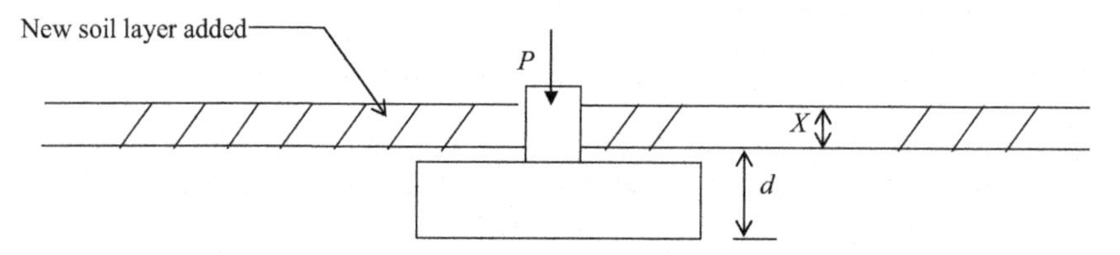

Figure 8.4 New soil layer added to increase the bearing capacity.

The higher the density of soil, the higher the ultimate bearing capacity. When "γ" increases, the ultimate bearing capacity also increases.

Effect of Friction Angle (φ'):

One may wonder where is the effect of the friction angle in the Terzaghi's bearing capacity equation. All bearing capacity coefficients (N_c, N_q, and N_γ) are obtained using the friction angle.

The higher the friction angle, the higher the bearing capacity factors. (see Table 8.1). Hence, the effect of the friction angle is incorporated into the Terzaghi's bearing capacity coefficients in an indirect way.

8.4 Bearing capacity in sandy soil (drained analysis)

Cohesion in sandy soils and nonplastic silts are considered zero. Design Example 8.1 shows the computation of bearing capacity in sandy soils.

Design Example 8.1: Column footing in a homogeneous sand layer

Find the ultimate bearing capacity of a 1.5 m × 1.5 m square footing placed in a sand layer (Figure 8.5). The density of the soil was found to be 18.1 kN/m^3 and friction angle was 30°.

Solution

Since the footing is placed in a sand layer, a drained analysis should be conducted. In drained analysis, cohesion in sand is zero.

$$q_{ult} = cN_c s_c + qN_q + 0.5BN_\gamma \gamma s_\gamma$$

Step 1: Find the Terzaghi's bearing capacity factors from Table 8.1, using $\varphi = 30°$.

$$N_c = 37.2, \quad N_q = 22.5, \quad N_\gamma = 19.7$$

Step 2: Find shape factors using Table 8.2.

For a square footing, $s_c = 1.3$ and $s_\gamma = 0.8$

Step 3: Find the surcharge (q).

$$q = \gamma d = 18.1 \times 1.2 = 21.7 \text{ kPa.}$$

Step 4: Apply the Terzaghi's bearing capacity equation.

$$q_{ult} = cN_c s_c + qN_q + 0.5BN_\gamma \gamma s_\gamma$$

$$q_{ult} = 0 \times 37.2 \times 1.3 + 21.7 \times 22.5 + 0.5 \times 1.5 \times 19.7 \times 18.1 \times 0.8$$

$$q_{ult} = 702.2 \text{ kPa}$$

Allowable bearing capacity ($q_{allowable}$) = q_{ult}/FOS = q_{ult}/3.0 = 234.1 kPa.

FOS = Factor of safety

Total load ($Q_{allowable}$) that can be safely placed on the footing

$$Q_{allowable} = q_{allowable} \times \text{area of the footing} = q_{allowable} \times (1.5 \times 1.5) = 526 \text{ kN.}$$

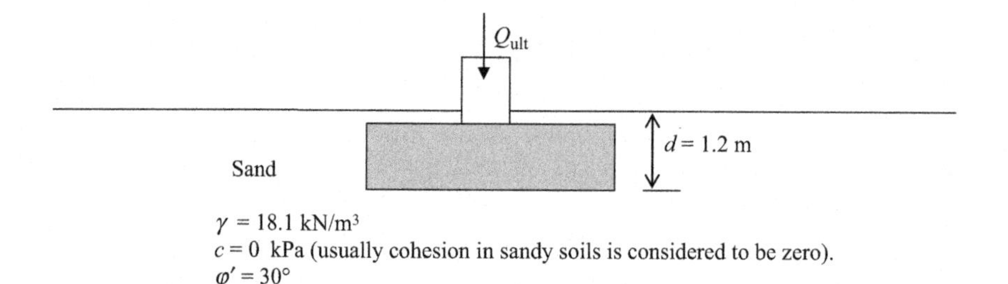

Figure 8.5 Column footing in a homogeneous sand layer.

In Design Example 8.1, we computed the bearing capacity of a column footing or a square footing placed in sandy soil. In Design Example 8.2, the computation of the bearing capacity of a wall footing placed in sandy soil is shown.

Design Example 8.2: Wall footing in a homogeneous sand layer

Find the ultimate bearing capacity of a 1.5 m wide wall (strip) footing placed in sandy soil (Figure 8.6). The density of the soil is found to be 18.1 kN/m^3 and friction angle to be 20°.

Solution

$$q_{ult} = cN_c s_c + qN_q + 0.5BN_\gamma \gamma s_\gamma$$

Step 1: Find the bearing capacity factors from Table 8.1.

$$N_c = 17.7, \qquad N_q = 7.4, \qquad N_\gamma = 5.0$$

Step 2: Find shape factors using Table 8.2.

For a wall footing, $s_c = 1.0$ and $s_\gamma = 1.0$

Step 3: Find the surcharge (q).

$$q = \gamma d = 18.1 \times 1.2 = 21.7 \text{ kPa.}$$

Step 4: Apply the Terzaghi's bearing capacity equation.

$$q_{ult} = cN_c s_c + qN_q + 0.5BN_\gamma \gamma s_\gamma$$

For sands, $c = 0$.

$$q_{ult} = 0 \times 17.7 \times 1.0 + 21.7 \times 7.4 + 0.5 \times 1.5 \times 5.0 \times 18.1 \times 1.0$$
$$q_{ult} = 228.5 \text{ kPa}$$

Allowable bearing capacity ($q_{allowable}$) = q_{ult}/FOS = q_{ult}/3.0 = 76.2 kPa.
Total load that can safely be placed on the footing:

$$\begin{aligned} Q_{allowable} &= q_{allowable} \times \text{area of the footing} = q_{allowable} \times (1.5 \times 1.0) \\ &= 114.2 \text{ kN / linear meter of footing.} \end{aligned}$$

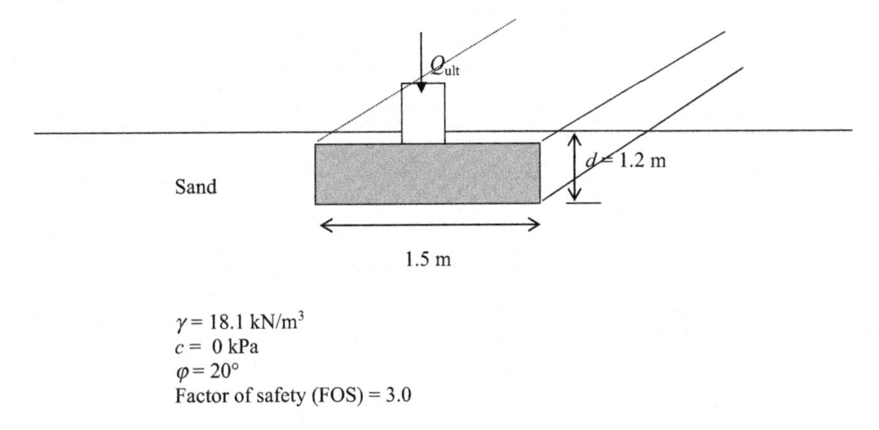

$\gamma = 18.1$ kN/m^3
$c =$ 0 kPa
$\varphi = 20°$
Factor of safety (FOS) = 3.0

Figure 8.6 Strip footing.

Note: In the case of wall footings, the area of the footing is taken as the width × 1 m or the capacity of the footing is given in terms of kN per 1 m length of the wall footing.

8.5 Bearing capacity in clay (undrained analysis)

I mentioned that during the construction of a foundation, the excess pore water would be dissipated in sandy soils. Therefore, a drained analysis is conducted. This is not the case with clay soils. Due to this reason, an undrained analysis should be conducted for clay soils. In clay soils, the cohesion term plays a major role in bearing capacity. Undrained friction angle is considered zero for clays and plastic silts. Note that drained friction angle is not zero for clay soils. Generally, many buildings are constructed fast enough not to allow drainage in clay soils. Due to this reason, assuming a friction angle for clay soils is not conservative.

Design Example 8.3: Column footing in a homogeneous clay layer

Find the ultimate bearing capacity of a 1.2 m × 1.2 m square footing placed in a clay layer (Figure 8.7). The density of the soil is found to be 17.7 kN/m^3 and the cohesion was found to be 20 kPa.

Solution

$$q_{ult} = cN_c s_c + qN_q + 0.5BN_\gamma \gamma s_\gamma$$

Step 1: Find the Terzaghi's bearing capacity factors from Table 8.1.
For clay soils, the friction angle (φ) is considered zero.

$$N_c = 5.7, \quad N_q = 1.0, \quad N_\gamma = 0.0$$

Step 2: Find shape factors using Table 8.2.
For a square footing $s_c = 1.3$ and $s_\gamma = 0.8$.

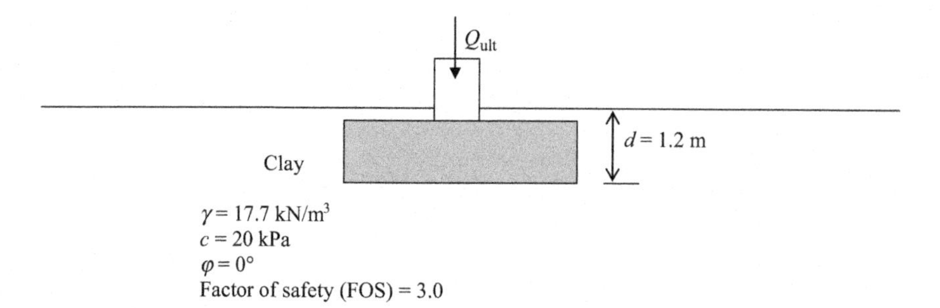

Figure 8.7 Column footing in a homogeneous clay layer.

Step 3: Find the surcharge (q).

$$q = \gamma d = 17.7 \times 1.2 = 21.2 \text{ kPa.}$$

Step 4: Apply the Terzaghi's bearing capacity equation.

$$q_{ult} = cN_c s_c + qN_q + 0.5BN_\gamma \gamma s_\gamma$$
$$q_{ult} = 20 \times 5.7 \times 1.3 + 21.2 \times 1.0 + 0$$
$$q_{ult} = 169.4 \text{ kPa}$$

Allowable bearing capacity ($q_{allowable}$) = q_{ult} /FOS = q_{ult} /3.0 = 56.5 kPa.
Total load ($Q_{allowable}$) that can be safely placed on the footing:

$$Q_{allowable} = q_{allowable} \times \text{area of the footing}$$
$$Q_{allowable} = q_{allowable} \times (1.2 \times 1.2) = 81.36 \text{ kN.}$$

Design Example 8.4

A 7 ft. tall dike is to be constructed on top of a soft clay layer using sandy material (Figure 8.8). The dike has to be constructed immediately.

Will the dike undergo a bearing failure? An unconfined compressive strength test of the clay sample yielded 700 psf.

Solution

Step 1: Pressure exerted on clay soil stratum due to the dike, needs to be computed.

- Volume of the dike per linear foot of length = $(4 + 6)/2 \times 7$ ft.^{3} = 35 ft.^{3}
- Weight of the dike per linear foot of length = $35 \times 120 = 4200$ lbs
  γ (soil density) = 120 pcf
- Pressure exerted on the clay stratum = 4200/6 = 700 psf

Step 2: Assume a factor of safety of 2.5.
Hence, the required bearing capacity of the soft clay layer = 2.5×700 psf = 1750 psf

Step 3: Bearing capacity of clay is given by the Terzaghi's bearing capacity equation.
Bearing capacity = $cN_c + qN_q + ½\gamma N_\gamma$
For clay soils $N_q = 1$ and $N_\gamma = 0$

Hence, the bearing capacity = $cN_c + qN_q$
where, q = overburden soil on the bearing stratum.
Since there is no overburden soil on the bearing stratum, $q = 0$.
Hence, the bearing capacity of the clay stratum = cN_c
From Table 8.1, for friction angle zero, $N_c = 5.7$.
Bearing capacity of the clay stratum = $5.7 \times c$
c = cohesion parameter

Step 4: Find the cohesion parameter of the clay layer.
Cohesion of the clay layer can be obtained by conducting an unconfined compressive strength test.
The load due to the dike is applied immediately and there is not enough time for the clay stratum to consolidate. The load is applied rapidly during an unconfined compressive strength test. For this scenario, the unconfined compressive strength test is appropriate.

Step 5: Unconfined compressive strength test yielded a value of 700 psf.
Hence $q_u = 700$ psf.
Unconfined compressive strength = $q_u = 700$ psf.
Shear strength (S_u) = c (cohesion)
Shear strength of clay soil (S_u) = $q_u/2 = 700/2 = 350$ psf. (Figure 8.9).
Since $\phi = 0$ for undrained clay soils, shear strength is solely dependant on cohesion.
(Drained clays have a ϕ value greater than zero)
Hence, $S_u = c + \sigma \tan \phi = c + 0 = 350$ psf
Bearing capacity of the clay layer = $5.7\,c = 5.7 \times 350 = 1995$ psf (from Step 3).
The bearing capacity required with a factor of safety of 2.5 = 1750 psf (from Step 2).
Hence, the soft clay layer will be safe against bearing failure.

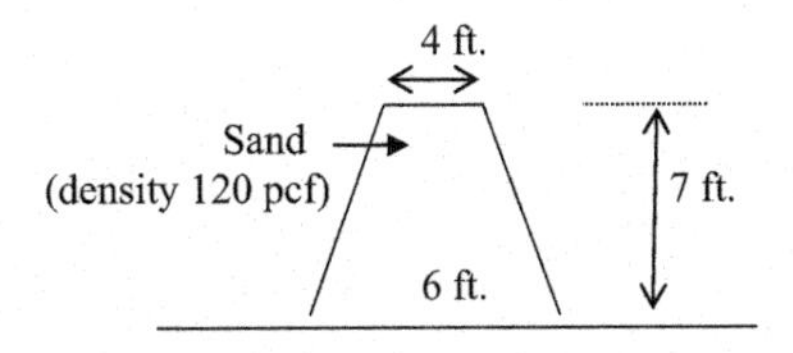

Figure 8.8 Dike built on soft clay.

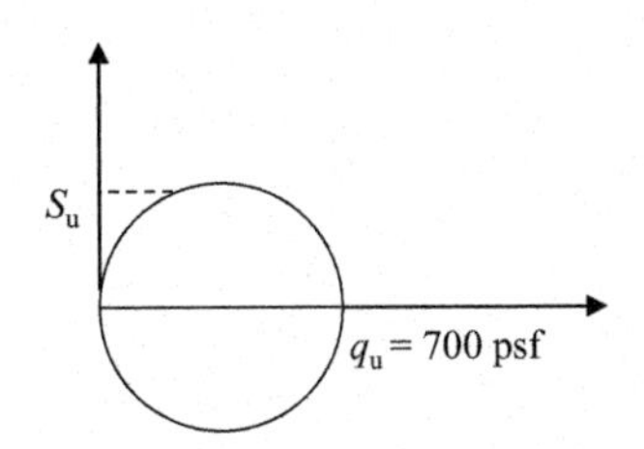

Figure 8.9 Mohr's circle.

8.6 Bearing capacity in layered soil

In reality, most sites contain more than one layer of soil. In such situations, the average values of friction angle and cohesion need to be obtained. This procedure can explained better with examples.

Average friction angle and cohesion within the pressure triangle needs to be obtained. The pressure triangle is drawn as shown in Figure 8.10, using the friction angle of the top layer. (φ_1).

Average cohesion within the pressure triangle is given by the following equation:

$$c_{\text{ave}} = \frac{c_1 h_1 + c_2 h_2}{H}$$

c_1 and c_2 are cohesion values of the first and second layers of soil, respectively. h_1 and h_2 are the thickness of soil layers measured as shown in Figure 8.10.

"H" can be found using trigonometry.

$$H = \frac{B}{2}\tan(45 + \phi_1/2)$$

Note that, H is equal to $(h_1 + h_2)$.

Note: Notice that the pressure triangle is found using φ_1, the friction angle of the top layer.

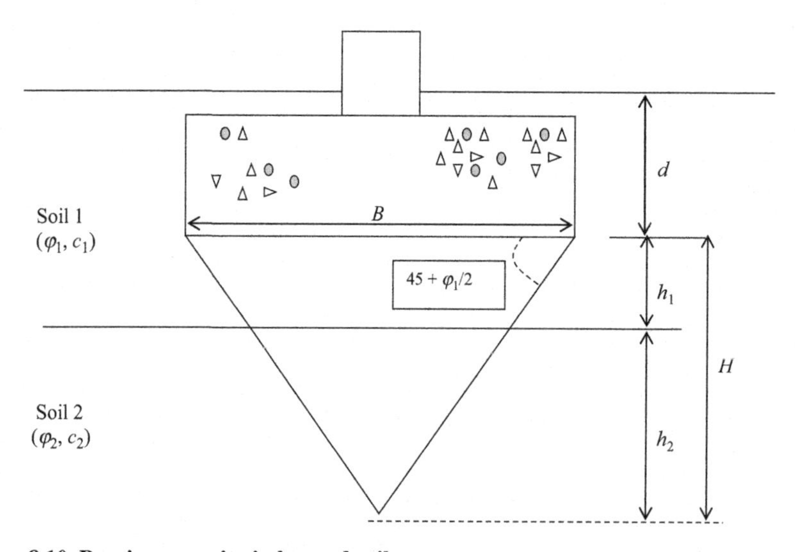

Figure 8.10 Bearing capacity in layered soil.

A similar procedure could be used to find the average friction angle within the pressure triangle.

$$\varphi_{\text{ave}} = \frac{\varphi_1 h_1 + \varphi_2 h_2}{H}$$

Average cohesion (c_{ave}) and average friction angle (φ_{ave}) are used in the Terzaghi's bearing capacity equation.

Design Example 8.5: Column footing in layered soil (clay and sand)

Find the ultimate bearing capacity of a 2 m × 2 m square footing placed in layered soils as shown in Figure 8.11. The density of all soils was found to be 17.7 kN/m^3 and the cohesion of clay was found to be 28 kPa. The friction angle of sand is 20°.

Solution

Use the Terzaghi's bearing capacity equation:

$$q_{\text{ult}} = cN_c s_c + qN_q + 0.5.\ BN_\gamma \gamma s_\gamma$$

Find the average friction angle and cohesion. To find the average friction angle and cohesion, the depth of the pressure triangle needs to be computed.

Step 1: Find the depth of the pressure triangle (H) (Figure 8.12).

$$H = \frac{B}{2}\tan(45 + \varphi_1/2)$$

B = width of the footing = 2 m.
φ_1 = Friction angle of the top layer = 0

$$H = \frac{2}{2}\tan(45 + 0) = 1\text{m}$$

h_1 is given to be 0.6 m and h_2 can be found.
$h_2 = H - h_1 = 1 - 0.6 = 0.4$ m

Step 2: Find average φ;

$$\phi_{\text{ave}} = \frac{\phi_1 h_1 + \phi_2 h_2}{H}$$

$$\phi_{\text{ave}} = \frac{0h_1 + 20h_2}{H}$$

Since $H = 1.0$ m, $h_1 = 0.6$ m, and $h_2 = 0.4$ m.

$$\varphi_{\text{ave}} = \frac{0 + 20^0 \times 0.4}{1} = 8^0$$

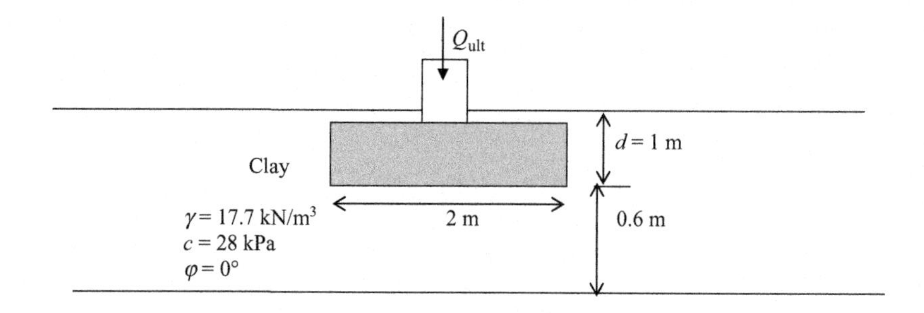

Figure 8.11 Shallow foundation on layered soil.

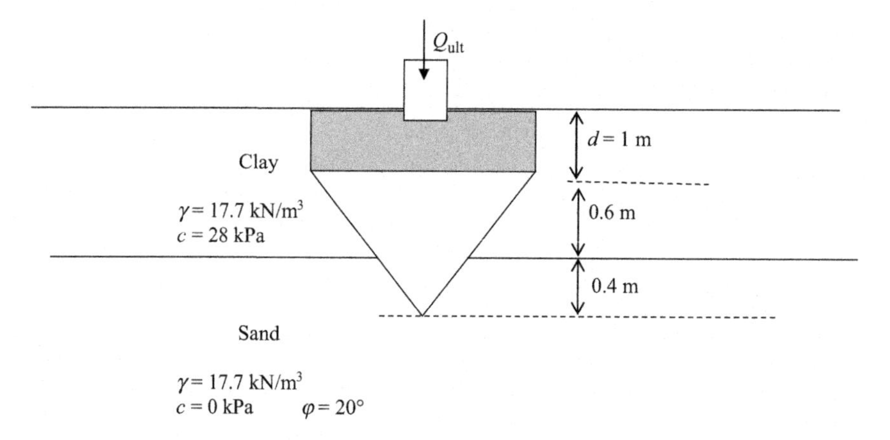

Figure 8.12 Effective depths for the bearing capacity equation.

Step 3:

$$c_{ave} = \frac{c_1 h_1 + c_2 h_2}{H}$$

Since, $H = 1.0$ m, $h_1 = 0.6$ m, $h_2 = 0.4$ m, $c_1 = 20$ kPa, and $c_2 = 0$.

$$c_{ave} = \frac{28.0.6 + 0}{1.0} = 16.8 \text{ kPa}$$

Use average "φ" and average "c" in the Terzaghi's bearing capacity equation to find the bearing capacity.

Step 4: Using an average "φ" of 8°, the Terzaghi bearing capacity factors can be found using Table 8.1.

$$N_c = 8.7; \quad N_q = 2.3; \quad N_\gamma = 0.9; \quad \text{(use interpolation)}.$$

From Table 8.2 for column (square) footings:

$$s_c = 1.3; \quad s_\gamma = 0.8$$

Step 5: Use the Terzaghi's bearing capacity equation to obtain q_{ult}.

$$q_{ult} = c_{ave}N_c s_c + qN_q + 0.5.\ BN_\gamma \gamma s_\gamma$$
$$q_{ult} = 16.8 \times 8.7 \times 1.3 + (17.7 \times 1) \times 2.3 + 0.5 \times 2 \times 0.9 \times 17.7 \times 0.8$$
$$q_{ult} = 191.1 + 40.7 + 12.7 = 244.5\ \text{kN/m}^2.$$
$$q_{allowable} = 244.5/\text{FOS} = 81.5\ \text{kN/m}^2 \text{ (assume a factor of safety (FOS) of 3.0).}$$

Step 6 Total allowable load on the footing:
Total allowable load on the footing (Q) = allowable bearing capacity × area of the footing

$$Q = q_{allowable} \times 4 = 326\ \text{kN}$$

Design Example 8.6: Column footing in layered soil (sand and sand)

Find the ultimate bearing capacity of a (2 m × 2 m) square footing placed in layered soils as shown in Figure 8.13. The density of all the soil layers was found to be 17.7 kN/m^3. The friction angles of sand layers are 15° and 20°, respectively.

Solution
Use the Terzaghi's bearing capacity equation

$$q_{ult} = cN_c s_c + qN_q + 0.5.\ BN_\gamma \gamma s_\gamma$$

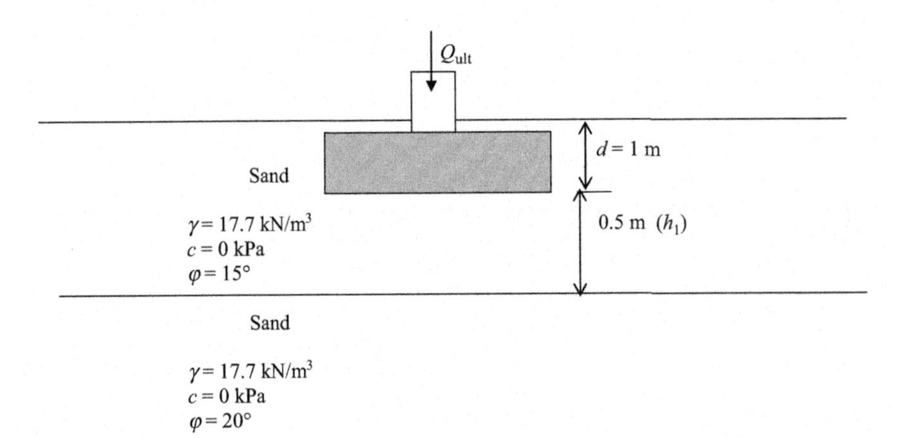

Figure 8.13 Shallow footing in two sand layers.

Step 1: Find the depth of the pressure triangle (H).

$$H = \frac{B}{2}\tan(45 + \varphi_1/2)$$

B = width of the footing = 2 m.
φ_1 = Friction angle of the top layer = 15°

$$H = \frac{2}{2}\tan(45 + 15/2) = 1.3\text{m}$$

h_1 is given to be 0.5 m and h_2 can be found. $h_2 = H - h_1 = 1.3 - 0.5 = 0.8$ m (Figure 8.14)

Step 2: Find average φ;

$$\varphi_{\text{ave}} = \frac{\varphi_1 h_1 + \varphi_2 h_2}{H}$$

$$\varphi_{\text{ave}} = \frac{15h_1 + 20h_2}{H}$$

Since, $H = 1.3$ m, $h_1 = 0.5$ m, and $h_2 = 0.8$ m.

$$\varphi_{\text{ave}} = \frac{15° \times 0.5 + 20° \times 0.8}{1.3} = 18°$$

Step 3:

$$c_{\text{ave}} = 0 \text{ since } c_1 \text{ and } c_2 \text{ are zero}$$

Use average "φ" and average "c" in the Terzaghi's bearing capacity equation to find the bearing capacity.

Step 4: Using an average "φ" of 18° degrees, the Terzaghi bearing capacity factors could be found.
Using Table 8.1 through interpolation, the following bearing capacity factors are obtained.

$$N_c = 15.8; \quad N_q = 6.2; \quad N_\gamma = 4.0;$$

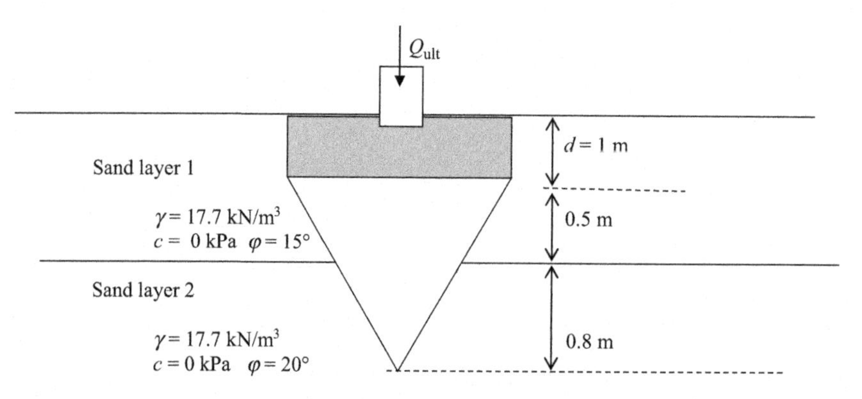

Figure 8.14 Effective depths of two sand layers.

From Table 8.2 for column (square) footings,

$$s_c = 1.3; \qquad s_\gamma = 0.8$$

Step 5: Use the Terzaghi's bearing capacity equation to obtain q_{ult}.

$$q_{ult} = c_{ave} N_c s_c + qN_q + 0.5BN_\gamma \gamma s_\gamma$$
$$q_{ult} = 0 + (17.7 \times 1) \times 6.2 + 0.5 \times 2 \times 4.0 \times 17.7 \times 0.8$$
$$q_{ult} = 0 + 109.7 + 24.6 = 134.4 \text{ kN/m}^2$$
$$q_{allowable} = 144.4/\text{FOS} = 44.8 \text{ kN/m}^2. \text{ (Assume a factor of safety (FOS) of 3.0)}$$

Step 6: Total allowable load on the footing;
Total allowable load on the footing (Q) = allowable bearing capacity × area of the footing

$$Q = q_{allowable} \times 4 = 179.2 \text{ kN}$$

What if the pressure triangle is located in one soil layer as shown in Figure 8.15?

In Figure 8.15, the effect of clay on the bearing capacity can be reasonably ignored. The designer should still look into the settlement aspects of the footing separately.

In the previous example, column footing placed on a soil strata consisting two different sand layers was considered. In this example, wall footing placed in two different clay layers is considered.

Design Example 8.7: Wall footing in layered soil (clay and clay)

Find the ultimate bearing capacity of a 2 m wide square wall footing placed in layered soils as shown in Figure 8.16. The density of the soil is found to be 17.7 kN/m^3 for all soils. The cohesion of the top clay layer was found to be 28 kPa and the cohesion of the bottom clay layer was found to be 20 kPa.

Solution

Use the Terzaghi's bearing capacity equation.

$$q_{ult} = cN_c s_c + qN_q + 0.5BN_\gamma \gamma s_\gamma$$

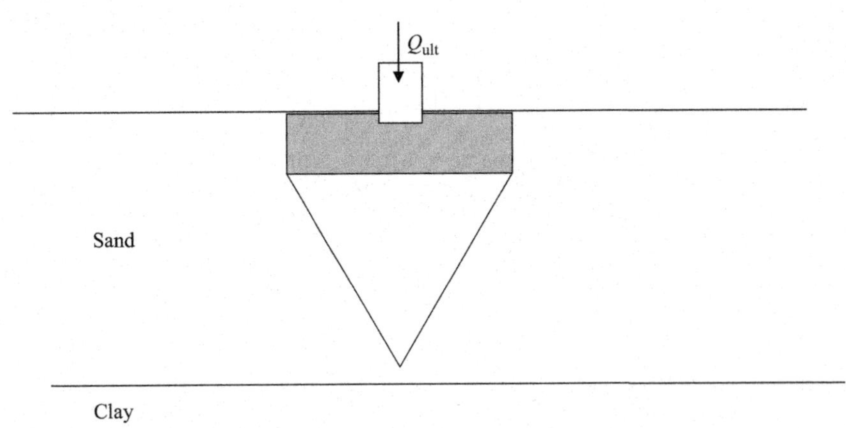

Figure 8.15 Shallow foundation in sand and clay layers.

Step 1: Find the depth of the pressure triangle (H).

$$H = \frac{B}{2}\tan(45 + \varphi_1/2)$$

B = width of the footing = 2 m.
φ_1 = friction angle of the top layer = 0.

$$H = \frac{2}{2}\tan(45 + 0) = 1\text{m}$$

h_1 is given to be 0.6 m and h_2 can be found. $h_2 = H - h_1 = 1 - 0.6 = 0.4$ m (Figure 8.17)

Step 2: Find average φ.
Average φ is zero since both the layers are clay.

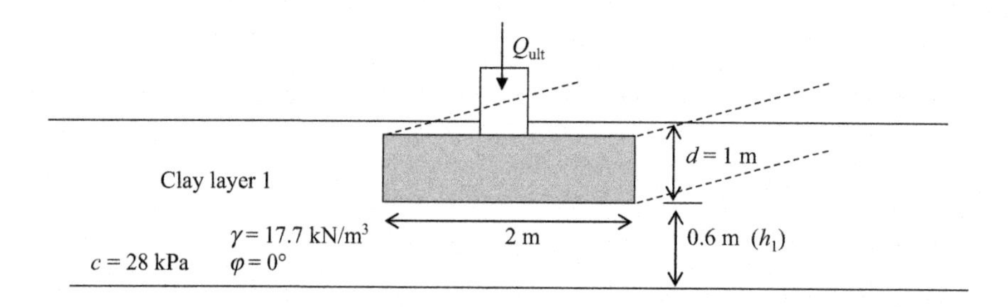

Figure 8.16 Shallow foundation in two clay layers.

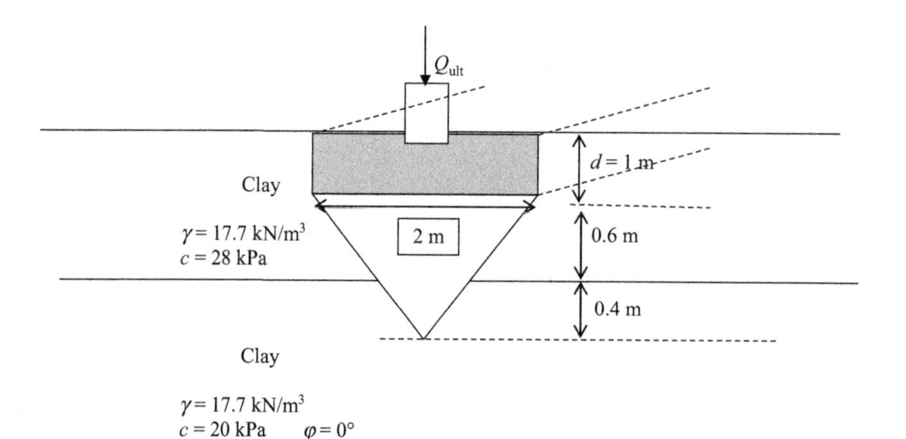

Figure 8.17 Pressure triangle.

Step 3: Find the average cohesion.

$$c_{ave} = \frac{c_1 h_1 + c_2 h_2}{H}$$

Since, $H = 1.0$ m, $h_1 = 0.6$ m, $h_2 = 0.4$ m, $c_1 = 28$ kPa, and $c_2 = 20$ kPa.

$$c_{ave} = \frac{28 \times 0.6 + 20 \times 0.4}{1.0} = 24.8 \text{ kPa}$$

Use average "φ" and average "c" in the Terzaghi's bearing capacity equation to find the bearing capacity.

Step 4: Using an average "φ" of 0°, the Terzaghi's bearing capacity factors can be found.
Using Table 8.1, the bearing capacity factors can be found.

$$N_c = 5.7; \quad N_q = 1.0; \quad N_\gamma = 0.0;$$

From Table 8.2, for wall (strip) footings:

$$s_c = 1.0; \quad s_\gamma = 1.0;$$

Step 5: Use the Terzaghi's bearing capacity equation to obtain the q_{ult}.

$q_{ult} = c_{ave} N_c s_c + q N_q + 0.5 B N_\gamma \gamma s_\gamma$
$q_{ult} = 24.8 \times 5.7 \times 1.0 + (17.7 \times 1) \times 1.0 + 0$
$q_{ult} = 141.4 + 17.7 = 159.1$ kN/m^2.
$q_{allowable} = 159.1/\text{FOS} = 53.0$ kN/m^2 (assume a factor of safety (FOS) of 3.0)

In a wall footing, the allowable load is given per linear meter of the footing.

$$q_{allowable} = 53.0 \times (\text{width of the footing})$$
$$= 53.0 \times 2 = 106 \text{ kN per linear meter of the footing.}$$

Design Example 8.8: Strip footing design in sand using SPT (N) values (No Groundwater Considered)

Find the allowable bearing capacity of the strip foundation shown in Figure 8.18. Sand was encountered below the footing and the SPT (*N*) values are as shown in Figure 8.18.
Assume the width of the footing to be 4 ft. for initial calculations.

Solution
Step 1:
- SPT (*N*) values obtained are given in Figure 8.18.

Step 2: Use Tables 2.2 and 2.3 to obtain φ (friction angle) and the total density using the average *N* value.

Step 3: In this case, $\varphi = 30$. (for fine sand with an average *N* value of 7 from Table 2.2).

Step 4:
- Terzaghi bearing capacity equation:

Ultimate bearing capacity (q_{ult}) = $cN_c s_c + qN_q + 0.5\gamma B N_\gamma s_\gamma$
c = cohesion (zero for sands).

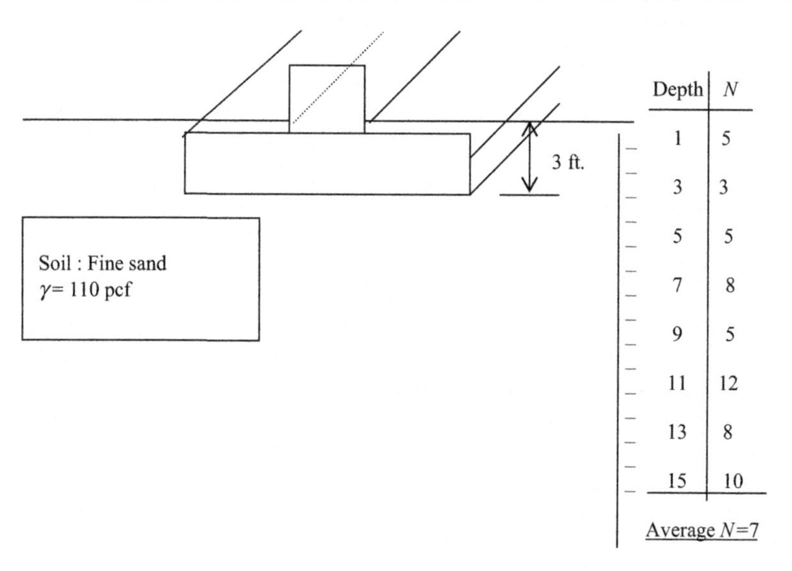

Figure 8.18 Strip footing in sand.

N_c, N_q, and N_γ Terzaghi's bearing capacity factors
For $\varphi = 30°$, obtain $N_c = 37.2$, $N_q = 22.5$, and $N_\gamma = 19.7$
γ = total soil density = 110 pcf.
[Note: γ' is equal to the total soil density(γ) above groundwater level. Below groundwater $\gamma' = \gamma - \gamma_w$; ($\gamma_w$ = density of water).]

q = effective stress at bottom of footing
$= 110 \times 3 = 330$ lbs/ft.^{2}

B = width of footing

$S_c = 1; \quad S_\gamma = 1$

Substitute all the values into the Terzaghi's bearing capacity equation:

Ultimate bearing capacity (q_{ult}) = $cN_cs_c + qN_q + 0.5\gamma' BN_\gamma s_\gamma$

$$q_{ult} = 0 + 330 \times 22.5 + 0.5 \times 110 \times 5 \times 19.7 \times 1.0 = 12{,}842 \text{ psf}$$

Allowable bearing capacity (q_a) = $q_{ult}/3.0$

$$q_{ult}/3.0 = 12{,}842/3.0 = 4{,}280 \text{ psf}$$

8.7 Bearing capacity when groundwater present

Groundwater reduces the density due to buoyancy. When groundwater is present, the density of soil needs to be modified. When a footing is loaded, the triangle of stress is developed below the footing (Figure 8.19). If groundwater is below the stress triangle, it does not affect the bearing capacity.

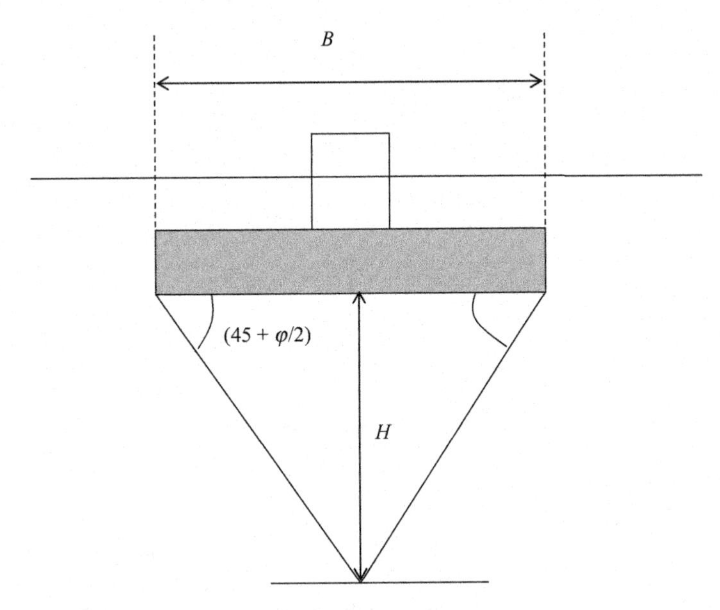

Figure 8.19 Pressure triangle below a footing.

It is assumed that the angle between the stress triangle and the footing is (45 + φ/2). Hence, from trigonometry we can obtain

$$H = B/2 \times \tan(45 + \varphi/2)$$

H = depth of the stress triangle; B = width of the footing.

Equivalent density of soil (γ_e):

If the groundwater is within the stress triangle, the equivalent density of soil is used (Figure 8.20).

Equivalent density of soil (γ_e) is given by

$$\gamma_e = \frac{(2H - d_w)}{H^2} \times d_w \times \gamma_{total} + \gamma' \times \frac{(H - d_w)^2}{H^2}$$

H, 0.5 B tan (45 + Φ/2); d_w, depth to groundwater measured from the bottom of footing; γ, buoyant density of soil = $\gamma - \gamma_w$; γ_w, density of water.

8.8 Groundwater below the stress triangle

If the groundwater level is below the stress triangle (Figure 8.21) ($d_w > H$), the effect due to groundwater is ignored.

When groundwater is below the stress triangle, γ (total soil density) is used in the bearing capacity equation.

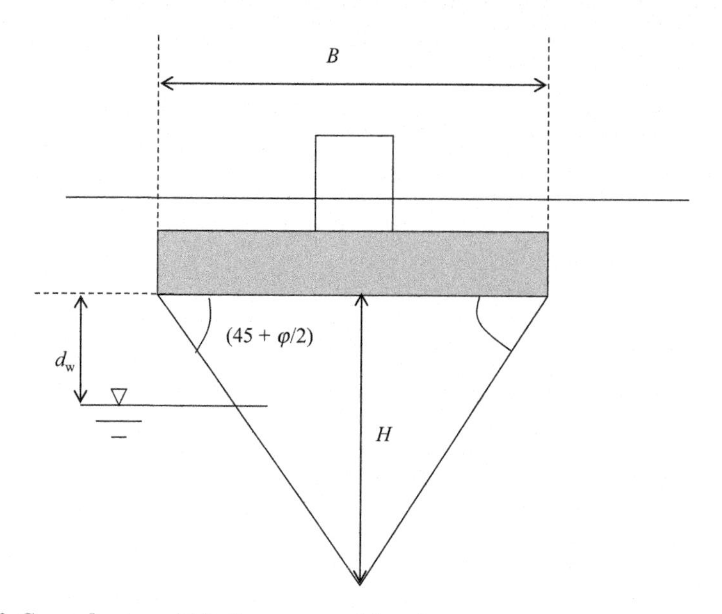

Figure 8.20 Groundwater within the pressure triangle.

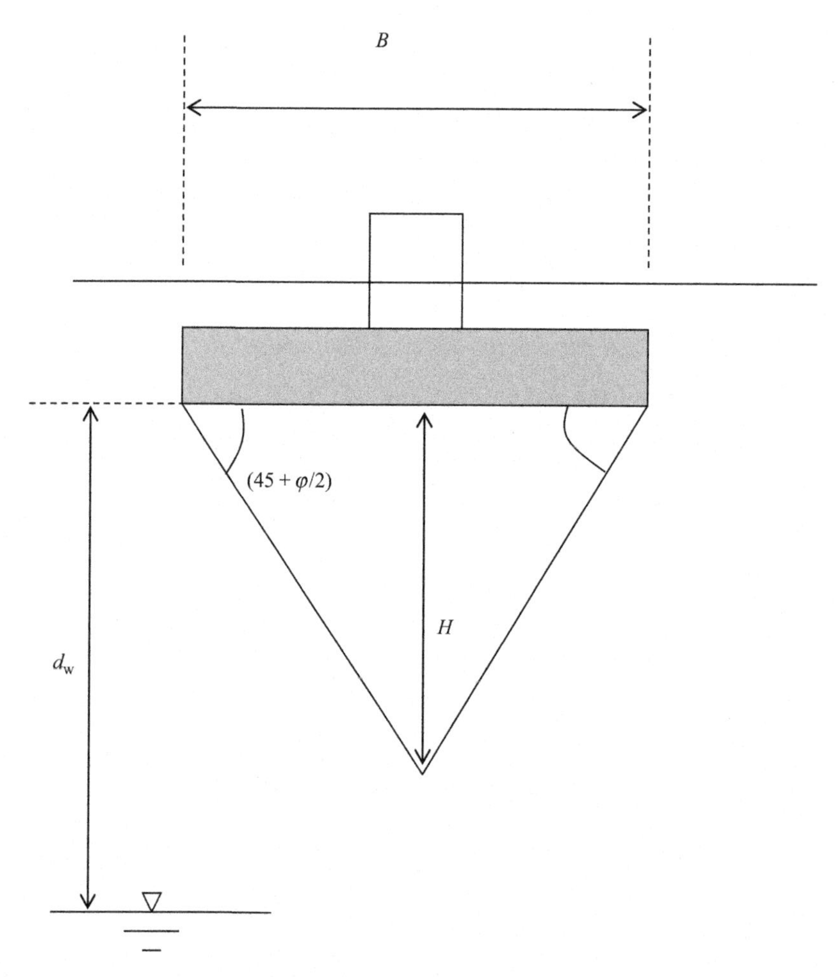

Figure 8.21 Groundwater is below the stress triangle.

8.9 Groundwater above the bottom of footing level

If the groundwater is above the bottom of footing (Figure 8.22), then γ' is used in the bearing capacity equation.

$$\gamma' = \gamma - \gamma_w$$

γ', buoyant density of soil; γ, density of soil; γ_w, density of water.

8.10 Groundwater at bottom of footing level

When the groundwater is at the bottom of footing level, d_w becomes zero.

In this case, the first term of the equation becomes zero and the second term becomes γ'.

$$\gamma_e = \frac{(2H - d_w)}{H^2} \times d_w \times \gamma_{total} + \gamma' \times \frac{(H - d_w)^2}{H^2}$$

Hence, when the groundwater is at bottom of footing level,

$$\gamma_e = \gamma'$$

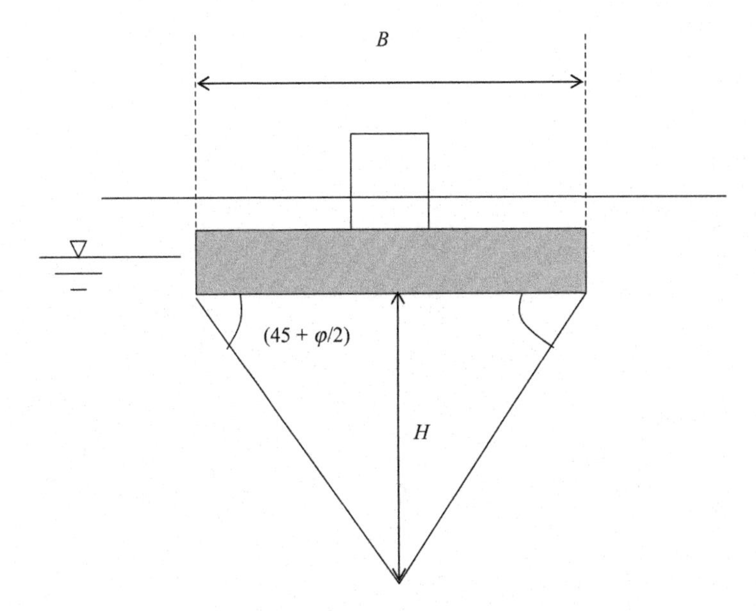

Figure 8.22 Groundwater above bottom of footing. If groundwater is above bottom of footing (γ') is used.

Design Example 8.9: Shallow foundation design (when groundwater is present)

The loading given by the structural engineer happens to be 15,000 lbs per linear foot of the wall. The average SPT (*N*) value is found to be 7, which relates to a friction angle of 30° for fine sands.
The groundwater is 4 ft. below the ground surface. The bottom of the footing is placed 3 ft. below the ground surface to provide protection against frost (Figure 8.23). Compute the bearing capacity of a 4 ft. wide strip footing.

Solution

Step 1: Width of the footing (B) = 4 ft.

- In this case, groundwater is below the bottom of footing.

$$\begin{aligned} H &= 0.5B\tan(45+\Phi/2) \\ &= 0.5\times 4\times \tan(45+30/2) \\ &= 0.5\times 4\times \tan(60) \\ &= 2\times 1.73 \\ H &= 3.46 \end{aligned}$$

$$\gamma_e = \frac{(2H-d_w)}{H^2}\times d_w \times \gamma_{total} + \gamma' \times \frac{(H-d_w)^2}{H^2}$$

d_w = depth to groundwater measured from the bottom of footing
d_w = 1 ft.
γ' = 110 − 62.4 = 47.6 pcf

$$\gamma_e = \frac{(2\times 3.46-1)}{(3.46)^2}\times 1\times 110 + 47.6\times \frac{(3.46-1)^2}{(3.46)^2}$$

$$\gamma_e = 54.4 + 24.1 = 78.5 \text{ pcf}$$

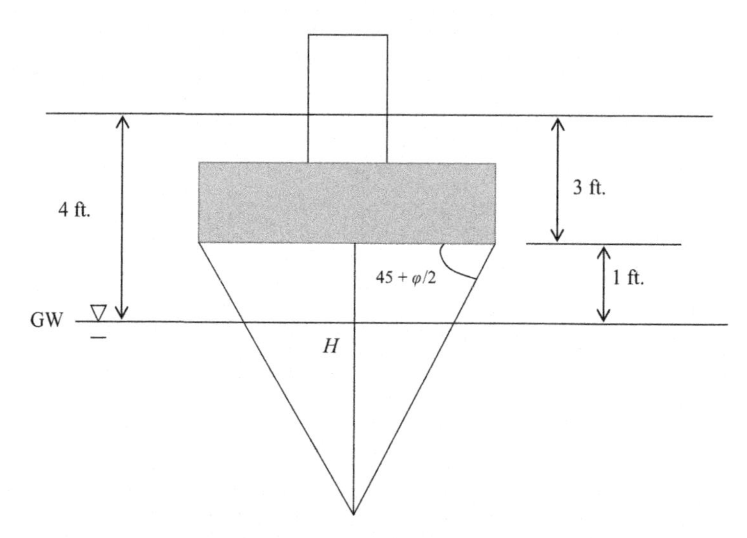

Figure 8.23 Groundwater within the pressure triangle.

Step 2: Terzaghi's bearing capacity equation:
Ultimate bearing capacity (q_{ult}) = $cN_cS_c + q'N_q + 0.5\gamma' BN_\gamma S_\gamma$

c = cohesion (zero for sands)
$\varphi = 30$, for $N = 7$, for fine sands. (Table 2.2)
N_c, N_q, and N_γ = Terzaghi's bearing capacity factors
For $\varphi = 30$,
Obtain $N_c = 37.2$, $N_q = 22.5$, $N_\gamma = 19.7$ (Table 8.1)
$S_\gamma = 1.0$, for strip footings (Table 8.2)

Step 3:

- The first term (cN_cS_c) in the Terzaghi's bearing capacity equation does not change due to groundwater elevation. (For sands "c" [cohesion] is zero).
- Second term (qN_q) will change depending on the elevation of the groundwater. "q" is defined as the effective stress at bottom of footing.

$$q = (110) \times 3$$
$$q = 330 \text{ psf}$$

$$qN_q = 330 \times 22.5$$
$$= 7425 \text{ psf}$$

Step 4: Apply the Terzaghi's bearing capacity equation.
Ultimate bearing capacity (q_{ult}) = $cN_cS_c + qN_q + 0.5\gamma_e \times B \times N_\gamma \times S_\gamma$

$$\text{Ultimate bearing capacity } (q_{ult}) = 0 + 7425 + 0.5 \times 78.5 \times 4 \times 19.7 \times 1$$
$$= 10518 \text{ psf}$$

$$\text{Allowable bearing capacity} = \text{ultimate bearing capacity/factor of safety}$$
$$= 10518/3.0$$
$$= 3506 \text{ psf}$$

Design Example 8.10: Strip and column footing design in sand (Groundwater effects considered)

The loading given by the structural engineer happens to be 15,000 lbs per linear foot of the wall. The average SPT (N) value is found to be 7, which relates to a friction angle of 30° for fine sands.
The groundwater is 2 ft. below the ground surface. The bottom of the footing is placed 3 ft. below the ground surface to provide protection against frost (Figure 8.24). Compute the bearing capacity of a 4 ft. wide strip footing.

Solution

Step 1: In this case, the groundwater is above the bottom of the footing.
Hence $\gamma_e = \gamma'$.

Step 2: Terzaghi's bearing capacity equation:
Ultimate bearing capacity (q_{ult}) = $cN_cS_c + q'N_q + 0.5\gamma' BN_\gamma S_\gamma$

c = cohesion (zero for sands)
$\varphi = 30°$ for $N = 7$, for fine sands. (Table 2.2)

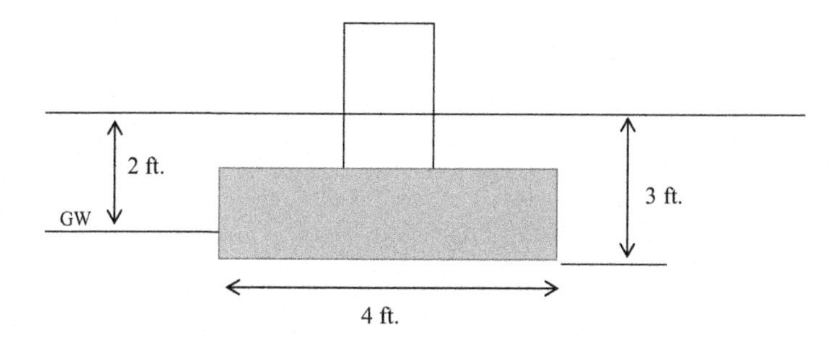

Figure 8.24 Shallow foundation (groundwater effect considered).

N_c, N_q, and N_γ = Terzaghi's bearing capacity factors.
For $\varphi = 30°$,
Obtain $N_c = 37.2$, $N_q = 22.5$, and $N_\gamma = 19.7$.
γ' = buoyant soil density
γ' is taken to be equal to $\gamma' = \gamma - \gamma_W$
(γ_W = density of water).

Step 3:

- The first term (cN_cS_c) in the Terzaghi's bearing capacity equation does not change due to the groundwater elevation. (For sands, "c" (cohesion) is zero).
- The second term ($q'N_q$) will change depending on the elevation of the groundwater. q' is defined as the effective stress at the bottom of the footing.

$$q' = (110) \times 2 + (110 - 62.4) \times 1 = 267.6 \text{ psf}$$
$$q'N_q = 267.6 \times 22.5 = 6021 \text{ psf}$$

The term (110 × 2) gives the effective stress due to soil above the groundwater level. The total density (γ) is used to compute the effective stress above the groundwater level.
The next term [(110 − 62.4) × 1] gives the effective stress due to soil below the groundwater table. Buoyant density (γ') is used to compute the effective stress below the groundwater level.

Step 4: γ_e is given by

$$\gamma_e = \frac{(2H - d_w)}{H^2} \times d_w \times \gamma_{total} + \gamma' \times \frac{(H - d_w)^2}{H^2}$$

When groundwater is above the bottom of the footing, $\gamma_e = \gamma'$

$$\gamma_e = (110 - 62.4) = 47.6 \text{ psf}$$

Apply the Terzaghi's bearing capacity equation.
Ultimate bearing capacity (q_{ult}) = $cN_cS_c + q'N_q + 0.5\gamma' BN_\gamma S_\gamma$

$$\text{Ultimate bearing capacity } (q_{ult}) = 0 + 6{,}021 + 0.5 \times 47.6 \times 4 \times 19.7 \times 1$$
$$= 7896 \text{ psf}$$

$$\text{Allowable bearing capacity} = \text{ultimate bearing capacity/factor of safety}$$
$$= 7896/3.0 = 2632 \text{ psf}$$

8.11 Meyerhof bearing capacity equation

So far we have studied the bearing capacity computation with Terzaghi's bearing capacity equation. In 1963, Meyerhof proposed another set of equations and some think that these set of equations are more accurate than the Terzaghi equations. Meyerhof gave two equations. One for vertical loads and another one for inclined loads.

8.11.1 Meyerhof equation for vertical loads

$$q_{ult} = \underbrace{cN_c s_c d_c}_{\text{Cohesion term}} + \underbrace{qN_q s_q d_q}_{\text{Surcharge term}} + \underbrace{0.5BN_\gamma \gamma s_\gamma d_\gamma}_{\text{Density term}}$$

q_{ult}, ultimate bearing capacity of the foundation (tsf or psf); c, cohesion of the soil; N_c, N_q, N_γ, Meyerhof bearing capacity factors; N_q, $e^{\pi \tan\varphi'}\tan^2[45 + \varphi'/2]$; N_c, $(N_q - 1)/\tan\varphi'$; N_γ, $(N_q - 1)\tan(1.4\varphi')$; s_c, s_q, s_γ, shape factors; s_c, $1 + 0.2\tan^2[45 + \varphi'/2] \times B/L$ for any φ'; s_q, $s_\gamma = 1 + 0.1\tan^2[45 + \varphi'/2] \times B/L$ for $\varphi' > 10°$; s_q, $s_\gamma = 1.0$, for $\varphi' = 0$, and $\varphi' < 10°$.

B is the shorter dimension of a rectangular footing and L is the longer dimension.

d_c, d_q, and d_γ = depth factors.

When the depth of the footing increases, the bearing capacity increases (Figure 8.25).

$d_c = 1 + 0.2\tan[45 + \varphi'/2] \times D/B$ for any φ'

$d_q = d_\gamma = 1 + 0.1\tan[45 + \varphi'/2] \times D/B$ for $\varphi' > 10°$.

Note that it is not $\tan^2[45 + \varphi'/2]$ as in shape factors.

$d_q = d_\gamma = 1$, for $\varphi' = 0$, and $\varphi' < 10°$.

8.11.2 Meyerhof equation for inclined loads

$$q_{ult} = \underbrace{cN_c d_c i_c}_{\text{Cohesion term}} + \underbrace{qN_q d_q i_q}_{\text{Surcharge term}} + \underbrace{0.5BN_\gamma \gamma d_\gamma i_\gamma}_{\text{Density term}}$$

As you can see, Meyerhof removed the shape factors for inclined loads.

Inclined loads would reduce the bearing capacity (Figure 8.26).

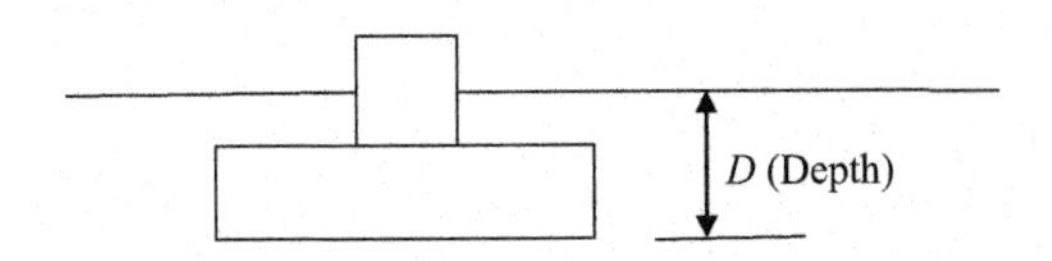

Figure 8.25 Shallow footing.

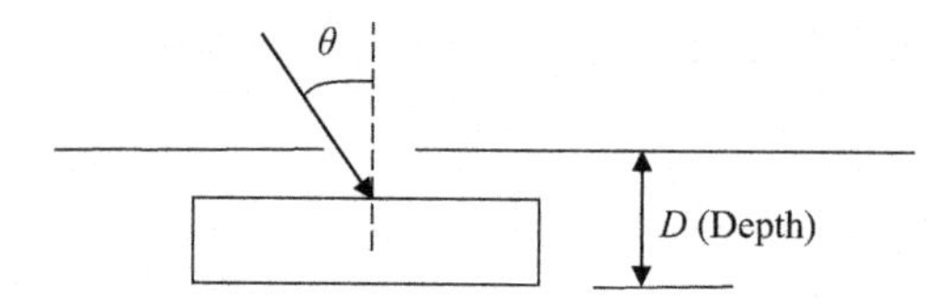

Figure 8.26 Inclined loading in footings.

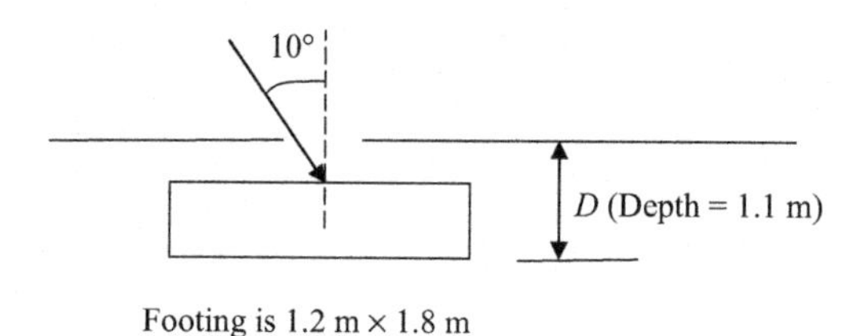

Figure 8.27 10° inclination.

$i_c = i_q = (1 - \theta/90)^2$ for any φ'
θ is in degrees.
$i_\gamma = (1 - \theta/\varphi)^2$ for $\varphi' > 0°$.
$i_\gamma = 0$ for $\varphi' = 0$.

Design Example 8.11

A column foundation with dimensions 1.2 m × 1.8 m was placed at a depth of 1.1 m from the ground surface (Figure 8.27). The soil profile is mainly medium sand with a friction angle of 35° and a soil unit weight of 17.1 kN/m^2. The inclination of the load is 10° from the vertical. What is the allowable load in kN on the footing if the desired factor of safety is 3.0?

Solution

Step 1: Write down the Meyerhof bearing capacity equation for inclined loads.

$$q_{ult} = cN_c d_c i_c + qN_q d_q i_q + 0.5BN_\gamma \gamma d_\gamma i_\gamma$$

$c = 0$ for medium sand
$q = \gamma D = 17.1 \times 1.1 = 18.81$ kN/m^2
$N_q = e^{\pi \tan\varphi'} \tan^2[45 + \varphi'/2]$
$N_q = e^{\pi \tan 35} \tan^2[45 + 35/2] = 9.023 \times 3.69 = 33.3$
$N_\gamma = (N_q - 1) \tan(1.4\, \varphi')$
$N_\gamma = (33.3 - 1) \tan(1.4 \times 35) = 37.16$
$d_q = d_\gamma = 1 + 0.1 \tan[45 + \varphi'/2] \times D/B$ for $\varphi' > 10°$.
$d_q = d_\gamma = 1 + 0.1 \tan[45 + 35/2] \times 1.1/1.2$ for $\varphi' > 10°$.
$d_q = d_\gamma = 1.17$
$i_c = i_q = (1 - \theta/90)^2$ for any φ'
$i_q = (1 - 10/90)^2 = 0.79$
$i_\gamma = (1 - \theta/\varphi)^2$ for $\varphi' > 0°$.
$i_\gamma = (1 - 10/35)^2 = 0.51$

Step 2: Apply the equation;

$$q_{ult} = cN_c d_c i_c + qN_q d_q i_q + 0.5BN_\gamma \gamma d_\gamma i_\gamma$$

$$q_{ult} = 0 + 18.81 \times 33.3 \times 1.17 \times 0.79 + 0.5 \times 1.2 \times 37.16 \times 17.1 \times 1.17 \times 0.51$$

$$q_{ult} = 0 + 578.9 + 227.5 = 806.4 \text{ kN/m}^2$$

$$q_{allowable} = 806.4/3.0 = 268.8 \text{ kN/m}^2$$

8.12 Eccentric loading

At some occasions loads are not placed on the center of gravity of the footing. In such situations, the bearing pressure under the footing is not uniform (Figure 8.28).

As you can see, when the load is acting at an eccentricity, the soil pressure is not uniform (Figure 8.29). The above footing is equated to a footing with a uniform pressure with width B'. B' is calculated using the following equation:

$$B' = B - 2e$$

It is important to mention here that $B' = B - 2e$ equation does not come from theory. It is a conservative simplification.

Similarly, eccentricity could be along the length (Figure 8.30). In that case, we have $L' = L - 2e$

Design Example 8.12

A 3 m × 5 m rectangular footing is loaded with an eccentricity of 0.5 m along the shorter dimension (Figure 8.31). The load on the footing is 50 kN. Find the soil pressure under the footing.

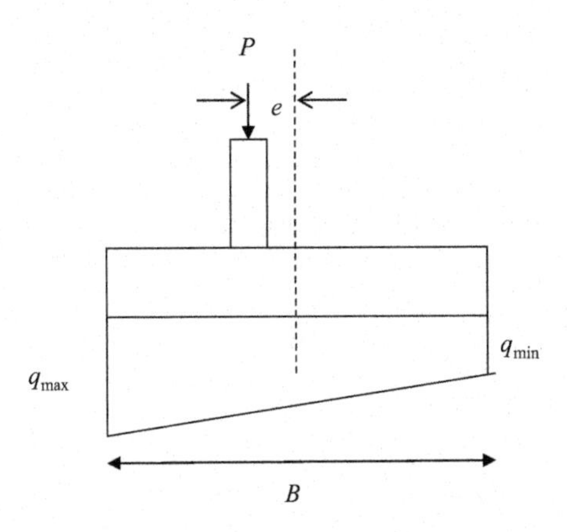

Figure 8.28 Eccentric loading.

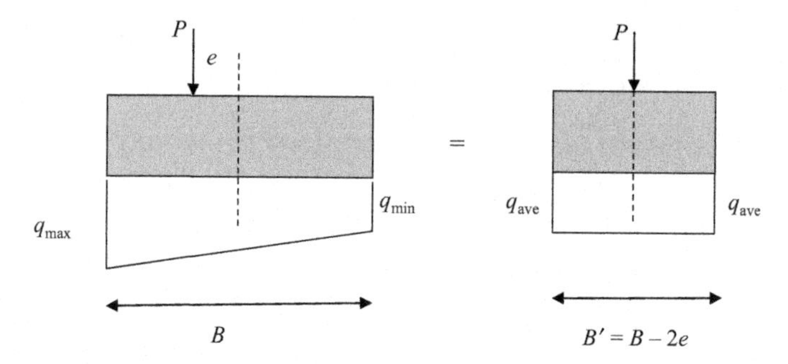

Figure 8.29 Eccentric loading and equivalent width.

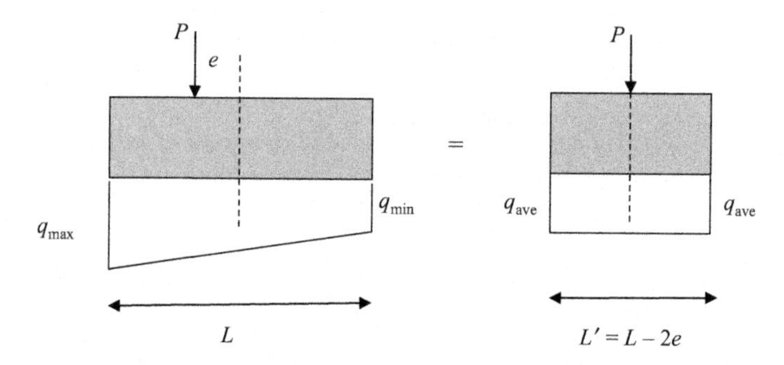

Figure 8.30 Eccentric loading and equivalent length.

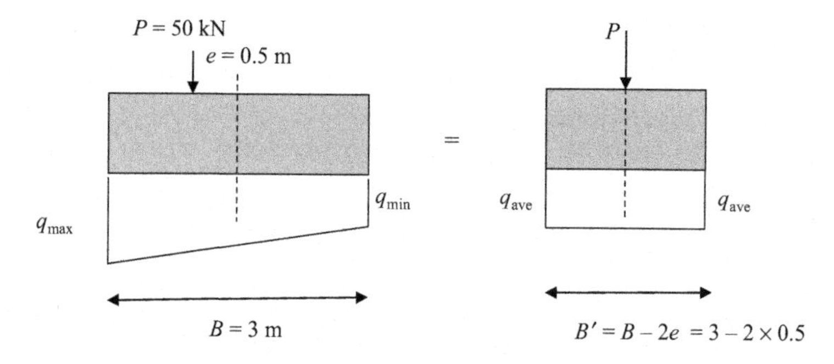

Figure 8.31 Eccentric loading with eccentricity 0.5 m.

Solution

$$B' = B - 2e = 3 - 2 \times 0.5 = 2\,\text{m}$$

$$\text{Pressure under the footing} = 50/(B' \times L)$$
$$= 50/(2 \times 5) = 5\ \text{kN/m}^2.$$

Design Example 8.13

A 3 m × 5 m rectangular footing is loaded with an eccentricity of 0.5 m along the longer dimension. The load on the footing is 50 kN (Figure 8.32). Find the soil pressure under the footing.

Solution

$$L' = L - 2e = 5 - 2 \times 0.5 = 4\ \text{m}$$

$$\text{Pressure under the footing} = 50/(B' \times L)$$
$$= 50/(3 \times 4) = 4.17\ \text{kN/m}^2.$$

Eccentricity on both directions:
Eccentricity can be on both the directions. Plan view of such a footing is shown in Figure 8.33.

$$B' = B - 2e_B$$
$$L' = L - 2e_L$$

Equivalent area $= A' = B' \times L' = (B - 2e_B) \times (L - 2e_L)$

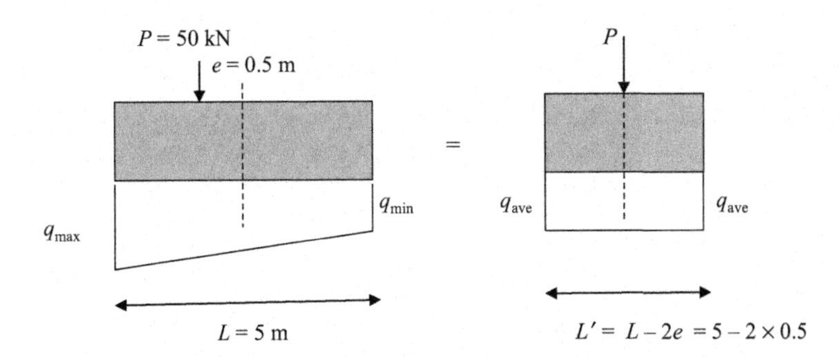

Figure 8.32 Eccentric loading with eccentricity 0.5 m along length.

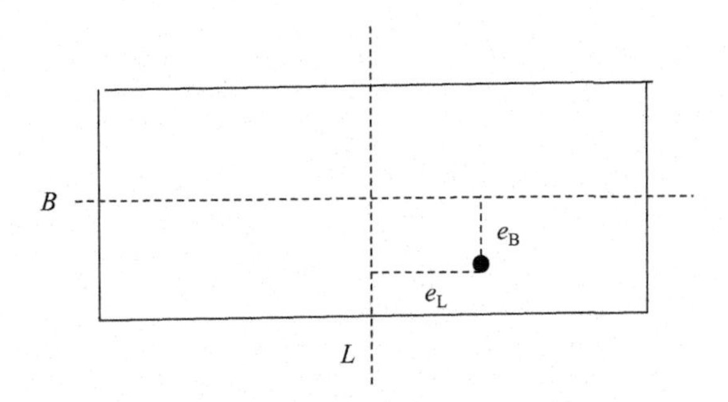

Figure 8.33 Plan view of eccentric loading in both directions width.

Design Example 8.14

A 3 m × 5 m rectangular footing is loaded with an eccentricity of 0.5 m along the longer dimension and 0.4 m along the shorter dimension. The load on the footing is 50 kN. Find the soil pressure under the footing.

Solution

$$L' = L - 2e_L = 5 - 2 \times 0.5 = 4\ \text{m}$$
$$B' = B - 2e_B = 3 - 2 \times 0.4 = 2.2\ \text{m}$$

A' = equivalent area = 4 × 2.2 = 8.8 m^2

$$\text{Pressure under the footing} = 50/(B' \times L') = 50/(4 \times 2.2) = 50/8.8 = 5.68\ \text{kN/m}^2.$$

Design Example 8.15

Find the ultimate bearing capacity of a (1.2 m × 1.8 m) rectangular footing placed in a clay layer (Figure 8.34). The density of the soil was found to be 17.7 kN/m^3 and the cohesion was found to be 20 kPa.

The following parameters are known:

$N_c = 5.7$, $N_q = 1.0$, and $N_\gamma = 0.0$

Shape factors: $s_c = 1.3$ and $s_\gamma = 0.8$

The load is placed with an eccentricity of 0.2 m along the longer direction (Figure 8.35).

Find the total load in kN that can be placed on this footing with a factor of safety of 3.0.

Solution

$$q_{ult} = cN_c s_c + qN_q + 0.5BN_\gamma \gamma s_\gamma$$

Step 1: Find the surcharge (q).

$$q = \gamma d = 17.7 \times 1.2 = 21.2\ \text{kPa}.$$

Step 2: Apply the Terzaghi's bearing capacity equation.

$$q_{ult} = cN_c s_c + qN_q + 0.5BN_\gamma \gamma s_\gamma$$
$$q_{ult} = 20 \times 5.7 \times 1.3 + 21.2 \times 1.0 + 0$$
$$q_{ult} = 169.4\ \text{kPa}$$

Allowable bearing capacity ($q_{allowable}$) = q_{ult}/FOS = q_{ult}/3.0 = 56.5 kPa.

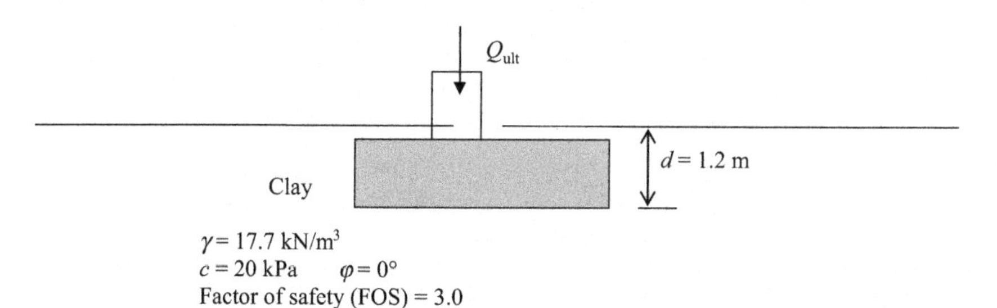

Figure 8.34 Column footing in a homogeneous clay layer with eccentricity 0.2 m.

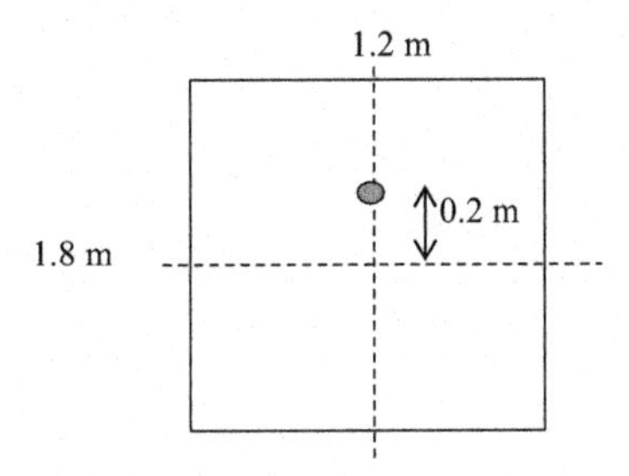

Figure 8.35 Plan view of eccentric loading.

Step 3: Find the equivalent area (A′).

$L' = L - 2.e = 1.8 - 2 \times 0.2 = 1.4$ m
$B' = B = 1.2$
$A' = B' \times L' = 1.2 \times 1.4 = 1.68$
$Q_{allowable} = q_{allowable} \times$ area of the footing
$Q_{allowable} = q_{allowable} \times (1.68) = 56.5 \times 1.68 = 94.92$ kN.

Design Example 8.16

A column foundation with dimensions 1.2 m × 1.8 m was placed at a depth of 1.1 m from the ground surface (Figure 8.36). The soil profile is mainly medium sand with a friction angle of 35° and a soil unit weight of 17.1 kN/m². The inclination of the load is 10° from the vertical. In addition, the load is applied at an eccentricity of 0.2 m along the length and 0.1 m along the width. ($e_L = 0.2$ m and $e_B = 0.1$ m). What is the allowable load in kN on the footing if the desired factor of safety is 3.0?
Ignore the moment due to inclined load.

Solution
Step 1: Write down the Meyerhof bearing capacity equation for inclined loads.

$q_{ult} = cN_c d_c i_c + qN_q d_q i_q + 0.5BN_\gamma \gamma d_\gamma i_\gamma$
$c = 0$ for medium sand
$q = \gamma D = 17.1 \times 1.1 = 18.81$ kN/m²
$N_q = e^{\pi \tan\varphi'} \tan^2[45 + \varphi'/2]$

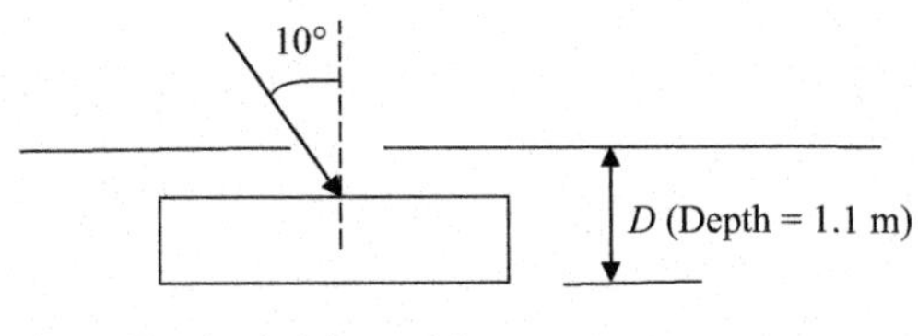

Figure 8.36 10° inclination of the load.

$N_q = e^{\pi \tan 35} \tan^2[45 + 35/2] = 9.023 \times 3.69 = 33.3$
$N_\gamma = (N_q - 1) \tan(1.4\ \varphi')$
$N_\gamma = (33.3 - 1) \tan(1.4 \times 35) = 37.16$
$d_q = d_\gamma = 1 + 0.1 \tan[45 + \varphi'/2] \times D/B$ for $\varphi' > 10°$.
$d_q = d_\gamma = 1 + 0.1 \tan[45 + 35/2] \times 1.1/1.2$ for $\varphi' > 10°$.
$d_q = d_\gamma = 1.17$
$i_c = i_q = (1 - \theta/90)^2$ for any φ'
$i_q = (1 - 10/90)^2 = 0.79$
$i_\gamma = (1 - \theta/\varphi)^2$ for $\varphi' > 0°$.
$i_\gamma = (1 - 10/35)^2 = 0.51$

Step 2: Apply the equation.

$q_{ult} = cN_c d_c i_c + qN_q d_q i_q + 0.5BN_\gamma \gamma d_\gamma i_\gamma$
$q_{ult} = 0 + 18.81 \times 33.3 \times 1.17 \times 0.79 + 0.5 \times 1.2 \times 37.16 \times 17.1 \times 1.17 \times 0.51$
$q_{ult} = 0 + 578.9 + 227.5 = 806.4\ kN/m^2$
$q_{allowable} = 806.4/3.0 = 268.8\ kN/m^2$

Step 3: Find the equivalent area (A').

$L' = L - 2e_L = 1.8 - 2 \times 0.2 = 1.4$ m
$B' = B - 2e_B = 1.2 - 2 \times 0.1 = 1$ m
$A' = B' \times L' = 1.4 \times 1.0 = 1.40$
$Q_{allowable} = q_{allowable} \times$ area of the footing
$Q_{allowable} = q_{allowable} \times (1.4) = 268.8 \times 1.4 = 376.32$ kN.

8.12.1 Tension under footing due to eccentric loading

When a footing is loaded with an eccentricity, one side of the footing will be loaded more as compared to the other side. When the eccentricity is increased, at some point, the loading on one side will be equal to zero. When the eccentricity is further increased, one side will be under tension (Figure 8.37).

Figure 8.37 shows high stress on one side and low stress on the other side. When the eccentricity is increased, the stress on one side becomes zero. Figure 8.37 shows negative pressure on one side due to very high eccentricity. It is good practice to make sure that negative pressure does not develop in footings.

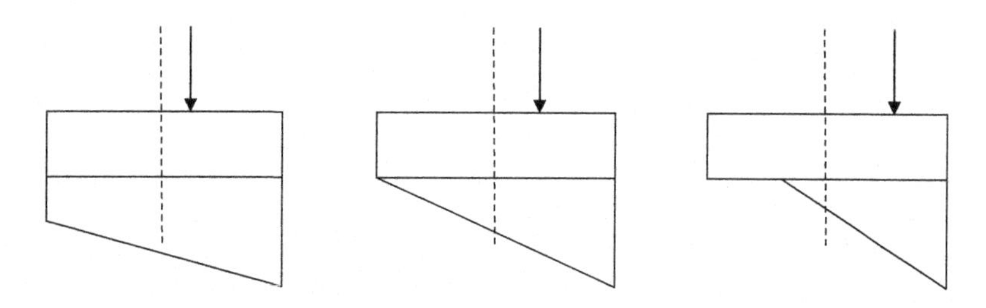

Figure 8.37 Eccentricity versus stress under the footing.

Figure 8.38 shows the case where tension is about to develop. Assume that the load is W, the eccentricity is "e", and the width of the footing is B.

Area of the stressed triangle under the footing = $qB/2$

This should be equal to the load W.

$$\text{Hence } W = qB/2 \tag{8.1}$$

In addition, one can take moments from point A.

$$W \times (B/2 + e) = [qB/2] \times 2B/3 \tag{8.2}$$

Replace W:

$[qB/2] \times (B/2 + e) = [qB/2] \times 2B/3$
$B/2 + e = 2B/3$
$e = 2B/3 - B/2 = B[4/6 - 3/6]$
$e = B/6$

When the eccentricity exceeds $B/6$, tension will develop under the footing. Either way, $B/6$ is the middle third of the footing (Figure 8.39).

If the load is within the middle third of the footing, the tensile forces will not develop in the footing.

8.13 Shallow foundations in bridge abutments

Shallow foundations in bridge abutments undergo heavy lateral loads in addition to vertical loads due to the approach fill (Figure 8.40).

Design Example 8.17

Find the safety factor against lateral forces acting on the foundation as shown in Figure 8.41. The soil and concrete interphase friction angle (δ) is found to be 20°, while the friction angle of soil

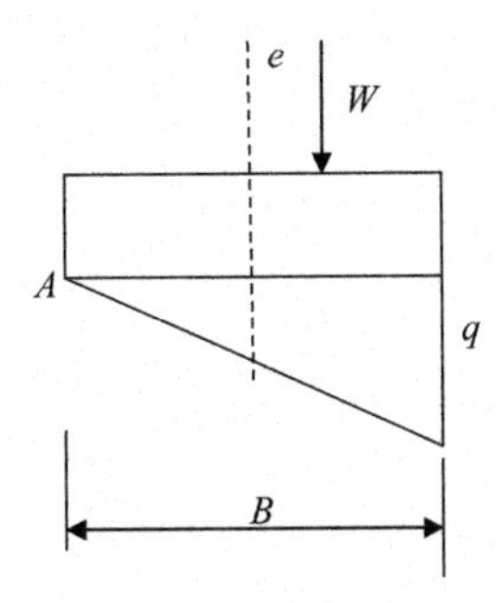

Figure 8.38 Tension about to be developed under the footing.

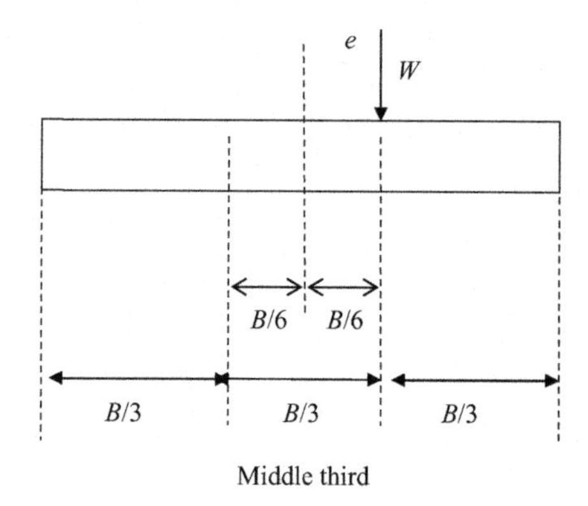

Figure 8.39 Middle third of the footing.

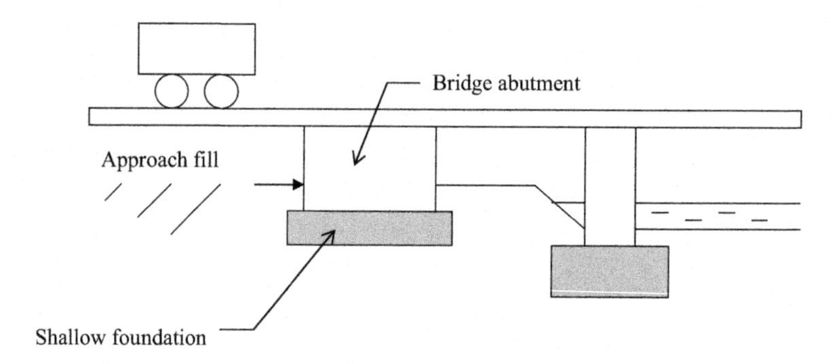

Figure 8.40 Shallow Foundation for a bridge abutment.

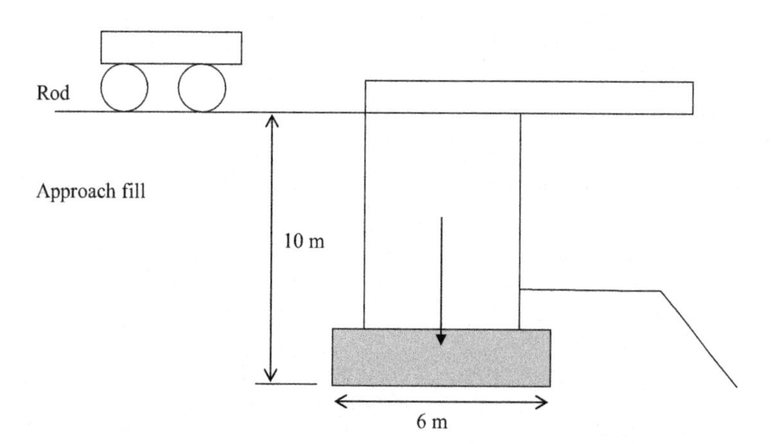

Figure 8.41 Bridge Abutment.

(φ) is 30°. The density of soil is 18.0 kN/m^3 and the weight of the abutment is 2000 kN per 1 m length of the abutment.

Solution

Step 1: Find the lateral earth pressure coefficient (K_a) (Figure 8.42).

$$K_a = \tan^2(45 - \varphi'/2)$$
$$K_a = \tan^2(45 - 30/2) = 0.333$$

Step 2:

$$\text{Stress at the bottom of the footing} = K_a \times \gamma \times h$$
$$= 0.33 \times 18 \times 10 = 59.4 \text{ kN/m}^2.$$

The passive pressure is ignored.

$$\text{Total lateral force} = \text{area of the stress triangle}$$
$$= 59.4 \times 10/2 = 297 \text{ kN}$$

Step 3:

Lateral resistance against sliding = weight of the abutment × tan (δ)

δ = friction angle between concrete and soil

δ = 20° and the weight of the abutment is given to be 2000 kN per 1 m length.

$$\text{Lateral resistance against sliding} = 2000 \times \tan(20°)$$
$$= 728 \text{ KN}$$

$$\text{Factor of safety} = \text{lateral resistance/lateral forces}$$
$$= 728/297 = 2.45$$

8.14 Bearing capacity computations (Eurocode)

Eurocode gives two equations to compute the bearing capacity. One equation for drained condition and the other one for the undrained condition. One should use the drained equation for sands and silts and undrained equation for clay soils and plastic silts.

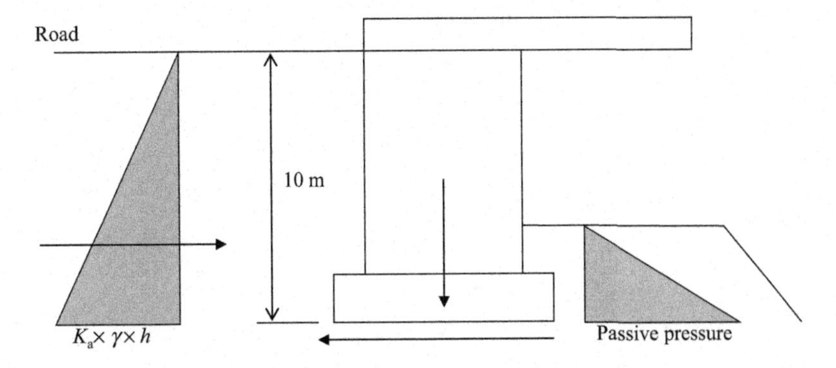

Figure 8.42 Bridge abutment stress diagram.

8.14.1 Drained equation for sands

In Figure 8.43, $R/A' = c'N_c b_c s_c i_c + q'N_q b_q s_q i_q + \frac{1}{2}\gamma' B' N_\gamma b_\gamma s_\gamma i_\gamma$.

R, load in kN; A', area of the footing. When the load is vertical and acting at the center of gravity of the footing, it is the exact area of the footing. If the load is inclined or not acting at the center of gravity of the footing, then A' will be different. This case would be explained later. c', Cohesion (kN/m^2); N_c, $(N_q - 1)\cot \varphi'$; N_q, $e.^{\pi \tan \varphi'} \tan^2 (45 + \varphi'/2)$; N_γ, $2 \times (N_q - 1) \tan \varphi'$.

8.14.1.1 Shape factors

s_q, $1 + B'/L' \sin \varphi'$ (for rectangular shapes); s_q, $1 + \sin \varphi'$ (for square and circular shapes); s_c, $(s_q N_q - 1)/(N_q - 1)$; s_γ, $1 - 0.3.(B'/L')$ (for rectangular shapes); s_γ, 0.70 (for square and circular shapes); B', $B - 2e_B$ = effective breadth of the footing; L', $L - 2e_L$ = effective length of the footing.

"b" factors are for inclination of the footing (Figure 8.44).

$$b_c = b_q - (1 - b_q)/(N_c \tan \varphi')$$
$$b_q = b_\gamma = (1 - \alpha \tan \varphi')^2 \text{ (α should be in radians)}$$

"i" factors are required when the load is applied at an angle (Figure 8.45).

$$i_q = [1 - 0.70.H/(V + A' \times c' \cot \varphi')]^m$$
$$m = m_B = [2 + B'/L']/[1 + B'/L']$$
$$m = m_L = [2 + L'/B']/[1 + L'/B']$$
$$m = m_\theta = m_L \cos^2\theta + m_B \sin^2\theta$$
$$i_c = (i_q N_q - 1)/(N_q - 1)$$
$$i_\gamma = [1 - H/(V + A' c' \cot \varphi')]^3$$

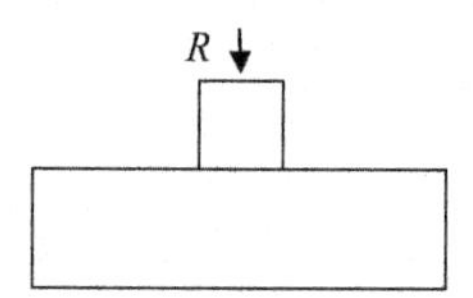

Figure 8.43 Footing with a load at the center of gravity.

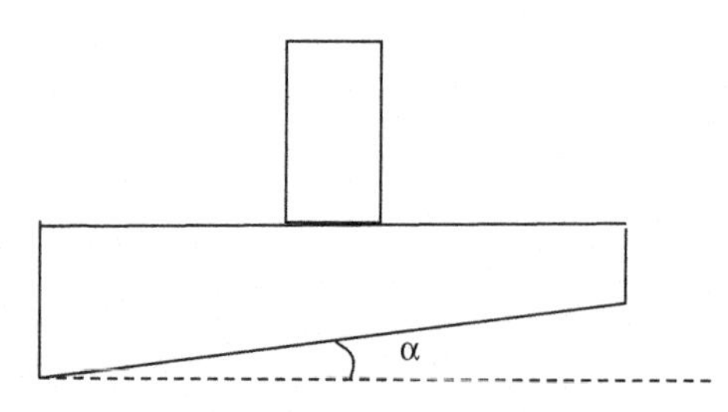

Figure 8.44 "b" factors are required when the bottom of the footing is not horizontal.

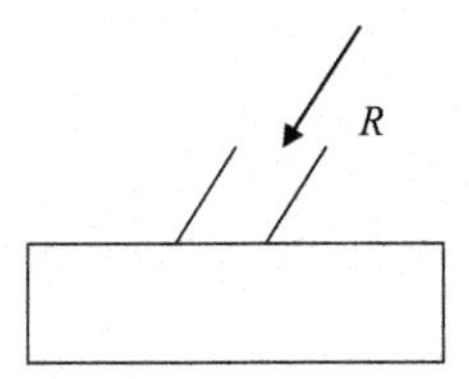

Figure 8.45 Footing with a load applied at an angle.

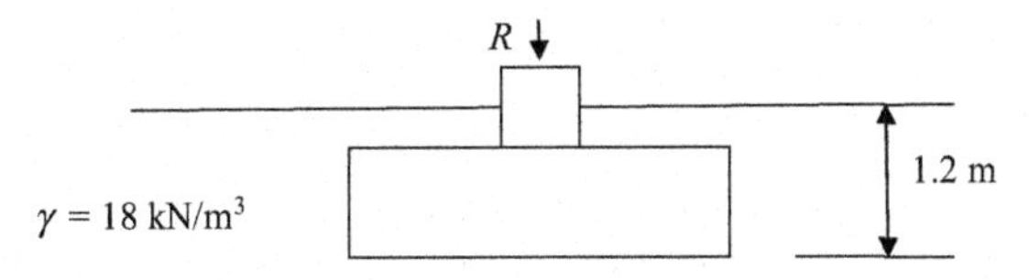

Figure 8.46 Footing with a load at the center of gravity.

Design Example 8.18

A shallow footing is placed 1.2 m below the ground level in sandy soil (Figure 8.46). The friction angle of the soil φ' is 30° degrees and the density of soil is 18 kN/m^3. The loading is vertical and the footing is not inclined. Length of the footing is 4 m and the width of the footing is 3 m. Assuming a factor of safety of 3.0, find the allowable load on the footing.

Solution

Step 1: Write down the bearing capacity equation for drained condition. For sandy soils, drained equation is used.

$R/A' = c'N_c b_c s_c i_c + q'N_q b_q s_q i_q + \frac{1}{2}\gamma' B' N_\gamma b_\gamma s_\gamma i_\gamma$

We can ignore the "*b*" factors and "*i*" factors. Hence, the equation can be simplified to the following equation:

$R/A' = c'N_c s_c + q'N_q s_q + \frac{1}{2}\gamma' B' N_\gamma s_\gamma$

$A' = A$ = area of the footing when the load is applied at the center of gravity of the footing vertically.

Step 2: Find the parameters.

For sandy soils $c' = 0$.

$q' = \gamma D = 18 \times 1.1 = 19.8$ kN/m^2.

$N_q = e^{\pi \tan \varphi'} \tan^2 (45 + \varphi'/2) = e^{\pi \tan 30} \tan^2 (45 + 30/2) = 6.13 \times 3.0 = 18.40$

$N_\gamma = 2 \times (N_q - 1) \tan \varphi' = 2 \times (18.40 - 1) \times 0.577 = 20.1$

Shape factors:

$s_q = 1 + \sin \varphi' = 1.5$

$s_c = (s_q N_q - 1)/(N_q - 1) = 26.6/17.4 = 1.53$

$s_\gamma = 1 - 0.3 (B/L) = 1 - 0.3 \times 3/4 = 0.775$

Step 3: Apply the bearing capacity equation.

$R/A' = c'N_c s_c + q'N_q s_q + \frac{1}{2}\gamma' B' N_\gamma s_\gamma$

$R/A' = 0 + 19.8 \times 18.40 \times 1.5 + \frac{1}{2} \times 18 \times 3 \times 20.1 \times 0.775 = 546.5 + 420.6 = 967.1$ kN/m^2.

R = area $\times$ 967.1 = 3 $\times$ 4 $\times$ 967.1 = 11,605.2

Allowable load on the footing = 11605/FOS = 11605/3.0 = 3868.4 kN

8.15 Undrained conditions

Undrained condition occurs in cohesive soils (clays and plastic silts). In cohesive soils, the drainage process takes a longer duration. Hence, it is reasonable to assume undrained conditions (Figure 8.47).

Equation for undrained condition:

$R/A' = (2 + \pi)\, c_u b_c s_c i_c + q$
R = load
$A' = B' \times L'$ ($B' = B - 2e$ and $L' = L - 2e$)
c_u = cohesion (kN/m^2)

"b" factors are for inclination of the footing.
$b_c = 1 - 2\alpha/(2 + \pi)$
α should be in radians.

s_c, shape factor; s_c, $1 + 0.2.\ B'/L'$ (for rectangular shapes); s_c, 1.2 (for square and circular shapes); i_c, $0.5 \times [1 + \{1 - H/(A'c_u)\}^{0.5}]$; H, horizontal component of the load; q, γd.

Design Example 8.19

A footing is placed 1.1 m below the ground surface (Figure 8.48). The density of soil is 17.5 kN/m^2. The soil is considered cohesive. The footing is 2 m $\times$ 3 m. The undrained cohesion of soil (c_u) is 50 kN/m^2. Find the ultimate bearing capacity of the footing.

Solution

Since, the soil is cohesive, use undrained equation.
Step 1: Write down the undrained equation.
$R/A' = (2 + \pi) c_u b_c s_c i_c + q$

c_u = cohesion (kN/m^2) = 50 kN/m^2
b_c = "b" factors are for inclination of the footing.

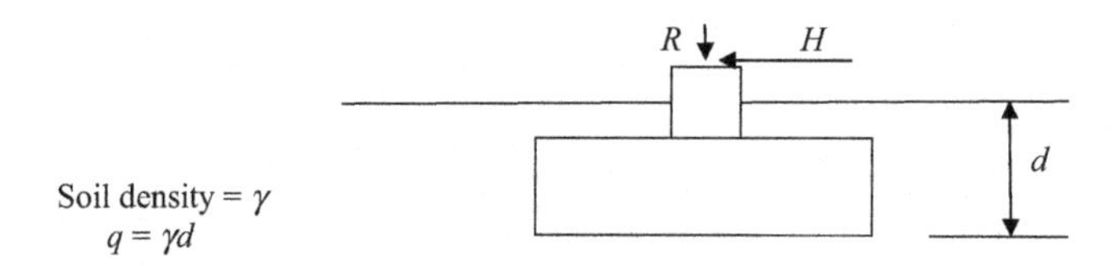

Figure 8.47 Footing with a load at the center of gravity (undrained condition).

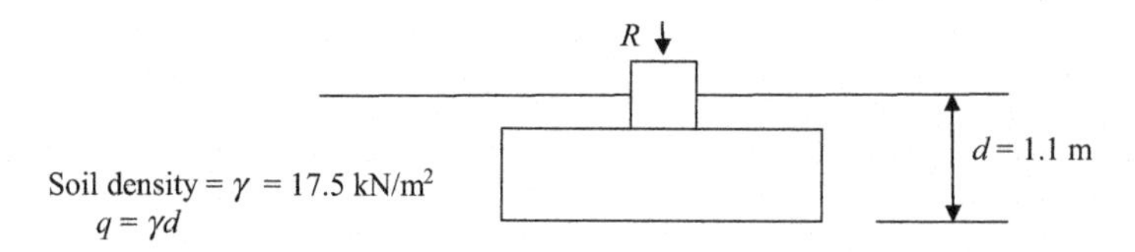

Figure 8.48 Footing with a load at the center of gravity (undrained condition example).

$b_c = 1 - 2\alpha/(2 + \pi)$
$\alpha = 0$; Hence, $b_c = 1.0$
s_c = shape factor
$s_c = 1 + 0.2.\ B'/L'$ (for rectangular shapes)
$s_c = 1 + 0.2 \times 2/3 = 1.13$

Since the load is vertical and also has no eccentricity $B' = B$ and $L' = L$

$i_c = 1.0$, since the load is vertical.
$q = \gamma d = 17.5 \times 1.1 = 19.25\ \text{kN/m}^2$.

Step 2: Apply the undrained equation.

$R/A' = (2 + \pi)c_u b_c s_c i_c + q$
$R/A' = (2 + \pi) \times 50 \times 1.0 \times 1.13 \times 1.0 + 19.25 = 309.75\ \text{kN/m}^2$.
$A' = A$ for footings with vertical loads acting at the center of gravity.
$R = 309.75 \times (2 \times 3) = 1858\ \text{kN}$

Elastic settlement of shallow foundations

9.1 Introduction

All soils undergo elastic settlement. Elastic settlement is immediate and much higher in sandy soils as compared to clay soils. Due to high Young's modulus values, elastic settlement is low in clay soils.

Elastic settlement, S_e is given by the following equation (AASHTO, 1998)

$$S_e = \frac{[q_0/(1-\nu^2)\times(A)^{1/2}]}{E_s\,\beta_z}$$

S_e = elastic settlement; q_0 = footing pressure (should be in same units as E_s); ν = Poisson's ratio (See Table 9.1 or the Poisson's ratio); A = area of the footing; E_s = elastic modulus or the Young's modulus of soil (See Table 6.1 to obtain E_s); β_z = factor to account for footing shape and rigidity (Table 9.1).

AASHTO provides another method to obtain the elastic modulus (Table 9.2).

A closer examination of Table 9.2 shows that elastic modulus of clay soils is significantly higher than sandy soils.

Elastic modulus, stress/strain; Strain, stress/elastic modulus.

Design Example 9.1

Find the immediate elastic settlement of the foundation shown in Figure 9.1. The dimensions of the foundation is 1.5 m × 1.5 m. The foundation is rested on medium dense coarse to medium sand. The column load is 1000 kN. The average SPT (*N*) value below the footing is 12.

Medium dense coarse to medium sand

γ = 18 kN/m^3

Average SPT (*N*) value = 12

Solution

STEP 1: Elastic settlement is given by the following equation:

$$S_e = \frac{[q_0/(1-\nu^2)\times(A)^{1/2}]}{E_s\,\beta_z}$$

S_e = elastic settlement; q_0 = footing pressure = 1000/(1.5 × 1.5) = 444 kN/m^2.

ν (Poisson's ratio) = Poisson's ratio is between 0.2 to 0.35 (From Table 9.2)

Assume ν = 0.25.

A = area of the footing = 1.5 × 1.5 = 2.25 m^2.

E_s = elastic modulus or the Young's modulus of soil

E_s = 670 N kPa (From Table 9.1, Use "fine to medium sand and slightly silty sand" category)

Table 9.1 Elastic modulus versus SPT (*N*) values

Soil type	E_s (ksf)	E_s (kN/m^2)
Silts and sandy silts	8 N	383 N
Fine to medium sand and slightly silty sand	14 N	670 N
Coarse sand and sands with little gravel	20 N	958 N
Sandy gravel and gravel	24 N	1150 N

Source: AASHTO (1998).

Table 9.2 Elastic modulus and Poisson's ratio for soils (AASHTO)

Soil type	Elastic modulus (ksf)	Elastic modulus (kN/m^2)	Poisson's ratio (ν)
Clay soils			
Soft clay	50–300	2,390–14,360	
Medium stiff clay	300–1,000	14,360–47,880	0.4–0.5
Stiff clay	1,000–2,000	47,880–95,760	(for all clays under undrained condition)
Very stiff clay	2,000 plus	95,760 plus	
Fine sand			
Loose fine sand	160–240	7,660–11,490	
Medium dense fine sand	240–400	11,490–19,150	0.25 for fine sand
Dense fine sand	400–600	19,150–28,720	
Coarse to medium sand			
Loose coarse to medium sand	200–600	9,570–28,720	
Medium dense coarse to medium sand	600–1,000	28,720–47,880	0.2–0.35 for coarse to medium sand
Dense coarse to medium sand	1,000–1,600	47,880–76,600	
Gravel			
Loose gravel	600–1,600	28,720–76,600	0.2–0.35
Medium dense gravel	1,600–2,000	76,600–95,760	0.2–0.35
Dense gravel	2,000–4,000	95,760–191,500	0.3–0.4
Silt	40–400	1,915–19,150	0.3–0.35

N = SPT (N) value

$E_s = 670 \times 12 = 8040$ kPa

β_z = factor to account for footing shape and rigidity

Assume $\beta_z = 1.0$

$$S_e = \frac{[444/(1-0.25^2)\times(2.25)^{1/2}]}{8040\times 1.0}$$

$$S_e = 0.088 \text{ m} \qquad (3.47 \text{ in.})$$

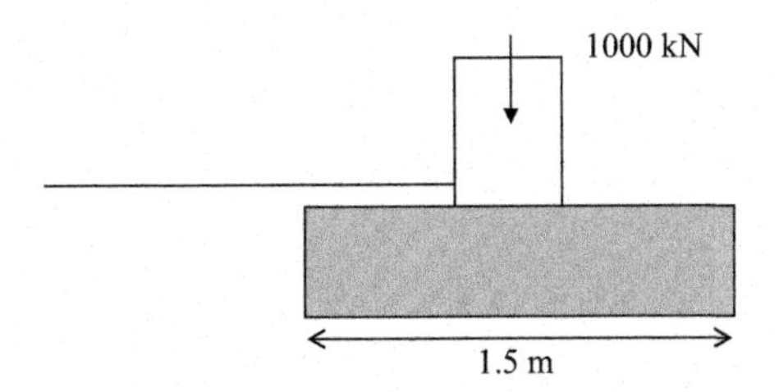

Figure 9.1 Elastic settlement of footing.

Reference

AASHTO, LRFD Bridge Design Specifications, 1998. American Association of State Highway and Transportation Officials.

Foundation reinforcement design 10

10.1 Concrete design (refresher)

10.1.1 Load factors

Load factors need to be applied as a safety measure against the uncertainties that could occur during the computation of loads. Loads of a building is computed by adding the weight of all the slabs, beams, columns, and roof elements.

Load factor for dead loads = 1.4
Load factor for live loads = 1.7
Load factor for wind loads = 1.7

10.1.2 Strength reduction factors (ϕ)

Concrete shear $\phi = 0.85$
Concrete compression $\phi = 0.70$
Concrete tension $\phi = 0.90$
Concrete beam flexure $\phi = 0.90$
Concrete bearing $\phi = 0.70$

Many parameters affect the strength of concrete. Water content, humidity, aggregate size and type, cement type, and the method of preparation. The strength reduction factors are used to account for variations that could happen during the preparation of concrete.

Different ϕ values are used for shear, tension, compression, and beam flexure as shown earlier.

Stress due to beam flexure is obtained using the following equation:

$$M/I = \sigma/y$$

10.1.3 How to find the shear strength?

If the compressive strength of a concrete is known, the following equation can be used to find the shear strength of the concrete:

$$\text{Shear Strength}\,(v_c) = 4 \times (f_c')^{1/2}$$

f_c', compressive strength of concrete; v_c, shear strength.

Shear strength value needs to be reduced by the strength reduction factor (ϕ).

$$\text{Shear Strength}\,(v_c) = 4 \times \phi \times (f_c')^{1/2}$$

$\phi = 0.85$ for concrete shear.

Geotechnical Engineering Calculations and Rules of Thumb

10.2 Design for beam flexure

Figure 10.1 shows, a beam subjected to bending. The beam is simply supported at two ends and loaded at the middle.

Upper concrete fibers are under compression and the lower steel is under tension.

The compression of concrete fibers is represented by a stress block as shown in the Figure 10.2. This is an approximate representation. The real stress distribution is a parabola.

The depth to reinforcements is "*d*" and the depth of the stress block is "*a*" ft.

The "*a*" value depends on concrete strength and steel strength.

By balancing forces:

$$f_y \times A_s = 0.85 f_c' \times a \times b$$

f_y, yield stress of steel; A_s, steel area; b, width of the beam; a, depth of stress block

In Figure 10.3,

Area of the stress block = $0.85 f_c' \times a \times b$

Take moments abut the steel

$$M = [0.85 f_c' \times a \times b]\ (d - a/2)$$
$$M/0.85 f_c' = a \times b \times (d - a/2) = Z_c$$

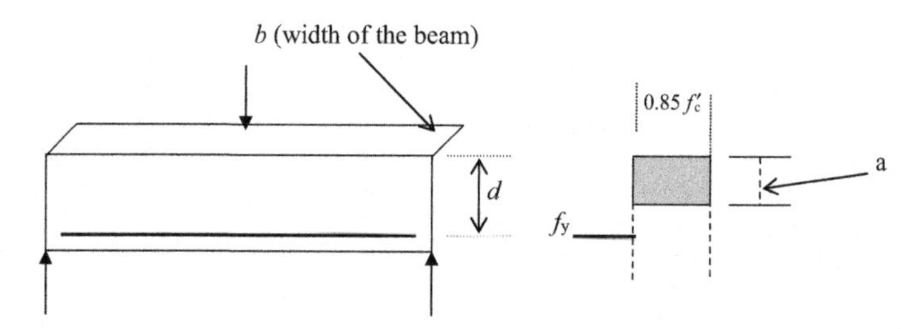

Figure 10.1 Beam subjected to bending.

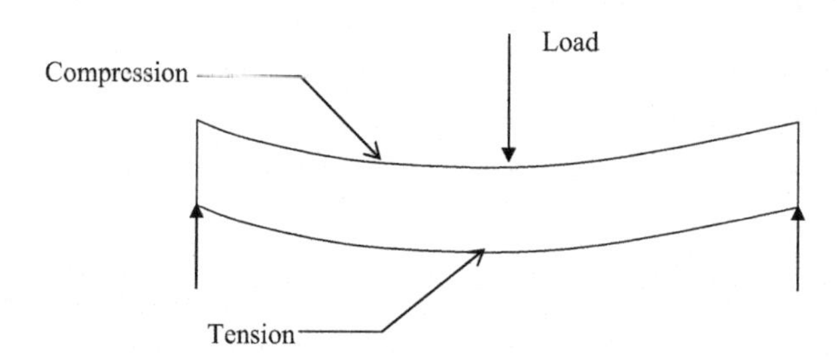

Figure 10.2 Top fibers are subjected to compression while the bottom fibers are subjected to tension.

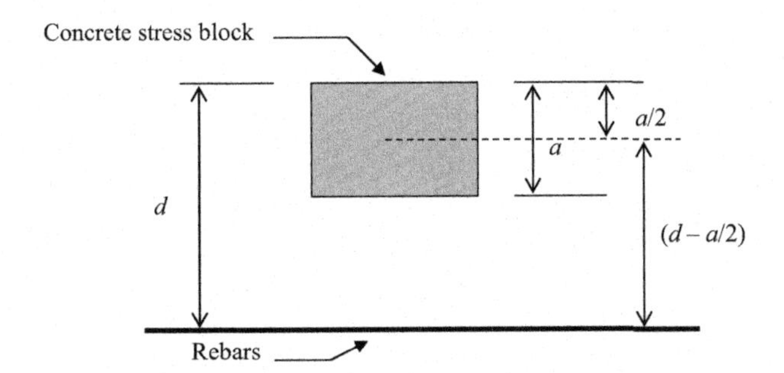

Figure 10.3 Concrete stress block and rebars.

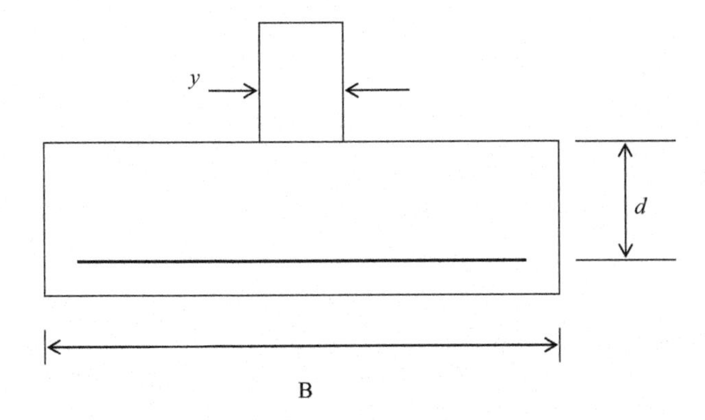

Figure 10.4 Shallow foundation with bottom rebars.

10.3 Foundation reinforcement design

10.3.1 Design for punching shear

Figure 10.4 shows, a shallow foundation with bottom rebars.

Punching failure occurs $d/2$ distance from the edge of the column (Figure 10.5).

10.3.2 Punching shear zone

Due to the load of the column, the footing could be punched through. Normally, the punching shear zone is taken to be at a distance of ($d/2$) from edge of the column.

"d" is the depth of the footing, measured to the top of the rebars from the top of the footing.

Hence, the dimensions of the punching shear zone is $(d + y) \times (d + y)$. For a punching failure to occur, this zone needs to fail.

$$\begin{aligned}\text{Dimensions of the punching shear zone} &= 4(y + d/2 + d/2)dv_c \\ &= 4(y + d)dv_c\end{aligned}$$

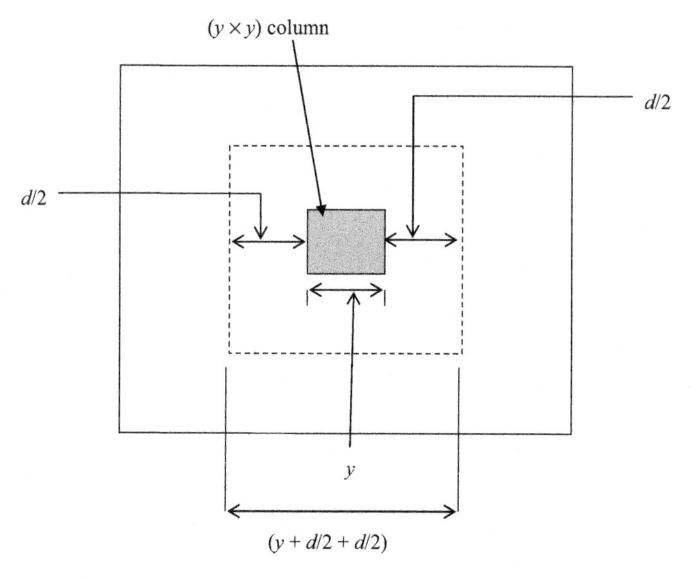

Figure 10.5 Punching shear zone in a footing.

$4 \times (y + d/2 + d/2)$ represents the perimeter of the zone.

"d" represents the depth or thickness of the footing.

The perimeter is multiplied by depth (d). This would give the resisting area for shear.

v_c is the shear strength per unit area of concrete.

The force that needs to be resisted = $[B^2 - (y + d)^2] \times q_{\text{allowable of soil}}$

In Figure 10.6,

Area of the hatched area outside the punching zone = $B^2 - (y + d)^2$

$$[B^2 - (y+d)^2] \times q_{\text{allowable of soil}} = 4 \times (y+d) \times d \times v_c$$

It is possible to find the depth of the footing required to avoid punching shear failure.

$$\text{Load from the column} = B^2 \times q_{\text{allowable of soil}}$$

- v_c is the shear strength of the concrete. It is computed using the following equation:

$$\text{Shear strength of concrete } (v_c) = 4\phi \times (f_c')^{1/2}$$

f_c' is the compressive strength of concrete. It normally varies from 3000 psi to 6000 psi.

For 3000 psi concrete, $v_c = 4 \times 0.85 \times (3000)^{1/2} = 186.2$ psi

Hence, "d" can be computed.

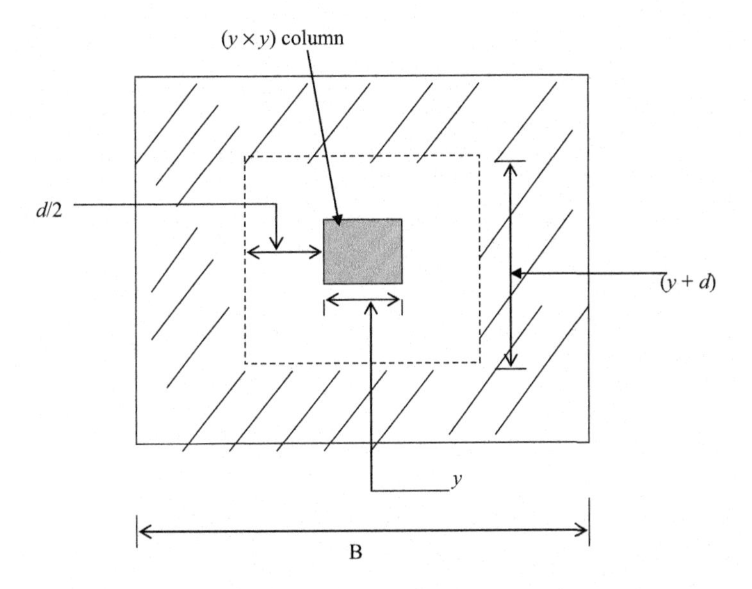

Figure 10.6 Stressed area of the footing.

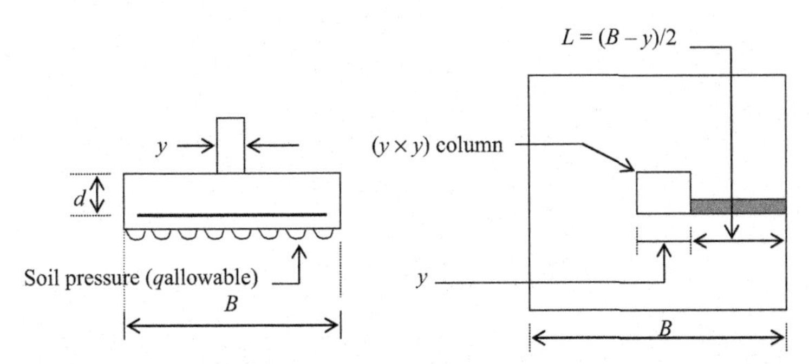

Figure 10.7 Soil pressure and rebars.

10.3.3 Design reinforcements for bending moment

From Figure 10.7, it is clear that a bending moment would be developed in the footing due to the load from the column and the soil pressure.

Assume a 1 ft. wide strip of the footing (shaded strip).

The forces acting on this strip are shown in Figure 10.8. The soil pressure acts from below. The footing is fixed to the column.

Footing/column boundary is considered to be cantilevered as shown in Figure 10.8. Now the problem has boiled down to a beam flexure problem.

$L = (B - y)/2$ (the strip is cantilevered)

$$\text{Bending moment}\,(M) = \frac{(q_{\text{allowable of soil}}) \times L^2}{2}$$

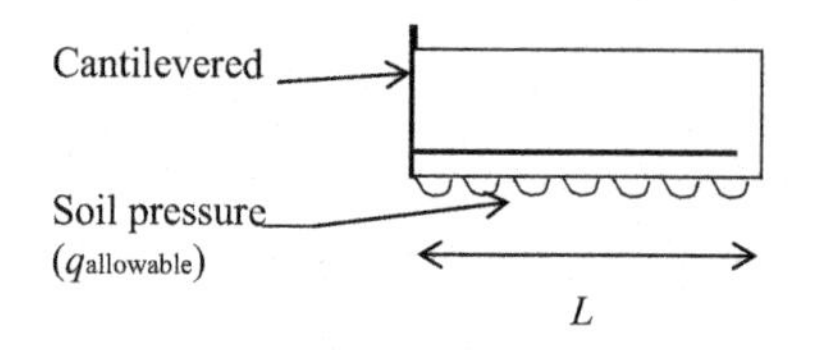

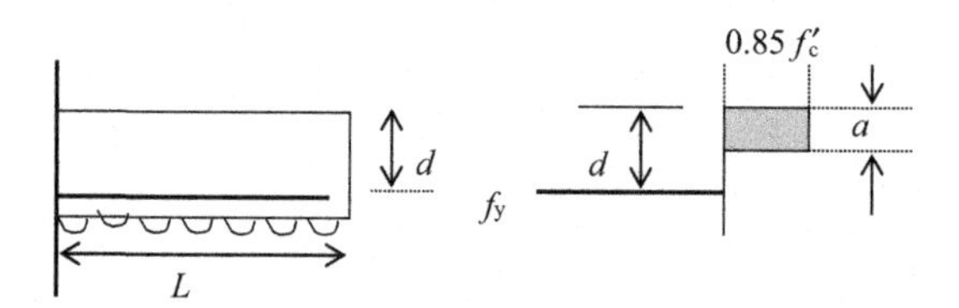

Figure 10.8 Cantilevered strip.

Bending moment = total load × distance to the center of gravity
Total load = $q_{\text{allowable of soil}} \times L$
Distance to center of gravity = $L/2$
Since we are considering a 1 ft. wide strip, width of the beam considered is 1 ft.

$$\text{Force in the steel} = A_s \times f_y$$

A_s, steel area; f_y, yield stress of steel.
This force in the steel should be equal to the force in concrete.
Hence, $0.85 f_c' \times b \times a = A_s \times f_y$
f_c', concrete compression strength; f_y, yield stress of steel; A_s, steel area; b, width of the beam; a, depth of the stress block.

Note: Above 0.85 represents the factor of safety against concrete workmanship. No factor of safety is assumed for steel.

$$a = (A_s \times f_y)/(0.85 f_c' \times b) \tag{10.1}$$

Moment arm = $(d - a/2)$
The distance between steel and the center of the stress block
Hence, resisting steel moment = $A_s \times f_y \times (d - a/2)$
This resisting moment should be equal to the moment induced due to loading.

$$\text{Moment induced due to soil pressure} = \frac{(q_{\text{allowable of soil}}) \times L^2}{2}$$

Hence,

$$\frac{(q_{\text{allowable of soil}}) \times L^2}{2} = A_s \times f_y \times (d - a/2) \tag{10.2}$$

In above equation $q_{\text{allowable}}$, L and f_y are known parameters.
"a" is obtained from Equation (10.1), in terms of A_s.

Table 10.1 **Reinforcing bars**

Bar number	Nominal diameter (in.)	Nominal diameter (mm)	Nominal weight (lb/ft.)
3	0.375 (3/8)	9.5	0.376
4	0.500 (4/8)	12.7	0.668
5	0.625 (5/8)	15.9	1.043
6	0.750 (6/8)	19.1	1.502
7	0.875 (7/8)	22.2	2.044
8	1.000	25.4	2.670
9	1.128	28.7	3.400
10	1.270	32.3	4.303
11	1.410	35.8	5.313
14	1.693	43	7.650

f_c', f_y, and b are known.

Only A_s is unknown in Equation (10.1).

Hence, in Equation (10.1), two parameters are unknown. They are required steel area, "A_s" and the required depth of the footing, "d".

Guess a reasonable footing depth and estimate A_s (Table 10.1).

Grillage design

11

11.1 Introduction

Typical load bearing concrete foundations are designed with steel reinforcements (Figure 11.1). It is not possible to design reinforcements for very high column loads. In such cases, grillages are used (Figure 11.2).

What is a grillage?

A grillage consists of two layers of "I" beams as shown in Figure 11.2.

The load from the column is transferred to the base plate and the base plate transfers the load to the concrete. The concrete transfers the load to the top layer of "I" beams and then to the bottom layer of beams.

The bottom layer of "I" beams would transfer the load to the concrete below and then to the rock underneath.

Design Example 11.1: Grillage design

A concrete encased steel column carrying 2000 tons needs to be supported. The allowable bearing capacity of the rock is 20 tsf. The steel column is supported on a 24 in. × 24 in. base plate. It is decided to have a grillage consisting of two layers of "I" beams. The engineer decides to have three "I" beams in the top layer and five "I" beams in the bottom layer.

1. Design the top layer of "I" beams.
2. Design the bottom layer of "I" beams.

Solution

Step 1: Size of the footing required = 2000 tons/20 tsf = 100 sq. ft.

Use a footing of 10 ft. × 10 ft.

Assumptions:

- Assume that the beams are 10 ft. long. (In reality, the beams are less than 10 ft., since the dimensions of the footing are 10 ft. × 10 ft.)
- Assume that the base plate is 24 in. × 24 in. and the load is transferred to the top layer of beams as shown in Figure 11.3.
- Assume that the load is transferred to a section of 30 in. (See Figure 11.4).

Step 2: Loads acting on top three beams are shown in Figure 11.4

Total load from the base plate = 2000 tons.

Since there are three "I" beams in the top layer, one I beam would take a load of 666.67 (2000/3) tons. This load is distributed in a length of 30 in.

Hence, the distributed load on the beam is 666.67/2.5 = 266.67 tsf.

Load from the top beams is distributed to the bottom layer of "I" beams.

Geotechnical Engineering Calculations and Rules of Thumb

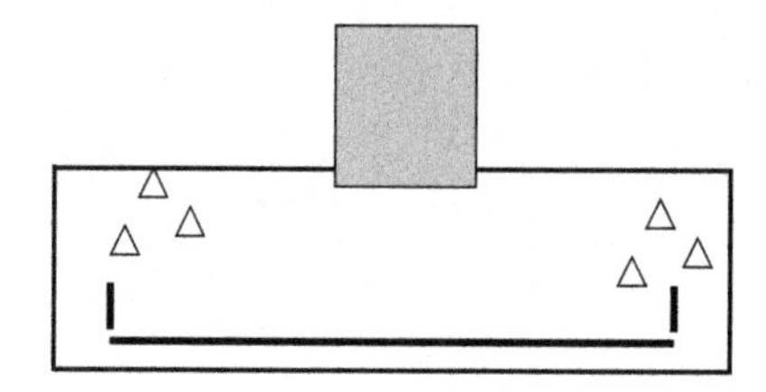

Figure 11.1 Regular footing with steel reinforcements.

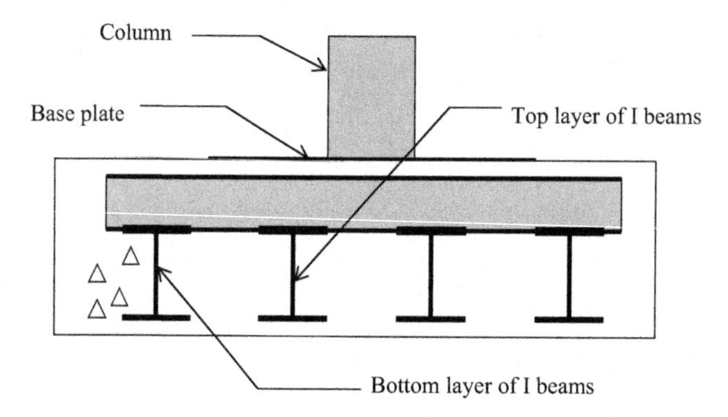

Figure 11.2 Grillage footing.

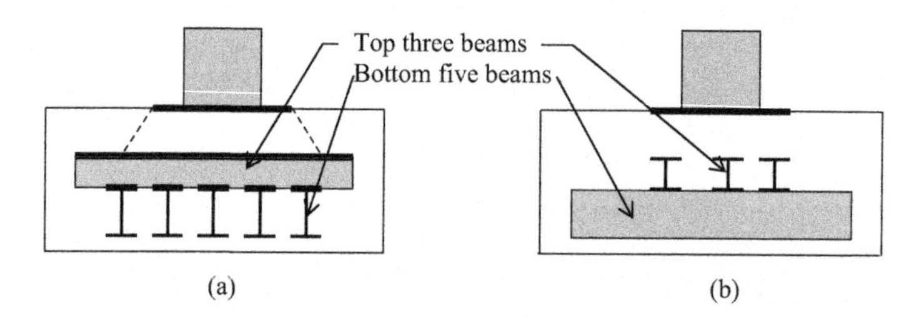

Figure 11.3 Grillage. (a) Front view. (b) Side view.

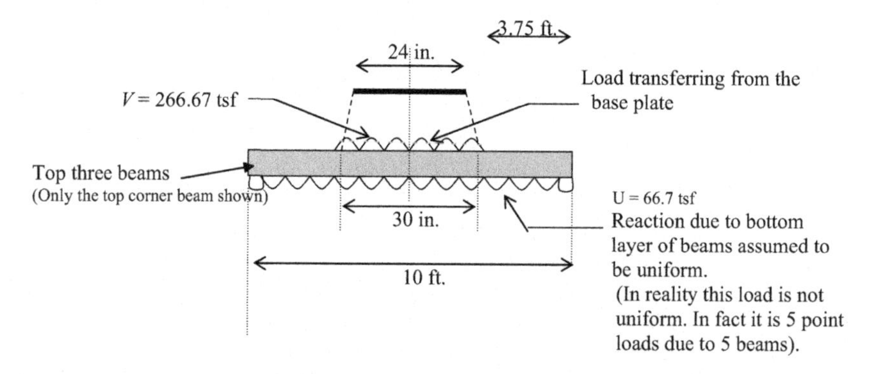

Figure 11.4 Loading on a grillage.

The bottom layer of "I" beams generate an upward reaction on the top "I" beams. This reaction is assumed uniform.

In reality, this upward reaction (U) is basically five concentrated reactions acting on the top layer of I beams. As you are aware, there are five beams in the bottom layer and each exerts a reaction.

Uniformly distributed load due to the reactions from the bottom layer = 666.67/10 = 66.7 tsf.

(Total load that needs to be transferred from one top beam is 666.67 tons and it is distributed over a length of 10 ft.).

Now the problem is to find the maximum bending moment occurring in the beam. Once maximum bending moment is found, "I" beam section can be designed.

Maximum bending moment occurs at the center of the beam. (See Figure 11.5).

Step 3: Find the maximum bending moment in the beam.

The reaction at the center point of the beam is taken to be "R".

Assume the bending moment at the center to be "M". For this type of loading, the maximum bending moment occurs at the center. (Take moments about the center point).

$$M = (66.67 \times 5 \times 2.5) - 266.67 \times 1.25 \times 1.25 / 2 = 625 \text{tons.ft.}$$

where, 66.67 × 5 represents the total load and 2.5 represents the distance to the center of gravity. Similarly 266.67 × 1.25 represents the total load and 1.25/2 represents the distance to the center of gravity.

The beam should be able to carry this bending moment. Select an "I" section that can carry a bending moment of 625 tons ft.

M = 625 tons. ft. = 2 × 625 = 1250 kip. ft.

$$M/Z = \sigma$$

M = Bending moment
Z = Section modulus
σ = Stress in the outermost fiber of the beam
Use an S – section with allowable steel stress of 36,000 psi.

σ = 36 ksi
$Z = M/\sigma$
Z = (1250 × 12) kip. in./36 ksi
Z = 417 in^3.

Use W36 × 135 section with section modulus 439 in^3.

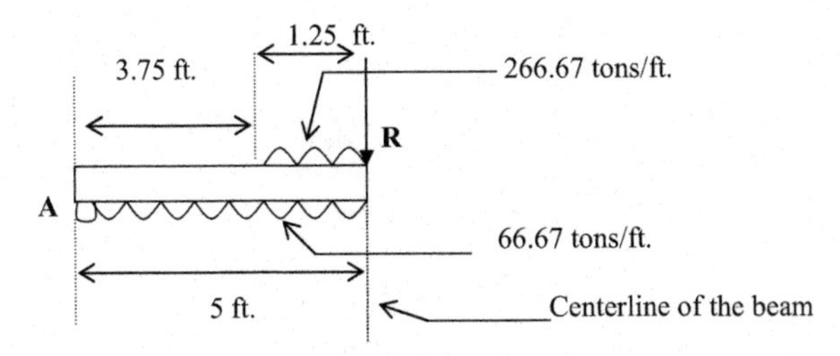

Figure 11.5 Half section of the grillage.

Step 4: Design the bottom layer of beams.

Three top beams rest on each bottom layer beam. Assume the top beams are 12 in. apart (Figure 11.6).

Each top layer beam carries a load of 666.67 tons. Each of the top layer beams rests on 5 bottom layer beams. Hence 666.67 tons is distributed over 5 bottom layer beams. Each bottom layer beam gets a load of 666.67/5 tons (=133.33) from each top beam.

There are three top layer beams.

Hence, each bottom layer beam carries a load of 3 × 133.33 = 400 tons.

All bottom layer beams sit on concrete. This load needs to be transferred to the concrete.

The reaction from concrete is considered to be uniformly distributed.

$$W = \text{Concrete reaction} = 400/10 = 40 \text{ tons per linear ft.}$$

Step 5: Find the maximum bending moment.

The maximum bending moment occurring in the bottom layer beams can be computed (Figure 11.7).

Cut the beam at the center. Then the concentrated load at the center needs to be halved since one half goes to the other section.

Take moments about point "C".

$$M = 40 \times 5 \times 2.5 - 133.33 \times 1 = 366.7 \text{ tons.ft}$$

Hence the maximum bending moment = 366.7 tons. ft.

M = 366.7 tons. ft. = 2 × 366.7 = 733.4 kip. ft.

$$M/Z = \sigma$$

M = bending moment; Z = section modulus; σ = stress in the outermost fiber of the beam.

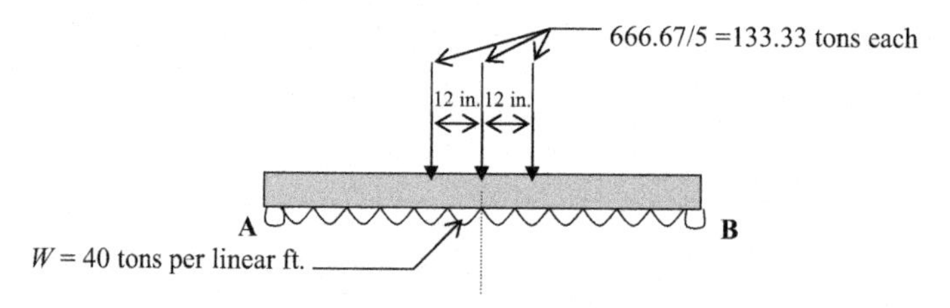

Figure 11.6 Forces on bottom layer of beams.

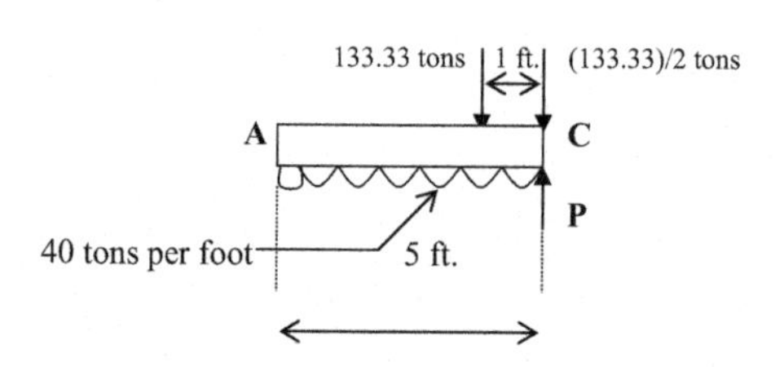

Figure 11.7 Half section of the grillage bottom layer.

Use a steel section with allowable steel stress of 36,000 psi.

$\sigma = 36$ ksi
$Z = M/\sigma$
$Z = (733.4 \times 12)$ kip. in./36 ksi
$Z = 244.4$ in^3.

Use *S* 24 × 121 section with section modulus 258 in^3.

Footings subjected to bending moment

12

12.1 Introduction

In addition to vertical loads, shallow footings are subjected to bending moments and horizontal forces (Figure 12.1). Foundation engineers need to address all the forces and moments in the design. Horizontal forces and bending moments are generated mostly due to wind loads.

Due to wind loads, the bending moment "M" will be acting on corner footings.

Wind load can be simplified to a resultant horizontal force and a bending moment at the footing level (Figure 12.2).

The wind load acting on the building is equated to a horizontal force (H) and a bending moment (M) (Figure 12.3). The lateral resistance (R) of the bottom of footing should be greater than the horizontal force (H).

"R" can be calculated by finding the friction between the bottom of the footing and soil. Friction "R" at the bottom of footing is given by

$$R = V \tan \delta$$

δ = friction angle between bottom of the footing and soil. Soil friction angle "φ" can be used as an approximation instead of "δ".

V = vertical force acting on the footing.

For stability, $R > H$.

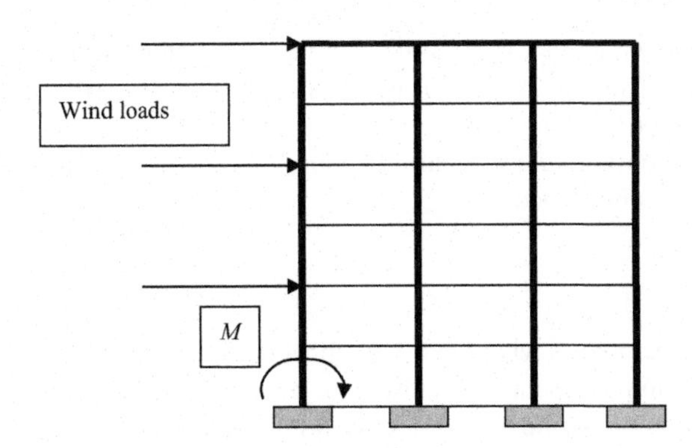

Figure 12.1 Loads on footings.

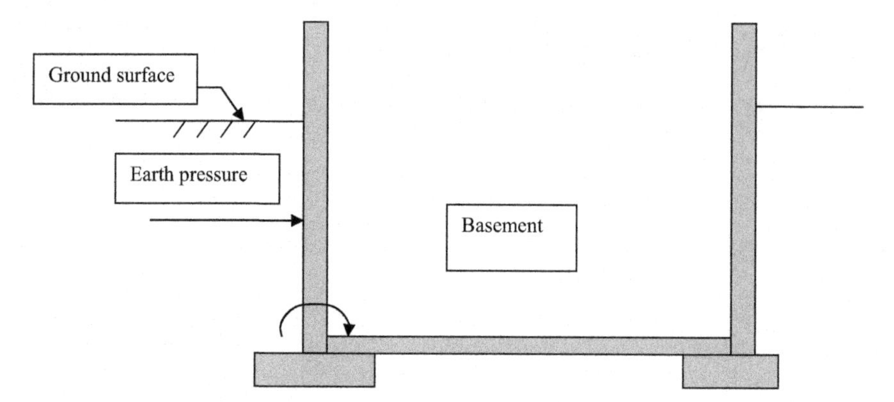

Figure 12.2 Building with a basement.

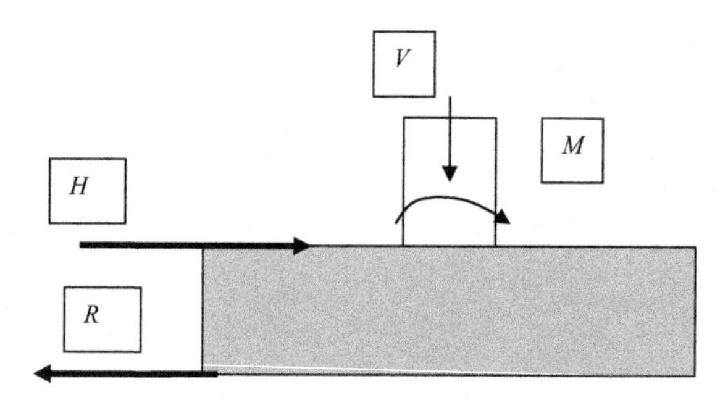

Figure 12.3 Forces and moments acting on a footing.

12.1.1 Representation of bending moment with an eccentric load

The bending moment acting on a footing can be represented by an eccentric vertical load (Figure 12.4). Vertical load that is not acting at the center of the footing will generate a bending moment.

Essentially, a footing subjected to a bending moment can be represented with a footing with an eccentric load.

$$M = Ve$$

M, bending moment due to wind load; *V*, vertical load.

When the bending moment is increased, the eccentricity also would increase.

Due to the bending moment, one side would undergo larger compression than the other side (Figure 12.5).

Figure 12.5a consists only of a vertical load. Figure 12.5b has a bending moment that has been represented by an eccentric vertical load.

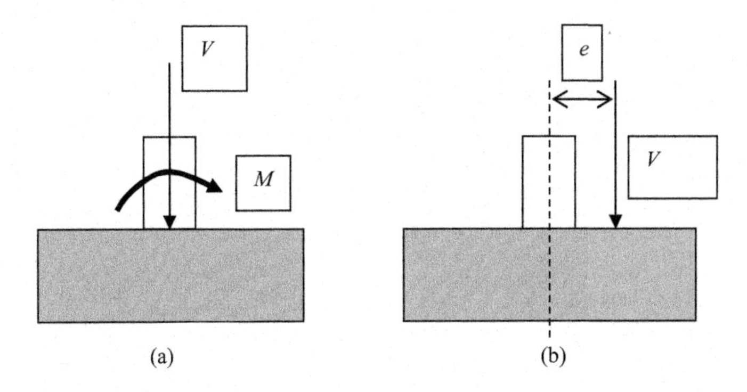

Figure 12.4 Equivalent footing (bending moment replaced with an eccentric load). (a) Footing with bending moment. (b) Bending moment is equated to an eccentric vertical load.

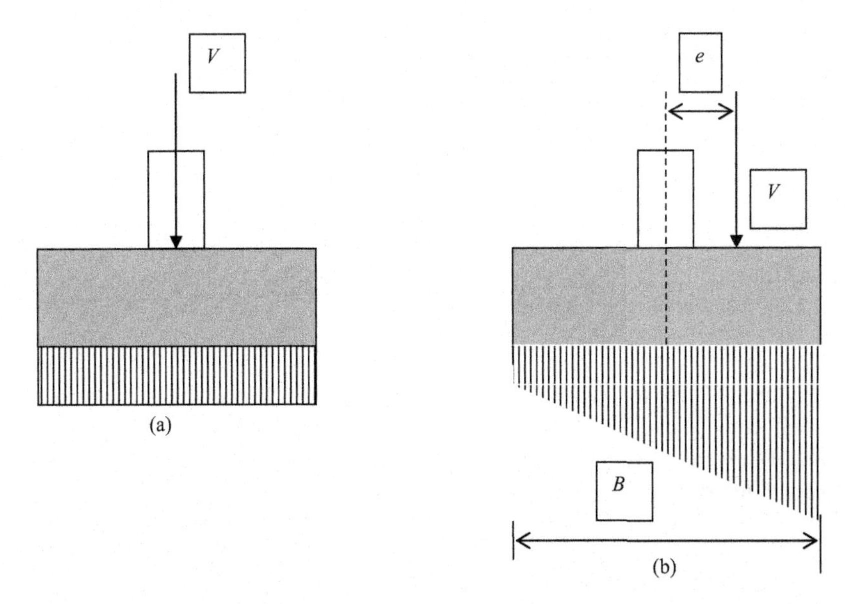

Figure 12.5 Stresses underneath footings. (a) Vertical load. (b) Eccentric vertical load.

When the bending moment is increased, the stress at the bottom of footing becomes uneven.

If the bending moment were to increase more, the pressure under footing in one side would become zero. When the bending moment is increased further, negative stresses start to develop. The foundation engineer should make sure that this does not happen.

In Figure 12.6, the load is applied at the center and the bearing pressure is uniform.

Bearing pressure $q = P/A$

A = area of the footing = BL (B = width, L = length)

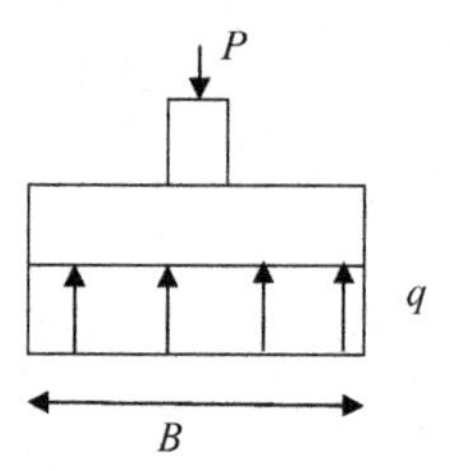

Figure 12.6 Uniform stress.

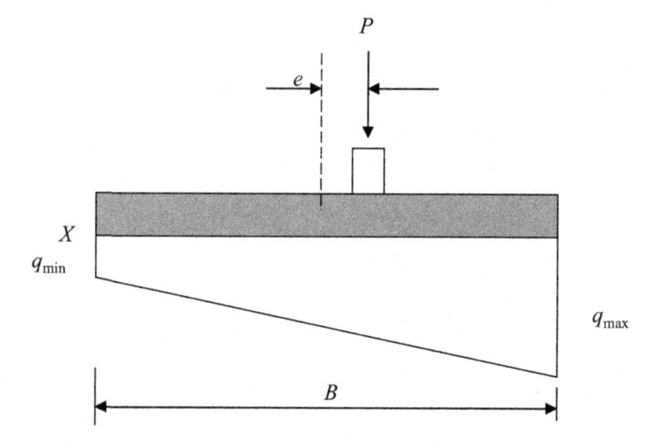

Figure 12.7 Load applied at an eccentricity.

In Figure 12.7, the load is applied with an eccentricity "*e*" and the bearing pressure is not uniform.

Area of the stress diagram = total load

Stress diagram is a trapezoid.

$$\text{Area of the stress trapezoid} = \frac{(q_{max} + q_{min})}{2} \times \text{B} \times \text{L}$$

Area of the stress diagram should be equal to the applied load.

$$\frac{(q_{max} + q_{min})}{2} \times \text{BL} = \text{P}$$

$$q_{max} = \frac{2\text{P}}{\text{BL}} - q_{min} \tag{12.1}$$

The stress diagram can be divided into a rectangle and a triangle (Figure 12.8). Take moments around point "*X*".

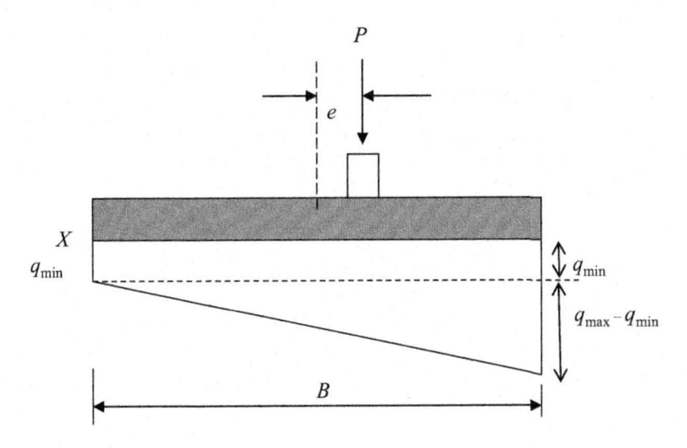

Figure 12.8 Eccentricity of the load.

Area of the rectangular portion $= [q_{min} \times BL]$
Moment due to rectangular portion $= [q_{min} \times BL] \times B/2 = q_{min} \times B^2 L/2$

Area of the triangular portion $= [q_{max} - q_{min}] \times (B/2) \times L$
Moment due to triangular portion $= [q_{max} - q_{min}] \times BL/2 \times 2B/3$
$= [q_{max} - q_{min}] \times B^2L/3$

$P \times (B/2 + e) =$ Moment due to stress rectangle + Moment due to stress triangle
$$P \times (B/2 + e) = q_{min} \times B^2L/2 + [q_{max} - q_{min}] \times B^2L/3$$
$$= q_{min} \times B^2L/2 + q_{max} \times B^2L/3 - q_{min} \times B^2L/3$$
$$= q_{min} \times 3B^2L/6 + q_{max} \times B^2L/3 - q_{min} \times 2B^2L/6$$

$$P \times (B/2 + e) = q_{min} \times B^2L/6 + q_{max} \times B^2L/3 \quad (12.2)$$

From Equation (12.1) $\quad q_{max} = 2P/(BL) - q_{min}$

$$P \times (B/2 + e) = q_{min} \times B^2L/6 + [2P/(BL) - q_{min}] \times B^2L/3$$

$$PB/2 + Pe = q_{min} \times B^2L/6 + 2P/(BL) \times B^2L/3 - q_{min} \times B^2L/3$$

$$PB/2 - 2PB/3 + Pe = q_{min} \times B^2L/6 - q_{min} \times 2B^2L/6$$

$$3PB/6 - 4PB/6 + Pe = -q_{min} \times B^2L/6$$
$$-PB/6 + Pe = -q_{min} \times B^2L/6$$

Multiply the whole equation by −1

$$PB/6 - Pe = q_{min} \times B^2L/6$$
$$q_{min} = [PB/6 - Pe] \times 6/(B^2L)$$
$$q_{min} = P/(BL) - 6Pe/(B^2L)$$

$$q_{min} = \frac{P}{BL} \times \frac{1-6e}{B} \tag{12.3}$$

If $6e/B > 1.0$, q_{min} becomes negative. In this case, a part of the foundation is not used. It is good practice to make sure that q_{min} is positive.

Insert q_{min} in Equation (12.1)

$$q_{max} = \frac{2P}{BL} - q_{min} \tag{12.1}$$

$$q_{max} = 2P/(BL) - P/(BL) \times [1 - 6e/B]$$
$$q_{max} = 2.P/(BL) - P/(BL) + P/(BL) \times 6e/B$$
$$q_{max} = P/(BL) + P/(BL) \times 6e/B$$
$$q_{max} = P/(BL)\ [1 + 6e/B]$$

Design Example 12.1

A footing is loaded with 6000 kN vertical load and 3300 kN.m moment (Figure 12.9). The width of the footing is 4 m and its length is 5 m.

1. What is the maximum and minimum stress under the given footing.
2. If the allowable bearing capacity of the footing is 500 kN/m^2, is this footing acceptable?

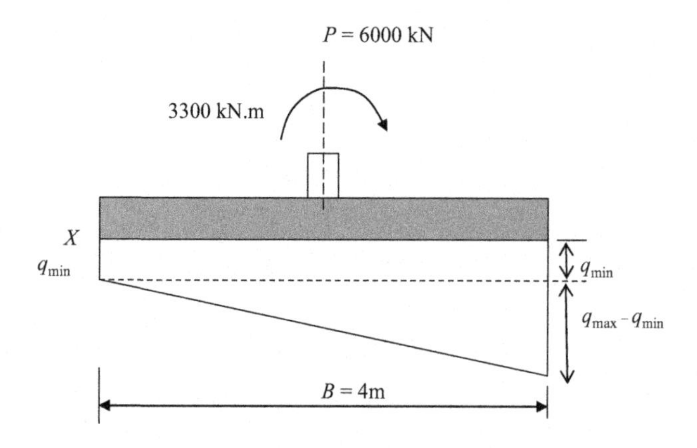

Figure 12.9 Bending moment in a footing.

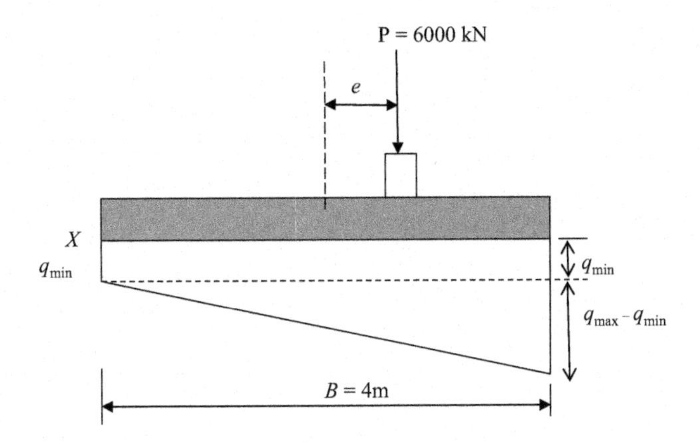

Figure 12.10 Bending moment replaced with an eccentric load.

Solution

The bending moment can be equated to an eccentricity (Figure 12.10).

Step 1: Find eccentricity (e):

$Pe = M = 3300$ kN.m

$e = 3300/6000 = 0.55$ m

Step 2: Investigate q_{min} is negative or positive.

$$q_{min} = \frac{P}{BL} \times \frac{1-6e}{B} \tag{12.3}$$

$q_{min} = P/(BL) \times [1 - 6 \times 0.55/4]$

$q_{min} = 6{,}000/(4 \times 5) \times 0.175 \text{kN/m}^2.$

$q_{min} = 52.5 \text{kN/m}^2.$

Step 3: Find whether q_{max} exceeds the bearing strength.

$q_{max} = P/(BL)[1 + 6e / B]$

$q_{max} = 6{,}000/(4 \times 5)[1 + 6 \times 0.55/4] = 547.5 \text{kN/m}^2.$

The allowable bearing capacity of soil is 500 kN/m^2.

Hence, the foundation is not acceptable. Increase the size of the footing.

In addition to the method used in Design Example 12.1, another method was described in Chapter 8. In that method, the B' and L' values were computed. $B' = (B' - 2e$ and $L' = L - 2e)$

Either method can be used.

Geogrids

13

Geogrids can be used to increase the bearing capacity of foundations in weak soils (Plate 13.1).

Geogrids distribute the load so that the factor of safety against bearing failure is increased (Figure 13.1).

The foundation load is better distributed, due to the geogrid. Hence, the pressure on the weak soil layer is reduced. Pressure reduction occurs since the load is distributed into a larger area.

Due to this reason, the settlement can be reduced and the bearing strength can be improved.

Geogrids distribute the load in a much even manner; however, it is not advisable to use geogrids for major settlement problems.

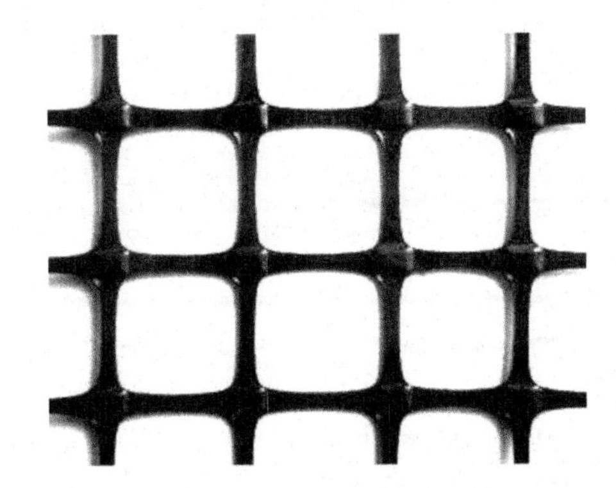

Plate 13.1 Geogrid.

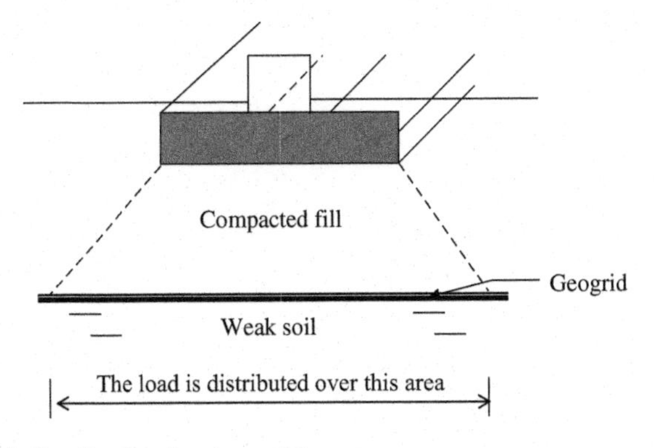

Figure 13.1 Load distribution in a geogrid.

13.1 Failure mechanisms

The foundation can fail within the compacted zone. This could happen if the soil is not properly compacted (Figure 13.2).

Foundation can fail due to failure of soil under the geogrid. Geogrids need to be large enough to distribute the load into a larger area (Figure 13.3).

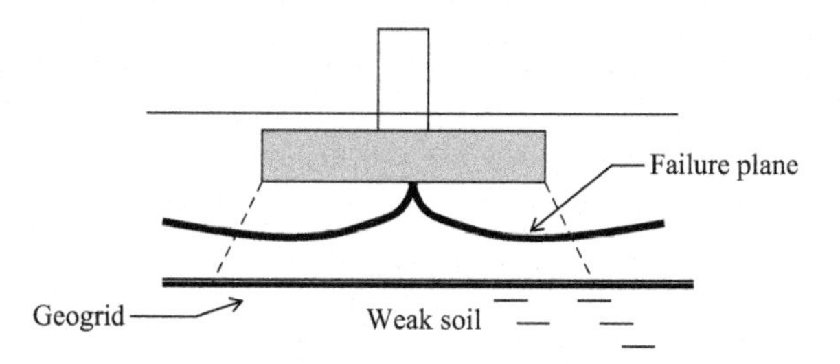

Figure 13.2 Failure in the compacted zone.

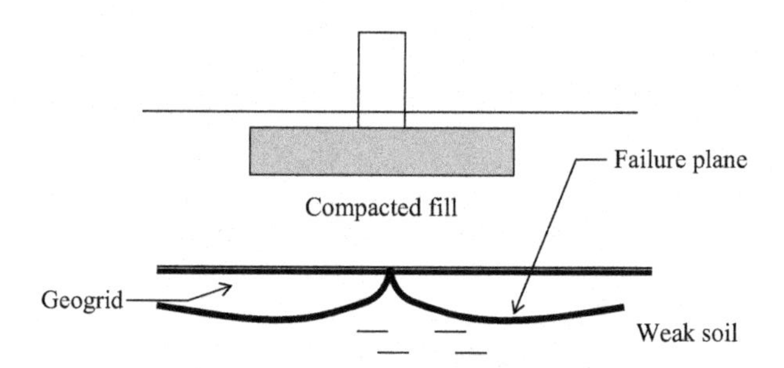

Figure 13.3 Failure within the weak soil layer.

Tie beams and grade beams

14

14.1 Tie beams

The purpose of tie beams is to connect the column footings together (Figure 14.1). Tie beams may or may not carry any vertical loads such as walls, etc.

14.2 Grade beams

Unlike tie beams, grade beams carry walls and other loads (Figure 14.2).

Grade beams are larger than tie beams, since they carry loads.

After the construction of pile caps or column footings, the next step is to construct the tie beams and grade beams (Figure 14.3). The supervision of rebars needs to be conducted by qualified personnel prior to concreting.

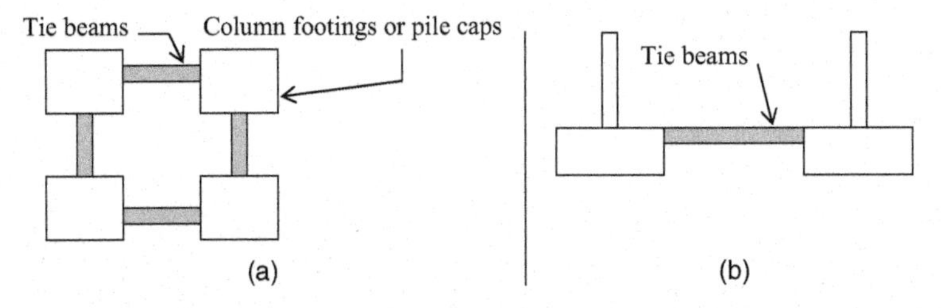

Figure 14.1 Tie beams.

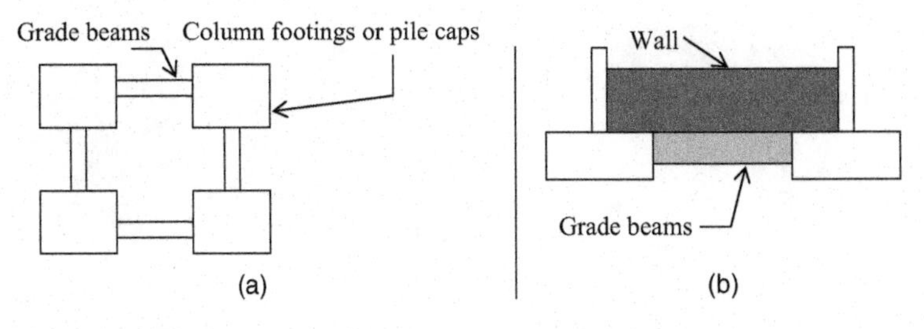

Figure 14.2 Grade beams.

Geotechnical Engineering Calculations and Rules of Thumb

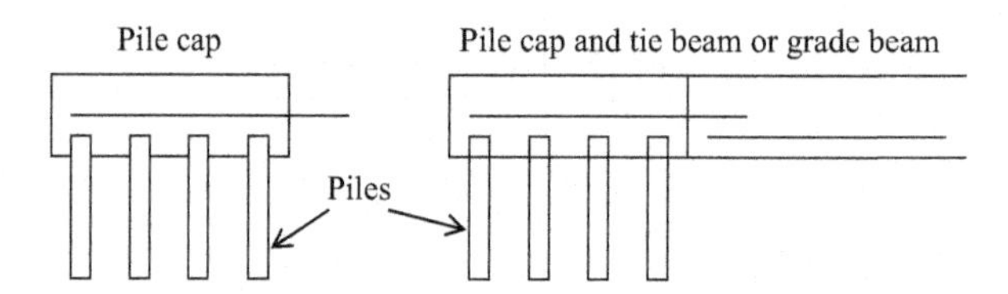

Figure 14.3 Pile cap and tie beams.

Note: The contractor should provide rebars jutting out prior to the concreting of pile caps.

This way when the grade beams are constructed, continuous rebars can be provided (Plates 14.1 and 14.2).

14.3 Construction joints

During the drying up process, the concrete would contract. If the construction joints are not provided, cracks could be generated due to the contraction process (Figure 14.4).

Construction Procedure

- Concrete a section (for example, concrete from "A" to "B" in Figure 14.4).

Plate 14.1 Pile cap before concreting (piles are seen at the bottom of the pile cap).

Plate 14.2 Pile cap after concreting.

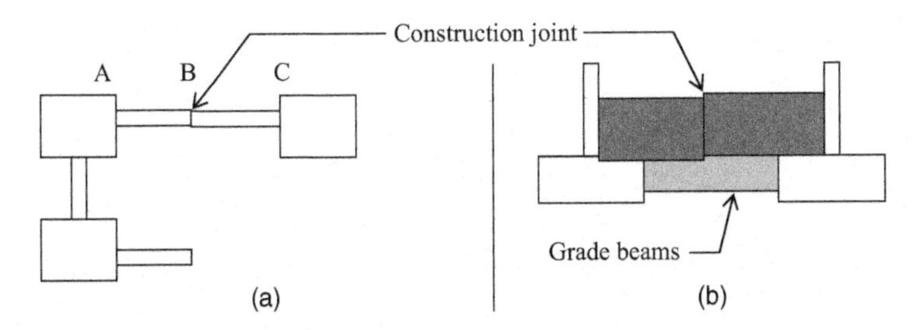

Figure 14.4 Construction joints.

- Wait for a reasonable time period for the concrete to contract.
- Concrete the next section.
- This way any contraction in the concrete would be eliminated.
- Typically, engineers recommend construction joints every 60–100 ft. of beams.

Drainage for shallow foundations

15.1 Introduction

Geotechnical engineers need to investigate the elevation of groundwater, since groundwater can create problems during the construction stage. The stability of the bottom may be affected by the presence of groundwater. In such situations, dewatering method needs to be planned. Maintaining sidewall stability is another problem that needs to be addressed (Figures 15.1 and 15.2).

Continuous pumping can be used to maintain a dry bottom. Stable side slopes can be maintained by providing shoring supports.

15.2 Dewatering methods

15.2.1 Well points

In some cases, it is not possible to pump enough water to maintain a dry bottom for concrete shallow foundations. In such situations, well points are constructed to lower the groundwater table (Figure 15.3).

15.2.2 Small scale dewatering for column footings (pump water from the excavation)

Groundwater level is lowered by constructing a small hole or a trench inside the excavation (as shown in Figure 15.4) and placing a pump (or several pumps) inside the excavation. For most column footing construction work, this method will be sufficient.

15.2.3 Medium scale dewatering for basements or deep excavations (pump water from trenches or wells)

For medium scale dewatering projects, trenches or well points can be used (Figure 15.5).

More pumps may be added, if necessary, to keep the excavation dry. A combination of submersible pumps and vacuum pumps can be used.

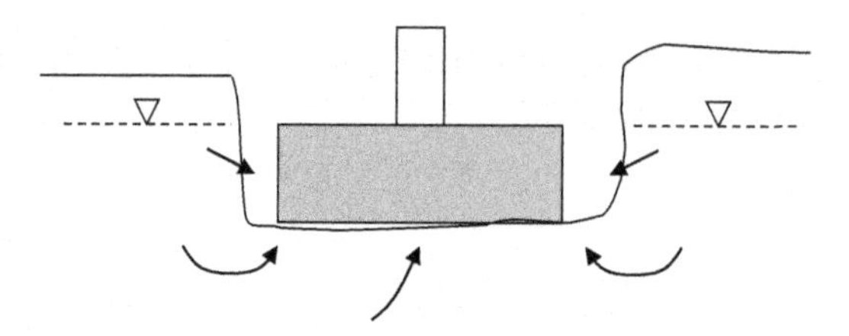

Figure 15.1 Groundwater migration during shallow foundation construction.

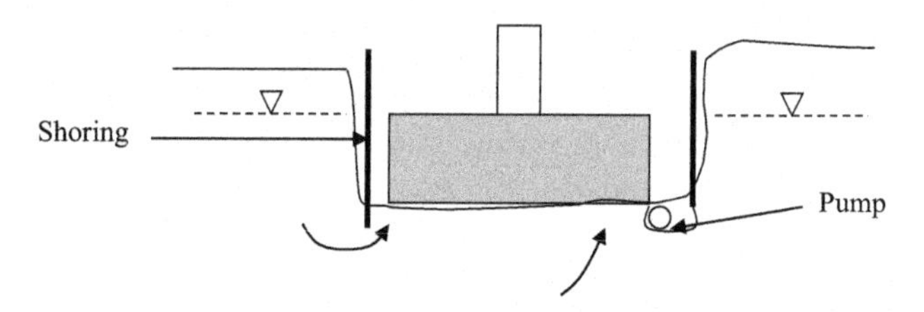

Figure 15.2 Pumping and shoring the sides.

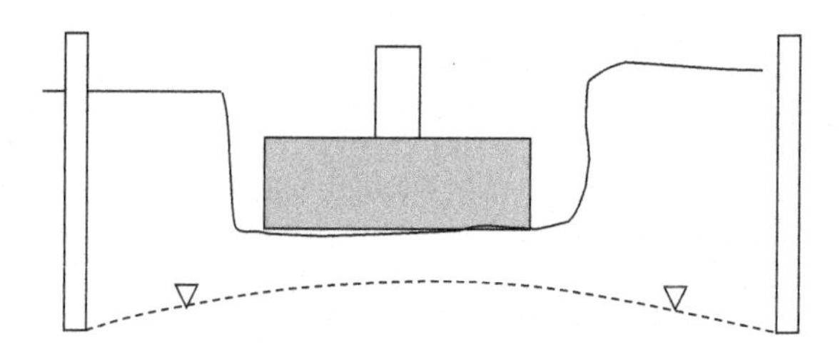

Figure 15.3 Use of well points to lower the groundwater table.

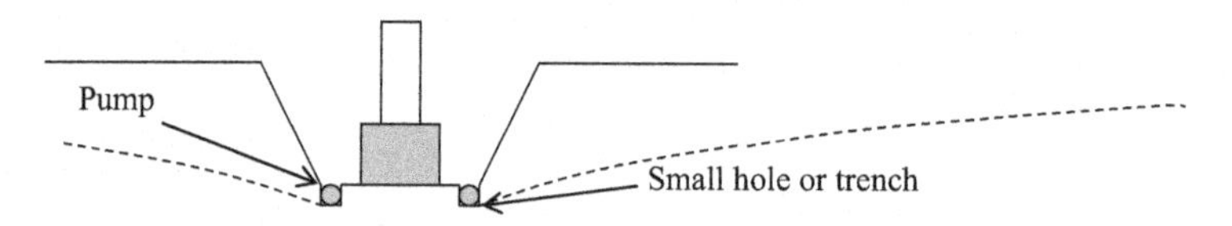

Figure 15.4 Dewatering of a column footing.

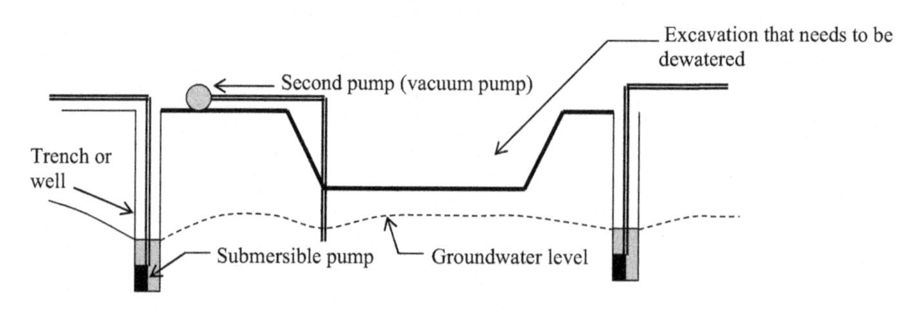

Figure 15.5 Groundwater lowering using well points.

15.2.4 Large scale dewatering for basements or deep excavations

15.2.4.1 Alternative 1

Well points or trenches are constructed. The main artery pipe is connected to each pump as shown in Figure 15.6. A strong high capacity pump would suck water out of all the wells as shown in Figure 15.6.

15.2.4.2 Alternative 2

A similar dewatering system can be designed using submersible pumps. In this case, instead of one pump, each well would get a submersible pump as shown in Figure 15.7.

Alternative 2 is more effective than alternative 1. The main disadvantage of alternative 2 is high maintenance. Since there is more than one pump, more maintenance work will occur compared to alternative 1. On the other hand, alternative 2 can be modified to include less well points since the pumping effort can be increased significantly.

The following points should be considered when a dewatering process is to be carried out.

- For most construction work, groundwater should be lowered at least 2 ft. below the excavation.
- A water quality assessment study needs to be conducted. If the groundwater is contaminated, discharge permits need to be obtained depending on the local regulations. In most cases, local and national environmental protection agencies need to be contacted and proper

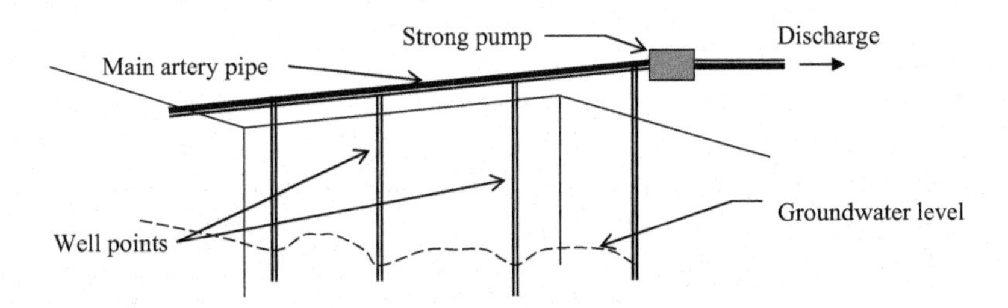

Figure 15.6 Well points in series with one large pump.

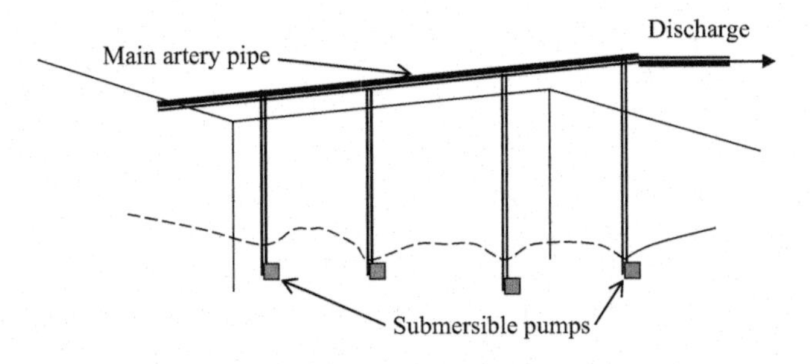

Figure 15.7 Well points in series with submersible pumps.

procedures should be followed. Discharging contaminated water into rivers, lakes, or storm water systems could be a major offense.

- If the water needs to be treated prior to discharge, dewatering may become extremely expensive. In that case "cutoff walls" or "ground freezing" options may be cheaper.

15.3 Design of dewatering systems

15.3.1 Initial study

Study the surrounding area and locate nearby rivers, lakes, and other water bodies. Groundwater normally flows toward surface water bodies such as rivers and lakes. If the site is adjacent to a major water body (such as a river or lake) the groundwater elevation will be same as the water level in the river.

15.3.2 Construct borings and piezometers

Soil conditions – Create soil profiles based on borings. Assess the permeability of the existing soil stratums. (Sandy soils will transmit more water than clayey soils).

Figure 15.8 shows the approximate permeability values of sands and gravel with respect to D_{10} values.

Step 1: Obtain the D_{10} value by conducting a sieve analysis.

Step 2: Use the graph to find the permeability.

Note: This graph should not be used for silts and clays.

The experimental values given in Table 15.1 were used to draw the graph in Figure 15.8.

D_{10} value = The size of the sieve that only 10% of the soil would pass through.

For example, if 10% of the soil would pass through a sieve that has an opening of 1.5 mm, then the D_{10} value of the soil is 1.5 mm. The D_{10} value is obtained by drawing a sieve analysis curve and then locating the 10% passing point in the curve. The corresponding sieve size is the D_{10} value.

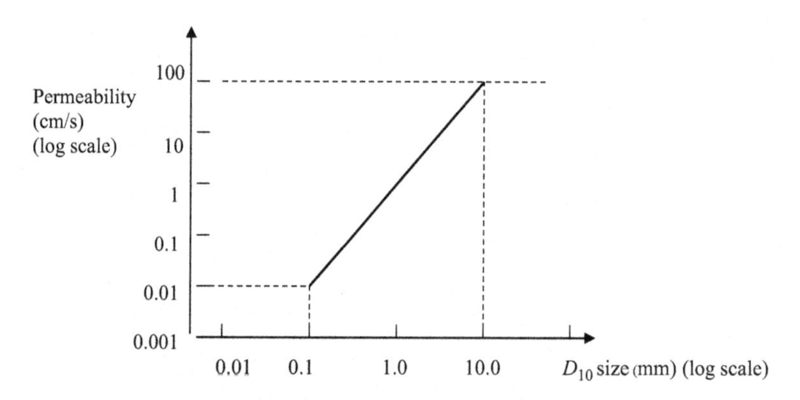

Figure 15.8 Permeability sandy soils.

Table 15.1 D_{10} value versus permeability

Soil type	D_{10} (mm)	Permeability (cm/s)
Sand	0.06	0.0036
Sand	0.01	0.01
Gravel	0.30	0.09
Gravel	1.5	2.3
Gravel	9.0	81

The graph in Figure 15.8 should be used with reservations. If highly variable soil conditions are found, Figure 15.8 should not be used to obtain the permeability. In such situations, a well pumping test needs to be conducted to obtain the permeability of soil.

- *Seasonal variations* – Groundwater elevation readings should be taken at regular intervals. In some sites, the groundwater elevation could be very sensitive to seasonal changes. The groundwater elevation during the summer could be drastically different from the winter.
- *Tidal effect* – Groundwater elevation changes with respect to tidal flow. In some sites, the groundwater elevation may show a very high sensitivity to high and low tides.
- *Artesian conditions* – Check for artesian conditions. Groundwater could be under pressure and well pumping or any other dewatering scheme could be a costly procedure.
- *Groundwater contamination* – If contaminated groundwater is found, then groundwater cutoff methods should be studied.

15.4 Ground freezing

Ground freezing can be used as an alternative to pumping. In some situations, the pumping of groundwater to keep excavations dry may be very costly. When groundwater is contaminated, the discharging of groundwater is a major problem. If contaminated groundwater is pumped out, that water needs to be treated prior to disposal. In such situations, ground freezing would be an economical way to control groundwater (Figure 15.9).

15.4.1 Ground freezing technique

Soil pores are full of moisture (Figure 15.10). The moisture in soil pores is frozen to produce an impermeable barrier.

Ground freezing is achieved by circulating calcium chloride brine. Calcium chloride brine is sent through one pipe and retrieved from two pipes (Figure 15.11).

Brine solution is sent through pipe “A” and removed from pipes “B” and “C”. During circulation, the brine solution extracts heat from the surrounding soil. When the heat is lost from the moisture in soil, ground freezing occurs.

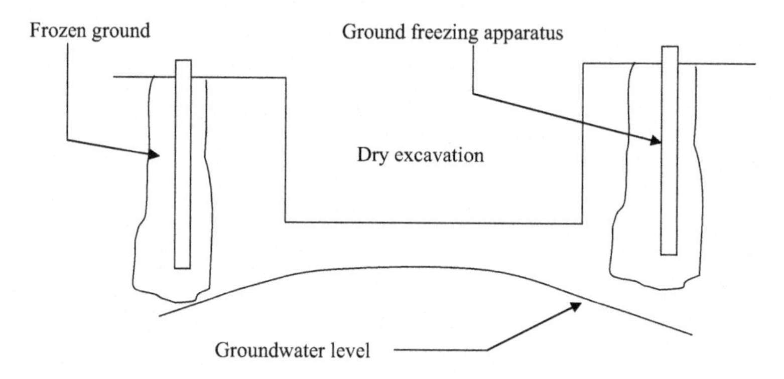

Figure 15.9 Ground freezing to keep excavations dry.

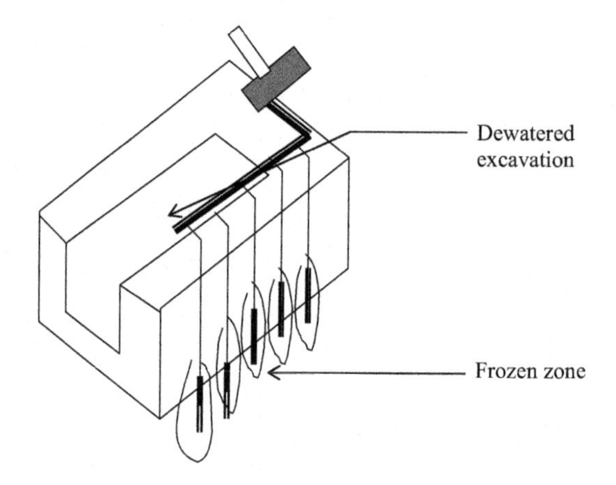

Figure 15.10 Ground freezing methodology.

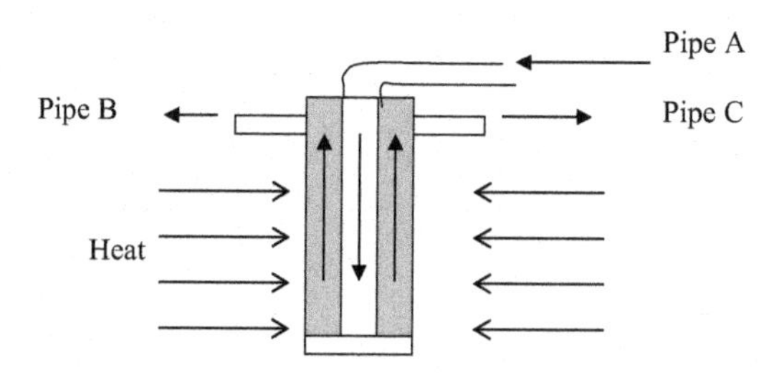

Figure 15.11 Brine flow for ground freezing.

Brine solution is removed from pipes "B" and "C", is pumped to the plant and energized, and sent back to the well.

A decade or so ago, ground freezing was considered to be an exotic method that no one knew exactly how to do. Recently, many projects have been successfully completed using the ground freezing technique. When construction work needs to be conducted in a contaminated groundwater media, the ground freezing technique becomes economical.

15.4.2 Ground freezing – practical aspects

Ground freezing is a complex operation. There are instances, where the method failed to cut off groundwater.

15.4.2.1 Sands

Ground freezing is easier in granular soils. Groundwater in granular soils occurs in the pores of the soil. The chemical bonds between soil grains and water are negligible. Hence, water would freeze quickly after the temperature is brought below the freezing point. (32°F or 0°C).

15.4.2.2 Clays

Ground freezing is unpredictable and difficult in clay soils. Water molecules are chemically bonded to clay particles. Due to this reason, water is not free to freeze. Substantially low temperatures are required to freeze clay soils.

15.4.2.3 Refrigeration plant size

Refrigeration plants are rated by tons (T_R).

$$1\ T_R = 12{,}000\ \text{BTU/h}$$

Typical refrigeration plants for ground freezing ranges from 100 to 200 T_R.

15.4.2.4 Liquid nitrogen (LN_2) versus brine

Liquid nitrogen is another chemical used for ground freezing. Extremely low temperatures can be achieved with liquid nitrogen. Cost of LN_2 plants can be higher than brine plants.

As mentioned earlier, clay soils may not be easy to freeze with chlorides. In such situations, LN_2 can be used. Ground freezing with LN_2 is much faster compared to brine.

15.4.2.5 Groundwater flow velocity

Groundwater flow velocity changes from site to site. High groundwater flow velocities create problems for the freezing process. It is a well known fact that more energy is needed to freeze moving water than still water.

Prior to any ground freezing project, the groundwater flow velocities should be assessed. If the flow velocity is high, the temperature may have to be brought down well below the freezing point to achieve success.

15.4.2.6 Ground heave

The ground can heave due to freezing. When water changes to ice, the volume increases approximately by 10% (Figure 15.12). The ground heave can cause problems for nearby buildings. In sandy soils, water freezes quickly and ground heave would be noticeable. In clay soils, the water freezes gradually. Hence, ground heave may not be apparent at the beginning.

15.4.2.7 Utilities

Ground freezing could freeze nearby utilities. If nearby utilities, such as gas lines or water lines, are present, proper authorities need to be consulted prior to recommending the ground freezing method to control groundwater.

15.4.2.8 Groundwater control near streams and rivers

Groundwater control near streams and rivers could be a challenging affair. In such situations, ground freezing may be more effective than well pumping.

15.4.2.9 Groundwater control in tunneling

Ground freezing method can also be utilized in tunnels. The fractured ground can be encountered unexpectedly during tunneling (Figure 15.13).

Holes are drilled ahead of the tunnel excavation and brine solution is injected. This would create a frozen zone in front of the tunnel excavation to freeze the groundwater.

15.4.2.10 Contaminant isolation

Ground freezing can be used for long periods to isolate contaminants (Figure 15.14). In such situations, durable piping and machinery need to be used. Contaminant isolation efforts could range from 10 to 30 years. During this time maintenance of the system is required.

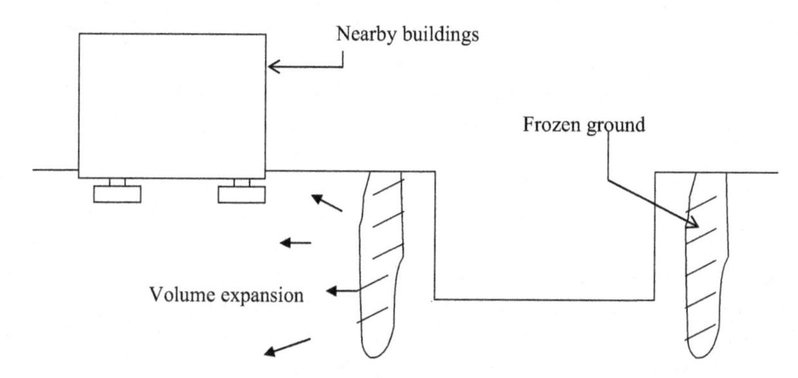

Figure 15.12 Volume expansion due to ground freezing could create problems for nearby buildings.

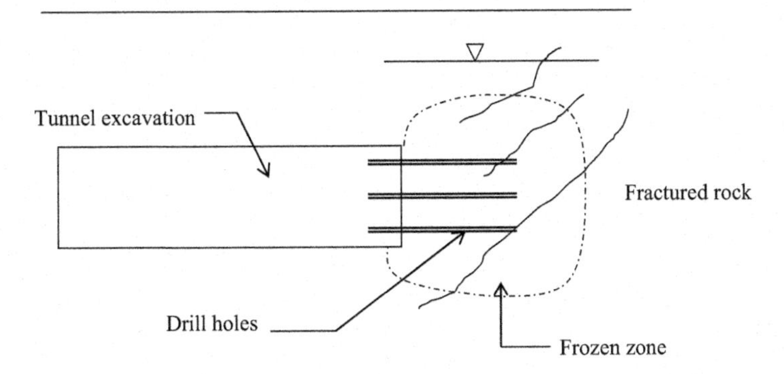

Figure 15.13 Ground freezing in tunneling.

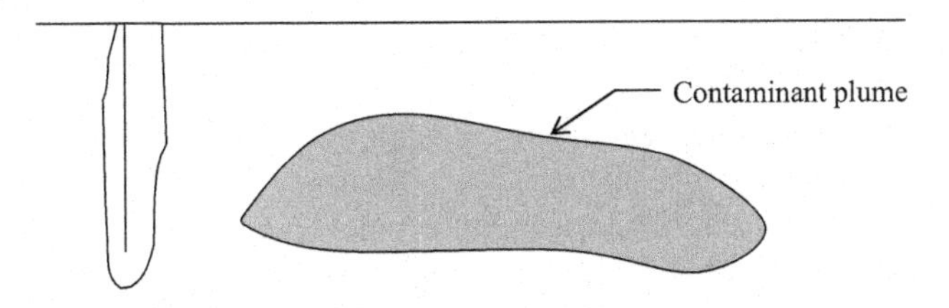

Figure 15.14 Ground freezing for contaminant isolation.

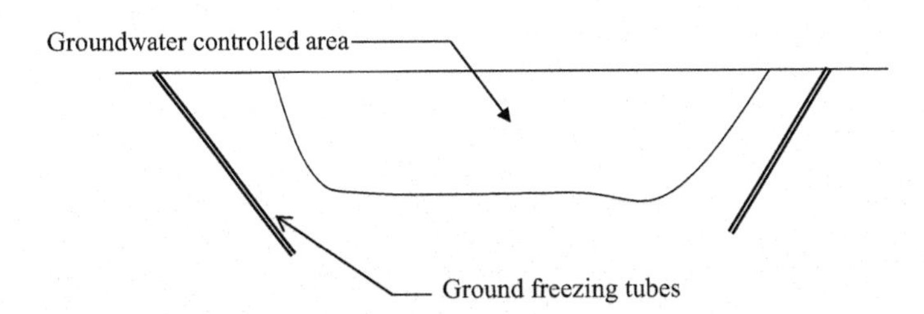

Figure 15.15 Ground freezing tubes installed in an angle.

15.4.2.11 *Directional ground freezing*

Due to the advancement of directional drilling techniques, the freezing tubes can be installed in an angle (Figure 15.15).

15.4.2.12 *Ground freezing for underpinning*

When excavations are planned near other buildings, underpinning of these buildings has to be conducted. Ground freezing can be used instead of underpinning in such situations (Figure 15.16).

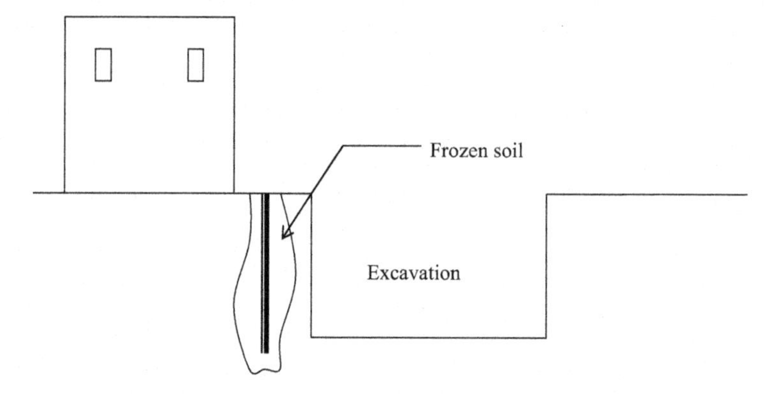

Figure 15.16 Stabilization of soil near existing buildings.

15.5 Drain pipes and filter design

Groundwater can create many problems for geotechnical engineers. Controlling the groundwater is a challenge in many situations. Drainage in shallow foundations, basements, and retaining walls are done using drain pipes (Figures 15.17 and 15.18).

Drain pipes tend to be clogged due to migration of silt and clay particles. It is important to filter the fine particles to avoid clogging.

Gravel filters have been one of the oldest techniques to filter fine particles. Gravel filters were very common prior to the arrival of geotextiles.

15.5.1 Design of gravel filters

Figure 15.19 shows the design of a gravel filter.

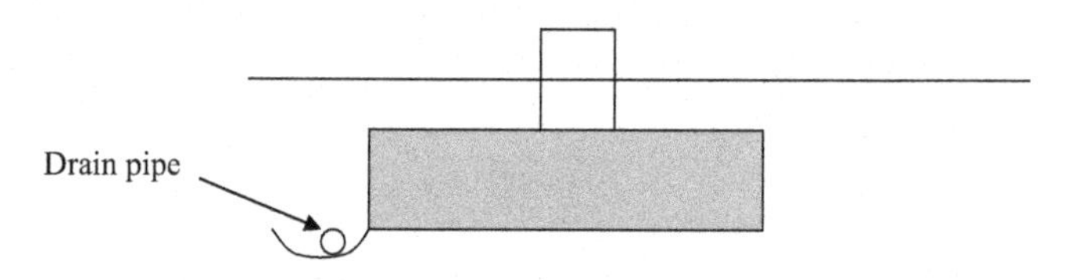

Figure 15.17 Drain pipe in a shallow foundation.

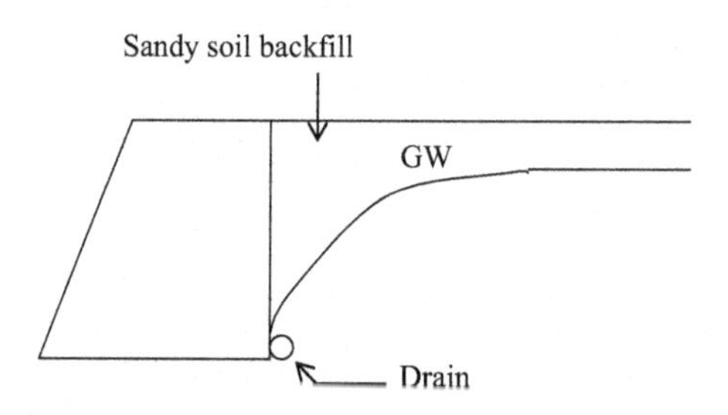

Figure 15.18 Drainage in retaining walls.

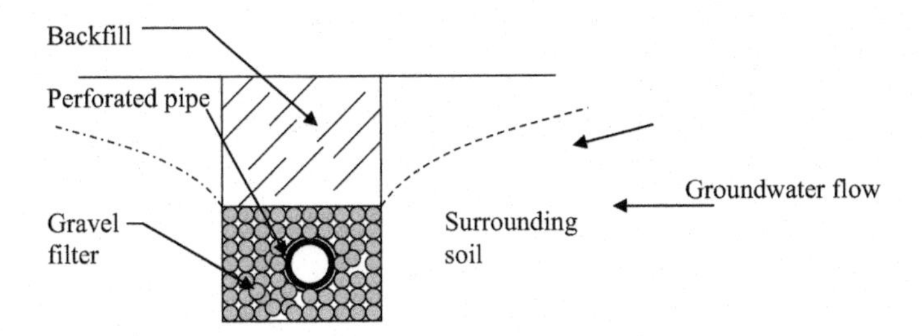

Figure 15.19 Gravel filter.

15.5.2 Purpose of gravel filters

Gravel filters would stop any soil particles from entering the pipe. At the same time, the gravel should allow ample water flow. If fine gravel were used, the water flow would be reduced. On the other hand, a large-size gravel would allow the soil particles to pass through and eventually clog the drain pipe.

Empirically, it has been found that if the gravel size were selected in accordance with the following two rules , very little soil would transport through the gravel filter and at thc same time the water would flow smoothly as well.

Rule 1: (*To block soil from entering the pipe*):

$$D_{15}\,(\text{Gravel}) < 5 \times D_{85}\,(\text{Soil})$$

D_{15} size = 15% of particles of a given soil or gravel would pass through the D_{15} size of that particular soil or gravel.

D_{85} size = 85% of particles of a given soil or gravel would pass through the D_{85} size of that particular soil or gravel.

How to obtain the D_{15} size for a gravel:

Conduct a sieve analysis test and draw the sieve analysis curve.

Draw a line across the 15% passing point.

Find the D_{15} size using the sieve analysis curve.

For the filter to stop the soil from washing away, the D_{15} value of the gravel should be smaller than $5 \times D_{85}$ of soil. However, the gravel size should not be too small. If the gravel size were too small, no water would pass through the gravel, which is not desirable.

Rule 2 is proposed to address that issue.

Rule 2: (*To let water flow*):

$$D_{50}\,(\text{Gravel}) > 25 \times D_{50}\,(\text{Soil}).$$

It has been found that the D_{50} value of gravel, should be larger than $25 \times D_{50}$ of soil, in order for the water to drain properly.

When selecting a suitable gravel to be used as a filter these two rules should be adhered to.

Design Example 15.1

The sieve analysis tests of surrounding soil gave the following D_{85} and D_{50} values.

$D_{85} = 1.5$ mm; $D_{50} = 0.4$ mm.

Find an appropriate gravel size for the filter.

Solution

Rule 1:

$$D_{15}\ (\text{Gravel}) < 5 \times D_{85}\ (\text{Soil})$$
$$D_{15}\ (\text{Gravel}) < 5 \times 1.5$$
$$D_{15}\ (\text{Gravel}) < 7.5$$

Hence, the D_{15} size of gravel should be less than 7.5 mm.

Rule 2:

$$D_{50}\ (\text{Gravel}) > 25 \times D_{50}$$
$$D_{50}\ (\text{Gravel}) > 25 \times 0.4$$
$$D_{50}\ (\text{Gravel}) > 20\,\text{mm}$$

Hence, select a gravel with a D_{15} value less than 7.5 mm and a D_{50} value greater than 20 mm.

15.6 Geotextile filter design

Geotextile filters are increasingly becoming popular in drainage applications. Ease of use, economy, and durability of geotextile filters have made it the number one choice of many engineers (Figure 15.20).

15.6.1 Geotextile wrapped granular drains (sandy surrounding soils)

Gravel is wrapped with a geotextile to improve the performance. The geotextile filters the water and the gravel act as the drain. There are two of types geotextiles (woven and nonwoven) available in the market. For sandy soils, both woven and nonwoven

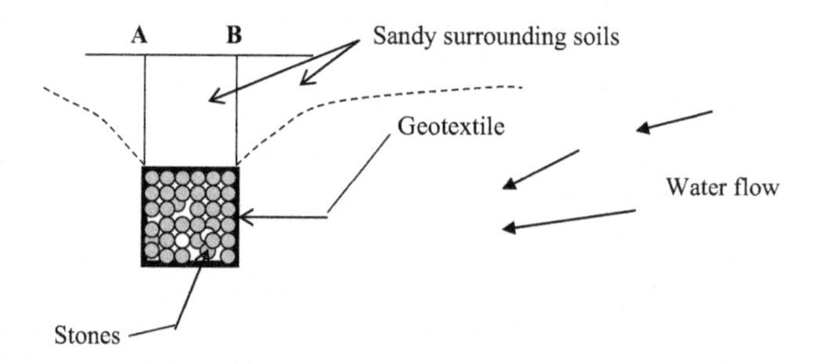

Figure 15.20 Geotextile wrapped drain filter.

gotextiles can be used. The geotextile is wrapped around stones and backfilled with original soil.

The stones act only as a medium to transport water. The task of stopping the soil from entering the drain is done by the geotextile.

Equations have been developed for two types of flows.

One-way flow

In this case, the flow is always in one direction across the geotextile. In Figure 15.20, the water enters the drain from sides of the drain.

Two-way flow (alternating flow)

In some instances, water goes through the geotextile in both directions. In case of heavy rain, the water enters the drain from the top (between points A and B in Figure 15.20) and flow into the drain and then go out of the drain to the surrounding soil. If the flow through the geotextile is possible in both directions, a different set of equations need to be used.

Design Example 15.2

Design a geotextile wrapped granular filter drain for a sandy soil with $D_{50} = 0.2$mm. It is determined that the flow across the geotextile is always in one direction.

Solution

Step 1:

H_{50} (Geotextile) $<$ 1.7 to 2.7 $\times$ D_{50} (soil) (for one way flow for sandy soils) (Zitscher, 1975)

The equation in Step 1 is valid for one way flow in a sandy soil surrounding.

H_{50} indicates that 50% of the holes in the geotextile are smaller than the H_{50} size.

D_{50} (soil) indicates that 50% of soil particles are smaller than the D_{50} size. Since, the D_{50} is found to be 0.2 mm, H_{50} can be computed.

H_{50} (Geotextile) $<$ (1.7 to 2.7) $\times$ 0.2 mm

H_{50} should be between 0.34 mm to 0.54 mm.

Hence, select a geotextile with a H_{50} size of 0.4 mm.

15.6.2 Alternating flow in sandy soils

In some situations, the flow can be in both directions. In such situations, the following procedure should be adopted.

Design Example 15.3

Design a geotextile wrapped granular filter drain for a sandy soil with $D_{50} = 0.2$ mm. It is determined that the flow across the geotextile could alternate.

Solution

For two-way flow

Step 1:

H_{50} (Geotextile) $<$ (0.5 to 1.0) $\times$ D_{50} (soil) (for two-way flow for sandy soils) (Zitscher, 1975)

Since, D_{50} was found to be 0.2 mm
H_{50} (Geotextile) < (0.5 to 1.0) × 0.2 mm
H_{50} should be between 0.1 mm and 0.2 mm.
Hence, select a geotextile with H_{50} size equal to 0.15 mm.

15.6.3 Geotextile wrapped granular drains (clayey surrounding soils)

Theory:

For cohesive soils, the direction of flow is not significant since the flow is not rapid as in sandy soils. Most Engineers prefer to use nonwoven geotextiles for cohesive soils.

Design Example 15.4

Design a Geotextile wrapped granular filter drain for a cohesive soil with D_{50} = 0.01 mm.

Solution

Step 1:

H_{50} (Geotextile) < (25 to 37) × D_{50} (soil) (one way and two way flow for clayey soils) (Zitscher, 1975)

- The equation in Step 1 is valid for cohesive surrounding soils for any type of flow.

H_{50} (Geotextile) < (25 to 37) × 0.01 mm
Since, D_{50} is equal to 0.01 mm, H_{50} should be between 0.25 mm and 0.37 mm.
Hence, select a geotextile with the H_{50} size equal to 0.30 mm.

15.6.4 Geotextile wrapped pipe drains

In some occasions flow through granular media alone is not enough. In such situations, a perforated pipe is used at the center as shown in Figure 15.21. The size of the pipe depends on the flow rate required.

The design of the geotextile in this case is exactly similar to the previous cases.

If the surrounding soil is sand and only one way flow is expected.

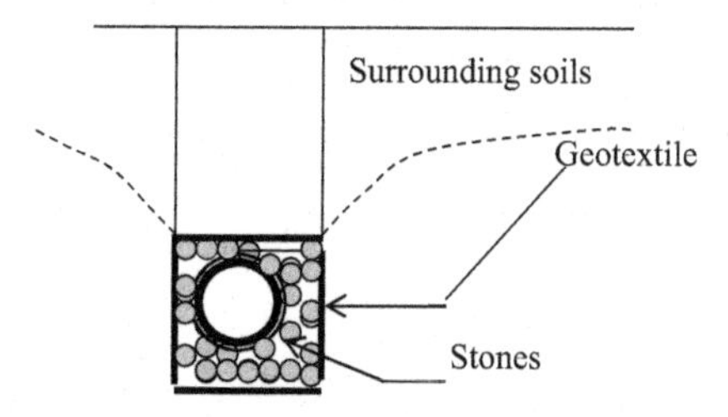

Figure 15.21 Geotextile wrapped pipe drain.

15.6.5 Summary

Sandy soils (one-way flow):
H_{50} (Geotextile) $<$ 1.7 to 2.7 $\times$ D_{50} (soil)

Sandy soils (two-way flow):
H_{50} (Geotextile) $<$ 0.5 to 1.0 $\times$ D_{50} (soil)

Clayey soils (one-way and two-way flow):
H_{50} (Geotextile) $<$ 25 to 37 $\times$ D_{50} (soil)

Reference

Zitscher, F.F., 1975. Recommendations for the Use of Plastics in Soil, vol. 52, Die Bautechnik, Berlin.

Selection of foundation type

16

There are a number of foundation types available for the geotechnical engineers.

16.1 Shallow foundations

Shallow foundations are the cheapest and the most common type of foundations (Figure 16.1).

Shallow foundations are ideal for the situations when the soil immediately below the footing is strong enough to carry the building loads. In some situations, the soil immediately below the footing could be weak or compressible. In such situations, other foundation types need to be considered.

16.2 Mat foundations

Mat foundations are also known as raft foundations. Mat foundations, as the name implies, spread like a mat. The building load is distributed in a large area (Figure 16.2) (Plate 16.1).

16.3 Pile foundations

Piles are used when the bearing soil is at a greater depth. In such situations, the load has to be transferred to the bearing soil stratum (Figure 16.3).

16.4 Caissons

Caissons are nothing but larger piles. Instead of a pile group few large caissons can be utilized. In some situations, caissons could be the best alternative (Figure 16.4).

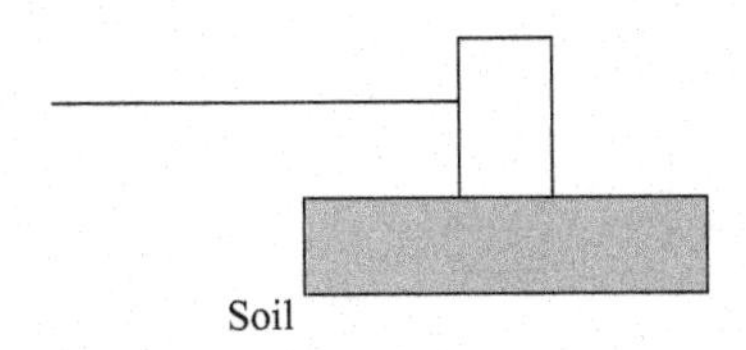

Figure 16.1 Shallow foundation.

Geotechnical Engineering Calculations and Rules of Thumb

Figure 16.2 Mat foundations.

Plate 16.1 Mat foundations.

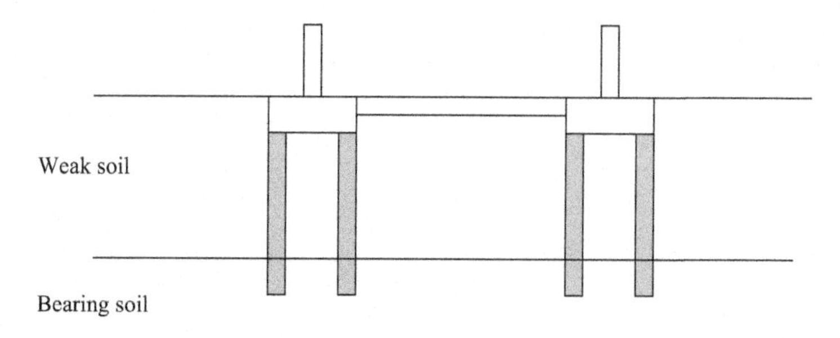

Figure 16.3 Pile foundations.

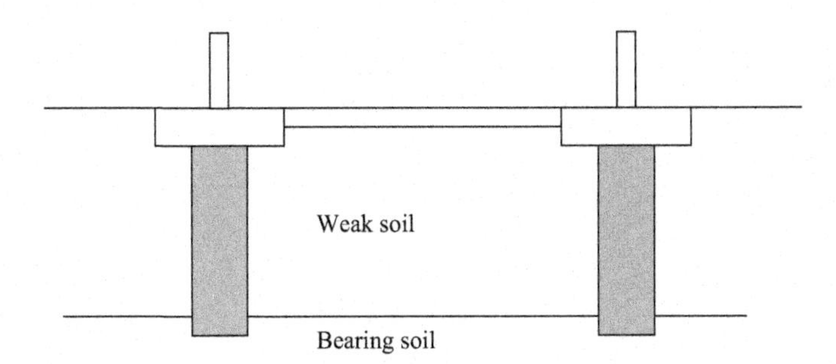

Figure 16.4 Caissons.

16.5 Foundation selection criteria

Normally, all attempts are made to construct shallow foundations. This is the cheapest and the fastest foundation type. The designer should look into the bearing capacity and settlement when considering shallow foundations.

The geotechnical engineer needs to compute the bearing capacity of the soil immediately below the footing. If the bearing capacity is adequate, the settlement needs to be computed. The settlement can be immediate or long term. Both immediate and long-term settlements should be computed.

Figure 16.5 shows a shallow foundation, mat foundation, pile group, and a caisson. The geotechnical engineer needs to investigate the feasibility of designing a shallow foundation due to its cheapness and the ease of construction. In Figure 16.5, it is clear that the weak soil layer just below the new fill may not be enough to support the shallow foundation. Settlement in soil due to loading of the footing also needs to be computed.

If shallow foundations are not feasible, then other options need to be investigated. Mat foundations can be designed to carry large loads in the presence of weak soils. Unfortunately, cost is a major issue with mat foundations. Piles can be installed, as

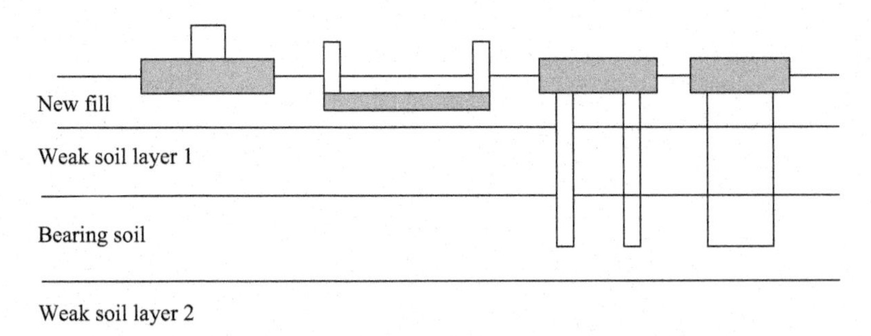

Figure 16.5 Different foundation types.

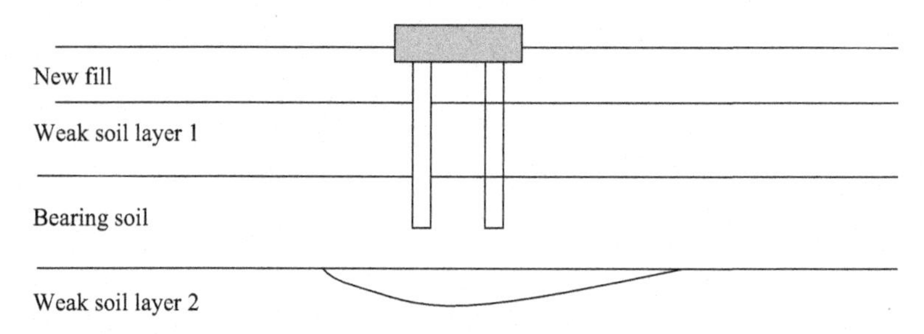

Figure 16.6 Punching failure (soil punching into the weak soil below due to pile load).

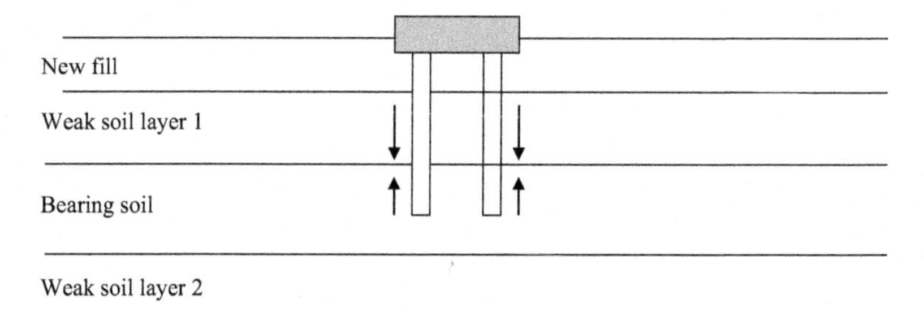

Figure 16.7 Negative skin friction.

shown in Figure 16.5, ending in the bearing stratum. In this situation, one needs to be careful of the second weak layer of soil below the bearing stratum. Piles could fail due to punching into the weak stratum (Figure 16.6).

The engineer needs to consider negative skin friction due to the new fill layer. The negative skin friction would reduce the capacity of piles (Figure 16.7).

Due to the new load of the added fill material, the weak soil layer 1 would consolidate and settle. The settling soil would drag down piles with it. This is known as negative skin friction or down drag.

Consolidation settlement of foundations

17

17.1 Introduction

What would happen when a foundation is placed on top of a clay layer? It is obvious that the clay layer would compress and, as a result, the foundation would settle.

Why would the foundation settle due to an applied load?

Clay is a mixture of soil particles, water, and air. When a clay soil is loaded, the water particles inside the soil mass are pressurized and tend to squeeze out (Figure 17.1).

The settlement due to the water squeezing out is known as primary consolidation.

Squeezing out of water is caused due to the dissipation of excess pore pressure. When all the water in the soil is squeezed out, one may declare that 100% primary consolidation has been achieved.

In reality, 100% primary consolidation usually can never be achieved. For all practical purposes, 90% primary consolidation is taken as the end of the primary consolidation process.

17.1.1 Secondary compression

Now, let us look at secondary compression.

After all the water is squeezed out, one would expect the settlement to stop. Interestingly, that is not the case. The settlement continues after all the excess pore water is squeezed out due to the fact that soil particles start to rearrange their orientation.

Let us see what this means (Figure 17.2).

These soil particles try to rearrange themselves to a more stable configuration (Figure 17.3).

The rearrangement of soil particles would create further settlement of the foundation. In reality, secondary compression starts as soon as the foundation is constructed. However, for simplicity, the usual practice is to compute the secondary compression after the primary consolidation is completed.

17.1.2 Summary of concepts learned

- Primary consolidation occurs due to the dissipation of excess pore pressure.
- In reality, the primary consolidation never ends. For all practical purposes, 90% consolidation is taken as the end of the process.
- Secondary compression occurs due to the rearrangement of soil particles, due to the load. Secondary compression has nothing to do with the pore water pressure.

Geotechnical Engineering Calculations and Rules of Thumb

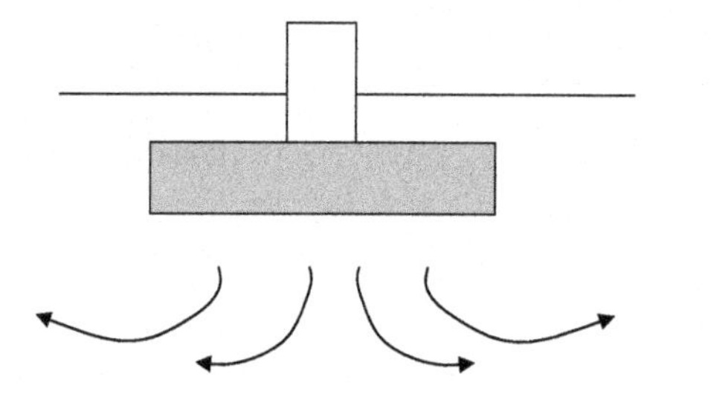

Figure 17.1 Water squeezing out from the soil.

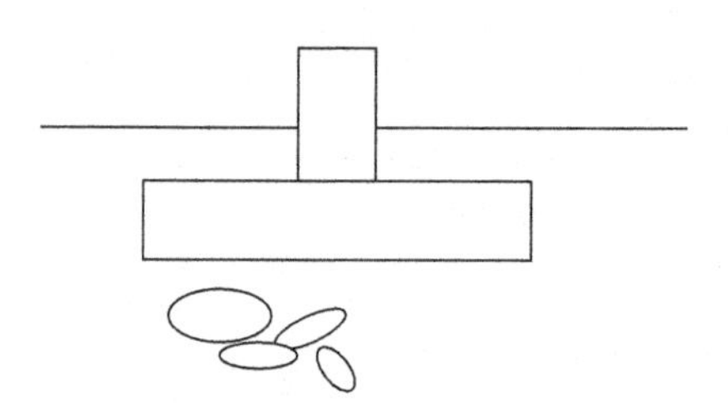

Figure 17.2 Soil particles below a footing.

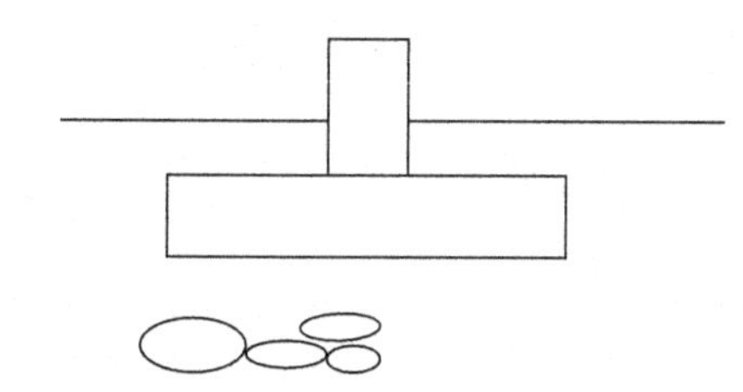

Figure 17.3 Stable rearrangement.

17.2 Excess pore pressure distribution

Figure 17.4 shows a group of piezometers installed closer to a shallow foundation.

Piezometer "A" is located at the same level as the bottom of the footing in the sand layer.

Piezometer A: Any excess pore pressure generated due to the footing load would be dissipated immediately since the sand is highly permeable.

Piezometer B: The bottom of piezometer B is at a lower depth than piezometer A. Excess pore pressure will not dissipate immediately since the water particles have to travel upward toward the sand layer. This would take some time depending upon the permeability of the clay layer.

Piezometer C, D, and E: Piezometers C, D, and E are lower than piezometer B. Hence, more time is required for the dissipation of pore pressure.

Piezometer F: The water level of piezometer F has dropped compared to piezometer E. This is because the water could dissipate to the sand layer below the clay layer.

Piezometer G: The excess pore pressure in piezometer G will dissipate immediately as in the case of piezometer A due to the closeness of the bottom sand layer.

Dissipation of excess pore pressure usually takes months or, in some cases, years to complete. Clays with low permeability would take a long time to complete the primary consolidation. As mentioned earlier, 100% primary consolidation usually would never be achieved.

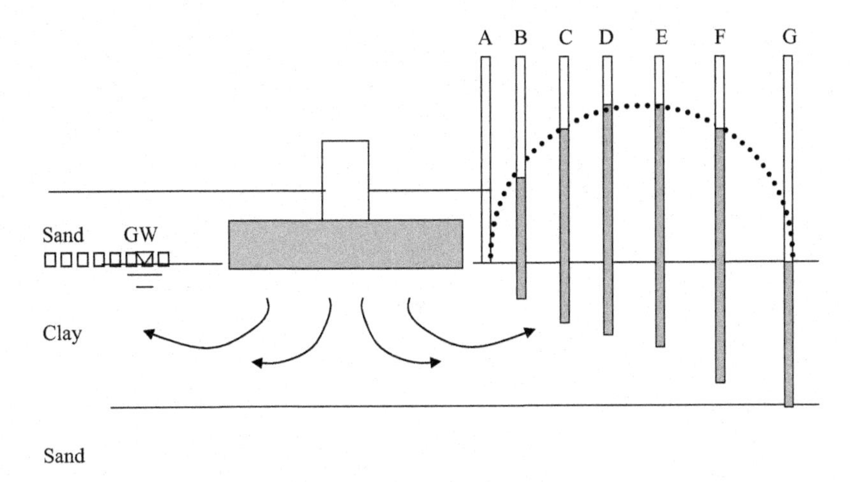

Figure 17.4 Pore pressure distribution near a loaded footing.

17.3 Normally consolidated clays and overconsolidated clays

Settlement due to consolidation occurs in clayey soils. Clay soils usually originate in lakebeds or in ocean floors. Sedimentation of clay particles occurs in calm waters (Figure 17.5).

Years down the road, the lake would dry up and the clay layer would be exposed.

The clay layer shown in Figure 17.6 had never been subjected to any loads other than the load due to the water body, such clays are known as normally consolidated clays. The earth is a dynamic planet. Changes occur in earth surface every day. Hurricanes, tsunamis, landslides, coming and going of glaciers, and volcanoes are some of the events that occur on the earth's surface. During the last ice age, a large area of northern hemisphere was under few hundred feet of ice. When glaciers are formed

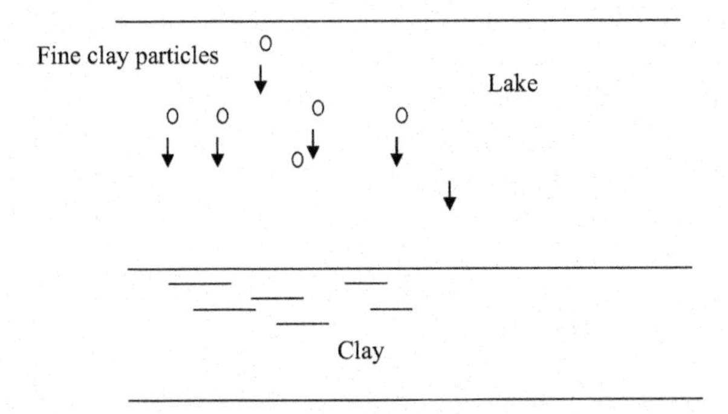

Figure 17.5 Formation of a clay layer due to sedimentation.

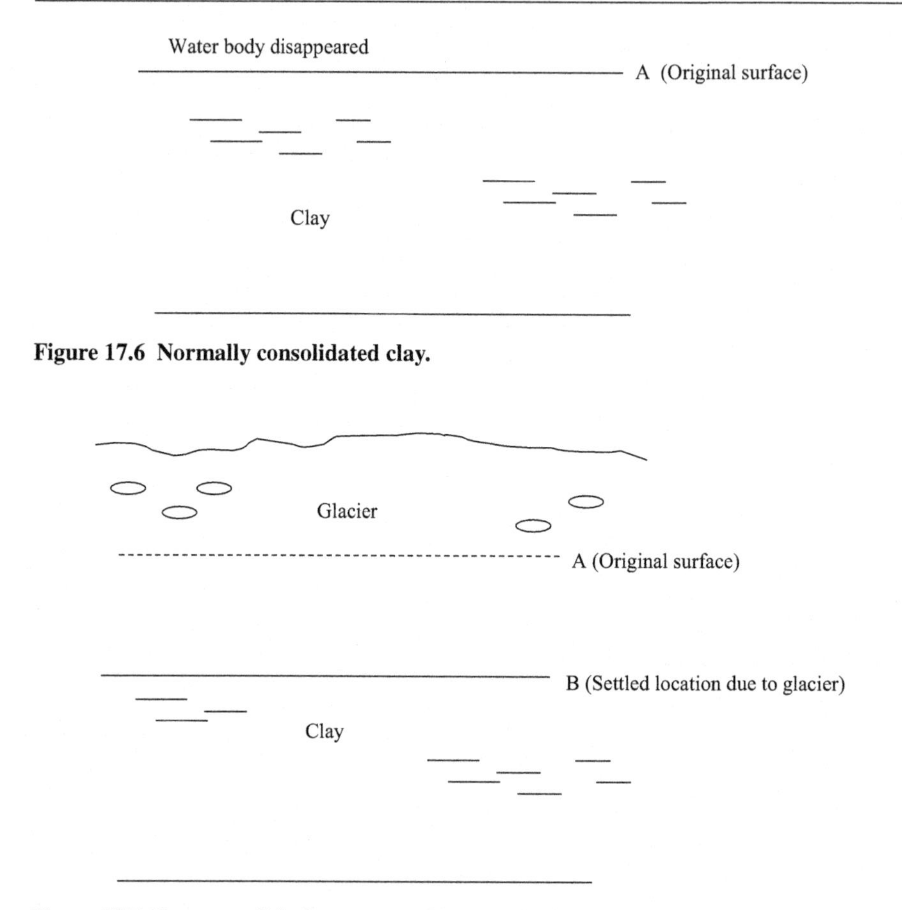

Figure 17.6 Normally consolidated clay.

Figure 17.7 Overconsolidation process. Clay layer would settle from point "A" to point "B" due to the glacier.

during an ice age, the clays in that region would be subjected to a tremendous load. Due to the ice load, the clay would undergo settlement (Figure 17.7).

Glaciers normally melt away and disappear at the end of ice ages. Today, glaciers can be seen only in north and south poles. After the melting of glaciers, the load on clay will be released. Hence, the clay layer would rebound. Clays that have been subjected to high pressures in the past are known as preconsolidated clays or over consolidated clays (Figure 17.8).

Now, let us assume that a footing was placed on the clay layer. The clay would settle due to the footing load (Figure 17.9).

The clay layer would undergo settlement due to the footing load. The new location of the top surface of the clay layer is shown in Figure 17.10.

The void ratio "e" versus log "p" graph is shown in Figure 17.11.

Now, let us assume that the footing load is increased. When the footing load is increased, the clay layer would further settle. If the load were large enough, the clay

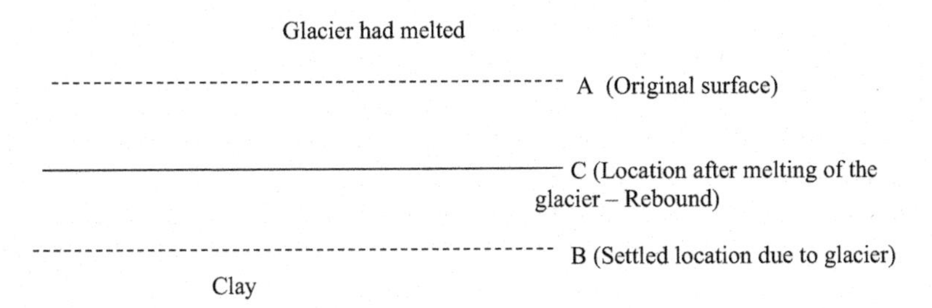

Figure 17.8 Overconsolidated clay. When the glacier is melted, the clay layer will rebound to point "C".

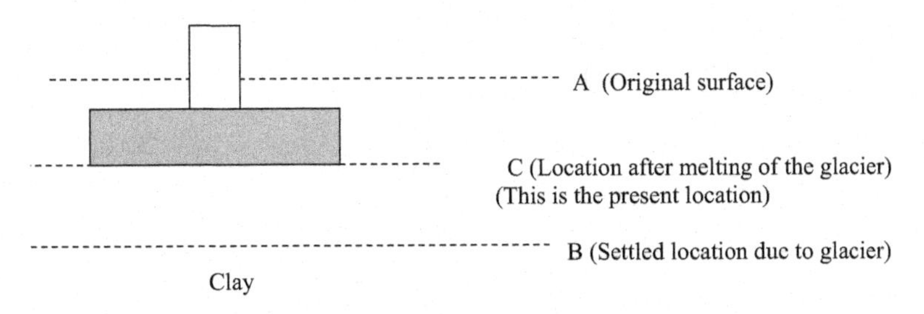

Figure 17.9 Footing load on overconsolidated clay. Footing was placed on top of the clay layer.

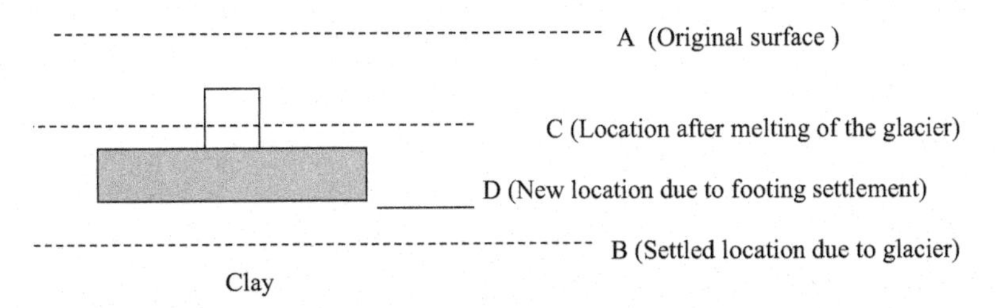

Figure 17.10 Footing settlement. Footing was placed on top of the clay layer.

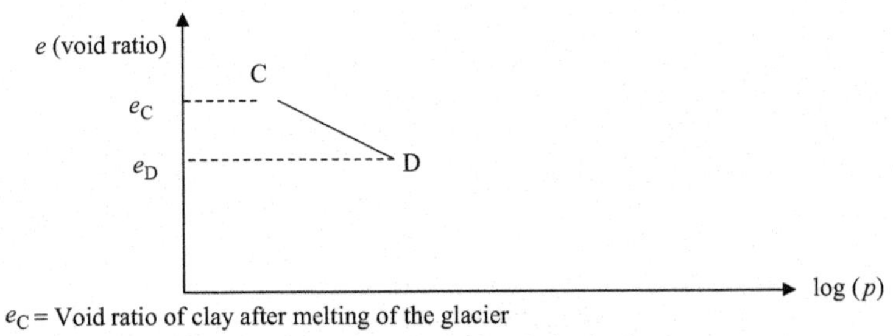

Figure 17.11 Change of void ratio.

layer would settle beyond point “B”. Point “B” is the surface of the clay layer during the period when the glaciers were present (Figure 17.12).

Void ratio “*e*” versus log “*p*” graph is shown in Figure 17.13.

When the load is increased, the void ratio of clay would decrease due to settlement.

In Figure 17.13, when the glacier was present, the void ratio of the clay layer was e_B. When the glacier melted away, the clay layer rebounded. Hence, the voids inside the clay increased. After the melting of the glacier, the void ratio increased from e_B to e_C.

This was the state of the clay layer until a footing was placed there.

When the footing load was increased, the void ratio decreased. As soon as the void ratio goes below e_B (void ratio when the glacier was present), the curve bends.

The top portion of the “*e*” versus log (*p*) curve is known as the recompression curve and the bottom portion is known as the virgin curve.

C to B – Recompression curve

B to D – Virgin curve

The gradient of the virgin curve is steeper than the recompression curve.

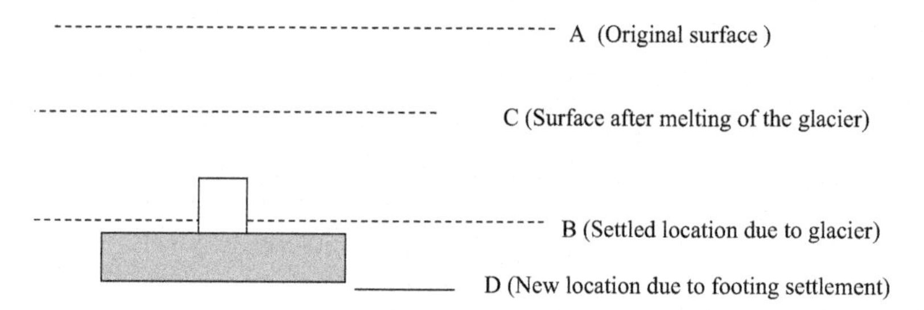

Figure 17.12 Settlement of footing beyond previous level. Settlement of footing beyond point “B”.

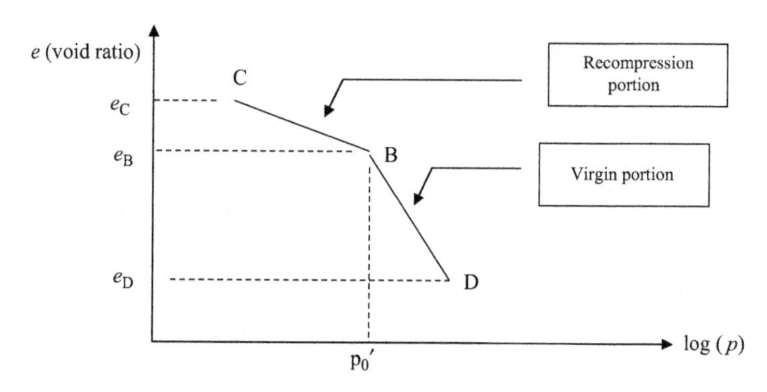

Figure 17.13 Void ratio after consolidation due to footing load.

The gradient of the recompression curve is usually denoted by C_r and the gradient of the virgin curve is denoted by C_c. The settlement is high for a given pressure within the virgin portion of the graph.

The pressure at point "B" is the largest pressure that the clay layer was subjected to prior to the foundation being placed. The pressure at point "B" is known as the preconsolidation pressure and is denoted by p_c'. The present stress prior to the placement of the footing is usually denoted by p_0'.

17.4 Total primary consolidation

When a clay layer is loaded, as we learnt in the previous section, it would start to consolidate and settle (Figure 17.14).

The settlement due to total primary consolidation in normally consolidated clay is given by the following equation. Please note that normally consolidated clays have never been subjected to a higher stress than what it is being subjected to now.

$$\Delta H = H \frac{C_c}{(1+e_0)} \log \frac{(p_0' + \Delta p)}{p_0'}$$

Description of terms: ΔH, total primary consolidation settlement; H, thickness of the clay layer; C_c, compression index of the clay layer; e_0, void ratio of the clay layer at mid point of the clay layer prior to loading; p_0', effective stress at the mid point of the clay layer prior to loading; Δp, increase of stress at the mid point of the clay layer due to the footing.

In Figure 17.15 the void ratio dimensions are given as: e_0, present void ratio; e_f, final void ratio after consolidation due to the footing load; p_0', present vertical effective stress at the mid point of the clay layer; p_f', final vertical effective stress at the mid point of the clay layer after placement of the footing; Δp, $p_f' - p_0'$; C_c, gradient of the curve (compression index).

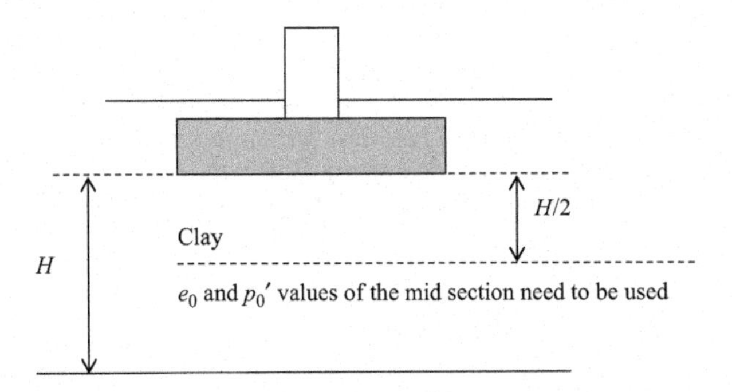

Figure 17.14 Shallow foundation on clay soil (properties of mid section).

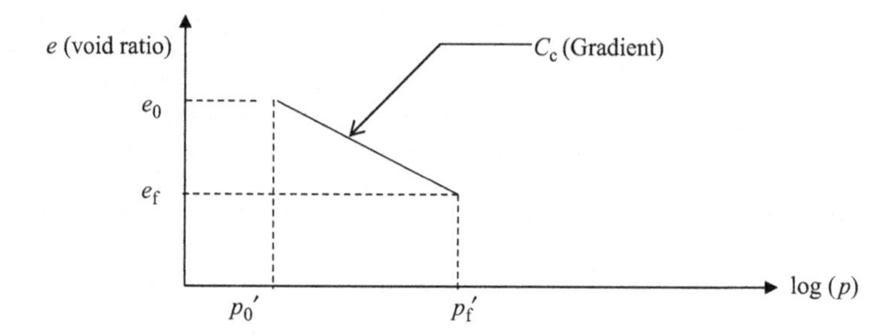

Figure 17.15 Change of void ratio for normally consolidated clay.

Design Example 17.1: Consolidation settlement in normally consolidated clay. No Groundwater

Find the settlement due to consolidation of a 3 m × 3 m column foundation with a load of 200 kN. The foundation is placed 1 m below the top surface and the clay layer is 9 m thick. There is a sand layer underneath the clay layer. The density of the clay layer is 18 kN/m^3, the compression index (C_c) of the clay layer is 0.32, and initial void ratio (e_0) of clay is 0.80. Assume that the pressure is distributed at 2:1 ratio and the clay is normally consolidated (Figure 17.16).

Solution

Step 1: Write down the consolidation settlement equation;

$$\Delta H = H\frac{C_c}{(1+e_0)}\log\frac{(p_0' + \Delta p)}{p_0'}$$

ΔH, total primary consolidation settlement; H, thickness of the compressible clay layer; C_c, compression index of the clay layer; e_0, void ratio of the clay layer at mid point of the clay layer prior to loading; p_0', effective stress at the mid point of the clay layer prior to loading.

Step 2: The clay layer is 9 m thick and the footing is placed 1 m below the surface. The top 1 m is not subjected to consolidation. A clay layer with a thickness of 8 m below the footing is subjected to consolidation due to the footing load.
Find the effective stress at the mid layer of the compressible clay stratum; (p_0');

$$p_0' = \gamma_{clay} 5$$

$$p_0' = 18 \times 5 = 90 \text{ kN/m}^2.$$

Step 3: Find Δp
Δp = increase of stress at the *mid point* of the clay layer due to the footing.
Total load of 100 kN is distributed at a larger area at the mid section of the clay layer.
Area of the mid section of the clay layer = (2 + 3 + 2) × (2 + 3 + 2) = 49 m^2.

$$\Delta p = 200/49 = 4.1 \text{ kN/m}^2.$$

Step 4: Apply values in the consolidation equation.

$$\Delta H = H\frac{C_c}{(1+e_0)}\log\frac{(p_0' + \Delta p)}{p_0'}$$

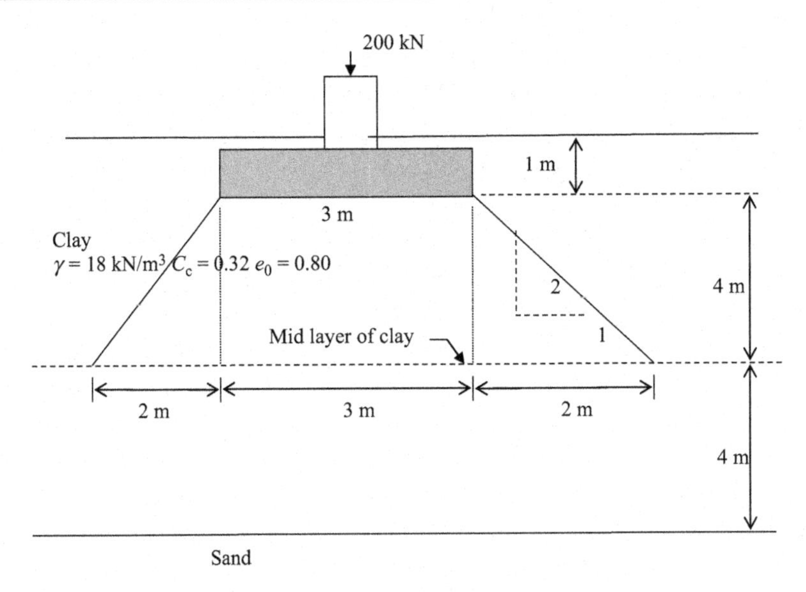

Figure 17.16 Footing on normally consolidated clay layer.

Following parameters are given: C_c, 0.32; e_0, 0.8.

$$\Delta H = 8.\frac{0.32}{(1+0.8)}\log\frac{(90+4.1)}{90}$$

$$\Delta H = 0.0275\ \text{m} \qquad (1\ \text{in.})$$

Note: "H" thickness of the clay layer should be calculated starting from the bottom of the footing. Although the total thickness of the clay layer is 9 m, the first 1 m of the clay layer is not compressed.

In Design Example 17.2, groundwater is considered.

Design Example 17.2: Consolidation settlement in normally consolidated clay. Groundwater is present

Find the settlement due to consolidation of a 3 m × 3 m) column foundation with a load of 180 kN (Figure 17.17). The sand layer is 4 m thick and the clay layer below is 10 m thick. The density of the sand layer is 17.5 kN/m³ and the density of the clay layer is 18 kN/m³. The groundwater is 1.5 m below the surface. The compression index (C_c) of the clay layer is 0.3 and initial void ratio (e_0) of clay is 0.8. Assume that the pressure is distributed at a 2:1 ratio and the clay is normally consolidated.

Solution

Step 1: Write down the consolidation settlement equation;

$$\Delta H = H\frac{C_c}{(1+e_0)}\log\frac{(p_0'+\Delta p)}{p_0'}$$

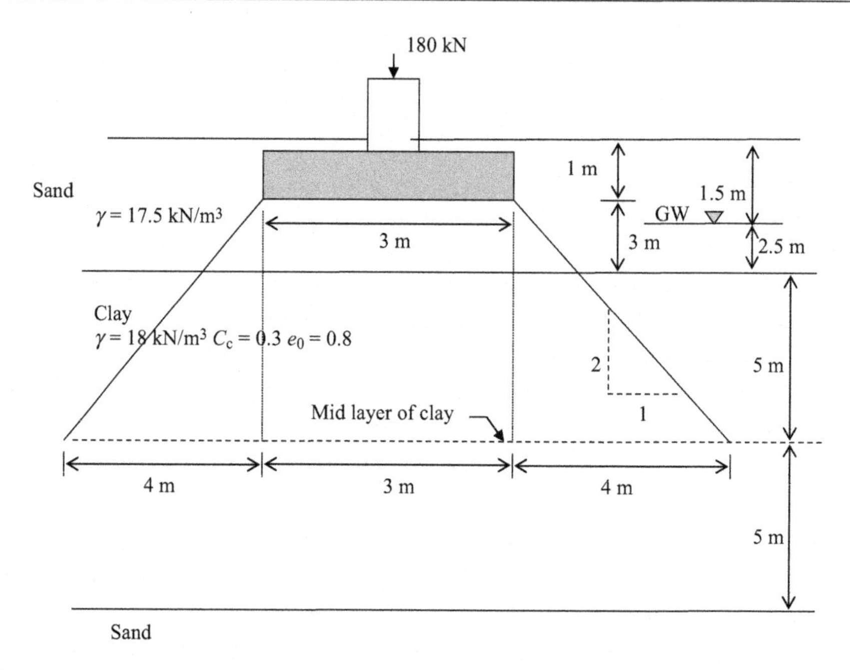

Figure 17.17 Shallow foundation on clay soil (parameters).

ΔH, total primary consolidation settlement; H, thickness of the clay layer; C_c, compression index of the clay layer; e_0, void ratio of the clay layer at mid point of the clay layer prior to loading; p_0', effective stress at the mid point of the clay layer prior to loading.

Step 2: Find the initial effective stress at the mid layer of the clay stratum (p_0');

$$p_0' = \gamma_{sand} \times 1.5 + (\gamma_{sand} - \gamma_{water}) \times 2.5 + (\gamma_{clay} - \gamma_{water}) \times 5$$

$$p_0' = 17.5 \times 1.5 + (17.5 - 9.807) \times 2.5 + (18 - 9.087) \times 5$$

$$p_0' = 86.5 \text{ kN/m}^2.$$

Step 3: Find Δp

Δp = Increase of stress at the mid point of the clay layer due to the footing.

Total load of 180 kN is distributed at a larger area at the mid section of the clay layer.

Area of the mid section of the clay layer = (4 + 3 + 4) × (4 + 3 + 4) = 121 m^2.

$$\Delta p = 180/121 = 1.5 \text{ kN/m}^2.$$

Step 4: Apply values in the consolidation equation.

$$\Delta H = H \frac{C_c}{(1+e_0)} \log \frac{(p_0' + \Delta p)}{p_0'}$$

$$\Delta H = 10 \frac{0.3}{(1+0.8)} \log \frac{(86.5 + 1.5)}{86.5}$$

$$\Delta H = 0.0124 \text{ m} \qquad (0.5 \text{ in.})$$

17.5 Consolidation in overconsolidated clay

The "e" versus log p curve for preconsolidated clay is shown in Figure 17.18.

Description of terms: e_0, initial void ratio; e_c, void ratio at pre consolidation pressure; e_f, final void ratio at the end of the consolidation process; p_0, initial pressure prior to the footing load; p_c', overconsolidation pressure (also known as preconsolidation pressure); p_f', pressure after the footing is placed; C_c, compression index; C_r, recompression index.

$$\Delta H = \underbrace{H\frac{C_r}{(1+e_0)}\log\frac{(p_0'+\Delta p_1)}{p_0'}}_{\substack{\text{Point B to C} \\ \text{Recompression portion}}} + \underbrace{H\frac{C_c}{(1+e_0)}\log\frac{(p_c'+\Delta p_2)}{p_c'}}_{\substack{\text{Point C to D} \\ \text{Virgin compression}}}$$

Point B: Point "B" indicates the present void ratio and the existing vertical effective stress.

Point C: Point "C" indicates the maximum stress that the clay had been subjected to in the past. Void ratio and the past maximum stress have to be obtained by conducting a laboratory consolidation test. The geotechnical engineer should obtain Shelby tube samples and send them to the laboratory to conduct a consolidation test.

Point D: Point "D" indicates the expected stress after the footing is constructed.

Unlike normally consolidated soils, in preconsolidated soils the settlement has to be computed in two parts.

Consolidation settlement from point B to C:

$$\Delta H_1 = H\frac{C_r}{(1+e_0)}\log\frac{(p_0'+\Delta p_1)}{p_0'}$$

$$\Delta p_1 = p_c' - p_0'$$

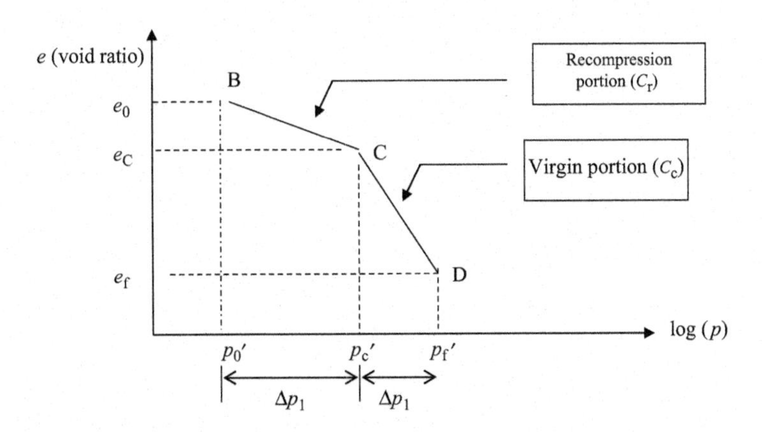

Figure 17.18 "e" versus log p curve for overconsolidated clay.

Consolidation settlement from point C to D:

$$\Delta H_2 = H\frac{C_c}{(1+e_0)}\log\frac{(p_c'+\Delta p_2)}{p_c'}$$

$$\Delta p_2 = p_f' - p_c'$$

Hence, the total settlement is given by

$$\Delta H = \Delta H_1 + \Delta H_2$$

Design Example 17.3: Consolidation settlement in overconsolidated clay. Groundwater not considered

Find the settlement due to consolidation in a 3 m × 3 m column foundation with a load of 270 kN (Figure 17.19). The foundation is placed 1 m below the ground surface and the clay layer is 9 m thick. There is a sand layer underneath the clay layer. The density of the clay layer is 18 kN/m^3 and the compression index (C_c) of the clay layer is 0.32, the recompression index C_r is 0.035, preconsolidation pressure (p_c') is 92 kN/m^2, and the initial void ratio (e_0) of clay is 0.80. Assume that the pressure is distributed at a 2:1 ratio and the clay is normally consolidated.

Solution

Step 1: Write down the consolidation settlement equation;

$$\Delta H = \underbrace{H\frac{C_r}{(1+e_0)}\log\frac{(p_0'+\Delta p_1)}{p_0'}}_{\text{Recompression portion}} + \underbrace{H\frac{C_c}{(1+e_0)}\log\frac{(p_c'+\Delta p_2)}{p_c'}}_{\text{Virgin compression}}$$

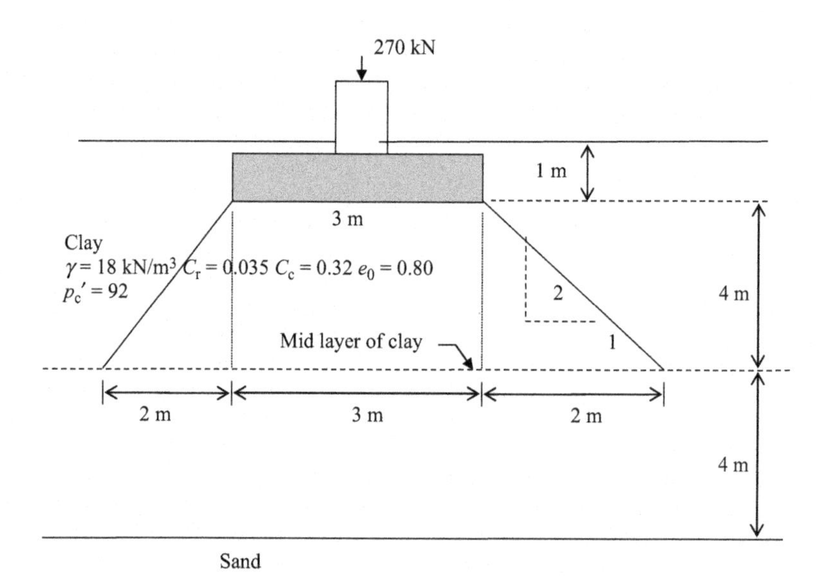

Figure 17.19 Settlement in overconsolidated clay.

ΔH, total primary consolidation settlement; H, thickness of the compressible clay layer; C_c, compression index of the clay layer; C_r, recompression index of the clay layer; e_0, void ratio of the clay layer at mid point of the clay layer prior to loading; p_0', effective stress at the mid point of the clay layer prior to loading; p_c', pre consolidation pressure; p_f', pressure after the footing is placed.

Step 2: The clay layer is 9 m thick and the footing is placed 1 m below the surface. The top 1 m is not subjected to consolidation. A clay layer with a thickness of 8 m below the footing is subjected to consolidation due to footing load.
Find the effective stress at the mid layer of the compressible clay stratum; (p_0');

$$p_0' = \gamma_{clay} \times 5$$

$$p_0' = 18 \times 5 = 90 \text{ kN/m}^2.$$

Step 3: Find Δp
Δp = Increase of stress at the *mid point* of the clay layer due to the footing.
Total load of 270 kN is distributed at a larger area at the mid section of the clay layer.
Area of the mid section of the clay layer = (2 + 3 + 2) × (2 + 3 + 2) = 49 m^2.

$$\Delta p = 270/49 = 5.5 \text{ kN/m}^2.$$

Existing stress (p_0') = 90 kN/m^2.
Preconsolidation stress = 92 kN/m^2

$$\text{Final pressure after foundation load } (p_f') = p_0' + \Delta p = 90 + 5.5 = 95.5 \text{ kN/m}^2$$

Step 4: Apply values in the consolidation equation.

$$\Delta H = H\frac{C_r}{1+e_0}\log\frac{p_0'+\Delta p_1}{p_0'} + H\frac{C_c}{(1+e_0)}\log\frac{p_c'+\Delta p_2}{p_c'}$$

p_0' = Effective stress at the mid point of the clay layer prior to loading = 90 kN/m^2.
(Δp_1) = Stress increase from initial stress (P_0') to preconsolidation pressure P_c'. Since p_0' is 90 and P_c' was given to be 92, Δp_1 would be 2 kN/m^2.
(Δp_2) = Stress increase from preconsolidation pressure (p_c') to final pressure P_f'. Since p_c' is 92 and final pressure (P_f') was found to be 95.5, Δp_2 would be 3.5 kN/m^2.

$$\Delta H = 8\frac{0.035}{1+0.8}\log\frac{90+2}{90} + 8\frac{0.32}{1+0.8}\log\frac{92+3.5}{92}$$

Figure 17.20 is a line diagram.

17.6 Computation of time for consolidation

In Chapter 16, we studied how to compute the total settlement due to primary consolidation. In this chapter, we would discuss how to compute the time taken for the consolidation process.

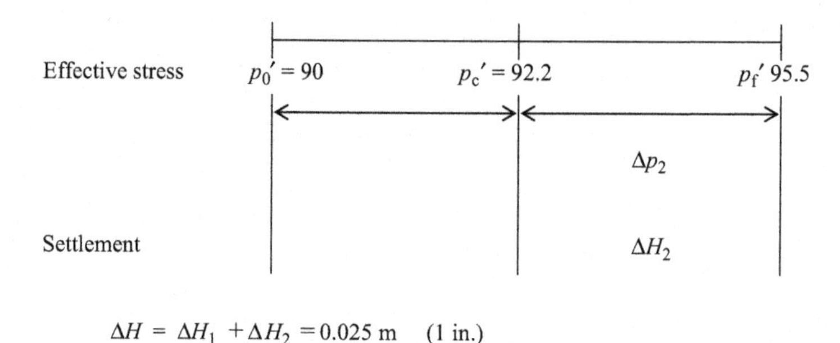

$$\Delta H = \Delta H_1 + \Delta H_2 = 0.025 \text{ m} \quad (1 \text{ in.})$$

Figure 17.20 Line diagram.

Table 17.1 ***U*% and T_v**

***U*% (percent consolidation)**	**T_v**
0	0.00
10	0.048
20	0.090
30	0.115
40	0.207
50	0.281
60	0.371
70	0.488
80	0.652
90	0.933
100	1.0 (approximately)

Time taken for primary consolidation is given by the following equation.

$$t = \frac{H^2 T_v}{c_v}$$

t, time taken for the consolidation process; H, thickness of the drainage layer (explained later); T_v, time coefficient; U%, percent consolidation.

T_v can be obtained from Table 17.1.

17.7 Drainage layer (H)

The thickness of the drainage layer is defined as the longest path that a water molecule has to take for drainage (Figure 17.21).

In the case of single drainage, drainage can occur only from one side. In Figure 17.21, water can not drain through the rock. Water can drain from the sand layer on top. In this case, the thickness of the drainage layer = H.

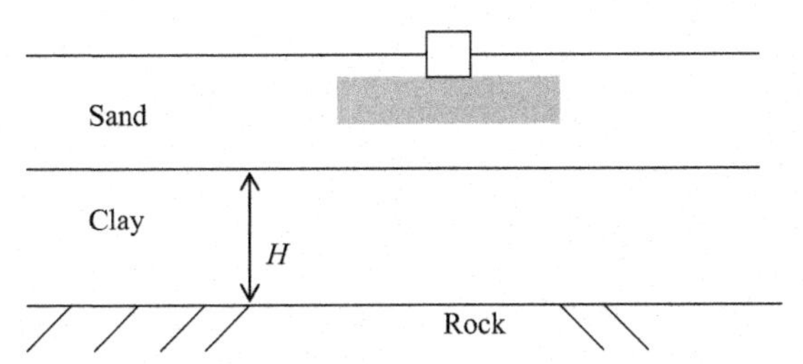

Figure 17.21 Thickness of the drainage layer (single drainage).

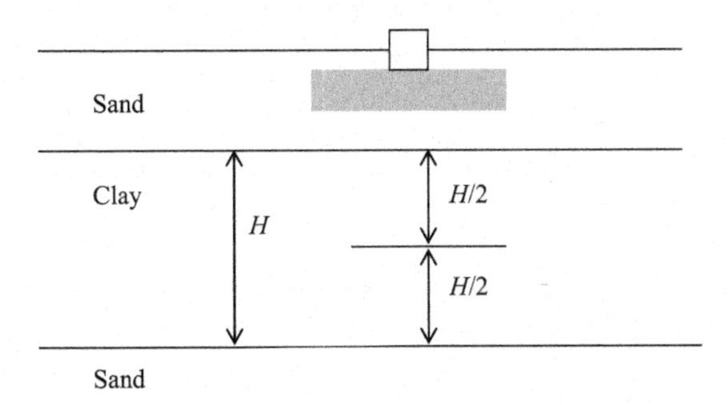

Figure 17.22 Thickness of the drainage layer (double drainage).

In Figure 17.22, the water molecule has the opportunity to either drain from the top or bottom. The longest drainage path for a water molecule is only *H*/2.

Design Example 17.4

Find the approximate time taken for 100% consolidation in the clay layer shown in Figure 17.23.

Solution

Step 1: Time taken for consolidation is given by

$$t = \frac{H^2 T_v}{c_v}$$

t, time taken for the consolidation process; *H*, thickness of the drainage layer; T_v, time coefficient.

The approximate T_v value for 100% consolidation is 1.0. (Table 17.1)

c_v is given to be 0.011 m²/day.

In Figure 17.23, the clay layer can drain from top and bottom. From the top it can drain to the surface and from bottom it can drain to the sand layer below. Hence, "*H*" should be half the thickness of the clay layer.

$$H = 2\,\text{m}.$$

$$t = \frac{2^2 \times 1.0}{0.011} = 363\ \text{days}$$

We can reasonably assume that all the settlement due to clay consolidation has occurred after 363 days.

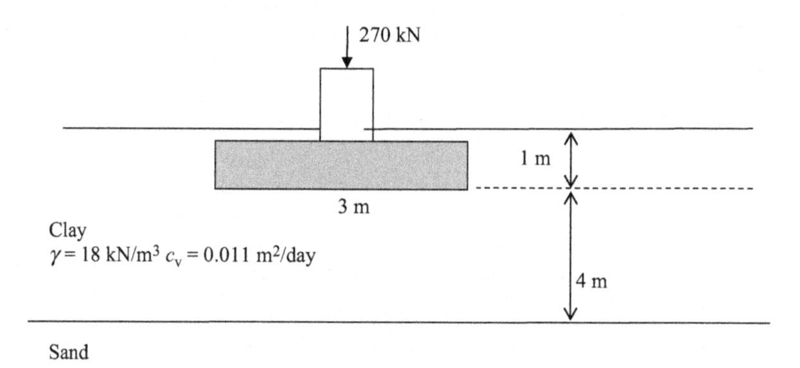

Figure 17.23 Consolidation in clay (double drainage).

Design Example 17.5

Find the settlement after 1 year in the clay layer shown in Figure 17.24 for the 3 m × 3 m column footing. Groundwater is 1.5 m below the surface. The following parameters are given for the clay layer: $\gamma = 18$ kN/m^3, $C_r = 0.032$, $C_c = 0.38$, $e_0 = 0.85$, $p_c' = 50$ kPa, $c_v = 0.011$ m^2/day.

Description of terms: ΔH, total primary consolidation settlement; H, thickness of the drainage layer; C_c, compression index of the clay layer in the virgin zone; C_r, recompression index of the clay layer in the recompression zone; e_0, void ratio of the clay layer at *mid point* of the clay layer prior to loading; p_0', effective stress at the mid point of the clay layer prior to loading; p_c', preconsolidation pressure; p_f', pressure after the footing is placed; c_v, consolidation coefficient.

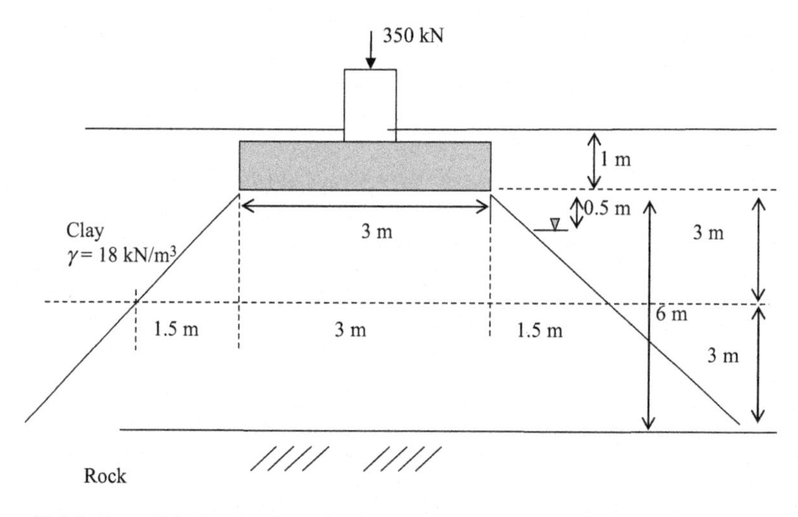

Figure 17.24 Consolidation in clay (single drainage).

Solution

Step 1: The first step is to find the percent consolidation ($U\%$) after 1 year.

$$t = \frac{H^2 T_v}{c_v}$$

$t = 1$ year = 365 days

H = Thickness of the drainage layer. In this case, drainage is possible only from top. Drainage is not possible from the bottom due to bedrock. This is a single drainage situation. Hence, $H = 6$ m. It should be mentioned here that in reality, the longest drainage path is 7 m, since the footing is placed 1 m below the surface.

T_v = Time coefficient.

Find T_v

$$T_v = \frac{c_v t}{H^2} = \frac{0.011 \times 365}{6^2} = 0.111$$

Step 2: From $U\%$ versus T_v (Table 17.1);

for ($U\%$) = 20%, $T_v = 0.009$,

for ($U\%$) = 30%, $T_v = 0.115$,

Through interpolation, find the $U\%$ for $T_v = 0.111$

$$\frac{U-20}{0.111-0.009} = \frac{30-20}{0.115-0.009}$$

$U = 29.6$, for $T_v = 0.111$

When t (time) = 1 year = $T_v = 0.111$, and $U\% = 29.6$

Step 3: Find the total settlement due to consolidation.

$$\Delta H = H\frac{C_r}{1+e_0}\log\frac{p_0' + \Delta p_1}{p_0'} + H\frac{C_c}{1+e_0}\log\frac{p_c' + \Delta p_2}{p_c'}$$

The clay layer is 7 m thick and the footing is placed 1 m below the surface. The top 1 m is not subjected to consolidation. The compressible portion of the clay layer is 6 m thick.

Find the effective stress at the mid layer of the compressible clay stratum (p_0');

$$p_0' = 1.5 \times \gamma_{clay} + 2.5 \times (\gamma_{clay} - \gamma_w)$$
$$p_0' = 1.5 \times 18 + 2.5 \times (18 - 9.807) = 47.5 \text{ kN/m}^2.$$

Step 4: Find Δp

Δp = Increase of stress at the mid point of the clay layer due to the footing.

Total load of 350 kN is distributed at a larger area at the mid section of the clay layer.

Area of the mid section of the clay layer = (3 + 1.5 + 1.5) × (3 + 1.5 + 1.5) = 36 m^2

$\Delta p = 350/36 = 9.72$ kN/m^2.

Step 5: Find the final pressure after application of the footing load;

p_f' = Initial pressure + pressure increase due to footing load at mid point

$p_f' = p_0' + \Delta p = 47.5 + 9.72 = 57.22$ kN/m^2.

Step 6: Apply values in the consolidation equation.

$$\Delta H = H\frac{C_r}{1+e_0}\log\frac{p_0' + \Delta p_1}{p_0'} + H\frac{C_c}{1+e_0}\log\frac{p_c' + \Delta p_2}{p_c'}$$

(Δp_1) = Stress increase from initial stress (p_0') to preconsolidation pressure p_c'.
Since p_0' is 47.5 and p_c' was given to be 50 kN/m^2, Δp_1 would be (50 – 47.5) kN/m^2.

$$\Delta p_1 = 2.5 \text{ kN/m}^2.$$

(Δp_2) = Stress increase from preconsolidation pressure p_c' to final pressure p_f'.
Since p_c' is 50 kN/m^2 and p_f' *is* was found to be 57.2, Δp_2 would be (57.22 – 50) kN/m^2.

$$\Delta p_2 = 7.22 \text{ kN/m}^2.$$

$$\Delta H = 6\frac{0.032}{1+0.85}\log\frac{p_0' + \Delta p_1}{p_0'} + 6\frac{0.38}{1+0.85}\log\frac{p_c' + \Delta p_2}{p_c'}$$
$$\Delta H = 0.1037\log\frac{47.5+2.5}{47.5} + 1.23\log\frac{50+7.22}{50}$$
$$\Delta H = 0.0023 + 0.072$$
$$\Delta H = 0.0743 \text{ m} \qquad (2.92 \text{ in.}).$$

Step 7: Find the settlement after 1 year;
Percent consolidation (U%) after 1 year is 29.6%. (see Step 2)
Settlement after 1 year = total settlement × percent consolidation
Settlement after 1 year = 0.0743 × 0.296 = 0.022 m (0.87 in.)

Secondary compression

18

After the primary consolidation process is completed, the secondary compression process starts. Secondary compression occurs due to the rearrangement of soil particles. This process is different from primary consolidation process. Primary consolidation occurs due to the dissipation of excess pore pressure (Figures 18.1 and 18.2).

Secondary compression is given by the following equation.

$$S = C_a/(1+e_0)L\times\log(t/t_p)$$

S, secondary compression; C_a, secondary compression coefficient; e_0, initial void ratio; L, thickness of the clay layer; t, time elapsed; t_p, time for completion of primary consolidation (t and t_p should be in same units).

Design Example 18.1

A 6 m thick clay layer is subjected to a loading. The secondary compression coefficient for the clay layer is 0.019. Computations show the primary consolidation to be completed in 10 years. The initial void ratio of the clay layer is 2.12. Find the secondary compression after 35 years.

Solution

Step 1: Gather all the data.

S, secondary compression; C_a, secondary compression coefficient = 0.019; e_0, initial void ratio = 2.12; L, thickness of the clay layer = 6 m; t, time elapsed = 35 years; t_p, time for completion of primary consolidation = 25 years.

Step 2: Apply the equation for secondary compression.

$$S = C_a/(1+e_0)L\times\log(t/t_p)$$
$$S = 0.019/(1+2.12)\ 6\log(35/25) = 0.0053\ \text{m} = 0.53\ \text{cm}$$

Design Example 18.2

Figure 18.3 shows a 3 m × 3 m column footing. The groundwater is 1.5 m below the surface. The following parameters are given for the clay layer: $\gamma = 18$ kN/m^3, $C_r = 0.032$, $C_c = 0.38$, $e_0 = 0.85$, $p_c' = 50$ kPa, $c_v = 0.011$ m^2/day, $C_a = 0.013$.

Find the total settlement after 10 years.

ΔH, total primary consolidation settlement; H, thickness of the drainage layer; C_c, compression index of the clay layer in the virgin zone; C_r, recompression index of the clay layer in the recompression zone; e_0, void ratio of the clay layer at *midpoint* of the clay layer prior to loading; p_0', effective stress at the midpoint of the clay layer prior to loading; p_c', preconsolidation pressure; p_f', pressure after the footing is placed; c_v, consolidation coefficient = 0.011 m^2/day; C_a, secondary compression coefficient.

Geotechnical Engineering Calculations and Rules of Thumb

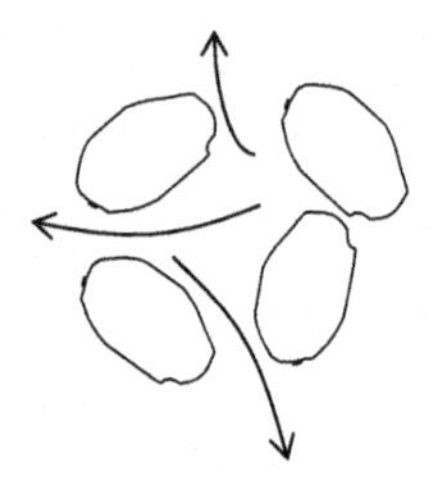

Figure 18.1 Primary consolidation is due to expulsion of water particles.

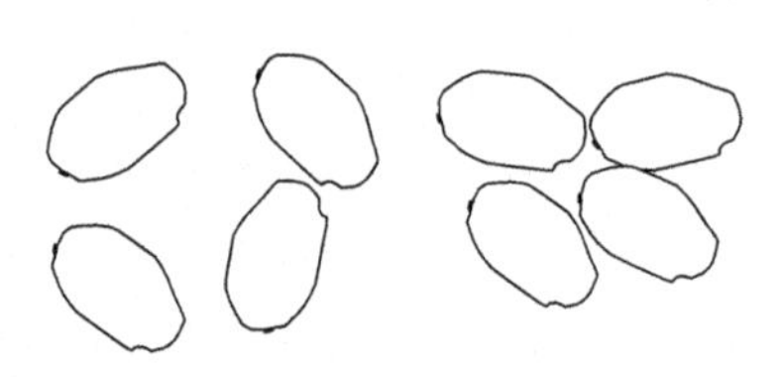

Figure 18.2 Secondary compression is due to rearrangement of soil particles.

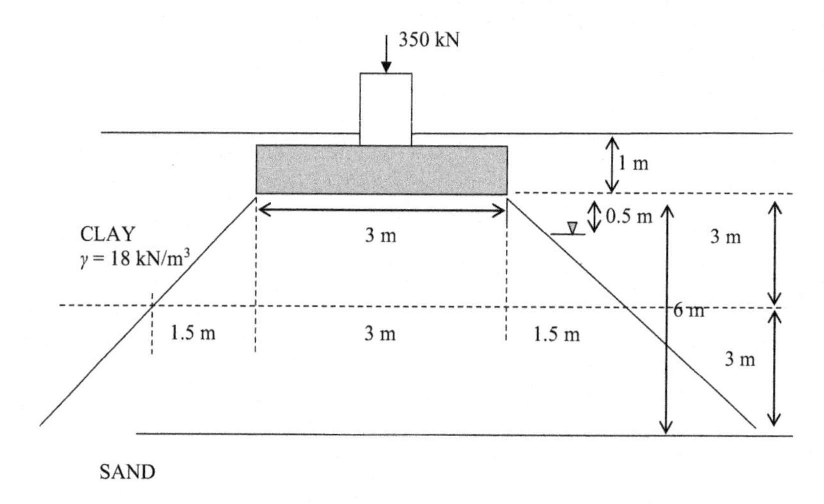

Figure 18.3 Consolidation in clay (single drainage).

Step 1: Find the total settlement due to consolidation.

$$\Delta H = H\frac{C_r}{1+e_0}\log\frac{p_0' + \Delta p_1}{p_0'} + H\frac{C_c}{1+e_0}\log\frac{p_c' + \Delta p_2}{p_c'}$$

The clay layer is 7 m thick and the footing is placed 1 m below the surface. The top 1 m is not subjected to consolidation. The compressible portion of the clay layer is 6 m thick.

Find the effective stress at the mid layer of the compressible clay stratum; (p_0');

$$p_0' = 1.5 \times \gamma_{clay} + 2.5 \times (\gamma_{clay} - \gamma_w)$$
$$p_0' = 1.5 \times 18 + 2.5 \times (18 - 9.807) = 47.5 \text{ kN/m}^2.$$

Step 2: Find Δp.

Δp = Increase of stress at the midpoint of the clay layer due to the footing.

Total load of 350 kN is distributed at a larger area at the mid section of the clay layer.

Area of the mid section of the clay layer = (3 + 1.5 + 1.5) × (3 + 1.5 + 1.5) = 36 m^2

$$\Delta p = 350/36 = 9.72 \text{ kN/m}^2.$$

Step 3: Find the final pressure after application of the footing load.

$p_f{}'$ = Initial pressure + pressure increase due to footing load at mid point

$p_f{}' = p_0{}' + \Delta p = 47.5 + 9.72 = 57.22\ \text{kN/m}^2$.

Step 4: Apply values in the consolidation equation.

$$\Delta H = H\frac{C_r}{1+e_0}\log\frac{p_0{}'+\Delta p_1}{p_0{}'} + H\frac{C_c}{1+e_0}\log\frac{p_c{}'+\Delta p_2}{p_c{}'}$$

(Δp_1) = Stress increase from initial stress ($p_0{}'$) to preconsolidation pressure $p_c{}'$.
Since $p_0{}'$ is 47.5 and $p_c{}'$ was given to be 50 kN/m^2, Δp_1 would be (50 − 47.5) kN/m^2.

$$\Delta p_1 = 2.5\ \text{kN/m}^2.$$

(Δp_2) = Stress increase from preconsolidation pressure $p_c{}'$ to final pressure $p_f{}'$.
Since $p_c{}'$ is 50 kN/m^2 and $p_f{}'$ *is* found to be 57.2, Δp_2 would be (57.22 − 50) kN/m^2.

$$\Delta p_2 = 7.22\ \text{kN/m}^2.$$

$$\Delta H = 6\frac{0.032}{1+0.85}\log\frac{p_0{}'+\Delta p_1}{p_0{}'} + 6\frac{0.38}{1+0.85}\log\frac{p_c{}'+\Delta p_2}{p_c{}'}$$
$$\Delta H = 0.1037\log\frac{47.5+2.5}{47.5} + 1.23\log\frac{50+7.22}{50}$$
$$\Delta H = 0.0023 + 0.072$$
$$\Delta H = 0.0743\ \text{m} \qquad (2.92\ \text{in.}).$$

Step 5: Find the time required for the completion of the primary consolidation;

Typically, 90% consolidation is assumed to be the end of primary consolidation.

$$t = \frac{H^2 T_v}{C_v}$$

$U = 90\%$;

From Table 17.1, find the T_v value.

$T_v = 0.933$

H = Thickness of the drainage layer. In this case, drainage is possible from both top and bottom. Sand layer is seen below the clay layer. This is a double drainage situation. Hence, $H = 3$ m.

Therefore, $t = 3^2 \times 0.933/0.011 = 763.4$ days

Since C_v is given in m^2/day, t will be in days.

Step 6: Find the settlement due to secondary compression.

$$S = C_a/(1+e_0)L \times \log(t/t_p)$$

S, secondary compression; C_a, secondary compression coefficient = 0.013; e_0, initial void ratio = 0.85; L, thickness of the clay layer = 6 m; t_p, time for completion of primary consolidation = 763.4 days (found in Step 5); t, time elapsed = 10 years − 763.4 days = 10 × 365 − 763.4 = 2886.6 days.

In this problem, we need to find the settlement after 10 years.

Primary consolidation starts from day 0 and ends on day 763.4.

Secondary compression starts from day 763.4.

$$S = C_a/(1+e_0)L \times \log(t/t_p)$$
$$S = 0.013/(1+0.85)6\log(2886.6/763.4) = 0.0243\ \text{m}$$

Total settlement = Total settlement due to primary consolidation + Total settlement due to secondary compression
Total settlement = 0.0743 + 0.0243 = 0.0986 m = 9.86 cm.

Seismic design of shallow foundations

19

Shallow footings could fail during earthquake events. Reduction of bearing capacity due to earthquakes was studied by Kumar and Mohan Rao (2002). They proposed to reduce Terzaghi's bearing capacity factors (N_c, N_q, and N_γ) based on horizontal ground acceleration. High ground accelerations would provide high reductions.

The following charts were provided by Kumar and Rao. Figures 19.1–19.3 show, bearing capacity factors plotted against horizontal earthquake acceleration coefficient $\alpha_{h.}$

Horizontal earthquake acceleration coefficient α_h = horizontal ground acceleration/g

g = gravitational acceleration = 9.81 m/s^2.

In Table 19.1 potential damage and α_h value has been discussed.

If you look at Figures 19.1–19.3, when α_h reaches 1.0, bearing capacity factors reach zero. This indicates severe loss of bearing capacity at this level of earthquake.

Design Example 19.1: Column footing in a homogeneous sand layer

Find the ultimate bearing capacity of a 1.5 m × 1.5 m square footing placed in a sand layer. The density of the soil is found to be 18.1 kN/m^3 and friction angle to be 30°. Foundation has to be designed for a horizontal ground acceleration of 1.6 m/s^2 (Figure 19.4).

Solution

Find α_h;

$\alpha_h = 1.6/g = 1.6/9.81 = 0.16$

Find the Terzaghi's bearing factors N_c, N_q and N_γ using an α_h value of 0.16.

$N_c = 32$

$N_q = 15$

$N_\gamma = 7$

Step 1: $q_{ult} = cN_c s_c + qN_q + 0.5BN_\gamma \gamma s_\gamma$

Step 2: Find shape factors using Table 8.2.

For a square footing $s_c = 1.3$ and $s_\gamma = 0.8$

Step 3: Find the surcharge (q).

$$q = \gamma d = 18.1 \times 1.2 = 21.7\ \text{kPa}.$$

Step 4: Apply the Terzaghi's bearing capacity equation.

$$q_{ult} = cN_c s_c + qN_q + 0.5BN_\gamma \gamma s_\gamma$$
$$q_{ult} = 0 \times 32 \times 1.3 + 21.7 \times 15 + 0.5 \times 1.5 \times 19.7 \times 7 \times 0.8$$
$$q_{ult} = 408.2\ \text{kPa}$$

Allowable bearing capacity ($q_{allowable}$) = q_{ult} /FOS = q_{ult} /3.0 = 136.1 kPa.

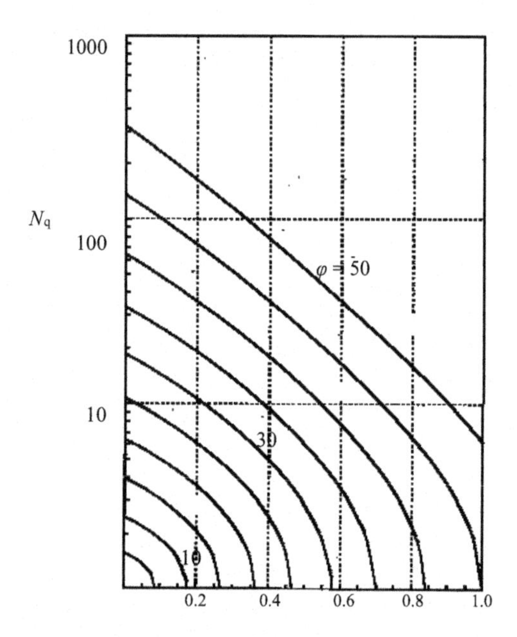

Figure 19.1 Horizontal earthquake acceleration coefficient α_h versus N_q.

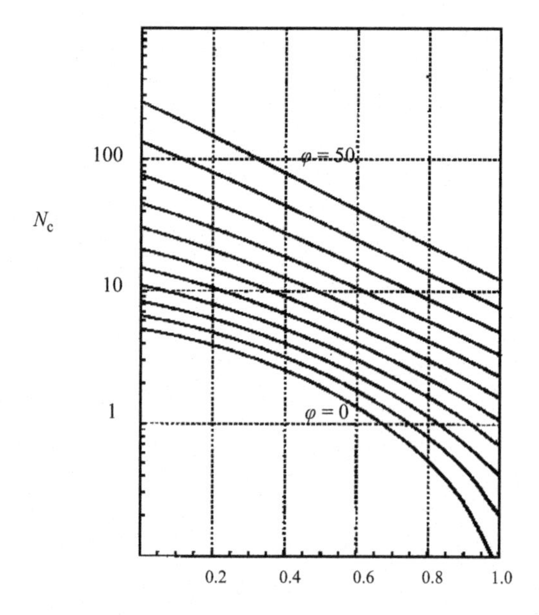

Figure 19.2 Horizontal earthquake acceleration coefficient α_h versus N_c.

FOS = factor of safety

Total load ($Q_{allowable}$) that could safely placed on the footing;

$$Q_{allowable} = q_{allowable} \times \text{area of the footing} = q_{allowable} \times (1.5 \times 1.5) = 306 \text{ kN}.$$

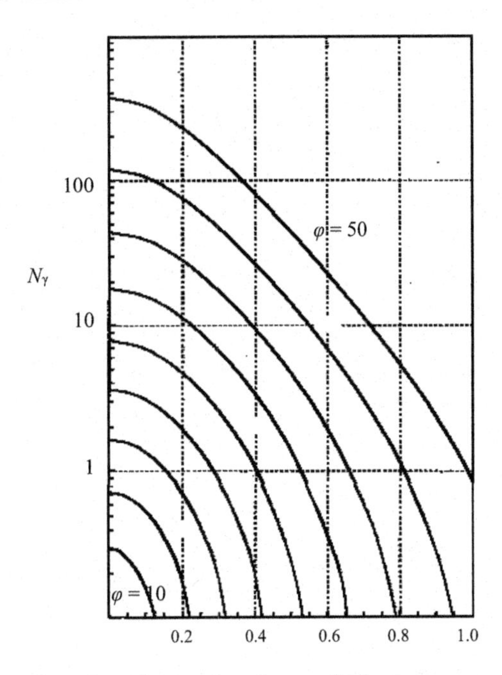

Figure 19.3 Horizontal earthquake acceleration coefficient α_h versus N_γ.

Table 19.1 **Potential damage and α_h value**

α_h value	Potential damage
0.04	Felt by people
0.25	Moderate damage to buildings
0.50	Considerable damage to buildings
1.0	Severe damage to buildings

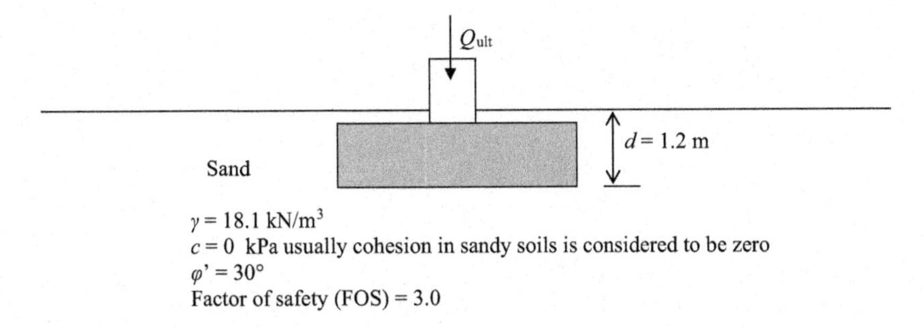

Figure 19.4 Column footing in a homogeneous sand layer.

19.1 Selection of α_h value for a given city

Let us say that a certain city has following probabilities.

Occurrence of α_h value of 1.0 = every 200 years

Occurrence of α_h value of 0.5 = every 90 years

Occurrence of α_h value of 0.25 = every 50 years

Occurrence of α_h value of 0.1 = every 20 years

If you are designing a hospital, then you should design the footing for an α_h value of 1.0. On the other hand, if you are designing a warehouse, α_h value of 0.1 or 0.25 may be sufficient. In the case of warehouse, loss of life would be minimal compared to a hospital. Normally local codes would provide guidance in selecting the design α_h value.

Reference

Kumar, J., Mohan Rao, V.B.K., 2002. Seismic bearing capacity factors for spread foundations. Géotechnique 52 (2), 79–88.

Part 3

Earth Retaining Structures

Earth retaining structures

20

20.1 Introduction

Earth retaining structures are an essential part of civil engineering. Retaining walls are everywhere, yet only few people notice them. Just take a drive for 20 min and count how many retaining walls you would see.

There are many types of retaining walls (Figure 20.1).

In the case of gravity walls, pure weight of the retaining wall holds the soil. Gravity walls are made of rock, concrete, and masonry. Presently concrete has been the most common material to construct gravity walls (Figures 20.2 and 20.3).

Gabions are baskets filled with rocks. These rock baskets can be used to construct retaining walls. Gabion walls are designed as gravity walls. Easy drainage through gabions is a major advantage. Gabion walls are in most cases cheaper than concrete gravity walls.

20.2 Water pressure distribution

Before we discuss the horizontal force due to soil, let us look at the horizontal force due to water.

Water pressure is same in all directions since it is a liquid. Vertical stress at a point inside water is same as the horizontal stress at that location (Figure 20.4).

$$P_h = P_v = \gamma_w h$$

P_h, horizontal pressure; P_v, vertical pressure; γ_w, density of water (usually taken as 62.4 pcf or 9.81 kN/m^3); h, depth to the point of interest.

Figure 20.5 shows the water pressure on a dam.

$$P_h = \text{horizontal pressure} = \gamma_w h$$
$$P_v = \text{vertical pressure} = \gamma_w h$$

In the case of water, horizontal pressure is equal to vertical pressure.

$$\gamma_w = \text{density of water} = 9.81\ \text{kN/m}^3 \qquad (62.4\ \text{pcf})$$

Total force (F) = area of the triangle

$$F = \gamma_w hh/2 = \gamma_w h^2/2$$

Geotechnical Engineering Calculations and Rules of Thumb

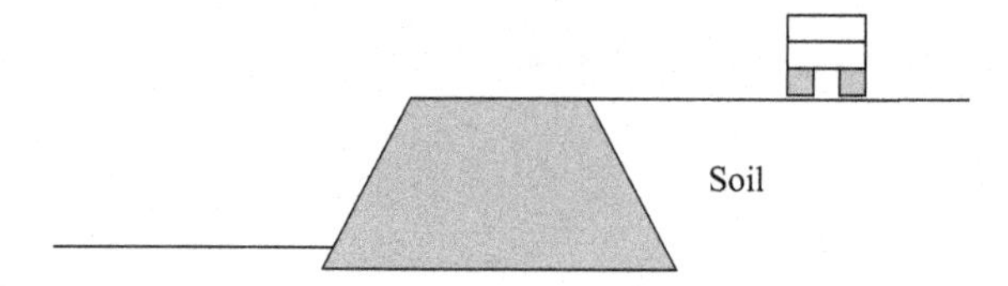

Figure 20.1 Gravity walls.

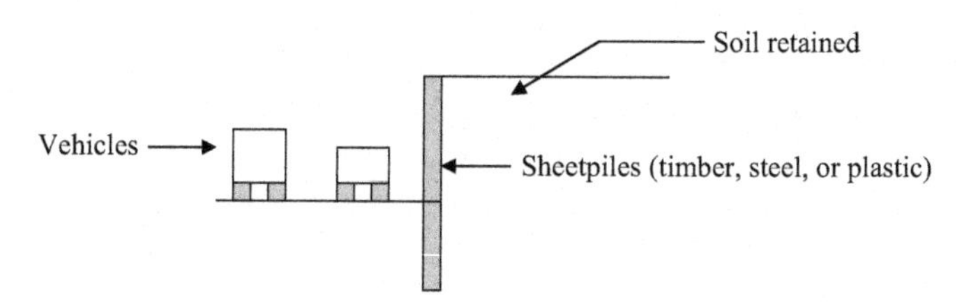

Figure 20.2 Sheetpile walls.

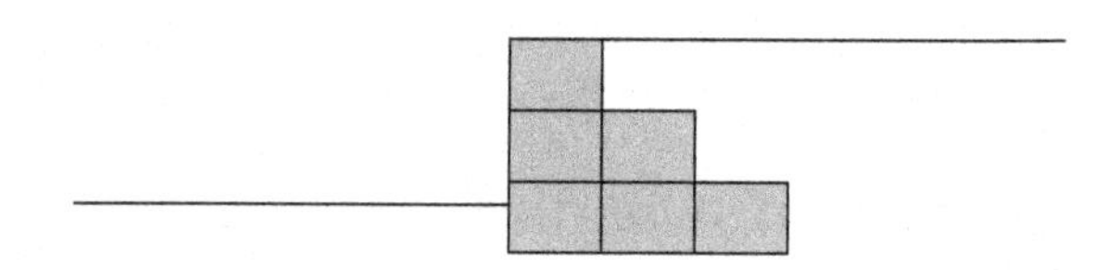

Figure 20.3 Gabion walls.

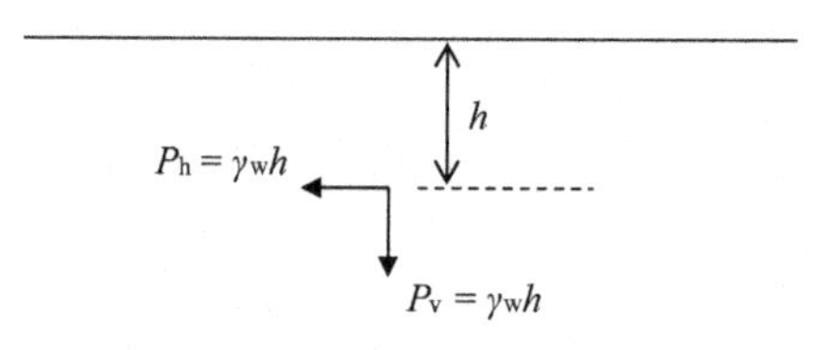

Figure 20.4 Pressure in water.

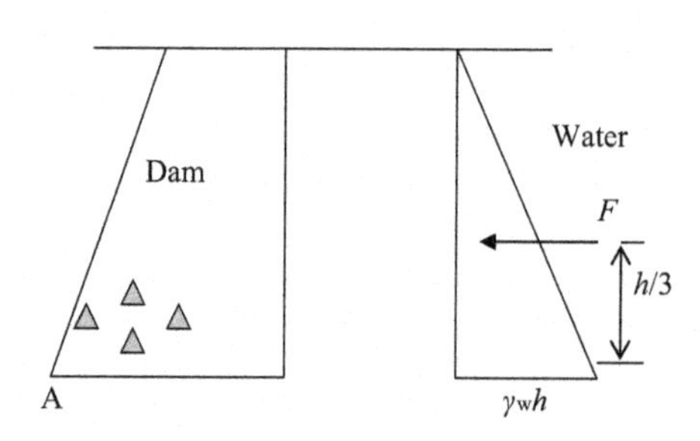

Figure 20.5 Water pressure on a dam.

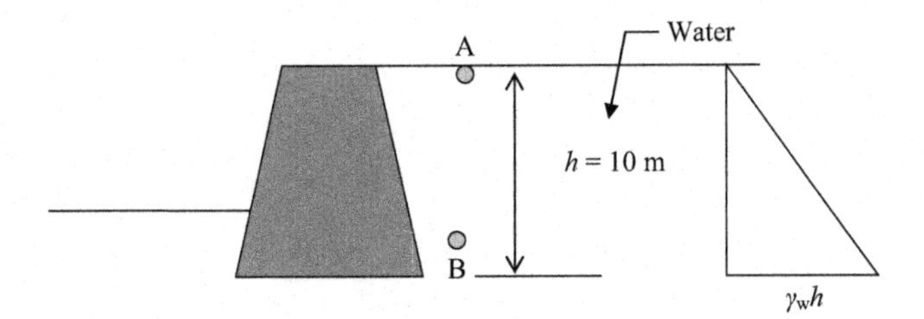

Figure 20.6 Dam subjected to water pressure.

The moment around point, "A" can be computed.

Total moment (M) = force × distance to the force

$$M = \gamma_w h^2/2 \times h/3 = \gamma_w h^3/6$$

The resultant pressure of the triangle acts $h/3$ distance from the bottom.

Design Example 20.1

Find the horizontal force and the overturning moment of the dam shown in Figure 20.6.

Solution

$$\gamma_w h = 10 \times 9.81 = 90.81\,\text{kN/m}^2.$$

Total pressure acting on the dam due to water is obtained by computing the area of the pressure triangle.

Total force acting on the dam = area of the pressure triangle $= \frac{1}{2} \times 10 \times 90.81$
$= 454.1\,\text{kN}$

In the case of water, pressure in all directions is the same.

20.2.1 Computation of horizontal pressure in soil

Following equations are used to compute the horizontal pressure in soils.

Vertical pressure in soil = density of soil × depth = $\gamma \times h$

This is when there is no groundwater.

Horizontal pressure in soil = lateral earth pressure coefficient × density of soil × depth
$= K \times \gamma \times h$

K = Lateral earth pressure coefficient

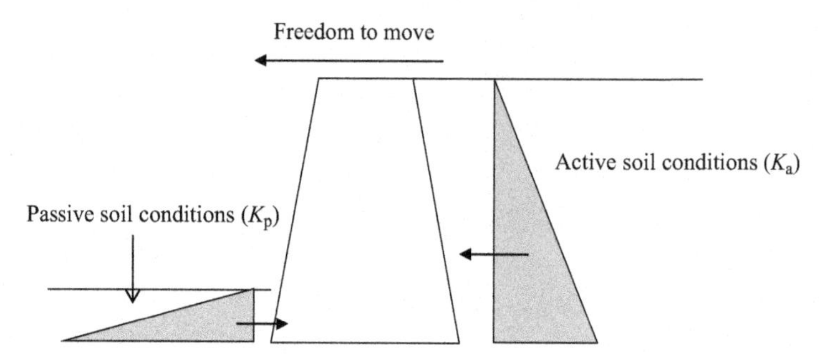

Figure 20.7 Freedom of movement of the retaining wall.

There are three lateral earth pressure coefficients.

- Active earth pressure coefficient (K_a)
- Passive earth pressure coefficient (K_p)
- Lateral earth pressure coefficient at rest (K_0)

20.3 Active earth pressure coefficient (K_a)

Active earth pressure coefficient is used when the retaining wall has the freedom to move (Figure 20.7).

The retaining wall is free to move to the left. When slight movement in the retaining wall occurs pressure on the right hand side will be reduced. On the other hand, pressure on the left hand side will be increased.

Hence, we can see that K_a is smaller than K_p.

Following equation is used to compute K_a and K_p.

$$K_a = \tan^2(45 - \phi/2)$$

$$K_p = \tan^2(45 - \phi/2)$$

Many designers do not consider the passive soil conditions in front of the retaining walls to be in the conservative side. Some codes require that erosion protection to be provided when passive earth pressure in front of a retaining wall is considered for the design.

20.4 Earth pressure coefficient at rest (K_0)

Following equation is used to compute K_0.

$$K_0 = 1 - \sin\varphi'$$

Earth pressure coefficient at rest is used for cantilever retaining walls (Figure 20.8).

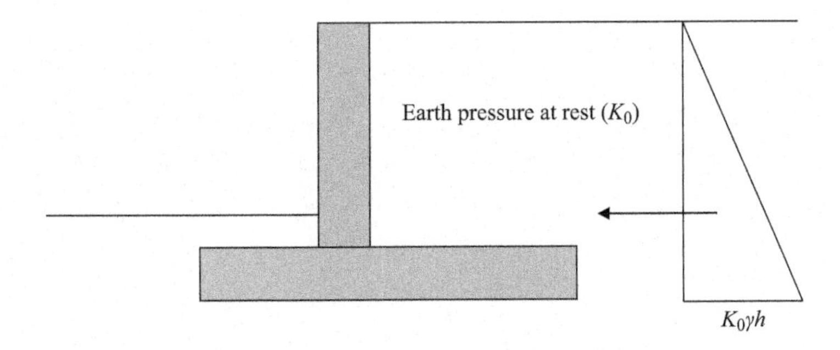

Figure 20.8 Cantilever retaining walls.

In a cantilever retaining wall, it is difficult to say whether the wall moves enough to create active conditions behind the wall. Hence, earth pressure coefficient at rest (K_0) is used for cantilever retaining walls. K_0 is larger than K_a.

$$K_a < K_0 < K_p$$

It should be noted that dense granular soil would have a higher K_0 value than loose soil (Plates 20.1 and 20.2).

Plate 20.1 Formwork construction for the stem of a cantilever wall.

Plate 20.2 Concreting the stem of a cantilever wall.

Gravity walls: sand backfill

21

21.1 Introduction

In gravity walls, soil pressure is restrained by the weight of the retaining wall.

In Figure 21.1, simple gravity retaining wall is shown. Modern gravity retaining walls are made of concrete.

The retaining wall can fail in three different ways.

1. The retaining wall can slide – sliding failure.
2. The retaining wall can overturn around the toe (point A).
3. Bearing failure of the foundation.

21.1.1 Resistance against sliding failure

For stability;

$$F_{(\text{floor})} > F$$

F, resultant earth pressure; $F_{(\text{floor})}$, friction between concrete and soil at the bottom face; $F_{(\text{floor})}$, $\tan\delta \times W$; W, weight of the retaining wall; δ, friction angle between concrete and soil.

Factor of safety against sliding failure = $F_{(\text{floor})}/F$

21.1.2 Resistance against overturning

The retaining wall can overturn around the toe (point A).

Overturning moment = $F \times h/3$

Resisting moment = $W \times a$

Factor of safety against overturning = resisting moment/overturning moment = =$3Wa/Fh$

How to find the resultant earth pressure?

Earth pressure at any given point is given by $K_a \gamma h$

K_a, active earth pressure coefficient = $\tan^2(45 - \varphi/2)$; γ, density of soil; h, height of the soil.

Figure 21.2, shows gravity retaining wall with soil pressure.

$$\begin{aligned}\text{Resultant earth pressure } (F) &= \text{area of the pressure triangle}\\ &= K_a \gamma h \times (h/2)\\ &= K_a \gamma h^2/2\end{aligned}$$

Geotechnical Engineering Calculations and Rules of Thumb

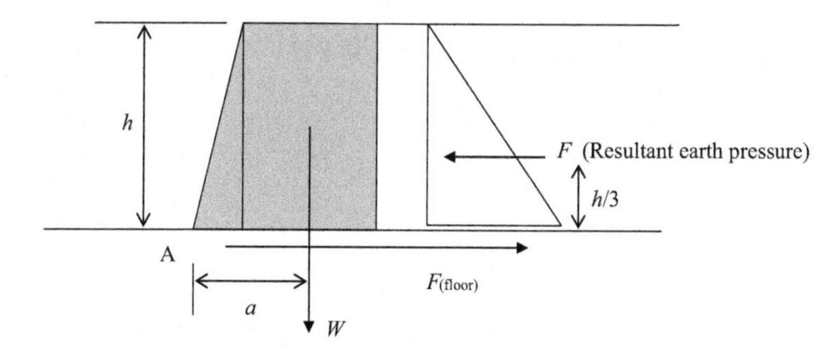

Figure 21.1 Gravity retaining wall.

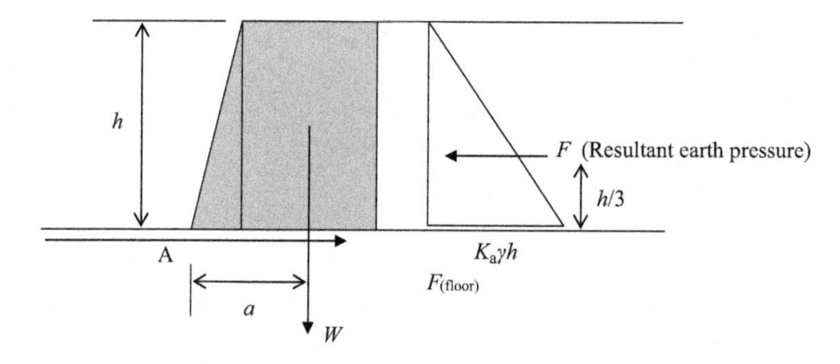

Figure 21.2 Gravity retaining wall with soil pressure.

The resultant earth pressure acts at the center of gravity of the pressure triangle, $h/3$ above the bottom.

Total moment (M) around point A = force (F) × distance to the force

$$M = (K_a\gamma h^2/2) \times h/3 = K_a\gamma h^3/6$$

21.1.3 Avoid tension in the base

It is important to avoid tension in the base of the retaining wall. This is done as explained in Design Example 21.1

Design Example 21.1: Gravity retaining wall with sand backfill – No groundwater

Find the factor of safety for the retaining wall shown in Figure 21.3. The height of the retaining wall (H) is 10 m, weight of the retaining wall is 2300 kN for 1 m length of the wall, acting at a distance

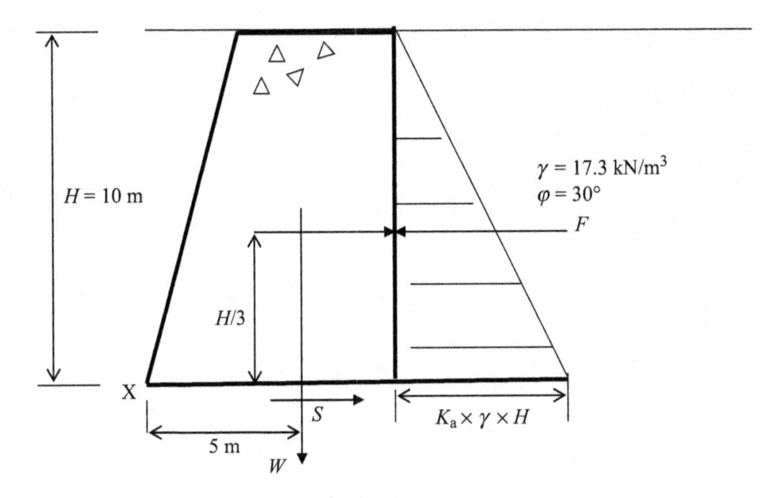

Figure 21.3 Gravity retaining wall with sand backfill.

of 5 m from the toe (point X). The friction angle of the soil backfill is 30°. The soil backfill mainly consists of sandy soils. The density of the soil is 17.3 kN/m^3. The resultant earth pressure force acts at $H/3$ distance from the bottom of the wall. The friction angle between soil and earth at the bottom of the retaining wall was found to be 20°.

Solution

Step 1: Find the resultant earth pressure

$$K_a = \tan^2(45 - \varphi/2) = \tan^2(45 - 30/2)$$
$$K_a = \tan^2(30) = 0.33$$

Earth pressure at the bottom of the base = $K_a \gamma H = 0.33 \times 17.3 \times 10 = 57.1$ kN/m^2.

Resultant earth pressure (F) = area of the pressure triangle = $57.1 \times 10/2$
= 285.5 kN per meter length of the wall

Step 2: Find the resistance against sliding at the base (S)

S = Weight of the wall $\times \tan(\delta)$
δ = Friction angle between concrete and soil at the bottom of the retaining wall
$S = W \times \tan(\delta) = 2300 \times \tan(20°) = 837$ kN per 1 m length of the wall
Weight of the retaining wall is given to be 2300 kN per 1 m length.
Factor of safety against sliding = 837/285.5 = 2.93

Step 3: Find the resistance against overturning (O)
Overturning will occur around point "X" of the retaining wall.
Resistance to overturning is provided by the weight of the retaining wall.
Overturning moment = $F \times H/3 = 285.5 \times 10/3$
= 951.7 kN.m (per 1 m length of the wall)
Resisting moment = $W \times a = 2,300 \times 5 = 11,500$ kN.m (per 1 m length of the wall)
Factor of safety against overturning = 11,500/951.7 = 12

21.2 Retaining wall design when groundwater is present

Groundwater exerts additional pressure on retaining walls when present. The following concept needs to be understood when drawing pressure triangles.

- Total density should be used above groundwater level to compute the lateral earth pressure. Below groundwater, the buoyant density of soil should be used.

$$\text{Buoyant density} = (\gamma - \gamma_w).$$

Pressure due to water should be computed separately.

Calculation of pressure acting on a gravity retaining wall when groundwater is present needs to be calculated in two parts.

- First, draw the pressure diagrams with effective stresses of soil.
- Secondly, draw the pressure diagrams for groundwater.
- Both the pressure diagrams are considered when computing the sliding force and the overturning moment (Figure 21.4).

$$\text{Earth pressure above groundwater} = K_a \times \gamma \times h_1$$

$$\text{Earth pressure below groundwater} = K_a(\gamma - \gamma_w)h_2$$

$$\text{Pressure due to water alone} = \gamma_w \times h_2$$

γ, total density of soil; γ_w, density of water; $(\gamma - \gamma_w)$, buoyant density of soil (or soil density below groundwater).

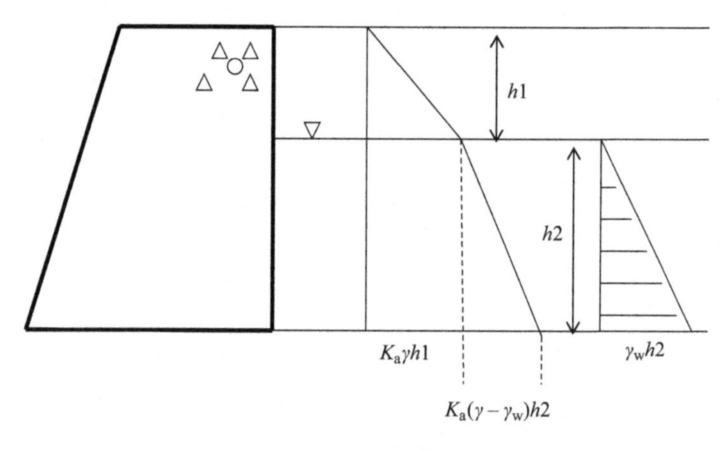

Figure 21.4 Earth pressure and pressure due to water.

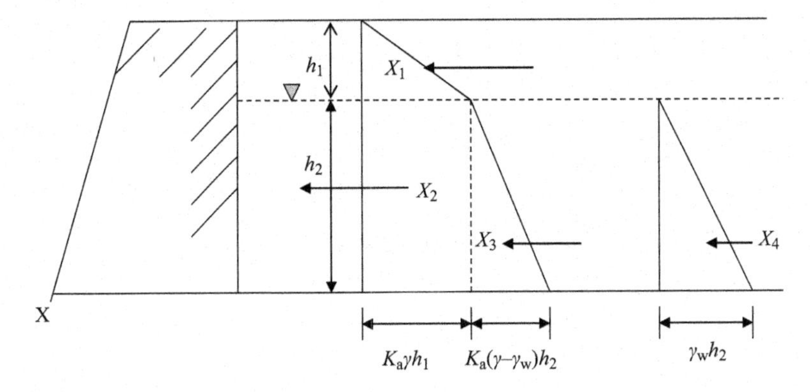

Figure 21.5 Forces acting on the retaining wall.

Figure 21.5 shows how to compute the factor of safety of a retaining wall when groundwater is present.

Total force = areas of all the triangles and rectangles

Area of the triangle, $X_1 = K_a \gamma h_1^2/2$

Area of the rectangle, $X_2 = K_a \gamma h_1 h_2$

Area of the triangle, $X_3 = K_a(\gamma - \gamma_w)\, h_2^2/2$

Area of the triangle, $X_4 = \gamma_w\, h_2^2/2$

$$\text{Total force} = X_1 + X_2 + X_3 + X_4$$
$$= K_a\gamma h_1^2/2 + K_a\gamma h_1 h_2 + K_a(\gamma - \gamma_w)h_2^2/2 + \gamma_w h_2^2/2$$

Total moment (M) around point X = moment of each area

Moment of triangle, $X_1 = K_a\gamma h_1^2/2 \times (h_2 + h_1/3)$

Moment of rectangle, $X_2 = K_a \gamma h_1 h_2 (h_2/2)$

Moment of triangle, $X_3 = K_a(\gamma - \gamma_w)h_2^2/2(h_2/3)$

Moment of triangle, $X_4 = \gamma_w\, h_2^2/2h_2/3$

$$\text{Total moment} = K_a\, \gamma h_1^2/2(h_2 + h_1/3) + K_a\, \gamma\, h_1 h_2 (h_2/2) +$$
$$K_a\, (\gamma - \gamma_w)\, h_2^2/2(h_2/3) + \gamma_w\, h_2^2/2h_2/3$$

Design Example 21.2: Gravity retaining wall with sand backfill – Groundwater Present

Find the factor of safety for the retaining wall shown in Figure 21.6. Height of the retaining wall (H) is 10 m, weight of the retaining wall is 2300 kN for 1 m length of the wall, acting at a distance of 5 m from the toe (point X). Friction angle of the soil backfill is 30°. The soil backfill mainly consists of sandy soils. The density of the soil is 17.3 kN/m^3. The resultant earth pressure force acts at H/3 distance from the bottom of the wall. Friction angle between soil and earth at the bottom of the retaining wall was found to be 20°. Groundwater is found to be at 4 m below the surface.

Solution

Step 1: Find the coefficient of earth pressure

$$K_a = \tan^2(45-\varphi/2) = \tan^2(45-30/2)$$
$$K_a = \tan^2(30) = 0.33$$

Step 2: Find the earth pressure at point B.
Earth pressure at point B (IB) = $K_a \gamma h = 0.33 \times 17.3 \times 4 = 22.8$ kN/m^2.

IB = ED

Step 3: Find the earth pressure at point C.
Earth pressure at point C = ED + DC
Distance "ED" was found to be 22.8 kN/m^2 in the Step 2.

$$DC = K_a \times (\gamma - \gamma_w) \times h = 0.33 \times (17.3 - 9.81) \times 6 = 14.8\text{kN/m}^2.$$

Earth pressure at point C = 22.8 + 14.8 = 37.6 kN/m^2.

Step 4: Find the pressure due to water at point G:
Water pressure at point G = $\gamma_w \times 6 = 9.81 \times 6 = 58.9$ kN/m^2.

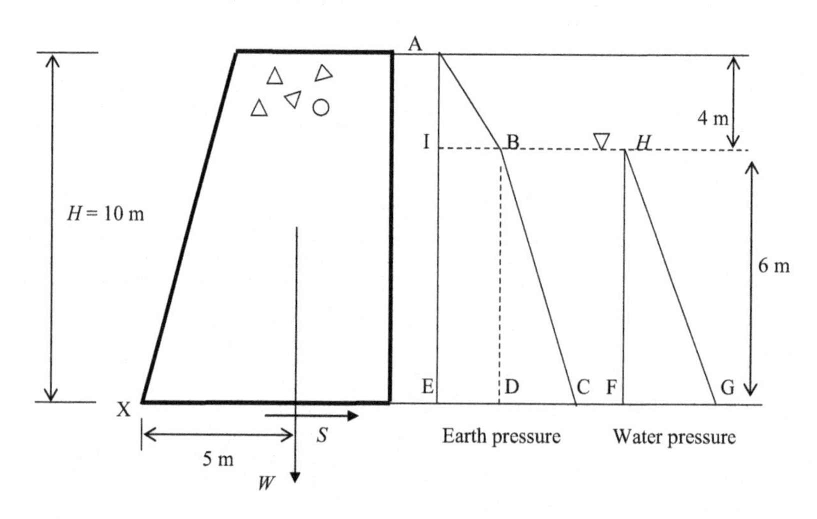

Figure 21.6 Overturning moment.

Step 5: Find the horizontal force due to soil and water (Figure 21.7):
Horizontal force due to soil and water can be computed by adding the areas of all triangles.
Area of triangle, ABI = 22.8 × 4 × ½ = 45.6 kN
Area of rectangle, IBED = 22.8 × 6 = 136.8 kN
Area of triangle, BCD = 14.8 × 6 × ½ = 44.4 kN
Area of triangle, HGF = 58.9 × 6 × ½ = 176.7 kN
Total area = 403.5 kN

Step 6: Find the resisting force against sliding:
S = Weight of the retaining wall × tan(δ)
δ = Friction angle between concrete and soil at the bottom of the retaining wall
$S = W \times \tan(\delta) = 2300 \times \tan(20°) = 837$ kN per 1 m length of the wall
Weight of the retaining wall is given to be 2300 kN per 1 m length.
Factor of safety against sliding = 837/403.5 = 2.07
Usually, a factor of safety of 2.5 is desired. Hence, it is advisable to increase the size of the retaining wall to increase the weight.

Step 7: Find the resistance against overturning (O) (Figure 21.8)
The retaining wall can overturn around point "X".
To find the overturning moment, we need to find the position of the forces in each unit.
Triangle ABI (F1) = The force is acting at the center of gravity of the triangle. The center of gravity of the triangle is 1/3 of the height of the triangle.
Distance to F1 from the bottom = L1 = 6 + 4/3 m = 7.33 m
Rectangle IBED (F2) = The force F2 is acting at the center of the rectangle.
Distance to F2 from the bottom = L2 = 6/2 m = 3 m
Triangle BCD (F3) = The force is acting at the center of gravity of the triangle. The center of gravity of the triangle is 1/3 of the height of the triangle.
Distance to F3 from the bottom = L3 = 6/3 m = 2 m
Triangle HGF (F4)
Distance to F4 from the bottom = L4 = 6/3 m = 2 m
Overturning moment = (F1 × L1) + (F2 × L2) + (F3 × L3) + (F4 × L4)

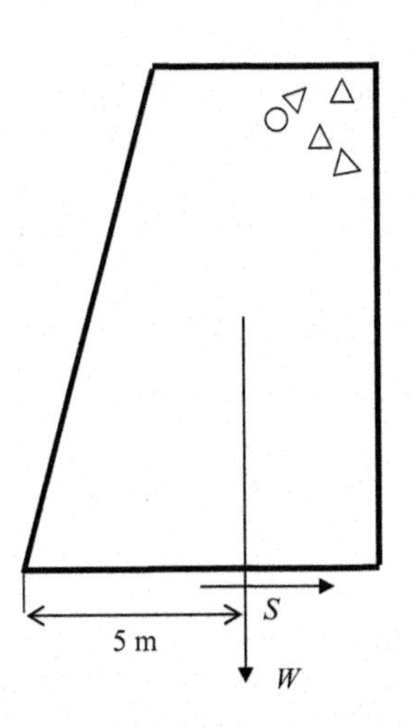

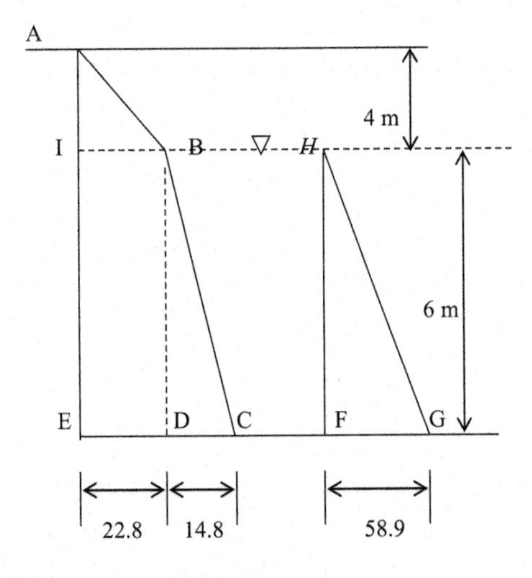

Figure 21.7 Forces and moments acting on earth retaining wall.

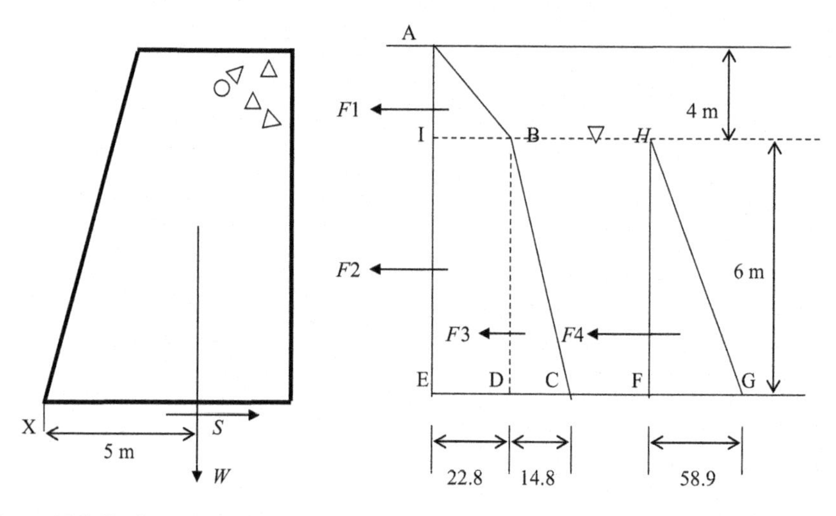

Figure 21.8 Resistance against overturning.

Overturning moment = 45.6 × 7.33 + 136.8 × 3 + 44.4 × 2 + 176.7 × 2
Overturning moment = 1186.8 kN.m per 1 m length of the wall.

Step 8: Find the resistance against overturning (*O*)
Resistance to overturning is provided by the weight of the retaining wall.
Resisting moment = $W \times a$ = 2,300 × 5 = 11,500 kN.m (per 1 m length of the wall)
Factor of safety against overturning = 11,500/1,186.8 = 9.7
Factor of safety of 2.5–3.0 is desired against overturning. Hence, the factor of safety against overturning is sufficient.

21.3 Retaining wall design in nonhomogeneous sands

The soil behind an earth-retaining structure may not be homogeneous. When there are different types of soil adjacent to a retaining wall, the earth pressure diagram would look different (Figure 21.9).

The lateral earth pressure at any location in soil is equal to the vertical effective stress multiplied by the lateral earth pressure coefficient.

Lateral earth pressure = vertical effective stress at that location multiplied by lateral earth pressure coefficient.

This statement can be represented mathematically as follows.

$$P_h = P_v \times K_a$$

P_h = horizontal stress
P_v = vertical effective stress (or simply effective stress)

When different types of sandy soils are encountered, K_a (lateral earth pressure coefficient) will be subjected to change. It is possible that different sandy soils are used

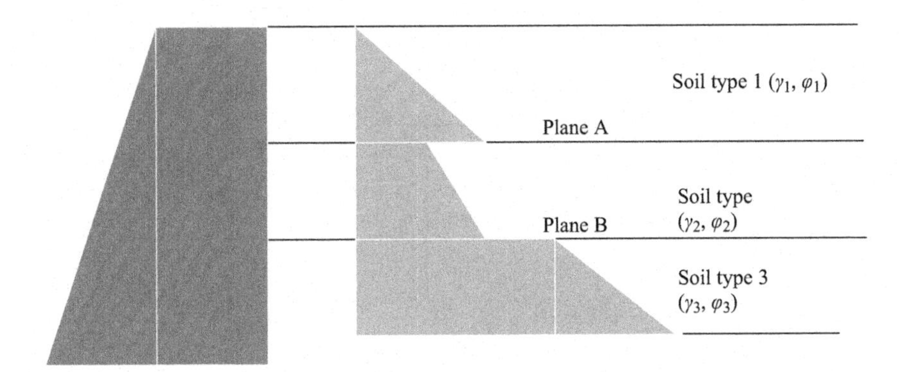

Figure 21.9 Lateral earth pressure in mixed soil.

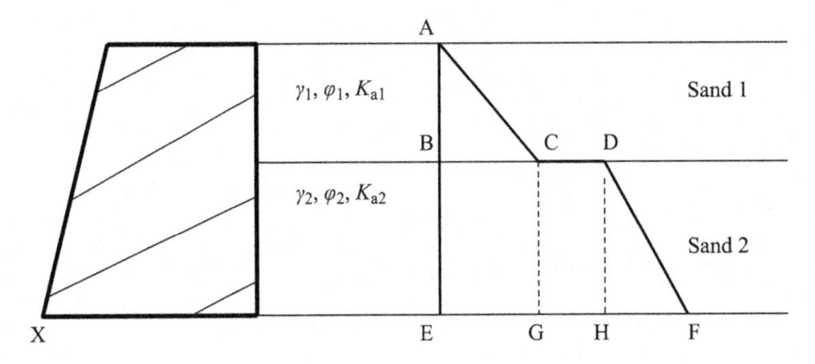

Figure 21.10 Lateral earth pressure diagram.

for backfilling. In some cases, the soil from the neighborhood is used for economic reasons (Figure 21.10).

Horizontal stress in sand layer 1 = $K_{a1} \times \gamma_1 \times h$

Horizontal stress in sand layer 2 = $K_{a2} \times \gamma_2 \times h$

Since K_{a1} and K_{a2} are different, there is a break at the start of the second layer.

These concepts can be better explained using an example.

Design Example 21.3: Gravity retaining wall in nonhomogeneous sandy soils

Find the factor of safety for the retaining wall shown in Figure 21.11. The height of the retaining wall (H) is 10 m, weight of the retaining wall is 2,300 kN for 1 m length of the wall, acting at a distance of 5 m from the toe (point X). Sandy backfill consists of two sand layers. The friction angle of the top sand layer is 30° and the density of the soil is 17.3 kN/m^3. The friction angle of the bottom sand layer is 20° with a density of 16.5 kN/m^3. The friction angle between soil and earth at the bottom of the retaining wall was found to be 20°.

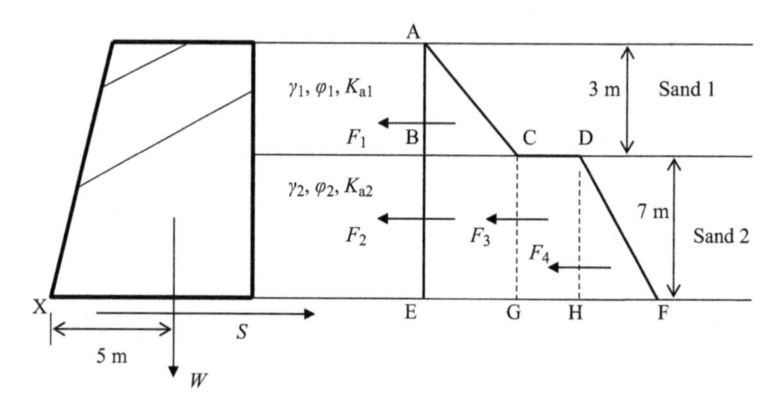

Figure 21.11 Forces due to mixed soil conditions.

Solution

Step 1: Find the lateral earth pressure coefficients of sand layers.

$$K_{a1} = \tan^2(45 - \varphi/2)$$

Sand layer 1 $K_{a1} = \tan^2(45 - 30/2) = 0.33$
Sand layer 2 $K_{a2} = \tan^2(45 - 20/2) = 0.49$

Step 2: Find the horizontal force acting on the retaining wall.

Length BC $= K_{a1} \times \gamma_1 \times h = 0.33 \times 17.3 \times 3 = 17.1\ \text{kN/m}^2$.
Length BD $= K_{a2} \times \gamma_1 \times h = 0.49 \times 17.3 \times 3 = 25.4\ \text{kN/m}^2$.

K_a value in sand 1 is 0.33 and the K_a value in sand 2 is 0.49.
Horizontal stress is found by multiplying the vertical effective stress with the lateral earth pressure coefficient (K_a).
In layer 1, the lateral earth pressure coefficient is K_{a1} and in layer 2, the lateral earth pressure coefficient is K_{a2}.

Length BD = length BC + length CD
25.4 = 17.1 + length CD
Length CD = 8.3 kN

The effective stress at both cases at point B is ($\gamma_1 \times h$) or (17.3×3). At point B, the density of soil layer 2 has no impact.
Find length EF:
First, find the vertical effective stress at point E.

Vertical effective stress at point E $= (\gamma_1 \times 3) + (\gamma_2 \times 7) = 17.3 \times 3 + 16.5 \times 7$
$= 167.4\ \text{kN/m}^2$.

Horizontal effective stress at point E = EF = $K_{a2} \times$ Vertical effective stress at point E
Horizontal effective stress at point E $= 0.49 \times 167.4 = 82\ \text{kN/m}^2$.

$$EF = EH + HF$$

$$EH = BD = 25.4$$
$$82 = 25.4 + HF$$

$$HF = 56.6$$

Step 3: Find the total horizontal force exerted on the retaining wall:
Total horizontal force exerted on the retaining wall = area of all triangles and rectangles.

Area of triangle, $ABC = \frac{1}{2} \times 3 \times 17.1 = 25.7\ \text{kN/m}^2$.

Area of rectangle, $BCEG = 17.1 \times 7 = 119.7\ \text{kN/m}^2$.

Area of rectangle, $CDGH = 8.3 \times 7 = 58.1\ \text{kN/m}^2$.

Area of triangle, $DHF = \frac{1}{2} \times 7 \times 56.6 = 198.1\ \text{kN/m}^2$.

Total horizontal force $= 25.7 + 119.7 + 58.1 + 198.1 = 401.6\ \text{kN/m}^2$.

Step 4: Find the resisting force against sliding:
S, weight of the retaining wall $\times \tan(\delta)$; δ, friction angle between concrete and soil at the bottom of the retaining wall.
$S = W \times \tan(\delta) = 2300 \times \tan(20°) = 837$ kN per 1 m length of the wall
Weight of the retaining wall is given to be 2300 kN per 1 m length.
Factor of safety against sliding = 837/401.6 = 2.08
Typically, a factor of safety of 2.5 is desired. Hence, increase the weight of the retaining wall.

Step 5: Find the resistance against overturning:
The retaining wall can overturn around point "X".
To find the overturning moment, we need to find the position of the forces in each unit.
Triangle ABC (F1) = The force is acting at the center of gravity of the triangle. The center of gravity of the triangle is acting at 1/3 of the height of the triangle.
Distance to F1 from the bottom = L1 = 7 + 3/3 m = 8 m
Rectangle BCEG (F2) = The force F2 is acting at the center of the rectangle.
Distance to F2 from the bottom = L2 = 7/2 m = 3.5 m
Rectangle CDHG (F3) = The force is acting at the center of gravity of the rectangle.
Distance to F3 from the bottom = L3 = 7/2 m = 3.5 m
Triangle DHF (F4)
Distance to F4 from the bottom = L4 = 7/3 m = 2.33 m
Overturning moment = (F1 × L1) + (F2 × L2) + (F3 × L3) + (F4 × L4)
Overturning moment = 25.7 × 8 + 119.7 × 3.5 + 58.1 × 3.5 + 198.1 × 2.33
Overturning moment = 1289.5 kN.m per 1 m length of the wall.

Step 6: Find the resistance against overturning:
Resistance to overturning is provided by the weight of the retaining wall.
Resisting moment = $W \times a$ = 2,300 × 5 = 11,500 kN.m (per 1 m length of the wall)
Factor of safety against overturning = 11,500/1,289.5 = 8.9
Factor of safety of 2.5–3.0 is desired against overturning. Hence, the factor of safety against overturning is sufficient.

21.3.1 General equation for gravity retaining walls

Retaining walls considered so far had a vertical surface and the backfill was horizontally placed; however, this may not be the case in some situations. The backfill could be at an angle β to the horizontal. The vertical surface of the retaining wall could be at an angle θ to the horizontal. The friction angle between soil and concrete could be δ.

In such situations, the K_a value is obtained from the following equation.

$$K_a = \frac{\sin^2(\theta+\varphi)}{\sin^2(\theta)\times\sin(\theta-\delta)\times[1+(p/q)^{1/2}]^2}$$

$p = \sin(\varphi+\delta)\times\sin(\varphi-\beta)$
$q = \sin(\theta-\delta)\times\sin(\theta+\beta)$
In Figure 21.12,
When $\theta = 90°$, $\beta = 0$, and $\delta \neq 0$:
$\sin(90-\varphi) = \cos(\varphi)$
$\sin\theta = 1.0$, for $\theta = 90°$
$\sin(90°-\delta) = \cos(\delta)$
$\sin(90+\beta) = \cos\beta$
$\cos\beta = 1.0$, for $\beta = 0°$.

$$K_a = \frac{\cos^2(\varphi)}{\cos(\delta)\times[1+(p/q)^{1/2}]^2}$$

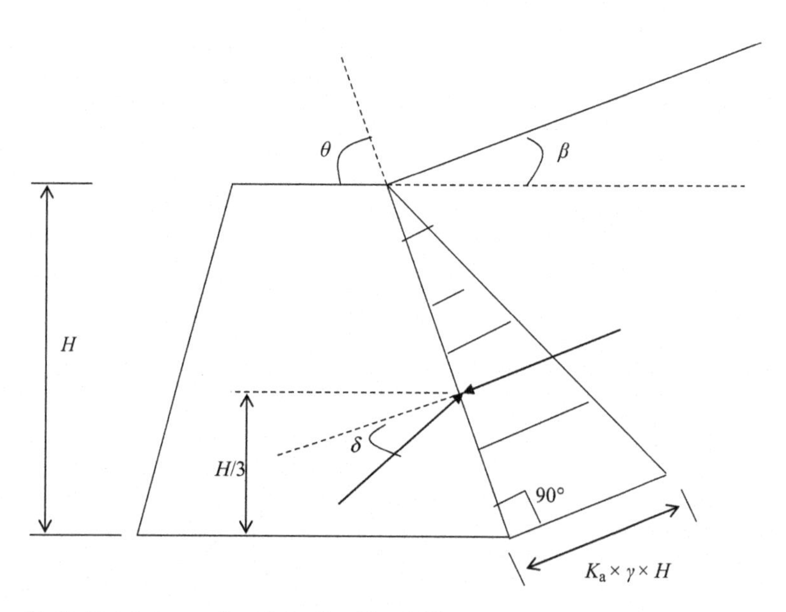

Figure 21.12 Retaining wall with inclined backfill.

$p = \sin(\varphi + \delta) \times \sin(\varphi)$
$q = \cos(\delta)$
In Figure 21.13,
When $\theta = 90°$, $\beta = 0$, and $\delta = 0$
$\sin(\theta + \varphi) = \cos(\varphi)$, for $\theta = 90°$.
$\sin 90° = 1.0$
$\sin(\theta - \delta) = \cos(\delta)$ for $\theta = 90°$.
$\sin(\theta + \beta) = \cos\beta$ for $\theta = 90°$.
$\cos\delta = 1.0$, for $\delta = 0°$

$$K_a = \frac{\cos^2(\varphi)}{\left[1 + (p/q)^{1/2}\right]^2}$$

$p = \sin^2(\varphi)$
$q = 1.0$

Hence, $K_a = \dfrac{\cos^2(\varphi)}{[1 + \sin\varphi]^2}$

This equation can be simplified to $K_a = \tan^2(45 - \varphi/2)$ (Figure 21.14).

21.3.2 Lateral earth pressure coefficient for clayey soils (active condition)

Earth pressure in clay soils is different from that in sandy soils. In the case of clay soils, lateral earth pressure is given by the following equation.

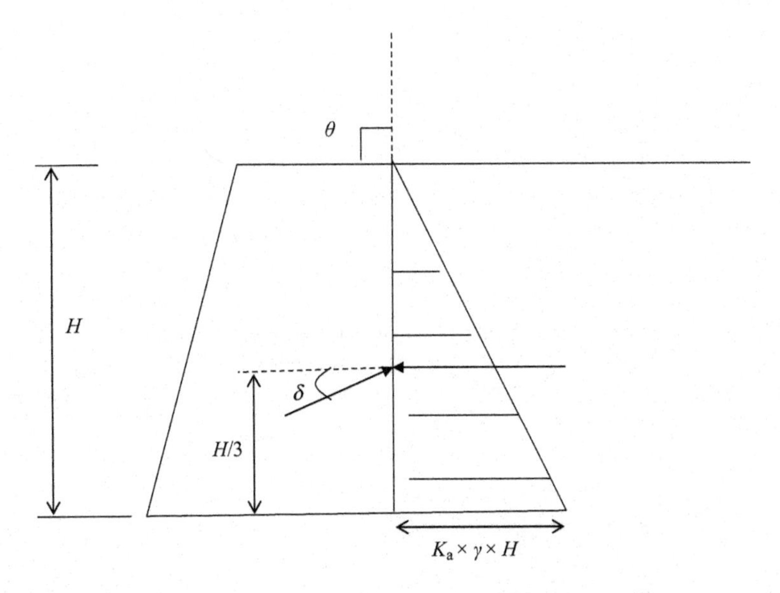

Figure 21.13 Retaining wall with nonzero concrete – soil friction angle.

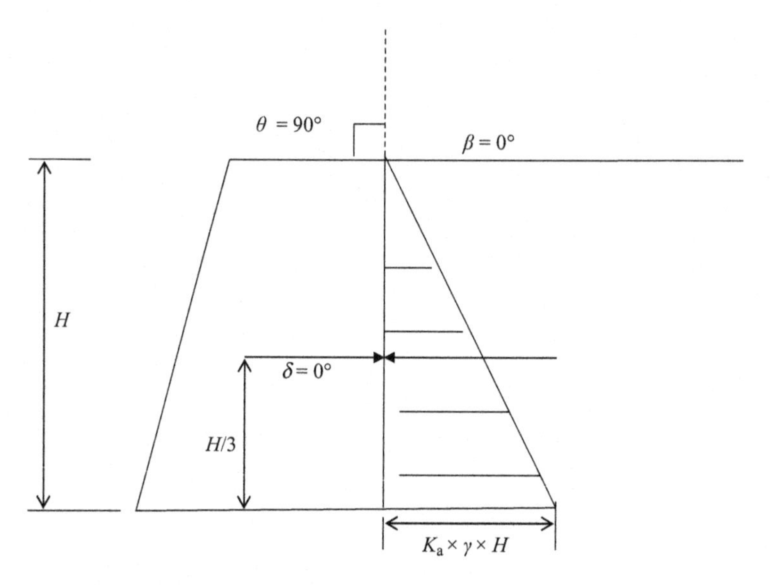

Figure 21.14 Gravity retaining wall and forces.

Lateral earth pressure = $\gamma h - 2c$

"c" is cohesion and "γ" is the density of the clay. When clay backfill is used, a portion of the clay layer will crack. The thickness of the cracked zone is given by $2c/\gamma$.

No active pressure is generated within the cracked zone (Figure 21.15).

Force (P) = area of the shaded portion of the triangle.

$$P_a = (\gamma h - 2c) \times h1/2$$
$$P_a = (\gamma h - 2c) \times (h - 2c/\gamma)/2$$

Usually, the cracked zone will be filled with water. Hence, the pressure due to water also needs to be accounted for (Figure 21.16).

Design Example 21.4: Gravity retaining wall with clay backfill

Find the factor of safety of the retaining wall shown in Figure 21.17 against sliding failure and overturning. Cohesion of the clay backfill is 20 kPa. Density of soil is $\gamma = 17.1$ kN/m^3. Assume that the cracked zone is not filled with water. Weight of the retaining wall is 2000 kN per meter run acting 3 m from the toe of the retaining wall (X) as shown in Figure 21.17. Height of the retaining wall is 8 m. Friction angle between base of the retaining wall and soil is 25°. Assume that the tension cracks are not filled with water.

Solution

Step 1: Find the thickness of the cracked zone:

Thickness of the cracked zone = $2c/\gamma = 2 \times 20/17.1 = 2.34$ m

Step 2: Find the pressure at the bottom:

Pressure at "B" = $(\gamma h - 2c) = (17.1 \times 8 - 2 \times 20) = 96.8$

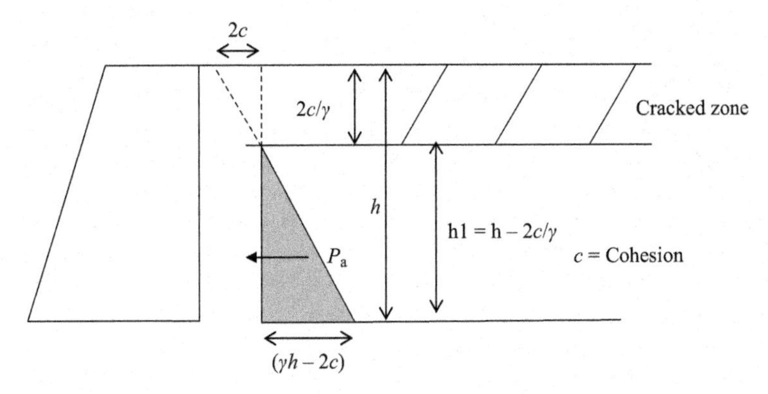

Figure 21.15 Gravity retaining wall in clay soil.

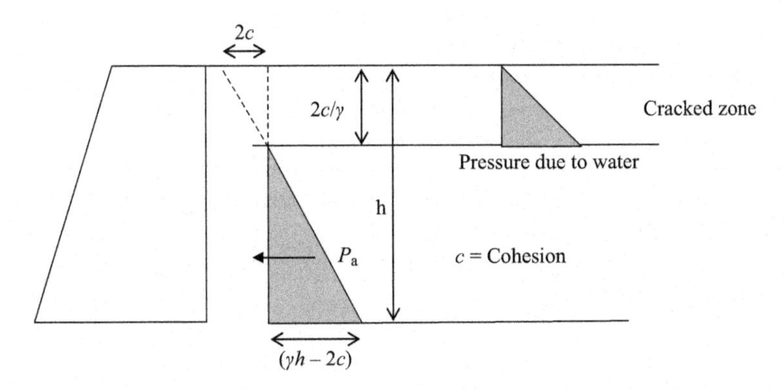

Figure 21.16 Forces due to clay soil.

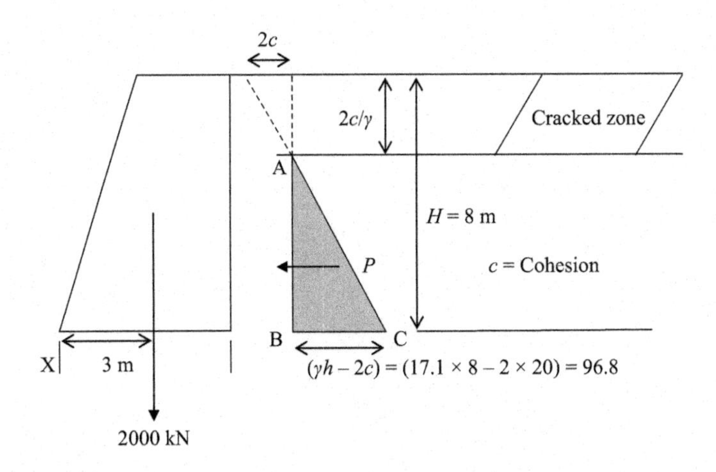

Figure 21.17 Force diagram.

Step 3: Find the lateral forces acting on the wall.
Lateral forces acting on the wall = area of triangle ABC
Note: In this case, it is assumed that the cracked zone is not filled with water.
Area of triangle, ABC = ½ × 96.8 × 8 = 387.2 kN

Step 4: Find the resistance to sliding:
Resistance to sliding = friction angle × weight of the retaining wall
Resistance to sliding = tan 25° × 2000 kN = 0.466 × 2000 = 932 kN.
Factor of safety against sliding = 932/387.2 = 2.4
Usually, a factor of safety of 2.5 is desired. Hence, increase the weight of the retaining wall.

Step 5: Overturning moment:
Resultant of the triangle ABC acts 1/3 distance from the base.
Distance to the resulting force = (8 − 2.34)/3 = 1.89 m
Overturning moment is obtained by obtaining moments around point "X"
Overturning moment = force × moment arm = 387.2 × 1.89 = 731.8 kN.m
Resistance to overturning is provided by the weight of the retaining wall.
Resistance to overturning = 2000 × 3 = 6000 kN.m
Factor of safety against overturning = 6000/731.8 = 8.2
Resistance against overturning is sufficient.

Design Example: 21.5: Gravity retaining wall with clay backfill and cracked zone filled with water

Find the factor of safety of the retaining wall shown in Figure 21.18 against sliding failure and overturning. Cohesion of the clay backfill is 20 kPa. Density of soil, $\gamma = 17.1$ kN/m^3. Assume that the cracked zone is not filled with water. Weight of the retaining wall is 2000 kN per meter run acting 3 m from the toe of the retaining wall (X) as shown. Height of the retaining wall is 8 m. The friction angle between the base of the retaining wall and soil is 25°. Assume that the tension cracks are filled with water.

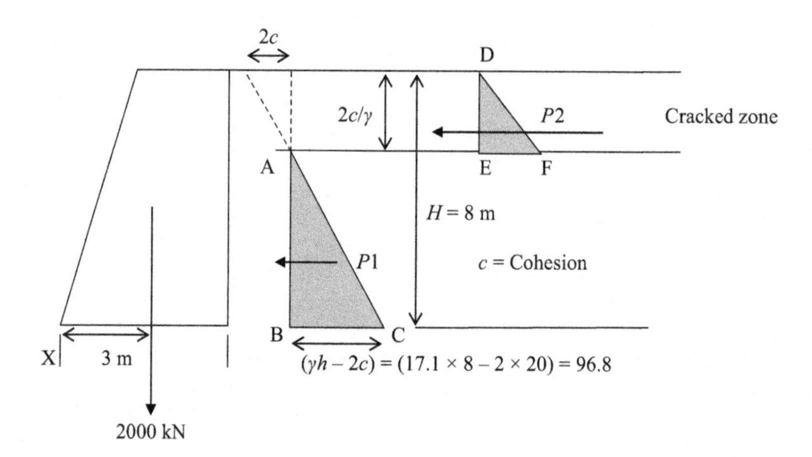

Figure 21.18 Water pressure due to cracked zone.

Solution

Step 1: Find the thickness of the cracked zone:
Thickness of the cracked zone = $2c/\gamma = 2 \times 20/17.1 = 2.34$ m

Step 2: Find the pressure at the bottom:
Pressure at "B" = $(\gamma h - 2c) = (17.1 \times 8 - 2 \times 20) = 96.8$
Pressure at "E" = $\gamma_w h = 9.81 \times 2.34 = 22.9$ kN

$$\gamma_w = \text{Density of water} = 9.81\,\text{kN/m}^3.$$

Step 3: Find the lateral forces acting on the wall.
Lateral forces acting on the wall = area of triangle ABC + area of triangle DEF
Note: In this case, it is assumed that the cracked zone is filled with water.

$$\text{Area of triangle ABC} = \tfrac{1}{2} \times 96.8 \times 8 = 387.2\,\text{kN}$$
$$\text{Area of triangle DEF} = \tfrac{1}{2} \times 22.9 \times 2.34 = 26.8\,\text{kN}$$
$$\text{Total sliding force} = 414\,\text{kN}$$

Step 4: Find the resistance to sliding:
Resistance to sliding = friction angle × weight of the retaining wall
Resistance to sliding = tan 25° × 2000 kN = 0.466 × 2000 = 932 kN.
Factor of safety against sliding = 932/414 = 2.25
Usually, a factor of safety of 2.5 is desired. Hence, increase the weight of the retaining wall.

Step 5: Overturning moment:
Resultant of the triangle ABC and DEF acts at 1/3 distance from the base.
Distance to the resulting force (ABC) from the bottom = (8 − 2.34)/3 = 1.89 m
Distance to the resulting force (DEF) from the bottom = (8 − 2.34) + 2.34/3 = 6.44 m
Overturning moment is obtained by obtaining moments around point "X"
Overturning moment of ABC = force × moment arm = 387.2 × 1.89 = 731.8 kN.m
Overturning moment of DEF = force × moment arm = 26.8 × 6.44 = 172.6 kN.m
Total overturning moment = 904.4
Resistance to overturning is provided by the weight of the retaining wall.
Resistance to overturning = 2000 × 3 = 6000 kN.m
Factor of safety against overturning = 6000/904.4 = 6.63
Resistance against overturning is sufficient.

21.3.3 Lateral earth pressure coefficient for clayey soils (passive condition)

Passive conditions occur in front of a retaining wall as shown in Figure 21.19. Due to pressure from the backfill, the retaining wall will try to move toward direction, Z. During the process, the soil in front of the retaining wall will be subjected to a passive condition. As mentioned earlier, if the soil is subjected to a force that loosens the soil, then active conditions would prevail. On the other hand, if the applied force subjects the soil to a denser condition, then passive conditions would prevail.

Lateral earth pressure (passive condition) = $\gamma h + 2c$

Here, "c" is cohesion and "γ" is the density of the clay.

$$\begin{aligned}\text{Total force} &= \text{area of the triangle} + \text{area of the rectangle}\\ &= \gamma h^2/2 + 2ch\end{aligned}$$

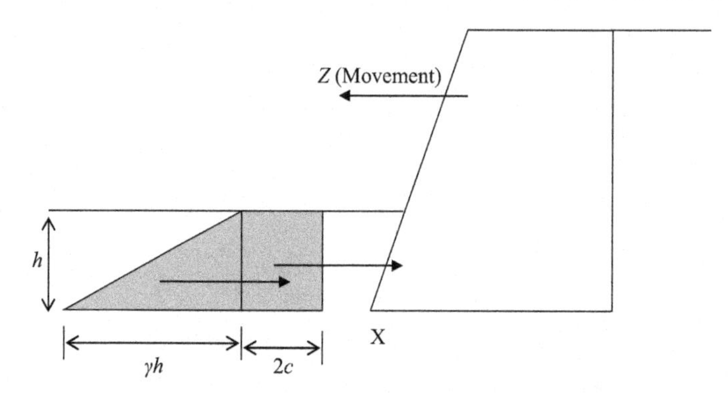

Figure 21.19 Passive earth pressure in clay soil.

Force due to triangle portion acts at $h/3$ distance above the bottom while the force due to the rectangular portion acts at $h/2$ distance from the bottom.

Moment around point "X" = moment of the triangle + moment of the rectangle

Moment of the triangle = $(\gamma h^2/2) \times h/3 = \gamma h^3/6$

Moment of the rectangle = $2ch^2/2$

Total moment around X = $\gamma h^3/6 + 2ch^2/2$

Design Example 21.6: Gravity retaining wall with clay backfill. Cracked zone is filled with water and passive pressure in front of the retaining wall considered.

Find the factor of safety of the retaining wall shown in Figure 21.20 against sliding failure and overturning. Cohesion of the clay backfill is 20 kPa. Density of soil (γ) is 17.1 kN/m^3. Weight of the retaining wall is 2000 kN per meter, acting 3 m from the toe of the retaining wall (X) as shown in Figure 21.20. Height of the retaining wall is 8 m. The friction angle between base of the retaining wall and soil is 25°. Assume that the tension cracks are filled with water. Assume that clay with cohesion 20 kPa is placed in front of the retaining wall and the passive pressure is not negligible. The depth of soil in front of the retaining wall is 1.5 m.

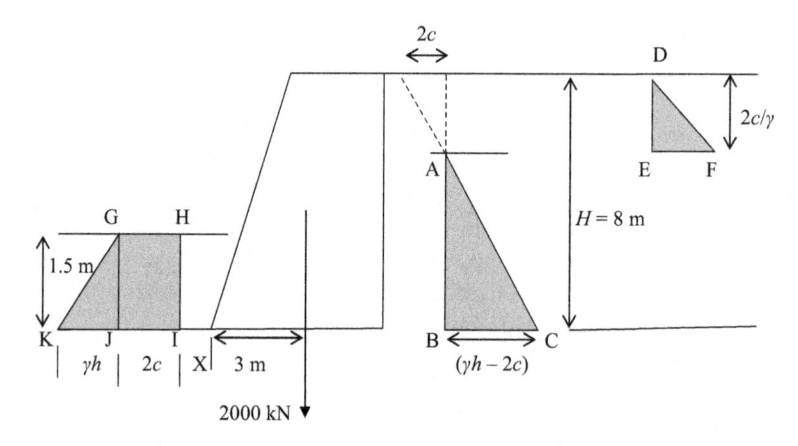

Figure 21.20 Force diagram including the passive earth pressure.

Solution

Step 1: Find the thickness of the cracked zone:
Height of the cracked zone (DE) = $2c/\gamma = 2 \times 20/17.1 = 2.34$ m

Step 2: Find the pressure at the bottom (BC):
Pressure at "B" = $(\gamma h - 2c) = (17.1 \times 8 - 2 \times 20) = 96.8$ kN/m^2.

Step 3: Find the pressure due to water
Pressure at "E" = $\gamma_w h = 9.81 \times 2.34 = 22.9$ kN/m^2.

Step 4: Find the pressure in the passive side
Pressure at "K" = KJ + JI = $\gamma h + 2c = 17.1 \times 1.5 + 2 \times 20 = 65.7$ kN/m^2.

Step 5: Find the lateral forces acting on the wall.

$$\text{Area of triangle, ABC} = \tfrac{1}{2} \times 96.8 \times (8 - 2.34) = 273.9\ \text{kN}$$

$$\text{Area of triangle, DEF} = \tfrac{1}{2} \times 22.9 \times 2.34 = 26.8\ \text{kN}$$

$$\begin{aligned}\text{Area of triangle, GKJ} &= \tfrac{1}{2} \times (\gamma h) \times 1.5 \\ &= \tfrac{1}{2} \times (17.1 \times 1.5) \times 1.5 \\ &= 19.2\ \text{kN}\end{aligned}$$

$$\begin{aligned}\text{Area of rectangle, GHIJ} &= 1.5 \times 2c \\ &= 1.5 \times 2 \times 20 \\ &= 60\ \text{kN}\end{aligned}$$

$$\text{Total sliding force} = 273.9 + 26.8 - 19.2 - 60\ \text{kN} = 221.5\ \text{kN}.$$

Step 6: Find the resistance to sliding:
Resistance to sliding = friction angle × weight of the retaining wall

$$\begin{aligned}\text{Resistance to sliding} &= \tan 250° \times 2000\ \text{kN} \\ &= 0.466 \times 2000 = 932.6\ \text{kN}.\end{aligned}$$

Factor of safety against sliding = 932.6/221.5 = 4.21
Usually, a factor of safety of 2.5 is desired. Hence, the factor of safety is sufficient.

Step 7: Overturning moment:
Resultants of triangles ABC, DEF, and GKJ act 1/3 distance from the base.
Distance to the resulting force (ABC) from the bottom = (8 − 2.34)/3 = 1.89 m
Distance to the resulting force (DEF) from the bottom = (8 − 2.34) + 2.34/3 = 6.44 m
Distance to the resulting force (GKJ) from the bottom = 1.5/3 = 0.5 m
Distance to the resulting force (GHJI) from the bottom = 1.5/2 = 0.75 m

21.3.4 Overturning moment is obtained by obtaining moments around point "X"

Overturning moment of ABC = force × moment arm = 273.9 × 1.89 = 517.7 kN.m
Overturning moment of DEF = force × moment arm = 26.8 × 6.44 = 172.6 kN.m
GKJ and GHJI are generating resisting moments.
Resisting moment of GKJ = force × moment arm = 19.2 × 0.5 = 9.6 kN.m

Resisting moment of GHJI = force × moment arm = 60 × 0.75 = 45 kN.m
Total overturning moment = 517.7 + 172.6 − 9.6 − 45 = 635.7 kN.m
Resistance to overturning is provided by the weight of the retaining wall.
Resistance to overturning = 2000 × 3 = 6000 kN.m
Factor of safety against overturning = 6000/635.7 = 9.4
Resistance against overturning is sufficient.

Design Example 21.7: Sand layer above clay layer (active condition) – no groundwater

The top layer of the backfill is sand with a friction angle (φ) of 30° (Figure 21.21). Density of the sand layer is 18 kN/m^3. The clay layer below the sand layer has a cohesion of 15 kPa and a density of 17.3 kN/m^3. Weight of the wall is 3000 kN/m and the center of gravity is acting 2 m from the toe. The friction angle between concrete and soil (δ) at the bottom of the retaining wall is found to be 25°.

Solution

Sand Layer

Step 1: Find the lateral earth pressure coefficient:

$$K_a = \tan^2(45 - \varphi/2)$$
$$K_a = \tan^2(45 - 30/2) = 0.33$$

Step 2: Find the lateral earth pressure

$$\text{Lateral earth pressure at point "B" (sand layer) (BC)} = K_a\,\gamma \times 6$$
$$= 0.33 \times 18 \times 6 \text{ kPa}$$
$$= 35.6 \text{ kPa.}$$

Note: Lateral earth pressure in sand and clay at the same depth has different values.

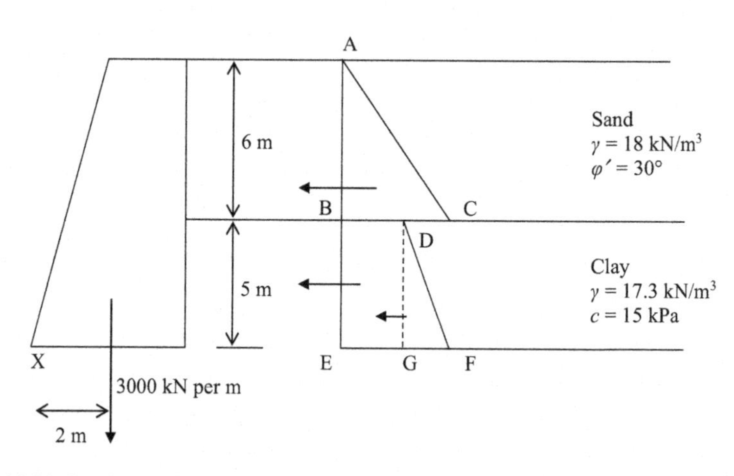

Figure 21.21 Sand layer underlain by a clay layer.

Clay Layer

Lateral earth pressure at point "B" (clay layer) (BD) $= \gamma h - 2c$
$= 18 \times 6 - 2 \times 15 = 78$ kPa

Lateral earth pressure at point "E" (clay layer) (EF) $= \text{BD} + \gamma h$
$= 78 + 17.3 \times 5$
$= 164.5$ kPa

Step 3: Find the forces due to lateral earth pressure:
Area of triangle, ABC = 0.5 × 35.6 × 6 = 106.8 kN
Area of rectangle, BDEG = 78 × 5 = 390 kN

Area of triangle, DGF $= 0.5 \times \text{GF} \times 5$
$= 0.5 \times (17.3 \times 5) \times 5$ kN
$= 216.3$ kN

Total sliding force $= 106.8 + 390 + 216.3$
$= 713.1$ kN

Step 4: Find the resistance against sliding:

Resistance against sliding $= \tan\delta \times$ weight of the retaining wall
$= \tan(25) \times 3000$
$= 1399$ kN

Factor of safety against sliding = 1399/713.1 = 1.96
Factor of safety of 2.5–3.0 is desired. Hence, increase the size of the wall.

Step 5: Find the height to center of gravity of forces:
Height to center of gravity of triangle, ABC = 5 + 6/3 = 7 m
Height to center of gravity of triangle, DGF = 5/3 = 1.67 m
Height to center of gravity of rectangle, BDEG = 5/2 = 2.5 m

Step 6: Find the overturning moment around the toe (point X)

Overturning moment due to triangle (ABC) $= 106.8 \times 7$
$= 747.6$ kNm

Overturning moment due to triangle (DGF) $= 216.3 \times 1.67$
$= 361.2$ kNm

Overturning moment due to rectangle (BDEG) $= 390 \times 2.5$
$= 975$ kN m

Total overturning moment $= 747.6 + 361.2 + 975$
$= 2083.8$ kNm

Total resisting moment $= 2 \times 3{,}000$ kNm
$= 6000$ kNm

Factor of safety against overturning = 6000/2083.8 = 2.88
Typically, a factor of safety of 2.5–3.0 is desired.

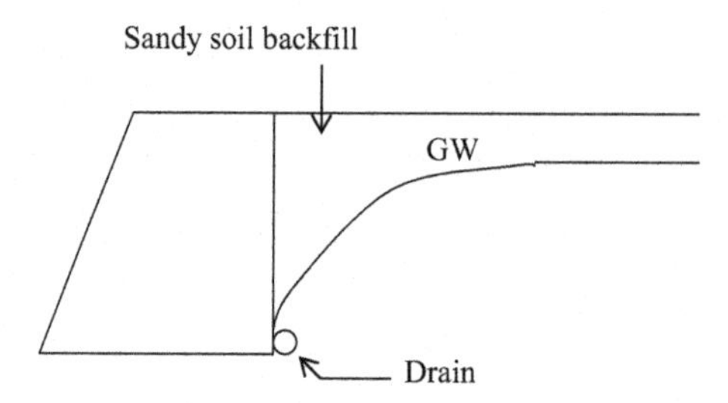

Figure 21.22 Drainage in retaining walls using sandy soils.

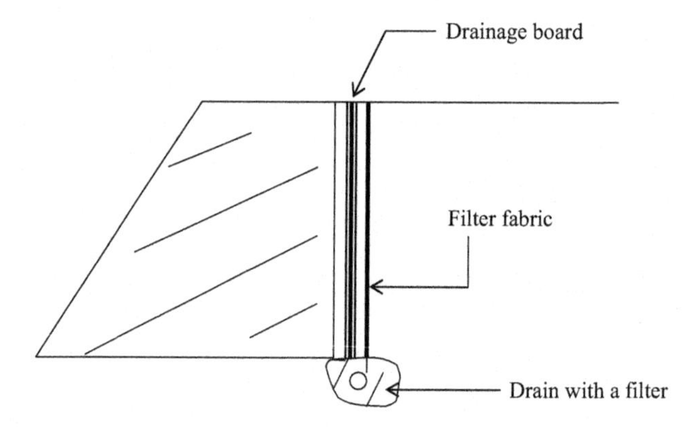

Figure 21.23 Drainage in retaining walls using geotextiles.

21.3.5 Earth pressure coefficients for cohesive backfills

Sandy soils (noncohesive) are the most suitable type of soils that can be used behind retaining walls. Sandy soils drain freely and therefore, the groundwater could be drained away using drains. This would reduce the pressure on the retaining wall (Figure 21.22).

Active earth pressure coefficient of K_a can be used for sandy soils.

21.3.6 Drainage using geotextiles

Drainage board can be used behind the wall to facilitate drainage. Filter fabric is used in front of the drainage board to filter any solid particles (Figure 21.23).

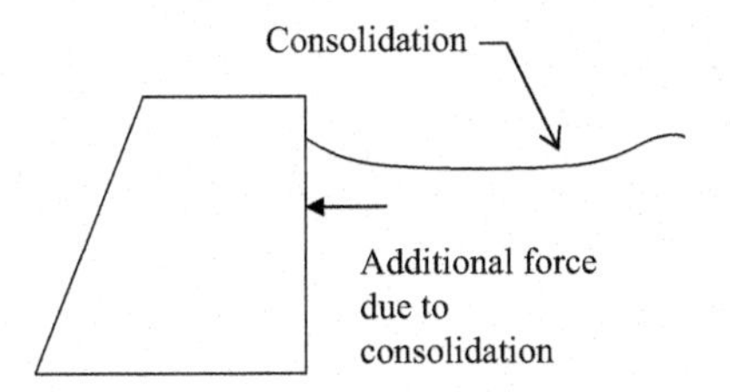

Figure 21.24 Thrust due to clay consolidation.

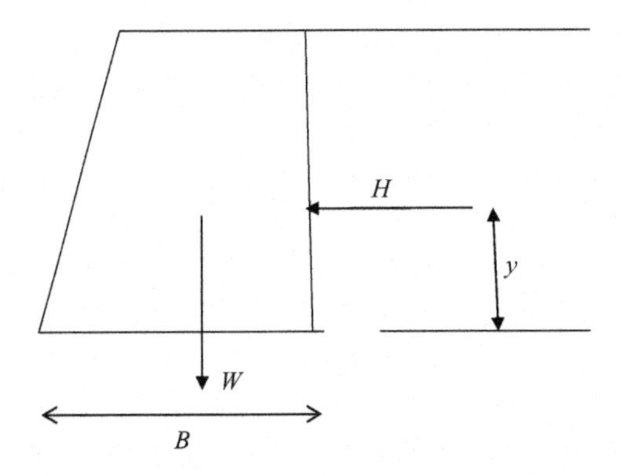

Figure 21.25 Forces action on a gravity retaining wall.

21.3.7 Consolidation of clayey soils

Clay soils are not desired as a backfill material for retaining walls. However, in some instances, clay soils are used to reduce costs. The geotechnical engineer needs to investigate the consolidation properties of clay soils. Clayey material tends to consolidate and generate an additional thrust on the retaining wall (Figure 21.24).

How to account for additional force due to clay consolidation?

Complex computer programs could be used to account for the additional thrust due to consolidation.

21.3.8 Avoiding tension in the base

It is important to make sure that tension is not developed in the base off the retaining wall. This is made sure by proportioning the retaining wall base so that the subgrade reaction acts within the middle third of the base (Figure 21.25).

Moment due to horizontal force is equated to an eccentricity. Eccentricity should be less than $B/6$.

Design Example 21.8

Check to see whether tension is developed under the base of the footing of the retaining wall. Density of backfill material is 17 kN/m^3. Lateral earth pressure coefficient (K_a) is 0.32. Density of concrete is 24 kN/m^3.

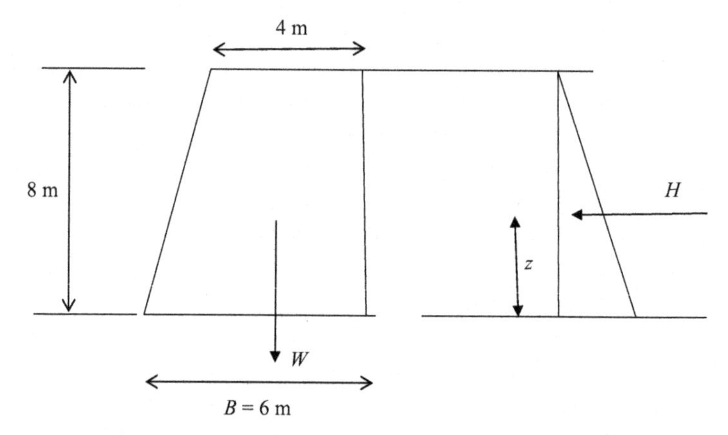

Solution

Step 1: Find the center of gravity of the retaining wall.

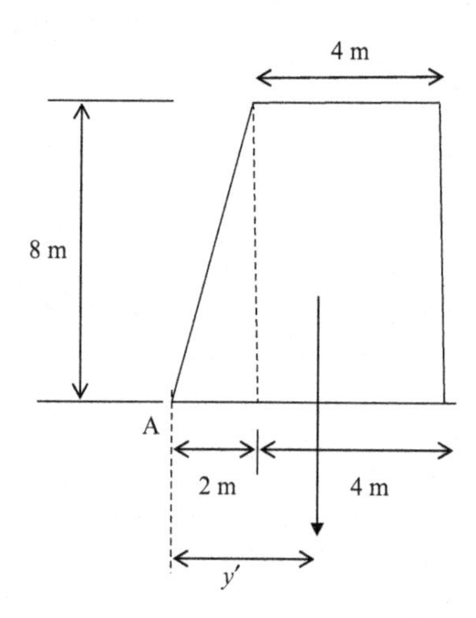

y' is the distance to center of gravity from point A.

The retaining wall can be divided into two pieces. One is a rectangular piece and the other one is a triangular piece.

Step 2: Find the location of the center of gravity (y'):

Weight of rectangular piece = $(4 \times 8) \times 24 = 768$ kN

Distance to center of gravity of the rectangular piece from point A = $2 + 4/2 = 4$ m

Weight of triangular piece = (2 × 8)/2 × 24 = 192 kN
Distance to center of gravity of the rectangular piece from point A = 2/3 m = 0.67 m
Total weight of the retaining wall = 768 + 192 = 960 kN
Apply the center of gravity equation;

$$960 \times y' = (768 \times 4) + (192 \times 0.67)$$
$$y' = 3.33$$

Step 3: Find the horizontal force due to earth pressure
Pressure at bottom of the base = $K_a \gamma h$
Pressure at bottom of the base = 0.32 × 17 × 8 = 43.52
Total force due to lateral earth pressure = H = area of the pressure triangle

$$H = 43.52 \times 8/2 = 174.1\,\text{kN}$$

Resultant of the lateral earth pressure acts at 8/3 distance from the bottom.
The moment due to lateral earth pressure = H × 8/3 = 174.1 × 8/3 = 464.2 kN

Step 4: Equate the moment to an eccentricity;

$$M = We$$
$$464.2 = 960 \times e$$

$$\text{Eccentricity } (e) = 0.484\,\text{m}$$

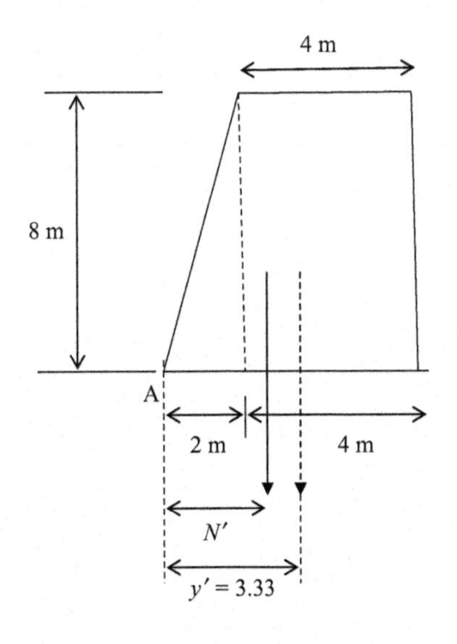

The dotted arrow in the previous figure shows the center of gravity. The solid arrow shows the location of the weight after accounting for the moment.

$$N' = 3.33 - e$$
$$N' = 3.33 - 0.484 = 2.836\,\text{m}$$

Width of the bottom of the base is 6 m.
One third of the base is 6/3 = 2 m.
N' is 2.836 m. Hence, the resultant is within the middle third of the retaining wall.

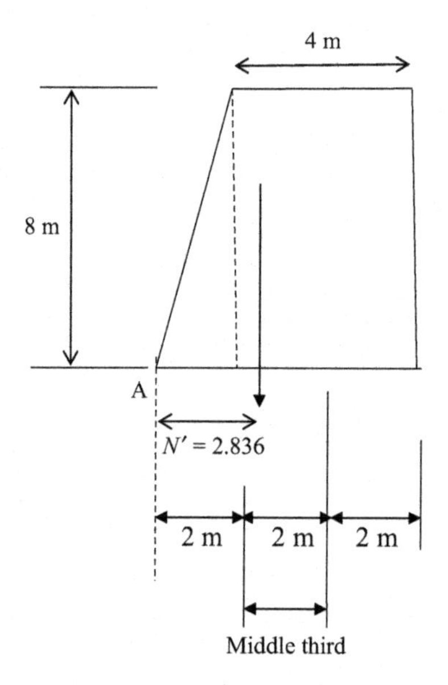

Since the resultant is within the middle third, tension will not be developed.

Cantilever walls

22

Cantilever walls are highly popular since they take less space than gravity walls. Also in most cases cheaper than gravity walls. Gravity walls require more concrete while cantilever walls require more steel to resist bending moment. General structure and forces in a cantilever wall is shown in Figure 22.1.

Description of terms: W_s, Weight of soil; $W1$, Weight of concrete stem; $W2$, Weight of concrete base; H_a, Horizontal force due to active soil pressure; H_p, Horizontal force due to passive soil pressure; F, Force due to friction between base and soil

Design Example 22.1

Cantilever wall is built with concrete density 24 kN/m^3. Density of soil backfill is 18 kN/m^3. Friction angle of soil backfill is 35° and adhesion coefficient between concrete and soil is 0.3.

1. Find the factor of safety against sliding;
2. Find the factor of safety against overturning;
3. Find whether there is tension under the footing;
4. Allowable bearing capacity of soil is 200 kN/m^2. Check to see whether there is a bearing failure.

Figure 22.2 shows cantilever wall and forces.

Solution

Step 1: Find lateral earth pressure coefficients;

$$K_a = \tan^2(45-\varphi/2) = \tan^2(45-35/2) = 0.271$$
$$K_p = \tan^2(45+\varphi/2) = \tan^2(45+35/2) = 3.69$$

Step 2: Find the weights and forces

Description of terms: W_s, Weight of soil; $W1$, weight of concrete stem; $W2$, weight of concrete base; H_a, horizontal force due to active soil pressure; H_p, horizontal force due to passive soil pressure; F, force due to friction between base and soil.

W_s = Weight of soil sitting on the retaining wall = $(2 \times 6) \times 18 = 216$ kN
$W1$ = Weight of concrete stem = $(0.5 \times 6) \times 24 = 72$ kN
$W2$ = Weight of concrete base = $(1 \times 3.5) \times 24 = 84$ kN
Total weight = $W1 + W2 + W_s = 372$ kN
H_a = Horizontal force due to active soil pressure = $(K_a \gamma H)\ H/2$
$H_a = (0.271 \times 18 \times 7) \times 7/2 = 119.5$ kN
H_p = Horizontal force due to passive soil pressure = $(K_p \gamma H)\ H/2$
$H_p = (3.69 \times 18 \times 1) \times 1/2 = 33.2$ kN
F = Force due to friction between base and soil
F = Adhesion coefficient × total weight of concrete and soil
$F = 0.3 \times 372 = 111.6$ kN

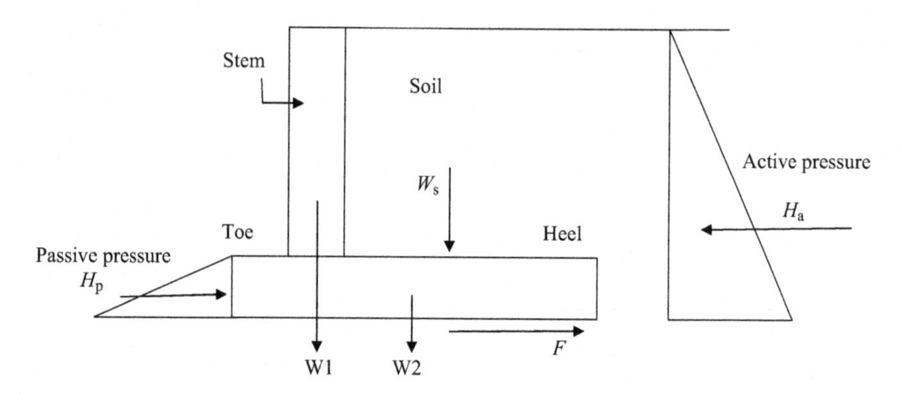

Figure 22.1 Cantilever wall.

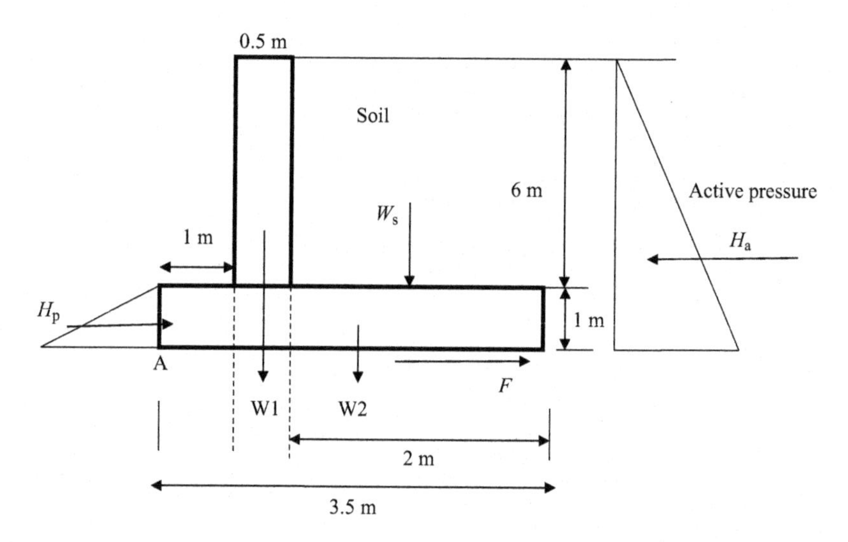

Figure 22.2 Cantilever wall and forces.

Step 3: Find the sliding force;
Sliding is due to horizontal force H_a
Sliding force = H_a = 119.5 kN

Step 4: Find the resisting force;
Resisting force = $H_p + F = 33.2 + 111.6 = 144.8$ kN
Factor of safety against sliding = 144.8/119.5 = 1.21
Factor of safety against sliding is not enough. Typically, a factor of safety of 2.0 or more is required.

Step 5: Find the overturning moment;
Overturning of the retaining wall occurs around point A in the retaining wall (Figure 22.2).
Overturning moment = $[(K_a\ \gamma H)\ H/2] \times H/3 = H_a \times H/3 = 119.5 \times 7/3 = 278.8$ kN.m

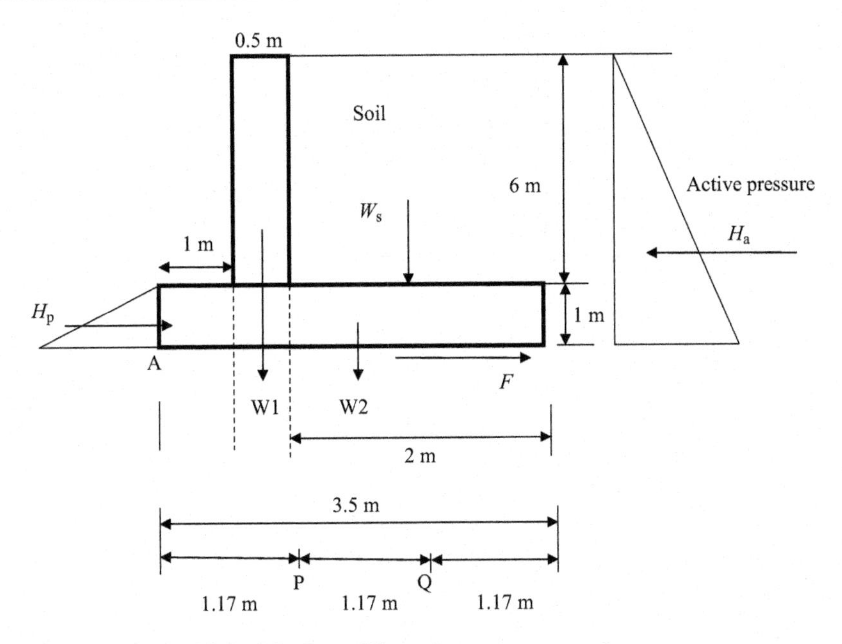

Figure 22.3 Middle third of the base. Dimensions are not to scale.

Step 6: Find the resisting moment;

Resisting moment = $W1 \times (1 + 0.5/2) + W2 \times (3.5/2) + W_s \times (1.5 + 1) + H_p \times H/3$

Resisting moment = $72 \times (1 + 0.5/2) + 84 \times (3.5/2) + 216 \times (1.5 + 1) + 33.2 \times 1/3 =$ 788.1 kN.m

Factor of safety against overturning = 777/278.8 = 2.83 (adequate)

Step 7: Check to see whether there is tension under the footing?

When the resultant force is acting at the middle third of the base, there will not be tension (Figure 22.3).

Divide the base into three parts. Middle third is between points P and Q.

Step 8: Find the center of gravity of the retaining wall;

Distance to center of gravity from point A = x'

Total weight of the retaining wall and backfill = $W = W1 + W2 + W_s = 72 + 84 + 216 = 372$ kN

$372.\ x' = W1 \times 1.25 + W2 \times 3.5/2 + W_s \times (1.5 + 1)$

$372.\ x' = 72 \times 1.25 + 84 \times 3.5/2 + 216 \times (1.5 + 1)$

$372.\ x' = 777$

$x' = 2.09$ ft.

Step 9: Convert overturning moment to an eccentricity;

Overturning moment was found to be 278.8 kN.m in Step 5.

Overturning moment = 278.8 kN.m

Overturning moment = 278.8 kN.m = $W e$

W is the total weight and "e" is the eccentricity. W was found to be 372 kN.

$278.8 = 372\ e$

$e = 0.75$

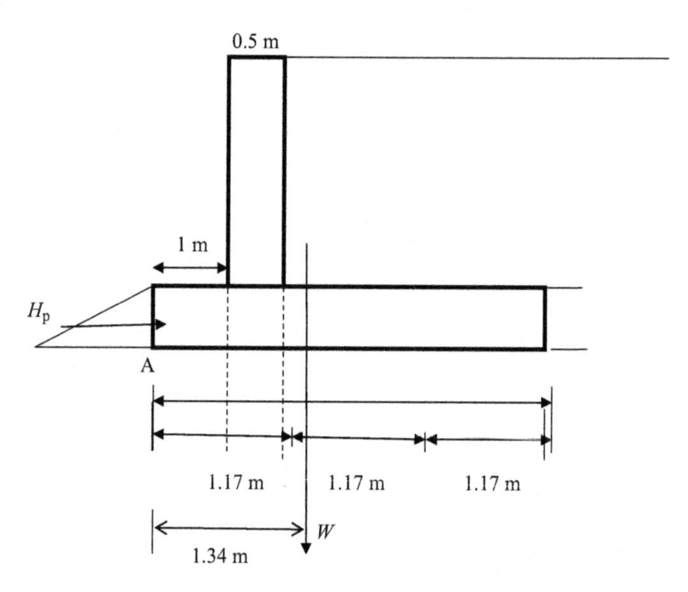

Figure 22.4 Eccentricity.

Step 10: Find the location of the resultant;
The center of gravity of the retaining wall and backfill is 2.09 ft. from point A. There is an eccentricity of 0.75 m. Hence the resultant is (2.09 − 0.75 = 1.34 m) from point A. Note that eccentricity is left to the center of gravity due to overturning moment (Figure 22.4).

The resultant is acting at the middle third of the base. Hence, tension will not develop in the base.

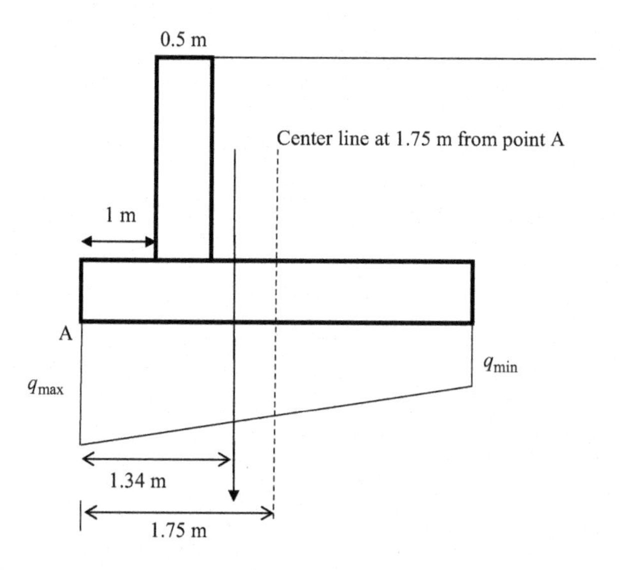

Figure 22.5 Eccentricity from centerline.

Step 11: Next question is to find the adequacy of bearing capacity.
Figure 22.5 shows the eccentricity from centerline.
Eccentricity measured from center line = 1.75 − 1.34 = 0.41 m
Find B' and L' (see Chapter 8 for complete description).
There is no eccentricity along the length. Hence $L = L$
Along the length, 1 m section is considered.

$$B' = B - 2e = 3.5 - 2 \times 0.41 = 2.68\ \text{m}$$

Soil bearing pressure = Total weight/($B' \times L'$) = 372/(2.68 × 1) = 138.8 kN/m^2.
Allowable bearing capacity of soil is 200 kN/m^2.
Hence, there is no bearing failure.

Gabion walls

23

23.1 Introduction

Computations involved in gabion walls are no different from regular gravity earth-retaining walls. Earth pressure forces are computed as usual along with stability of the wall with respect to rotation and sliding.

Gabion baskets are manufactured in different sizes. Typical basket is approximately 3 ft. in size. Smaller baskets are easier to handle. At the same time smaller baskets would have more seams to be connected. Not all gabion baskets are perfect cubes. Some baskets are made with elongated shapes (Plate 23.1).

Gabion baskets are connected to build a wall (Plate 23.2).

Design Example 23.1

An embankment of 15 ft. high has to be contained. 5 ft. Gabion baskets are placed as shown in Figure 23.1 as proposed. Find the factor of safety of the gabion wall.
Soil density (γ) = 120 pcf,
soil friction angle (φ) = 30°.
Friction angle between gabion baskets and soil (δ) = 20°.
Assume all groundwater is drained.

Solution
Step 1: Find the lateral earth pressure coefficient

$$K_a = \tan^2(45 - \varphi/2) = \tan^2(45 - 30/2) = 0.333$$

Step 2: Compute horizontal force due to earth pressure
Horizontal force (H) = Pressure due to soil

$$BC = K_a \times \gamma \times h = 0.333 \times 15 \times 120 = 600 \text{ psf}$$

$$\text{Area }(ABC) = \tfrac{1}{2} \times 15 \times BC = \tfrac{1}{2} \times 15 \times 600 = 4{,}500 \text{ lbs}$$

Total horizontal force (H) = 4,500 lbs

Step 3: Compute the weight of the gabion wall
Assume the density of stones to be 160 pcf. There are 6 gabion baskets.
W = Weight of gabion baskets = 6 × (5 × 5) × 160 = 24,000 lbs
There is soil sitting on gabion baskets.
Weight of soil = 3 × (5 × 5) × 120 = 9000 lbs
Total weight of the gabion wall = 24,000 + 9,000 = 33,000 lbs

Geotechnical Engineering Calculations and Rules of Thumb

Plate 23.1 Gabion baskets.

Plate 23.2 Gabion wall completed.

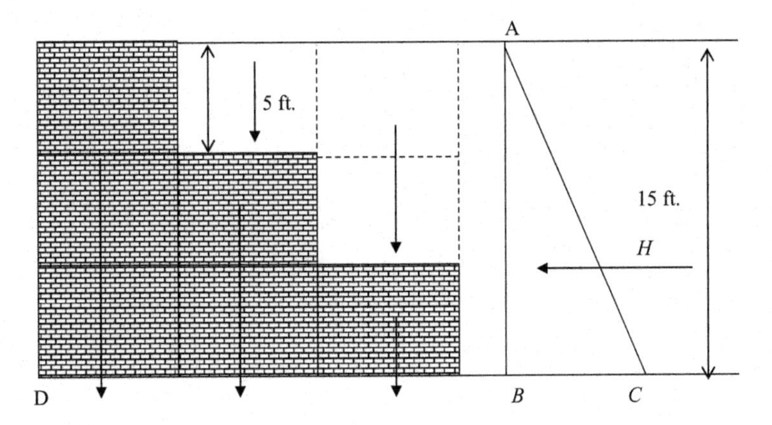

Figure 23.1 Gabion wall and forces.

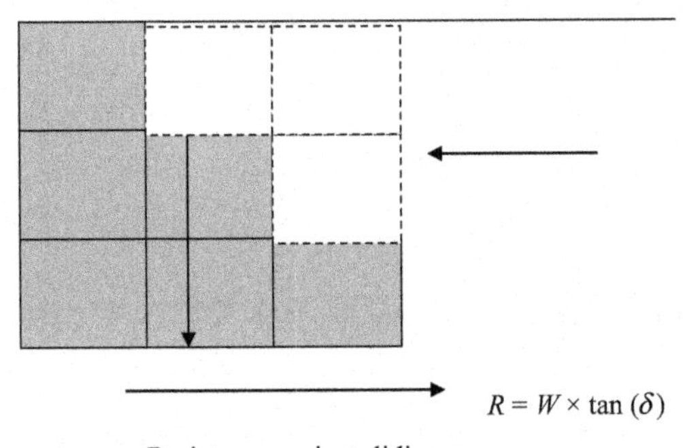

Figure 23.2 Forces acting on a gabion wall.

Step 4: Resistance against sliding (Figure 23.2)

Resistance against sliding = Weight of the gabion wall × tan (δ)

$$R = 33{,}000 \times \tan(20°)$$
$$= 12{,}011 \text{ lbs}$$

Factor of safety against sliding = resistance against sliding/horizontal force
= 12,011/4,500
= 2.67

Step 5: Overturning moment

Resultant force acts 1/3 of the length of the triangle.

Obtain moments around point "D".

Overturning moment = $H \times 1/3 \times 15$
= $4{,}500 \times 5$
= 22,500 lbs ft.

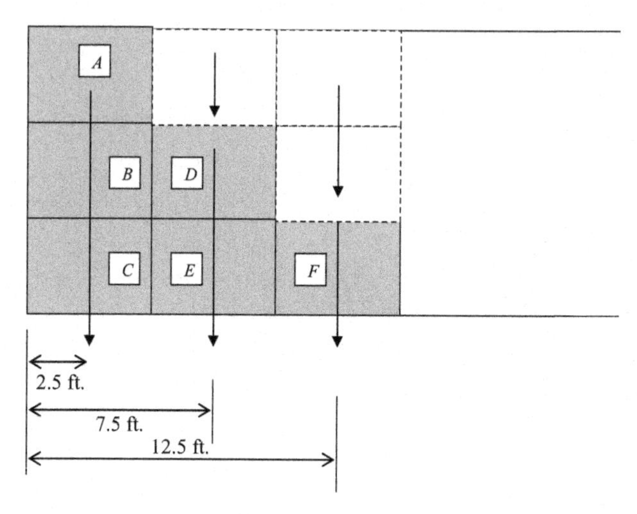

Figure 23.3 Weight due to gabion baskets.

Step 6: Resisting moment

There are six rock blocks and three soil blocks.

There are three blocks in the far left and one block at far right. There are two blocks in the middle (Figure 23.3).

Resisting moment due to gabion baskets

$$\text{Resisting moment due to three blocks in the left } (ABC) = 3 \times (5 \times 5) \times 160 \times 2.5 = 30{,}000 \text{ lbs ft.}$$

$$\text{Resisting moment due to two blocks } (DE) = 2 \times (5 \times 5) \times 160 \times 7.5 = 60{,}000 \text{ lbs ft.}$$

$$\text{Resisting moment due to one block } (F) = 1 \times (5 \times 5) \times 160 \times 12.5 = 50{,}000 \text{ lbs ft.}$$

Resisting moment due to soil sitting on top of gabion baskets

$$= 1 \times (5 \times 5) \times 120 \times 7.5 + 2 \times (5 \times 5) \times 120 \times 12.5$$

$$= 22{,}500 + 75{,}000$$

Resisting moment = 97,500 lbs ft.

Step 7:

$$\text{Factor of safety against overturning} = \text{resisting moment/overturning moment} = 97{,}500/22{,}500 = 4.33$$

23.2 Log retaining walls

Timber logs are arranged to build log walls. These type of retaining walls are rare today but can be used in wetland mitigation and for temporary structures (Figures 23.4 and 23.5).

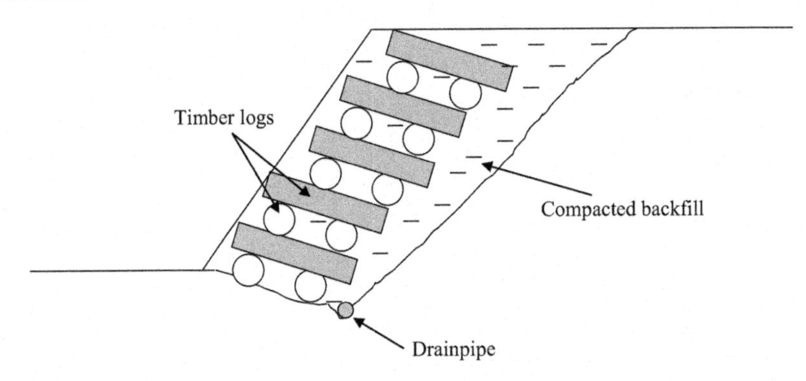

Figure 23.4 Log retaining wall (side view).

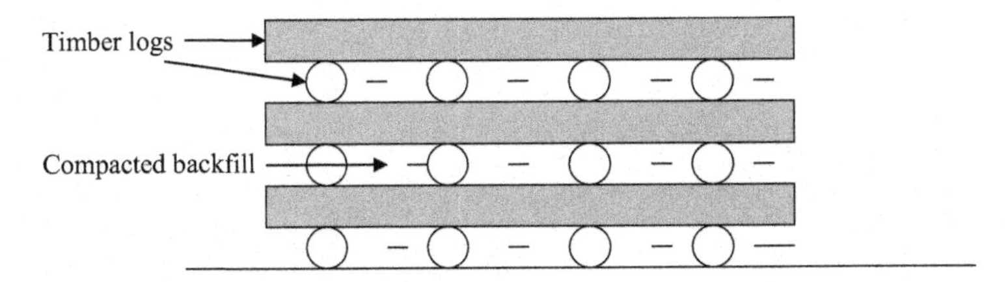

Figure 23.5 Log retaining wall (Frontal view).

Log walls can be constructed cheaply and still widely used in areas where timber is common.

23.2.1 Construction procedure of log walls

Figures 23.6 and 23.7 show placement of timber logs and placement of backfill between logs and compact soil, respectively.

- Place the next layer of logs and backfill.
- Continue the process.

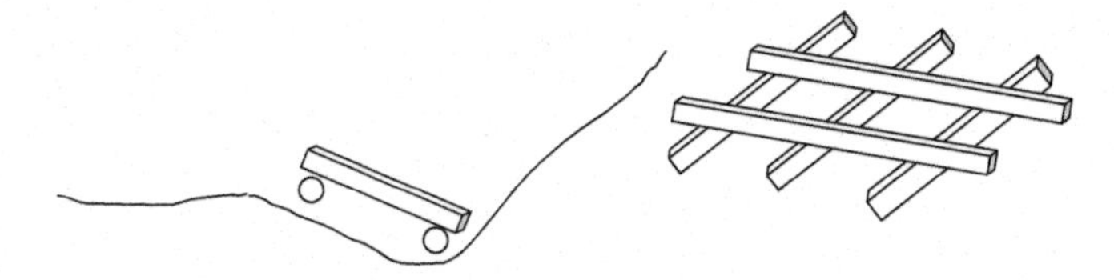

Figure 23.6 Placement of timber logs.

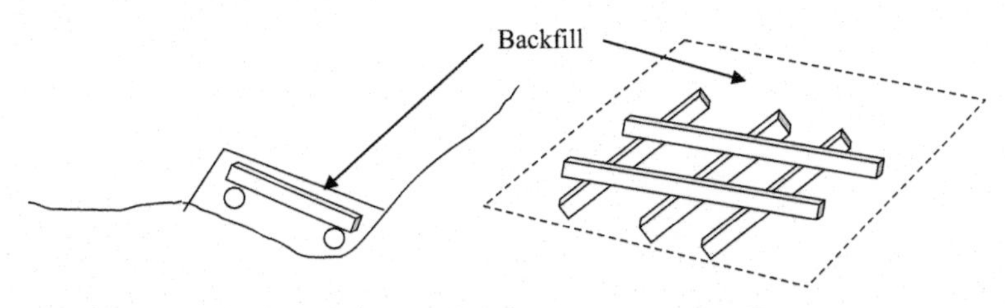

Figure 23.7 Placement of backfill between logs and compact soil.

Reinforced earth walls

24

Reinforced earth walls are becoming increasingly popular. Reinforced earth walls are constructed by holding the lateral earth pressure through metal strips (Figure 24.1).

Soil pressure in reinforced earth walls are resisted by metal strips (Figure 24.2).

Following equations can be developed to compute the lateral earth force acting on a facing unit.

Equations to compute the horizontal force on the facing unit (H): σ_v', Vertical effective stress; σ_h', $K_a \times \sigma_v'$ = horizontal stress; K_a, lateral earth pressure coefficient = $\tan^2(45 - \varphi/2)$; lateral soil pressure (F), $(K_a \times \sigma_v') \times$ area of the facing unit

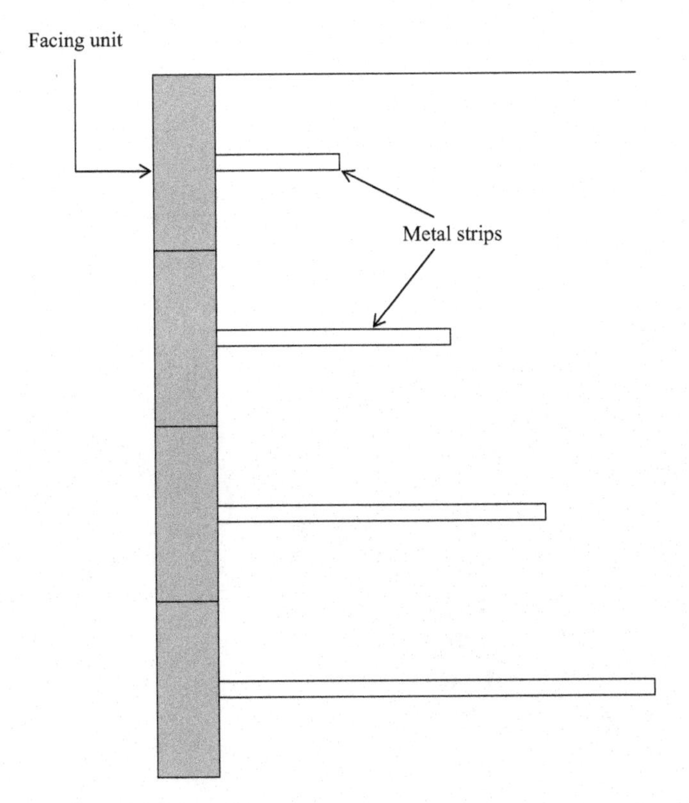

Figure 24.1 Reinforced earth wall.

Geotechnical Engineering Calculations and Rules of Thumb

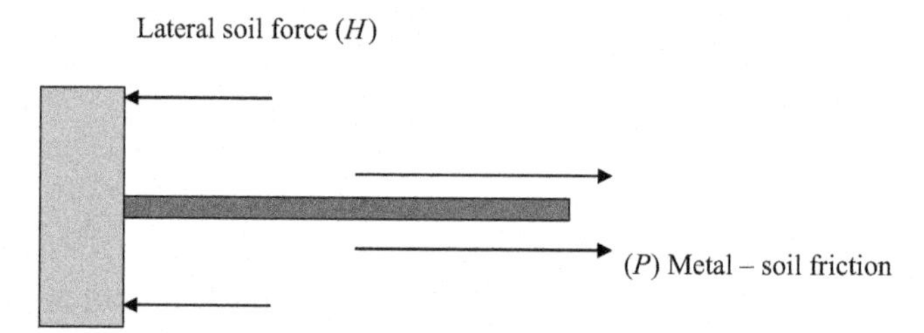

Figure 24.2 Forces acting on facing units and metal strips.

Equations to compute the metal – soil friction (P): The metal soil friction depends on the vertical effective stress acting on the strip, area of the metal strip and the friction angle between metal and soil.

Metal – soil friction (P) = $2 \times (\sigma_v' \times \tan \delta) \times$ area of the strip
δ, friction angle between metal and soil; σ_v', vertical effective stress.
The quantity is multiplied by 2, since there are two surfaces, top and bottom in the metal strip.
Factor of safety (FOS) = P/H
Plate 24.1 shows construction of metal strips.

Plate 24.1 Construction of metal strips.
Source: California DOT.

Design Example 24.1

Find the length of metal strips A, B and C in Figure 24.3. The density (γ) of soil is found to be 17.5 kN/m^3 and the friction angle (φ) of soil is 25°. Facing units are 1 m × 1 m and the metal strips are 0.5 m wide. Find the length of the metal strips if required factor of safety is 2.5.

Solution

Step 1: Find the vertical effective stress at the center of the facing unit
Facing unit "A"
Find the vertical effective stress at the center of the facing unit (σ_v');

$$\sigma_v' = \gamma \times h = 17.5 \times 6\ \text{kN/m}^2.$$
$$= 105\ \text{kN/m}^2$$

Find the lateral earth pressure coefficient (K_a);

$$K_a = \tan^2(45 - \varphi/2)$$
$$= \tan^2(45 - 25/2)$$
$$= 0.406$$

Find the horizontal stress at the center of the facing unit (σ_h')

$$\sigma_h' = K_a \times \sigma_v'$$
$$= 0.406 \times 105\ \text{kN/m}^2.$$
$$= 42.6\ \text{kN/m}^2.$$

Area of the facing unit "A" = Width × length = 1 × 1 m^2 = 1 m^2.

Total horizontal force on the facing unit = horizontal stress × area of the facing unit
= 42.6 kN

Step 2: Find the metal – soil friction in facing unit "A";

$$\text{Metal – soil friction } (P) = 2 \times (\sigma_v' \times \tan\delta) \times \text{area of the strip}$$
$$= 2 \times 105 \times \tan 20 \times 0.5L$$
$$= 38.2L$$

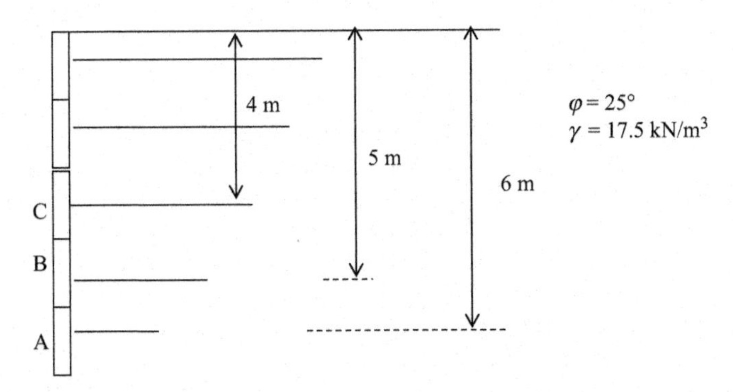

Figure 24.3 Reinforcing strips.

Required factor of safety (FOS) = 2.5

$$\text{FOS} = \text{Metal soil friction/horizontal force on the facing unit}$$
$$2.5 = 38.2L / 42.6$$
$$L = 2.79 \text{ m}$$

The length of the metal strip within the active soil zone does not contribute to the strength. Hence, the metal strip should extend 2.79 m beyond the active failure zone (Figures 24.4–24.6).

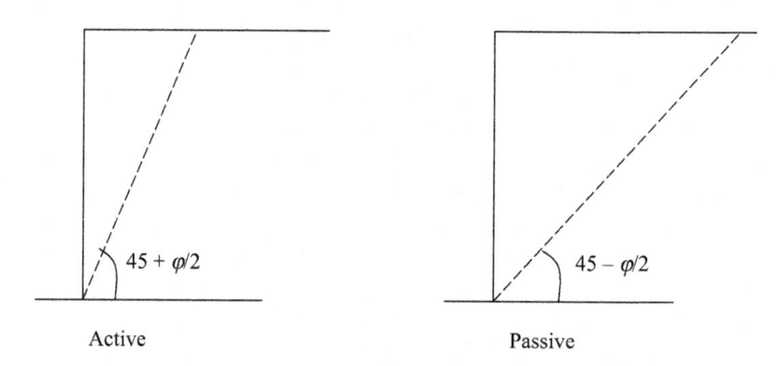

Figure 24.4 Active and passive failure planes.

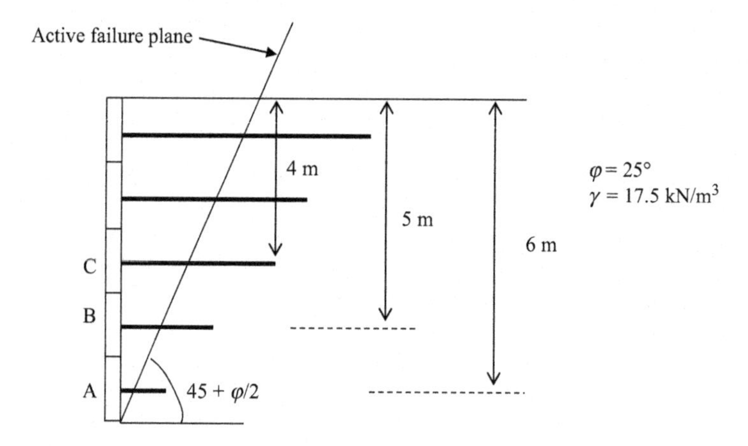

Figure 24.5 Metal strips and active failure plane.

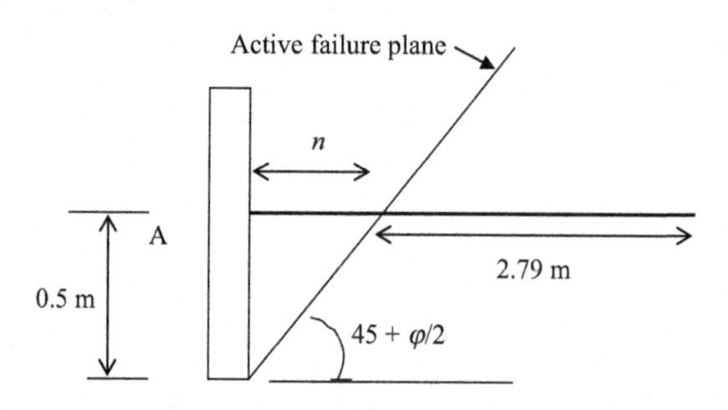

Figure 24.6 Facing unit "A" and the active failure plane.

Figure 24.6 shows the facing unit "A" and the active failure plane.
Distance "n" can be computed since the distance to the center of the facing unit is 0.5 m.

$$n = 0.5 \times \tan(45 - \varphi/2)$$

$$n = 0.5 \times \tan(45 - 25/2)$$

$$n = 0.32\text{m}$$

Total required length of the strip = 0.32 + 2.79 = 3.11 m.

Structural design of retaining walls

25

Gravity walls do not need structural design. Cantilever walls require structural design since they undergo tension. Concrete is good in resisting compression but not so good under tensile forces. When concrete undergoes tension, reinforcement bars need to be included.

Figure 25.1 shows a cantilever retaining wall. Figure 25.2 shows how the retaining wall would deflect due to earth pressure. Please note that the drawing is an exaggerated figure.

Structural rebars are required in the stem on the soil side. Also structural rebars are needed at the bottom of the toe and top of the heel.

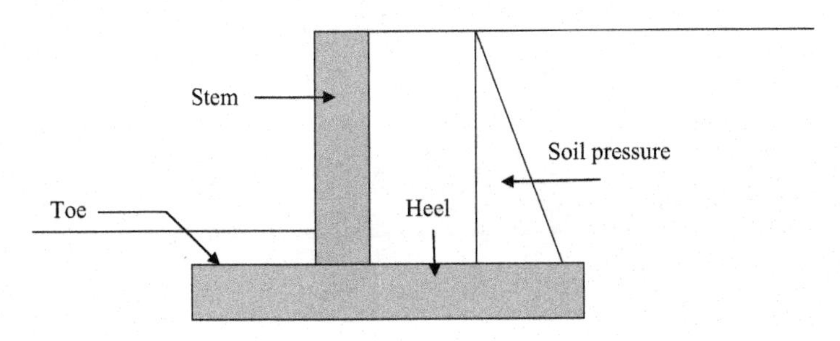

Figure 25.1 Cantilever retaining wall.

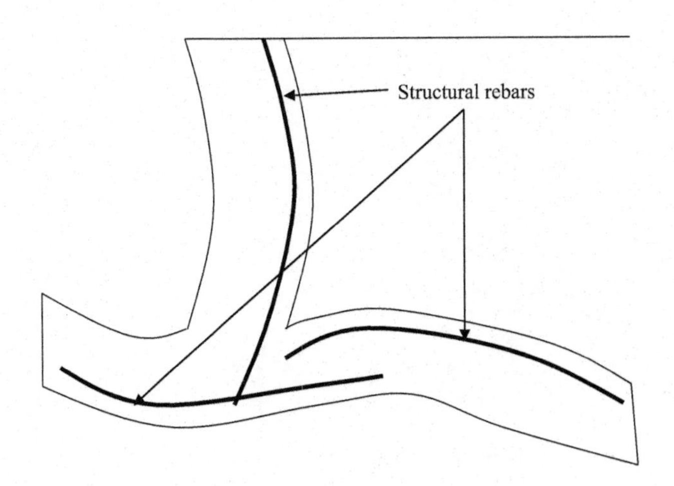

Figure 25.2 Bending of stem, toe, and heel (exaggerated diagram).

Geotechnical Engineering Calculations and Rules of Thumb

Finding bending moment in the stem for reinforcement design:
Figure 25.3 shows the bending moment acting on the stem.
Bending moment acting on the stem = $(K_a \gamma H) \times H/3$
Reinforcement bars need to resist this bending moment.
Plate 25.1 shows rebars for a cantilever retaining wall.

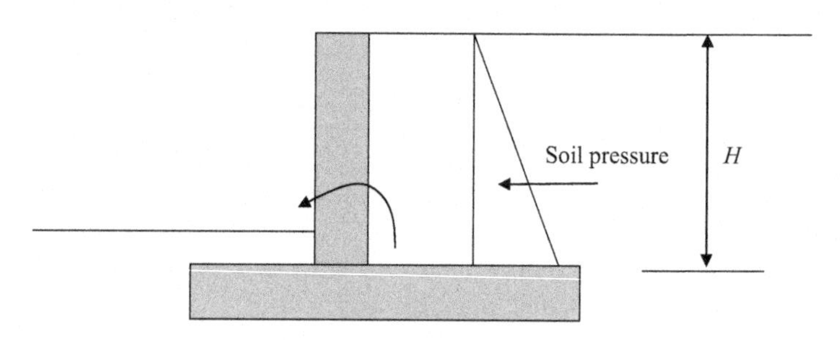

Figure 25.3 Bending moment acting on the stem.

Plate 25.1 Rebars for a cantilever retaining wall.

Part 4

Geotechnical Engineering Strategies

Geotechnical engineering software

26

There is a saying in computer science, "Garbage in garbage out". This means that one can have the best software in the world, but if you don't have soil data, then you will not get a good design. A software is a computer program that conducts computations. It is the responsibility of the design engineer to provide soil data, SPT/CPT data, soil density data, soil strata identifications, cohesion values, and friction angles.

The following procedure needs to be followed for any computer program:

1. Input soil strata data – The program should have a method to acquire soil strata data. Identification of soil strata and the thickness of each strata has to be given as an input to the program.
2. Input cohesion and friction angle values for each strata
3. Loading data

Once all the loading data and soil data are provided the program would be able to output a design for a footing.

There are many geotechnical engineering software available in the market. In this chapter, an introduction to some of the more famous software will be discussed.

26.1 Shallow foundations

26.1.1 SPT foundation

SPT Foundation is a program available at www.geoengineer.org/sptprogram.html.

The program computes the bearing capacity of shallow foundations using Terzaghi's bearing capacity equation. The SPT blow count in the soil is used to obtain the friction angle to be used in the Terzaghi's bearing capacity equation. The program uses the correlation provided by Hatanaka and Uchida (1996).

The program is capable of conducting the settlement of foundations as well. The average SPT (N) value within the depth of influence below the footing is used for settlement computations. A portion of the program is available without charge. As per authors of the website, they use the method proposed by Burland and Burbidge (1984).

26.1.2 ABC bearing capacity computation

Free programs from the following web site are capable of conducting bearing capacity computations:

http://www-civil.eng.ox.ac.uk/people/cmm/software/abc/.

ABC can be used to solve bearing capacity problems with many layers of soil. The program can be used to solve rigid foundations resting on a cohesive-frictional soil mass, loaded to failure by a central-vertical force. The program provides option for the cohesion to vary linearly with depth.

In reality, such situations may be rare. The program authors claim that the computations are done using a finite element grid using stress analysis without resorting to approximations. The program is fully documented and provided with a user manual. It is a freeware at present, but will probably be commercially sold soon.

26.1.3 Settle 3D

Settle 3D is a program by www.rocscience.com for analysis of consolidation and settlement under foundations, embankments, and surface excavations. The program is capable of conducting a 3D analysis of foundation settlement.

26.1.4 Vdrain – consolidation settlement

Vdrain is a free program available from www.cofra.com to calculate settlement and consolidation of soft soils.

26.1.5 Embank

Embank is a program that computes settlement under embankment loads. Embankments are common for bridge abutments. The program computes vertical settlement due to embankment loads. For the case of a strip symmetrical vertical embankment loading, the program superimposes two vertical embankment loads. For the increment of vertical stresses at the end of fill, the program internally superimposes a series of 10 rectangular loads to create the end of fill condition. This program can be downloaded free of charge from the FHWA website.

26.2 Slope stability analysis

26.2.1 Reinforced soil slopes (RSS)

This program is capable of designing and analyzing reinforced soil slopes (RSS). The programs are based on the FHWA manual, *Reinforced Soil Structures Volume I – Design and Construction Guidelines* (FHWA-RD-89-043).

This program analyzes and designs soil slopes strengthened with horizontal reinforcement. The program also analyzes unreinforced soil slopes. The analysis is performed using a two-dimensional limit equilibrium method. The program is a predecessor of a very popular STABL computer program of the 1990s (Federal Highway Administration, USA, http://www.fhwa.dot.gov/index.html).

26.2.2 *Mechanically stabilized earth walls (MSEW)*

The program MSEW is capable of conducting design and analysis of mechanically stabilized earth walls. The program is available from http://www.msew.com/downloads.htm.

26.3 Bridge foundations

26.3.1 *FB-MultiPier*

The FB-MultiPier is capable of analyzing bridge piers connected to each other. Each pier is considered to be supported on piles or caissons. The settlement and load bearing capacity of foundations can be computed using this program. The authors state that this program conducts a nonlinear structural finite element analysis and a nonlinear static soil models for axial and lateral soil behavior.

The product needs to be licensed prior to service and can be downloaded from http://bsi-web.ce.ufl.edu/products/.

26.4 Rock mechanics

This is a freeware available from Southern Illinois University, Carbondale. The program can be downloaded from their website: http://www.engr.siu.edu/mining/kroeger/.

The program considers rock joints and water pressure to compute the stability (Figure 26.1).

26.4.1 *Wedge failure analysis*

Wedge failure is important for tunneling engineers and rock slope stability computations. The formation of wedges due to joints needs to be analyzed. The free computer

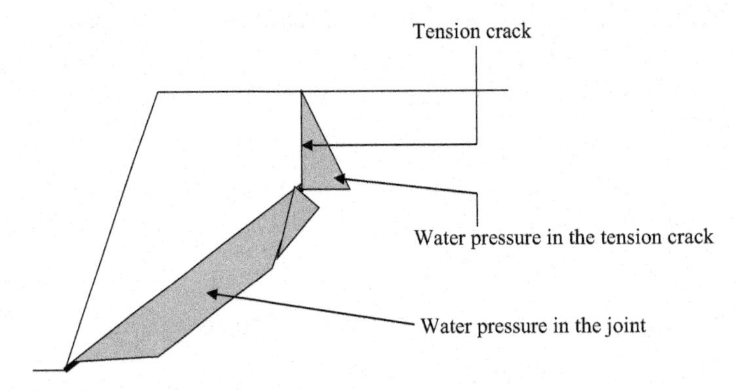

Figure 26.1 Rock slope stability.

program by http://www.engr.siu.edu/mining/kroeger/ can be used to analyze the formation of wedges in a rock formation.

26.4.2 Rock mass strength parameters

RocLab is a software program available from http://www.rocscience.com/ for determining rock mass strength parameters. The user can use the program to visualize the effects of changing rock mass parameters, on the failure envelopes. Rocscience.com has collaborated with E. Hoek, one of the most distinguished authorities in the rock mechanics field and it provides very valuable information on rock mechanics.

26.5 Pile design

26.5.1 Spile

Spile is a versatile program to determine the ultimate vertical static pile capacity. The program is capable of computing the vertical static pile capacity in clayey soils and sandy soils. The program is designed based on the equations presented by Nordlund (1979), Meyerhof (1976), and Tomlinson (1985).

26.5.2 Kalny

This software conducts pile group analysis. This program provides pile forces for regular and irregular pile groups for multiple load combinations in a single spreadsheet. Check governing forces for "corner" piles or review forces on all piles in a two dimensional plane.

In certain situations, certain piles may have to be discarded due to dog legging or damage during driving. In such situations, new piles have to be installed.

26.6 Lateral loading analysis – computer software

When a lateral load (P) is applied as shown in Figure 26.2, the following resistances are developed:

PS1, Passive soil resistance due to pile cap on one side of the pile cap; $S1$, skin friction at the base of the pile cap; $P1$ and $P2$, lateral soil resistance of piles.

If the pile cap is connected to other pile caps with tie beams, there would be resistance due to tie beams as well.

26.6.1 Lateral loading analysis using computer programs

Input parameters to computer programs are twofold.

1. Pile parameters
2. Soil parameters

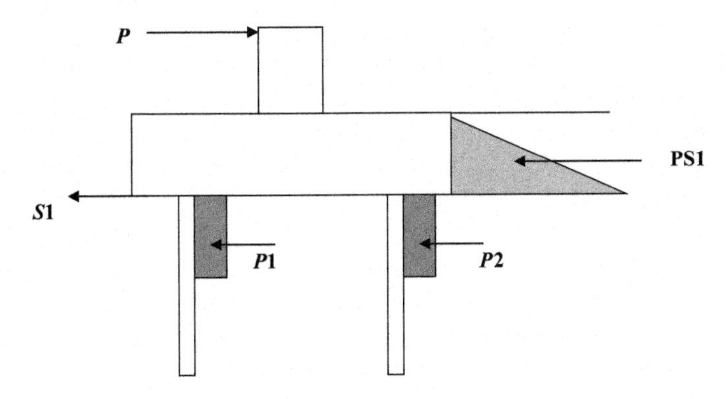

Figure 26.2 Lateral loading analysis.

26.6.2 Pile parameters

- Pile diameter
- Center to center spacing of piles in the group
- Number of piles in the group
- Pile cap dimensions

26.6.3 Soil parameters for sandy soils

Soil parameters should be provided to the computer for each strata.

- Strata thickness
- ϕ' value of the strata
- Coefficient of subgrade reaction (k)

Note: ϕ' value of sandy soil can be calculated using the following equation:
$\phi' = 53.881 - 27.6034.e^{-0.0147N}$ (Peck et al., 1974)
N, average SPT value of the strata.

Note: The coefficient of subgrade reaction (k) can be obtained using Table 26.1. Table 26.1 shows the coefficient of subgrade reaction (k) versus N (SPT).

Similarly, the soil parameters for other strata also need to be provided.

Table 26.1 Coefficient of subgrade reaction (k) versus N (SPT)

SPT (N)	8	10	15	20	30
k (kN/m^3)	2.67 E-6	4.08 E-6	7.38 E-6	9.74 E-6	1.45 E-6

Source: Johnson and Kavanaugh (1968).

Table 26.2 **Coefficient of subgrade reaction versus undrained shear strength**

Coefficient of subgrade reaction	Average undrained shear strength (tsf)		
	0.5–1 tsf	1–2 tsf	2–4 tsf
k_s (static) lbs/in.^{3}	500	1000	2000
k_s (cyclic) lbs/in.^{3}	200	400	800

Source: Reese (1975).

26.6.4 Soil parameters for clayey soils

Soil parameters required for clayey soils are as follows:

- S_u (undrained shear strength. "S_u" is obtained by conducting unconfined compressive strength tests)
- ε_c (strain corresponding to 50% of the ultimate stress. If the ultimate stress is 3 tsi, then "ε_c" is the strain at 1.5 tsi).
- k_s (coefficient of subgrade reaction)

Coefficient of subgrade reaction for clay soils is obtained from Table 26.2. Table 26.2 shows the coefficient of subgrade reaction versus undrained shear strength.

26.7 Finite element method

Most geotechnical engineering software are based on the finite element method.

Finite element method is considered to be the most powerful mathematical method that exists today for solving piling problems.

- Any type of soil condition could be simulated using the finite element method.
- A complicated soil profile is shown in Figure 26.3. Nodes in each of the finite element, is given the soil properties of that layer such as ϕ, γ (density), cohesion, SPT (N) value, etc.
- Nodes of elements in layer 1 is given the soil properties of layer 1. Similarly, nodes on layer 2 will be given the soil properties of layer 2.
- Due to this flexibility, the isolated soil pockets can also be effectively represented.

26.7.1 Representation of time history

- Capacity of a pile is dependent upon the history of loading. A pile that was loaded gradually would have a higher capacity than a pile that was loaded rapidly.
- Assume that a developer is planning to construct a 10-story building in 5 years. In this case, a full building load on piles would gradually develop in a time period of 10 years.
- On the other hand, if the developer changes his mind and decides to construct the 10-story building in 2 years, the full load on piles would develop in 2 years. If the piles were to be fully loaded in 2 years, the capacity of piles would be less than the first scenario.
- In such situations, the finite element method could be used to simulate the time history of loading.

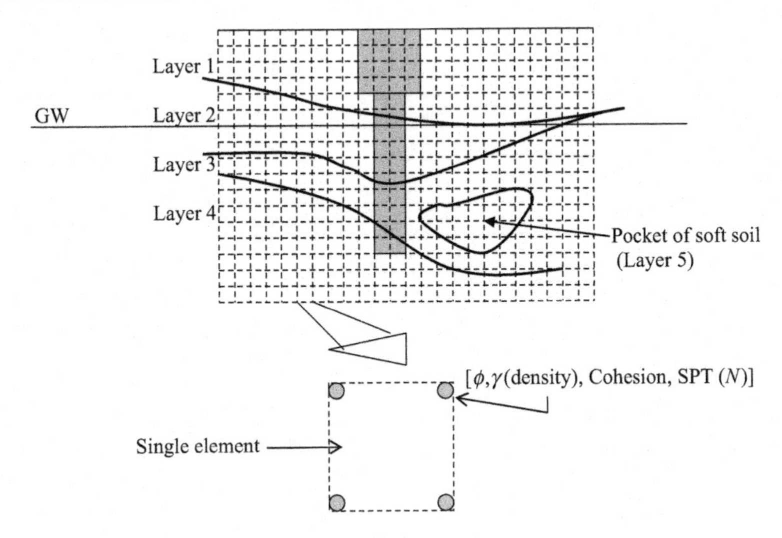

Figure 26.3 Finite element grid.

26.7.2 Groundwater changes

Change of groundwater conditions also affect the capacity of piles. Change of groundwater level can be simulated easily in finite element method.

26.7.3 Disadvantages

The main disadvantage of the finite element method is its complex nature. In many cases, the engineers may wonder whether it is profitable to perform a finite element analysis.

26.7.4 Finite element computer programs

Computer programs are available with finite element platforms. These programs can be used to solve a wide array of piling problems. The user is expected to have a working knowledge of finite element analysis to use these programs. More specialized computer programs are also available in the market. These programs do not require the knowledge of finite element analysis.

26.8 Boundary element method

- Boundary element method is a simplified version of the finite element method. In this method, only the elements at boundaries are considered.
- Only the elements at soil pile boundary are represented .

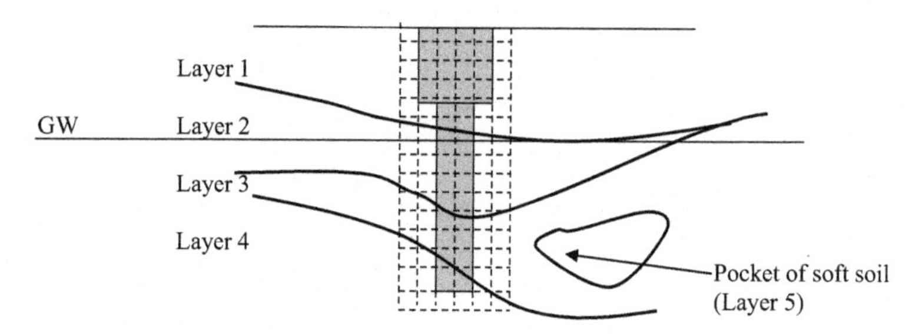

Figure 26.4 Boundary element method.

- In this method, a full soil profile is not represented. As one can see in Figure 26.4, the isolated soft soil pocket is not represented.
- On the other hand, less elements will make the computational procedure much simpler than the finite element method.

References

Burland, J.B., Burbidge, M.C., 1984. Settlement of foundations on sand and gravel. Proceedings of the Institution of Civil Engineers, Part 1. pp. 1325–1381, 1985, Dec., 78.

Hatanaka, M., Uchida, A., 1996. Empirical correlation between penetration resistance and effective friction of sandy soil. Soils Found. 36 (4), 1–9.

Johnson, S.M., Kavanaugh, T.C., 1968. The Design of Foundations for Buildings. McGraw Hills, NY.

Meyerhof, G.G., 1976. Bearing capacity and settlement of pile foundations. J. Geotech. Eng. 102 (GT3), 195–228, paper 11962.

Nordlund, R.L., 1979. Point bearing and shaft friction of piles in sand. In: Fifth Annual Fundamentals of Deep Foundation Design, University of Missouri, Rolla.

Peck, et al., 1974. Foundation Engineering. John Wiley and Sons, NY.

Reese, L.C., 1975. Field testing and analysis of laterally loaded piles in stiff clay. In: Proceedings – Offshore Technology, 1975. Conference, vol. II, Houston, TX.

Tomlinson, M.J., 1985. Foundation design and construction, Longman Scientific and Technical, Essex, England.

Geotechnical instrumentation 27

27.1 Inclinometer

Assume that you are driving piles on a slope. There is a concern that the slope may fail due to the pile-driving operation. In such situations, inclinometers can be installed to monitor any movement in the slope. Generally, prior to any major slope failure there are small movements. These slight movements can be detected by inclinometers. Slight movements in slopes occur due to various natural reasons such as wetting and drying of soils, small earthquakes, freezing and thawing of soil, etc. All these slight movements in slopes can be detected by inclinometers depending upon the accuracy of the inclinometer.

Now let us look at how an inclinometer works. First, a borehole is drilled and a casing is installed. The casing is grouted around the annulus so that it is thoroughly fixed to the ground. The casing will not move unless that ground moves.

Figure 27.1 shows a borehole and the casing, inclinometer, and a data logger. Inclinometer is a probe with two rollers. When the inclinometer is inserted into the borehole, it would provide the vertical profile of the borehole. The inclinometer is inserted into the borehole and it travels along the casing. The inclinometer would send signals to the data logger regarding the verticality of the borehole. If the borehole is vertical, it would give a vertical line as shown on the right-hand side of Figure 27.1. Figure 27.2 shows the same borehole and casing after slope movement.

When there is a movement in the slope, the borehole and the casing would deform. When the inclinometer is inserted, it would draw a profile of the borehole in the data logger. As you can see in this case, the profile is not vertical. By looking at the data, one can tell whether the slope has moved. In addition, if the slope has moved, the inclinometer data can be used to obtain the extent of the movement.

Figure 27.3 shows a typical inclinometer data set. As you can see, the leftmost line is almost vertical. On Day 2, it has moved. On Day 3, it has moved further. It is obvious that this is an unstable slope.

27.1.1 In-place inclinometers

The main disadvantage of portable inclinometers is that they do not provide instant information. We would not know any slope movement until someone lowers the inclinometer into the borehole. If instant information is to be recorded, in-place inclinometers should be installed.

In-place inclinometers are shown in Figure 27.4. In Figure 27.4, two in-place inclinometers are shown installed at different depths. These two inclinometers are connected to a data logger. When there is movement, the inclinometers would immediately log the data. Hence, it is possible to know the time of the slope movement. This information is important. For instance, if there is a movement when there is pile

Geotechnical Engineering Calculations and Rules of Thumb

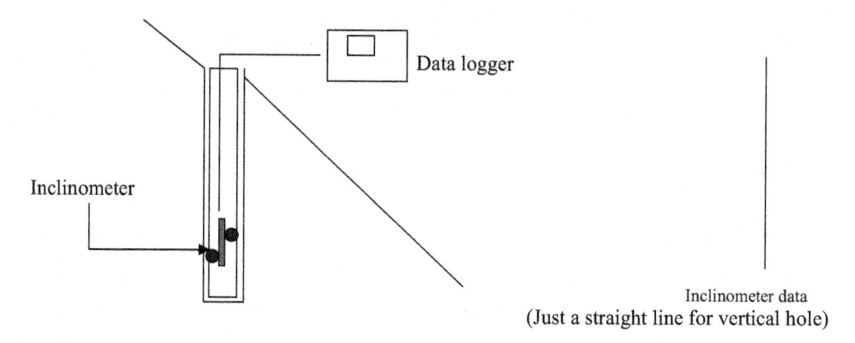

Figure 27.1 Inclinometer.

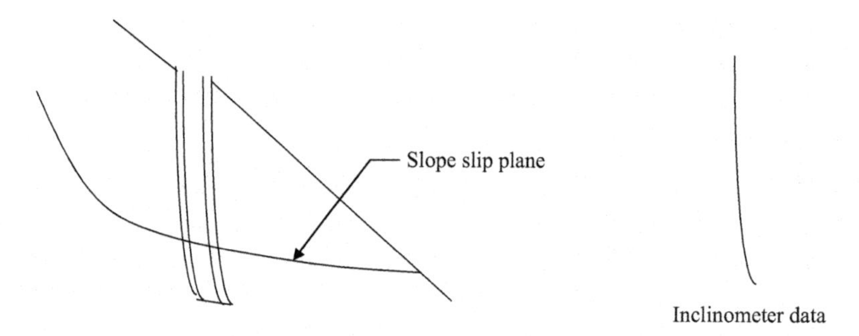

Figure 27.2 Slope movement.

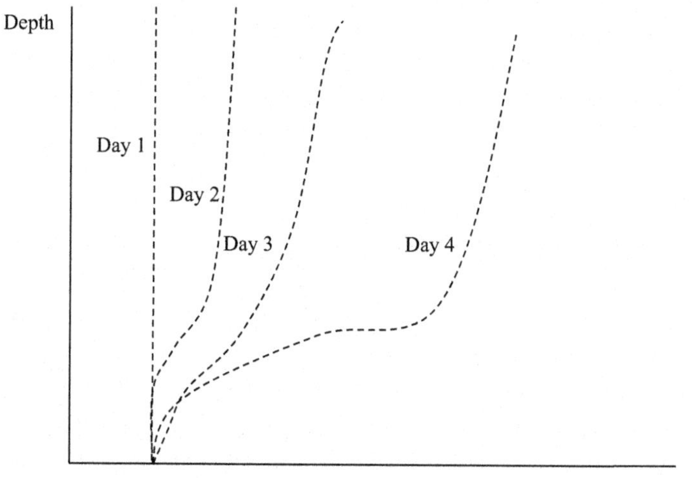

Figure 27.3 Inclinometer data in different days.

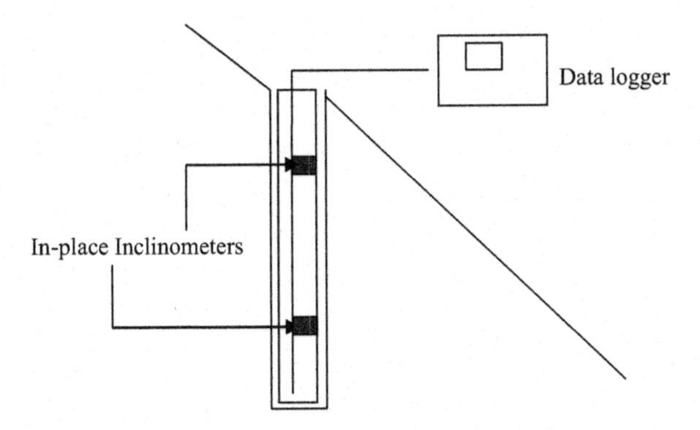

Figure 27.4 In-place inclinometer.

driving or any other construction activities we can deduce that the slope movement is occurring due to such activities.

Practice Problem

Borings were to be done in a slope. There is a concern that due to drilling operation, the slope would be unstable. Provide a method to measure the slope movement.

Solution

Install inclinometers extending below the potential slip plane as shown in Figure 27.5.

If the slope moves along the assumed slip plane, the inclinometer profiles would be as shown in Figure 27.6.

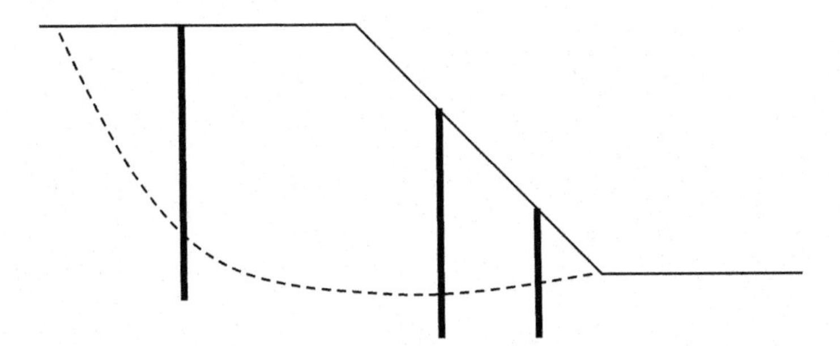

Figure 27.5 Inclinometers in a slope.

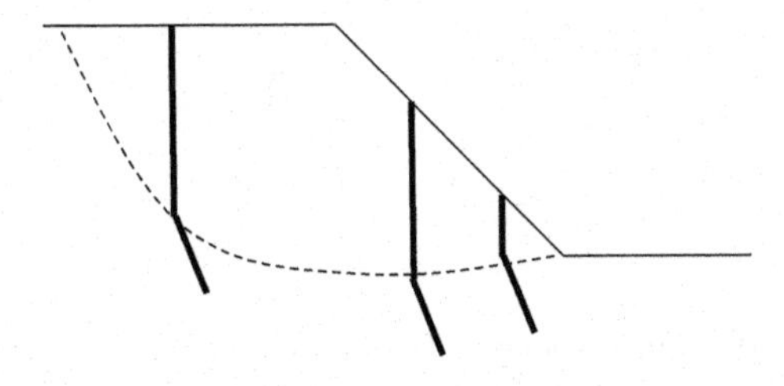

Figure 27.6 Movement of slope.

Inclinometers can be used to predict the slip plane. A slight movement of soil has been observed prior to most landslides, earthquakes, and slope failures. Some believe that insects, horses, and pigs feel these slight movements. An inclinometer is designed to measure the slight ground movements that occur prior to a slope failure (Figure 27.7).

- An inclinometer is shown in Figure 27.8. The inclinometer is lowered into the well. The wheels are fixed at the side to guide through the well casing.
- The pendulum has an electrical potential. When the inclinometer is vertical, both detectors (detector "A" and "B") feel the same potential.
- When the inclinometer is inclined as shown in Figure 27.8b, by an angle of "α", the pendulum is closer to detector "A" than to detector "B". Hence, detector "A" would record a higher reading than detector "B". This electrical potential difference can be utilized to obtain the angle of inclination (or angle "α").

27.1.2 Procedure

- The well is installed.
- Most inclinometers are approximately 2 ft. long. The inclinometer is inserted into the well and a reading (angle ϕ) is taken from 0 ft. to 2 ft. If the well casing is vertical, the reading should be 90°.
- Next, the inclinometer is inserted 2 ft. more and another reading is taken. The process is continued to the bottom of the well.

Obtained readings would look like as shown in Table 27.1.

- The next set of readings were taken a day or two later. If the soil had moved, the inclinometer would record different readings. A typical set of readings are shown in Table 27.2.

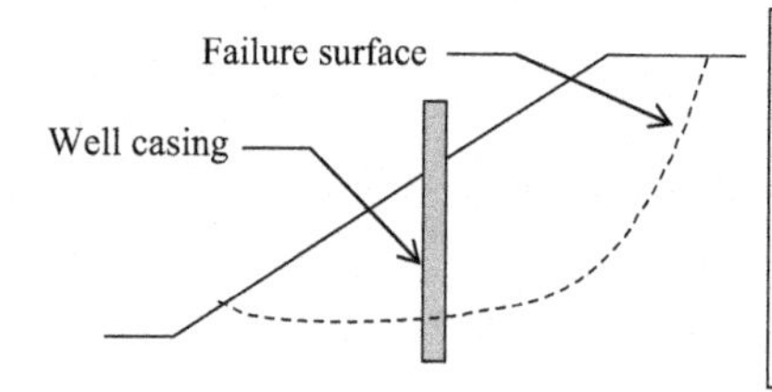

- First, a well is installed. The well is constructed vertically.
- Next, the inclinometer is lowered.
- If the well is properly installed, the inclinometer readings would show that the well is vertical.

Figure 27.7 Inclinometer.

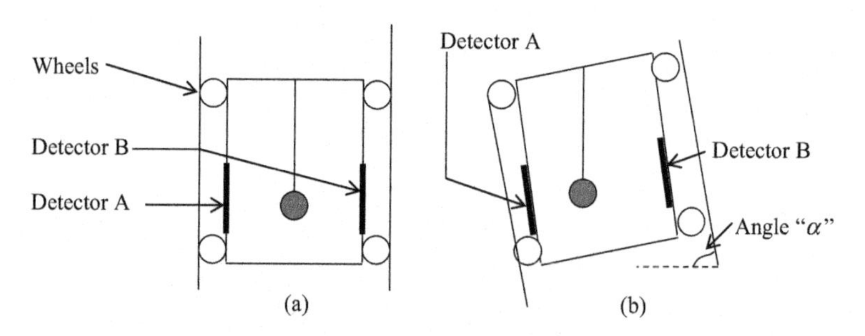

Figure 27.8 Working of the inclinometer. (a) Vertical inclinometer, (b) inclined inclinometers.

Table 27.1 **Inclinometer readings**

Depth (ft.)	0–2	2–4	4–6	6–8	8–10	10–12	12–14	14–16	16–18	18–20	20–22	22–24	24–26	26–28
Angle "α"	90.0	90.0	90.2	90.3	90.3	90.4	90.1	90.5	90.3	90.5	90.1	90.2	90.7	90.3

Table 27.2 **Second set of inclinometer readings**

Depth (ft.)	0–2	2–4	4–6	6–8	8–10	10–12	12–14	14–16	16–18	18–20	20–22	22–24	24–26	26–28
Angle "α"	90.0	90.0	90.1	90.2	88.3	88.1	87.3	87.2	89.1	88.5	88.3	89.1	89.0	89.2

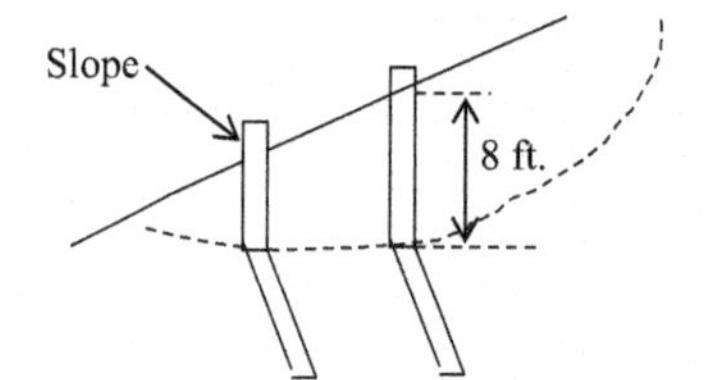

- At a depth of 8 ft., the inclinometer has bent as shown. This would indicate that the slope is not stable.
- The slope indicator results could be used to identify the *failure plane.*

Figure 27.9 Measuring soil movement.

As you can see, at a depth of 8–10 ft., the inclinometer has bent since the angle had deviated from 90°.

It is possible to identify the slip circle, by installing many inclinometers (Figure 27.9).

27.2 Extensometers

Extensometers are used to check the movement of rock in rock joints. A hole is drilled in rock and the extensometers are fitted (Figure 27.10). Any movement in the rock will be recorded by extensometers.

In Step 1, a borehole is drilled in rock. In Step 2, the anchors are inserted. In step 3, the anchors are expanded to lock in with the surrounding rock. There are many expanding mechanisms. There are snap ring types where a ring is snapped and the anchors expand and hug the rock. Then there are hydraulic types (Figure 27.11).

In Step 4, the rock joint moves and the distance between the two anchors changes due to movement in the rock joint. The change of distance between the two anchors will be recorded in the data logger. Hence, the movement at the rock joint can be assessed.

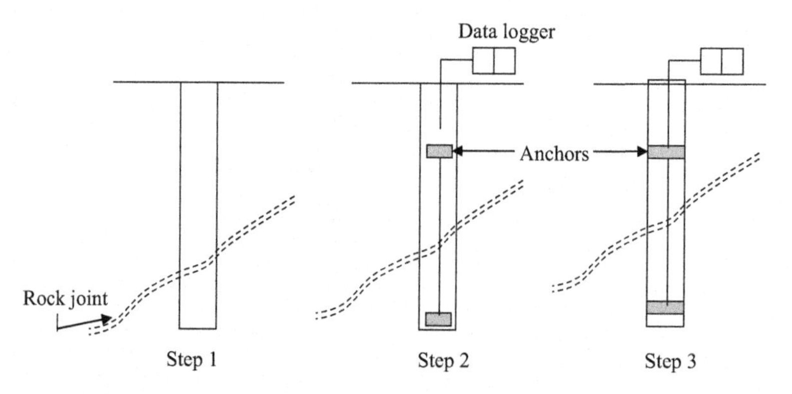

Figure 27.10 Extensometer installation procedure. Step 1: Drill a hole in rock. Step 2: Insert extensometers. Step 3: Expand the anchors.

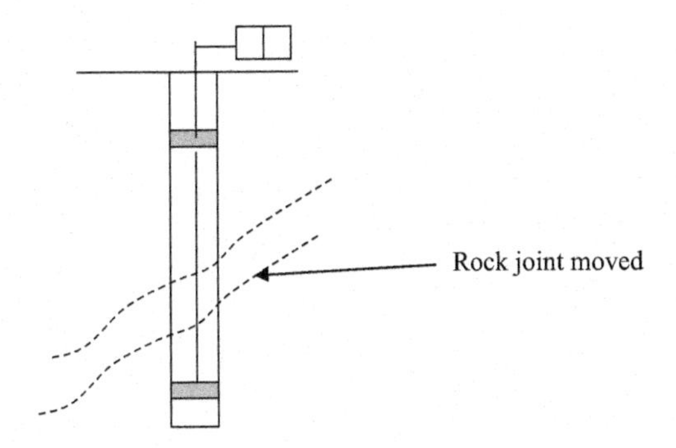

Figure 27.11 Extensometer in action.

27.3 Rock pressure gauge

Slight rock movement can be measured by using a pressure gauge (Figure 27.12). A pressure gauge is designed with an inflatable soft pack. The soft pack material can be expanded. Once expanded, the pressure gauge will be tightly wedged between rocks. Any rock movement would increase or decrease the pressure in the gauge.

27.4 Settlement plates

Settlement is an important parameter that needs to be monitored. Settlement monitoring is required during soil surcharging. Soil surcharging is done to compress and consolidate the existing clay soil. During consolidation, the ground would settle. Settlement of the ground needs to be monitored to assess the consolidation of the clay layer. In addition, there are situations where cracks develop in a building. Cracks in a building could be due to weak concrete, rusting of rebars, lack of expansion joints, or settlement of footings. A settlement plate is a very simple apparatus. It has a plate on the bottom and a rod. The rod is placed inside a PVC tube (Figure 27.13).

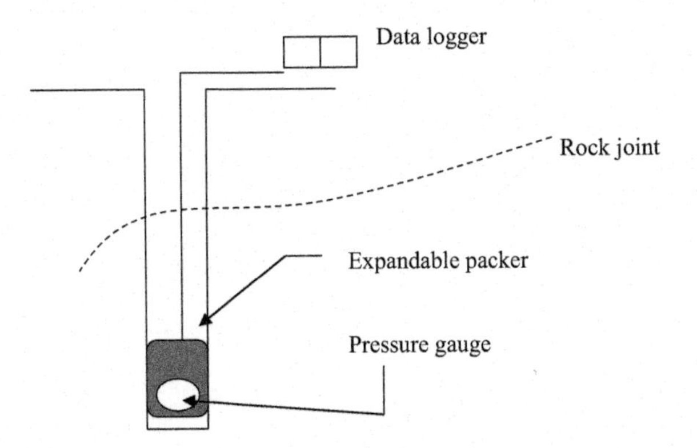

Figure 27.12 Rock pressure gauge.

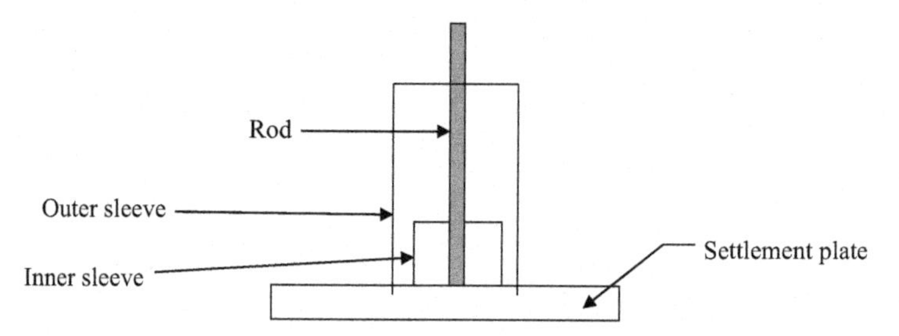

Figure 27.13 Settlement plate.

The rod and the inner sleeve are attached to the settlement plate. Hence, when the plate settles the rod and inner sleeve also settle. The amount of settlement can be obtained by measuring the settlement of the rod.

Assume that you need to obtain the settlement of ground due to an embankment. In this case, a settlement plate can be installed prior to constructing the embankment.

Figure 27.14 shows original ground and a settlement plate. The rod is welded to the settlement plate. Now construct the embankment (Figure 27.15).

When the ground settles due to the embankment, the plate also would settle. The settlement can be measured by shooting the elevation of the top of rod.

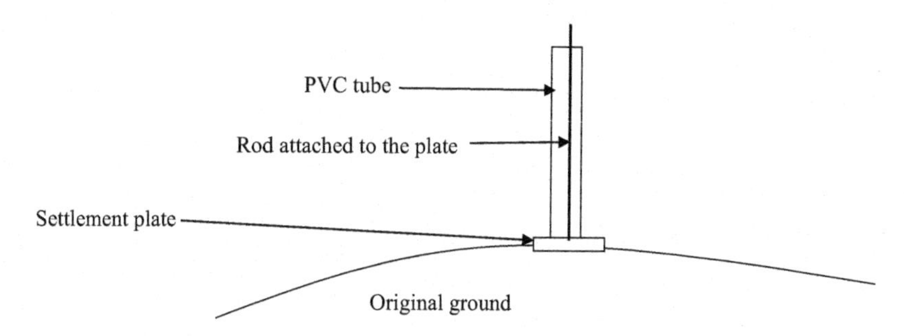

Figure 27.14 Settlement plate placed on ground.

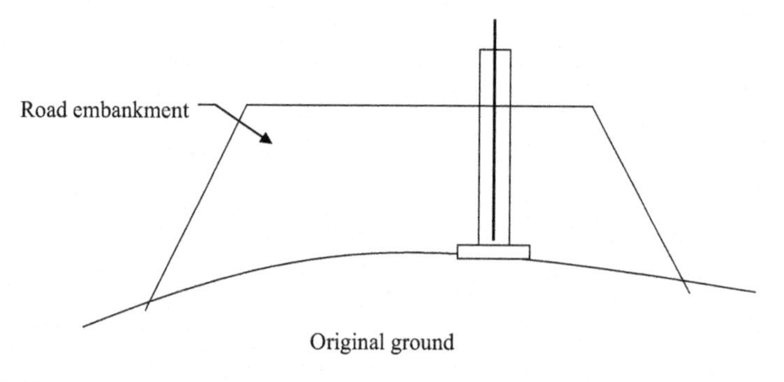

Figure 27.15 Embankment and settlement plate.

27.5 Borros anchors (settlement monitoring)

Borros anchors are installed deep inside the ground to monitor the settlement. Following is the procedure to install the Borros anchors (Figure 27.16):

Step 1: Drill a borehole.
Step 2: Install a PVC casing and backfill or grout the annulus. Annulus is the space between PVC pipe and borehole.
Step 3: Install the Borros anchor.
Step 4: Extend the prongs.

The Borros anchor is anchored to the soil. Hence, any settlement in the soil at the anchor level can be measured. This type of anchor is well suited for clay soils. Loose sandy soils may not move the anchor with soil.

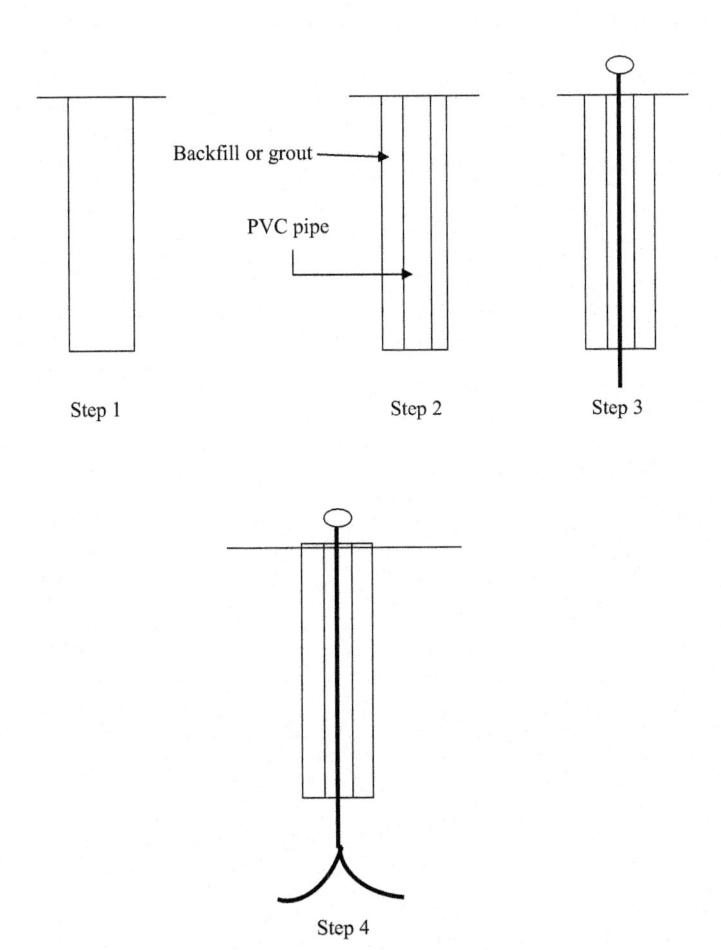

Figure 27.16 Installation of Borros anchors. Step 1: Drill a borehole. Step 2: Insert a PVC pipe and backfill. Step 3: Install the Borros anchor. Step 4: Extend the prongs.

Practice Problem

A building is constructed as shown in Figure 27.17. The top layer is medium dense sand followed by red clay and gray clay. Cracks started to appear in the concrete walls of the house. The concrete strength of concrete walls was measured and no problem was found. In addition, the rebars seem to be in good condition. Geotechnical engineers suspect that the settlement could be the reason for cracks. Devise a plan to identify the settling soil strata.

Solution

Install the three Borros anchors as shown in Figure 27.18.

Borros anchors can be installed as shown in the Figure 27.18. This setup can be used to identify the settling layer. In addition, a multianchor borehole can be used.

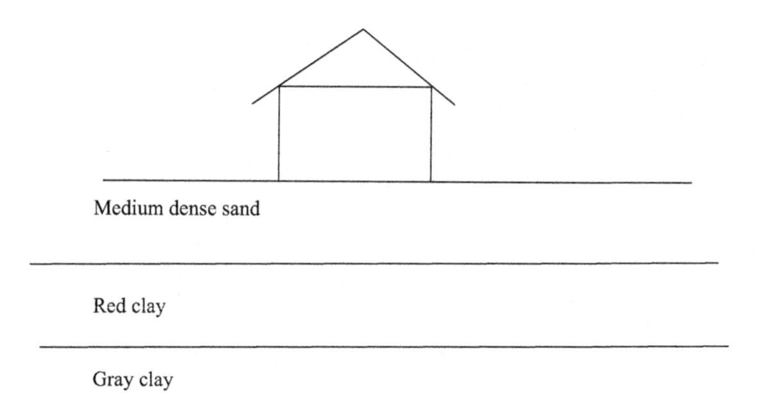

Figure 27.17 Structure on layered soil.

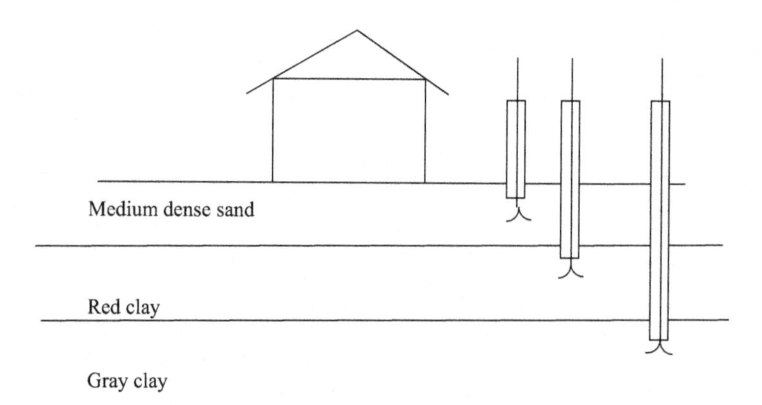

Figure 27.18 Installation of Borros anchors in layered soil.

27.6 Tiltmeter

- The operation of the tiltmeter is similar to a level. In a regular level, air bubble is used to identify the tilt. If the air bubble is at the center, then the tilt is zero.
- Tiltmeters are filled with an electrically sensitive liquid instead of water. When the tiltmeter is tilted, the bubble moves.
- The electrical signal coming from the tiltmeter is dependant upon the location of the bubble.
- Tiltmeters are used to measure the tilt in buildings, bridges, tunnels, etc.

27.6.1 Procedure

- Tilt plates are attached to the building wall, as shown in Figure 27.19.
- A tiltmeter is temporarily fixed to the plate. The tilt is measured. The tiltmeter is removed and the tilt in the next plate is measured. One tiltmeter can be used to measure the tilt of many tilt plates.

Horizontal tiltmeters can be installed to investigate the soil movement due to foundations or embankments (Figure 27.20).

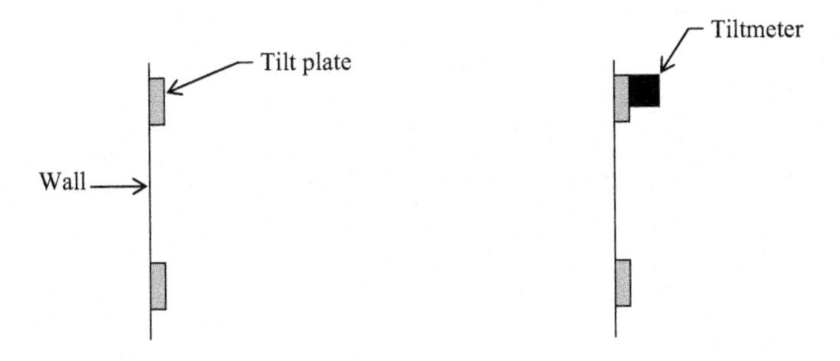

Figure 27.19 Tilt plates.

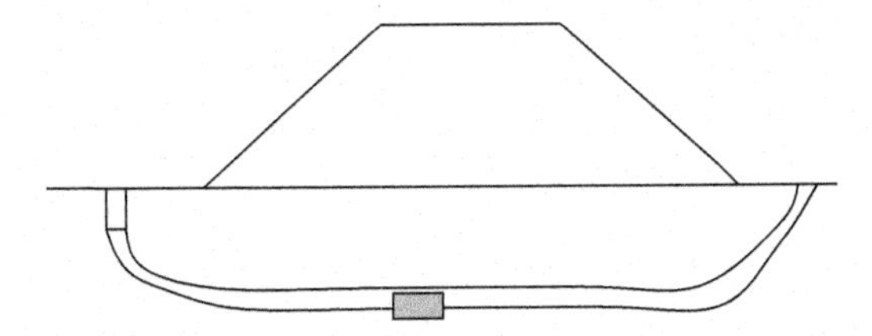

Figure 27.20 Horizontal tiltmeters.

Unbraced excavations

28

28.1 Introduction

Excavations less than 15 ft. deep can be constructed without any supporting system under most soil conditions. Excavations constructed without a supporting system are known as unbraced excavations. The excavations need to be properly sloped in unbraced excavations.

28.1.1 Unbraced excavations in sandy soils (heights less than 15 ft.)

Design Example 28.1:

Recommend permanent sloping angles for the soil types shown in Figure 28.1. Use Table 28.1 to obtain β.

Table 28.1 does not recommend a sloping angle for such a situation. Bracing is needed.

Figure 28.1 Unbraced excavations. (a) Rounded coarse sand, (b) sand with water emerging from slope.

Table 28.1 Sloping angles for unbraced excavations in sands

Soil type	Temporary slope β (°)	Permanent slope β (°)
1. Gravel with boulders	53	
2. Sandy gravel	39	34
3. Angular sand	39	34
4. Rounded coarse sand	34	30
5. Rounded fine sand	30	27
6. Sand with water emerging from slope	22–16	

Source: Koerner (1984).

Table 28.2 Sloping angles for unbraced excavations in clay

Soil type	Plasticity index	Depth of excavation (ft.)	β (°)
Clayey silt	<10	0–10 ft.	39
		10–15 ft.	32
Silty clay	10–20	0–15 ft.	39
Plastic clay	>20	0–10 ft.	39

Source: Koerner (1984).

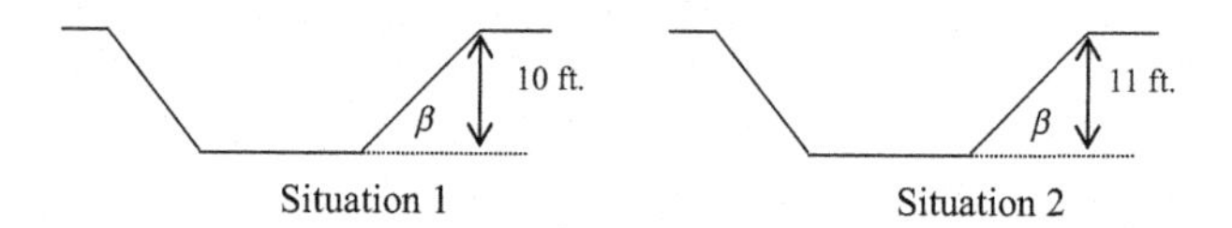

Figure 28.2 Embankments. Situation 1: silty clay (PI = 12) and use table 20.2 to obtain $\beta = 39°$. Situation 2: clayey silt (PI = 8) and use table 20.2 to obtain $\beta = 32°$.

28.1.2 Unbraced excavations in cohesive soils (heights less than 15 ft.)

High plastic silts and clays are considered as cohesive soils. Table 28.2 shows the sloping angles for unbraced excavations in clay.

Design Example 28.1

Recommend sloping angles for the soil types shown in Figure 28.2.

Reference

Koerner, R., 1984. Construction and Geotechnical Methods in Foundation Engineering. McGraw-Hill, New York.

Braced excavations

29

In some cases, sloping may not be feasible due to nearby buildings, roads, and various other obstructions. In such situations, shoring is done. Timber shoring is still very popular and is probably the cheapest option.

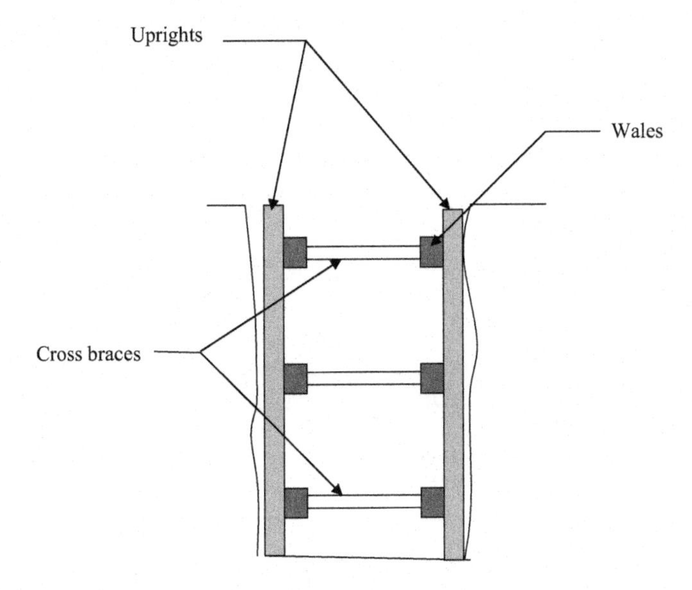

The elements of a timber shoring system are

- Cross braces (also known as struts),
- uprights,
- wales (sometimes not needed), and
- sheathing (sometimes not needed).

Cross braces, uprights, and wales are timber elements. Sheathing is timber boards or plywood.

29.1 Design of cross braces

Table 29.1 can be used to design cross braces.

Table 29.1 Design of cross braces

Depth of trench	Horizontal spacing	Width of trench					Vertical spacing (ft.)
		Up to 4 ft.	**Up to 6 ft.**	**Up to 9 ft.**	**Up to 12 ft.**	**Up to 15 ft.**	
5–10 ft.	Up to 6 ft.	4 × 4	4 × 4	4 × 6	6 × 6	6 × 6	4
	Up to 8 ft.	4 × 4	4 × 4	4 × 6	6 × 6	6 × 6	4
	Up to 10 ft.	4 × 6	4 × 6	4 × 6	6 × 6	6 × 6	4
10–15 ft.	Up to 6 ft.	4 × 4	4 × 4	4 × 6	6 × 6	6 × 6	4
	Up to 8 ft.	4 × 4	4 × 4	6 × 6	6 × 6	6 × 6	4
	Up to 10 ft.	6 × 6	6 × 6	6 × 6	6 × 8	6 × 8	4
15–20 ft.	Up to 6 ft.	6 × 6	6 × 6	6 × 6	6 × 8	6 × 8	4
	Up to 8 ft.	6 × 6	6 × 6	6 × 6	6 × 8	6 × 8	4
	Up to 10 ft.	8 × 8	8 × 8	8 × 8	8 × 8	8 × 10	4

Source: OSHA C.1.1

Design Example 29.1

A trench with a depth of 18 ft. needs to be constructed. The width of the trench is 10 ft. What is the size of the cross bracing required? The contractor would like to place cross bracing every 10 ft.

Solution

The depth of the trench is 18 ft. Hence, locate the row with depth 15–20 ft. The width is 10 ft. Hence, locate the column "up to 12 ft." Now, the contractor has three choices. The contractor wishes to place cross bracing every 10 ft.

Hence, the size of the cross bracing required = 8 × 8

Nominal size 8 × 8 (8 × 8 in.) is not the actual size. Actual size is 7.5 × 7.5 in.

29.1.1 Design of uprights

Table 29.2 shows the design of uprights and wales.

Table 29.2 Design of uprights

Depth of trench	Wale size	Wale vertical spacing (ft.)	Maximum allowable horizontal spacing of uprights and size				
			Close	**4 ft.**	**5 ft.**	**6 ft.**	**8 ft.**
5–10 ft.	Not required	–				2 × 6	
	Not required	–					2 × 8
	8 × 8	4			2 × 6		
	8 × 8	4				2 × 6	
10–15 ft.	Not required	–					3 × 8
	8 × 8	4			2 × 6		
	8 × 10	4				2 × 6	
	10 × 10	4					3 × 8

Source: OSHA Table C.1.1

29.1.2 Aluminum hydraulic shoring

Instead of timber, there is another alternative – Aluminum hydraulic shoring. These shoring can be reused many times. Though the initial cost is high, the cost per given year may be less with this type of shoring.

Raft design

30.1 Introduction

Rafts are known as mat foundations or floating foundations (Plate 30.1).

Rafts are constructed in situations where shallow foundations are not feasible. The foundation engineer has the choice of picking either rafts or piles depending on the situation. Most engineers usually prefer to select piles not necessarily due to merit but due to the familiarity of piles compared to rafts.

30.2 Raft design in sandy soils

30.2.1 Method proposed by Peck et al.

Three prominent foundation engineers, Peck et al. (1974) proposed the following method to design raft foundations.

Plate 30.1 Rebars for a raft.

Peck et al. (1974) gave the following equation to find the allowable pressure in a raft.

$$q_{(\text{allowable})\,\text{Raft}} = 0.22 \times N \times C_{\text{w}} + \gamma\, D_{\text{f}}$$

$q_{(\text{allowable})\,\text{Raft}}$ = allowable average pressure in the raft is given in tsf.
N = average SPT (N) value to a depth of $2B$ (B = lesser dimension of the raft).

$$C_{\text{w}} = 0.5 + 0.5 \frac{D_{\text{w}}}{(D_{\text{f}} + B)}$$

D_{w}, depth to groundwater measured from the ground surface (ft.); D_{f}, depth to bottom of the raft measured from the ground surface (ft.); γ, total density of the soil (measured in tons per cubic feet).

Note: units should match the units of $q_{(\text{allowable})\,\text{Raft}}$ which is in tsf.
Hence the total density (γ) should be in tcf.

Design Example 30.1

Find the capacity of the raft shown in Figure 30.1.
The following information is given.

- The raft dimensions are 100 ft. × 100 ft.
- Average SPT (N) value of the soil is 15.
- Total density of soil is 115 pcf.

Solution

Step 1: Write down the equation for the allowable bearing capacity of the raft allowing a settlement of 2 in. [$q_{(\text{allowable})\,\text{Raft}}$].

$$q_{(\text{allowable})\,\text{Raft}} = 0.22 \times N \times C_{\text{w}} + \gamma D_{\text{f}}$$

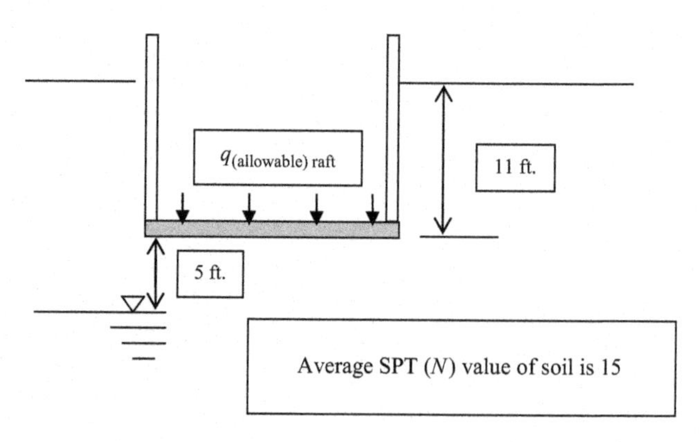

Figure 30.1 Basement raft.

Step 2: Find the correction factor for groundwater (C_W)

$$C_w = 0.5 + 0.5\frac{D_w}{(D_f + B)}$$

D_W, depth to groundwater measured from the ground surface (ft.) = 16 ft.; D_f, depth to bottom of the footing measured from the ground surface = 11 ft.; B, width of the raft = 100 ft.

$$C_w = 0.5 + 0.5\frac{16}{11+100} = 0.572$$

Step 3: Use the equation in Step 1 to find $q_{\text{(allowable) Raft}}$.

$$q_{\text{(allowable) Raft}} = 0.22 \times N \times C_w + \gamma D_f$$

$N = 15$
$C_W = 0.572$
$\gamma = 115$ pcf = 0.0575 tcf (γ needs to be converted to tcf since the equation is unit sensitive)
$D_f = 11$ ft.

$$q_{\text{(allowable) Raft}} = 0.22 \times N \times C_w + \gamma D_f$$

$$q_{\text{(allowable) Raft}} = 0.22 \times 15 \times 0.572 + 0.0575 \times 11 = 2.52 \text{ tsf}$$

Step 4:

Total load that can be carried by the raft = 2.52 × (100 × 100 ft.) tons
= 25,200 tons

Reference

Peck, R.B., Hanson, W.E., Thornburn, T.H., 1974. Foundation Engineering. John Wiley and Sons, New York.

Rock mechanics and foundation design in rock

31

31.1 Introduction

Very high-load foundations, caissons, and piles are carried down to rock to increase the capacity. Rocks can provide much higher bearing capacity than soil. Unweathered rock can have a bearing capacity in excess of 60 tsf. On the other hand, weathered rock or rocks subjected to chemical attack could be unsuitable for foundation use.

31.2 Brief overview of rocks

When a rock is heated to a very high temperature, it melts and becomes lava. The lava coming out from a volcano cools down and becomes rock. The earth's diameter is measured to be approximately 8000 miles and the bedrock is estimated to be only 10 miles. If you scale down the 10-mile thick bedrock to 1 in., then the earth will be a 66 ft. diameter sphere. The earth has a solid core with a diameter of 3000 miles and the rest is all lava, known as the mantle (Figure 31.1).

Occasionally, lava comes out of the earth during volcanic eruptions, cools down, and becomes rock (Figure 31.2). Such rocks are known as igneous rock.

Some of the common igneous rocks are

- Granite,
- Diabase,
- Basalt, and
- Diorite.

Volcanic eruptions are not the only process of rock origin. Soil particles constantly deposit in lakebeds and ocean floor. Many million years later, these depositions solidify and convert into rock. Such rocks are known as sedimentary rocks (Figure 31.3).

Some of the common sedimentary rocks are

- Sandstones,
- Shale,
- Mudstone,
- Limestone, and
- Chert.

Other than these two rock types, geologists have discovered another rock type. The third rock type is known as *metamorphic rocks*. When a tadpole becomes a completely different creature during its development process, we call it a frog.

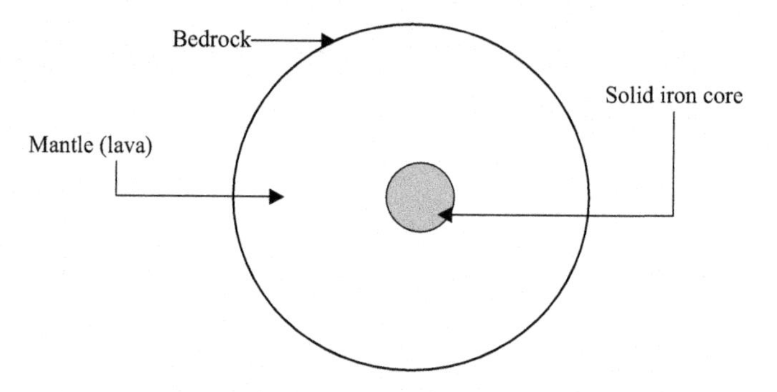

Figure 31.1 Earth.

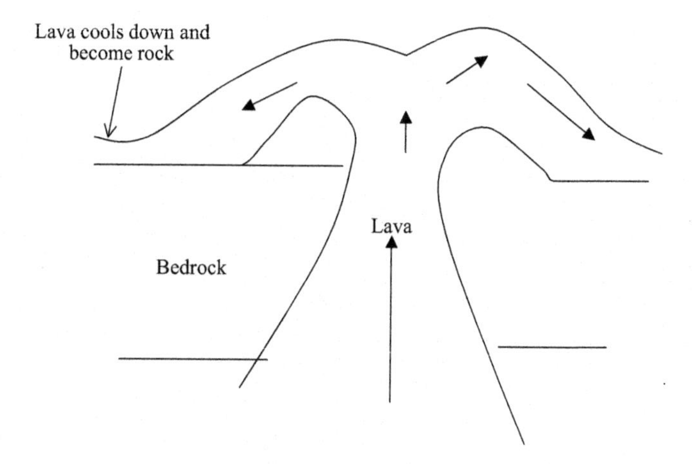

Figure 31.2 Volcanic eruptions and lava flow.

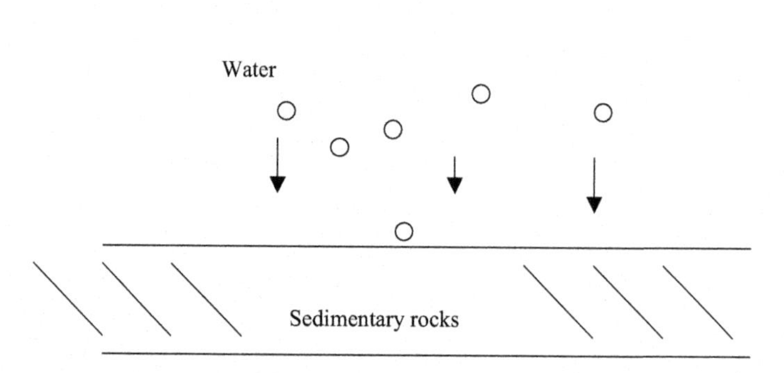

Figure 31.3 Sedimentary process.

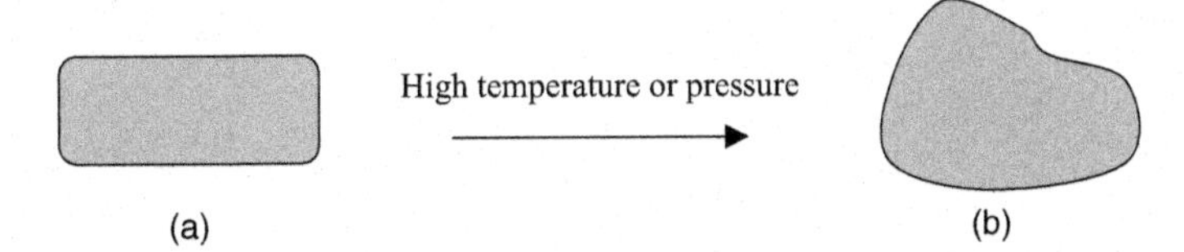

Figure 31.4 Formation of metamorphic rocks. (a) Igneous or sedimentary rock. (b) Metamorphic rock.

Similarly, either sedimentary rocks or igneous rocks could change and becomes a completely new rock type.

How could a sedimentary rock or an igneous rock change into a new rock type? This happens when a rock is subjected to extreme pressure or temperature. Unthinkably, large pressures are originated during the plate tectonic movements. Other events such as earthquakes, meteors, and volcanic eruptions also generate large pressures and temperatures (Figure 31.4).

Some of the common metamorphic rocks are:

- Gneiss,
- Schist,
- Marble, and
- Slate.

Gneiss is usually formed when granite or similar igneous rock is subjected to heat or pressure. In many instances, shale could be the parent rock of slate and limestone could be the parent rock of marble.

31.3 Rock joints

Identification of rock type is important for the foundation or tunneling engineer. Yet in most situations, suitability of a rock for foundation is dependant upon rock joints. Rock joint is a fracture in the rock mass. Most rocks consist of joints. Joints could occur in the rock mass due to many reasons.

- *Earthquakes* – Major earthquakes could shatter the bedrock and create joints.
- *Plate tectonic movements* – Continents move relative to each other. When they collide, the bedrock would fold and joints would be created.
- Volcanic eruptions.
- Generation of excessive heat in the rock.

Depending on the location of the bedrock, a number of joints in a core run could vary. Some core runs contain few joints while some other core runs may contain dozens of joints.

Joint Set: When a group of joints is parallel to each other, that group of joints is called a "*joint set*" (Figure 31.5).

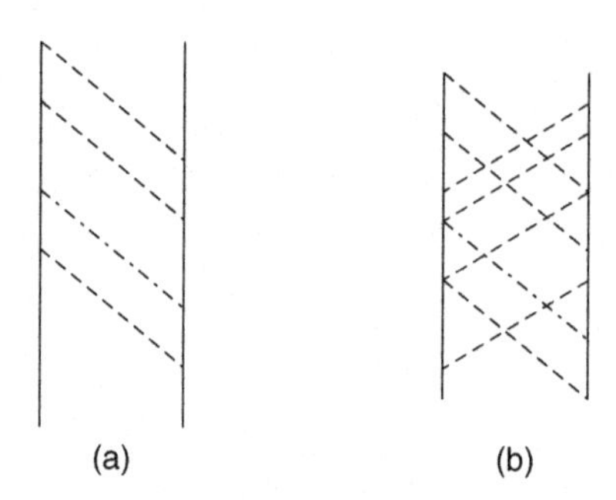

Figure 31.5 Joint sets. (a) Core run with one set of joints. (b) Core run with two sets of joints.

31.3.1 Foundations on rock

The foundation shown in Figure 31.6 can be considered unstable. If the foundation is already constructed, the engineer should consider the following factors that could reduce the friction in joints.

Joint smoothness: If the joints are smooth, the friction between joints can be considered minimal.

Water in joints: If water is flowing through joints, friction in joints can be reduced.

Presence of clay particles in joints: It is important to investigate the presence of clay particles in joints. Clay particles are deposited in joints due to groundwater movement. Clay particles in joints tend to reduce the joint friction.

Earthquakes: If the building is located in an earthquake-prone region, the risk of failure will be high.

Rock bolts can be installed to stabilize the rock (Figure 31.7).

In Figure 31.7, rock joints are stable and rock bolts may not be necessary.

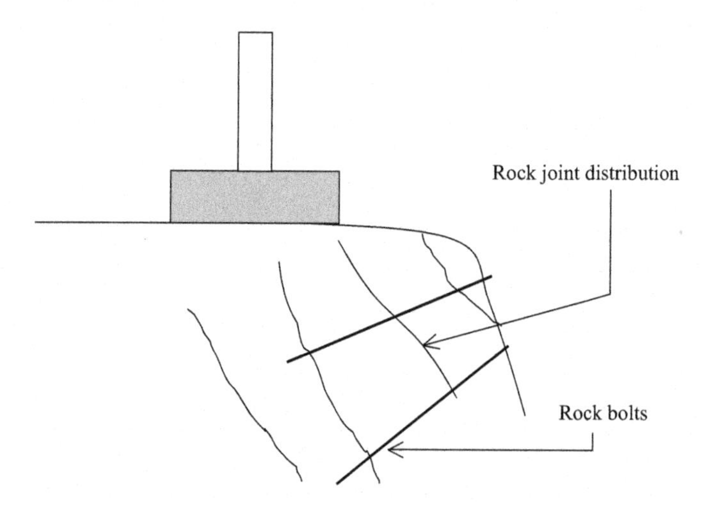

Figure 31.6 Foundation placed on rock with unstable joints.

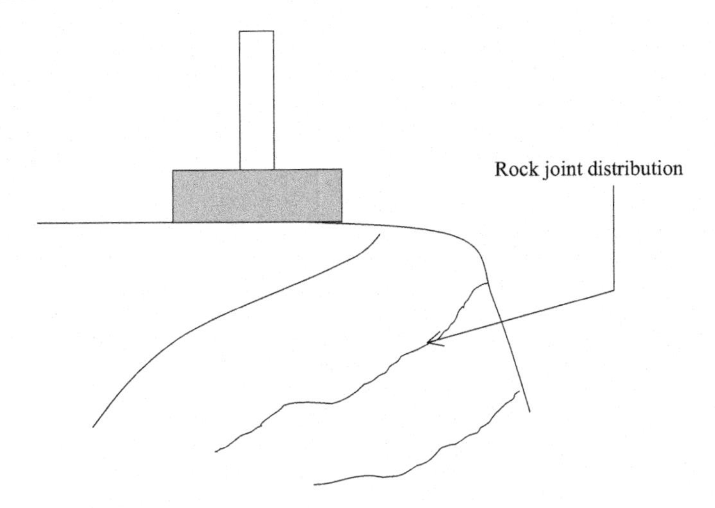

Figure 31.7 Foundation placed on rock with stable joints.

31.4 Rock coring and logging

The information regarding rock formations is obtained through rock coring. Rock coring is a process where a rotating diamond cutter is pressed to the bedrock. Diamond being the hardest of all material cuts through rock and a core is obtained. During the coring process, water is injected to keep the core bit cool and to remove the cuttings.

A typical rock core is 5 ft. in length. Rock cores are safely stored in core boxes made of wood (Figure 31.8).

A typical core box can store 4 rock cores, each with a length of 5 ft. (Figure 31.9). Core recovery is measured and noted for each rock core.

$$\text{Recovery percentage} = \frac{\text{Core recovery}}{60 \times 100}$$

For example, if the core recovery were 40, the recovery percentage would be 66.7%.

31.4.1 RQD (rock quality designation)

Rock Quality Designation (RQD) is obtained through the following process.

- Arrange all rock pieces as best as possible to simulate the ground conditions.
- Measure all rock pieces greater in length than 4 in.
- RQD is given by;

$$\text{RQD} = \frac{\text{Total length of all rock pieces greater than 4 in.}}{60 \times 100}$$

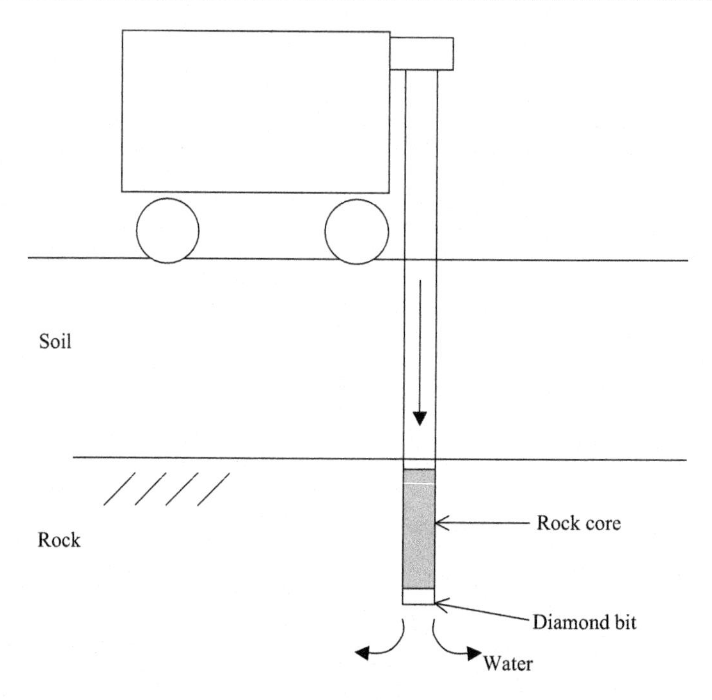

Figure 31.8 Rock coring.

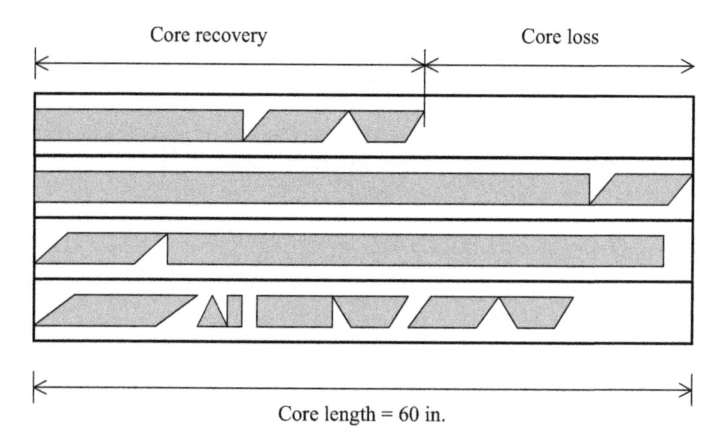

Figure 31.9 Core box.

Design Example 31.1

Lengths of rock pieces in a 60 in. rock core is measured to be
2, 3, 1, 4.5, 5.5, 2, 6, 8, 3, 2, 15
Find the core recovery percentage and RQD.

Solution

Step 1: Total length of all the pieces is added to be 52 in.

$$\text{Recovery percentage} = \frac{52}{60} \times 100$$
$$= 86.7\%$$

Step 2: Find all the pieces greater than 4 in.

4.5, 5.5, 6, 8, 15

Total length of pieces greater than 4 in. = 39 in.

$$\text{RQD} = \frac{39}{60} \times 100$$
$$= 65\%.$$

31.4.2 Joint filler materials

Some joints are filled with matter. The typical joint filler materials are

- *Sands*: Sands occur in joints where there is high-energy flow (alternatively, high velocity).
- *Silts*: Silts indicate that the flow is less energetic.
- *Clays*: Clay inside joints indicates stagnant water in joints.

Joint filler material information would be very useful in interpreting the Packer test data. It is known that filler material could clog up joints and reduce the flow rate with time.

31.4.3 Core loss information

It has been said by experts that core loss information is more important than the rock core information. Core loss location may not be obvious in most cases. Coring rate, color of return water, and arrangement of core in the box can be used to identify the location of the core loss (Figure 31.10).

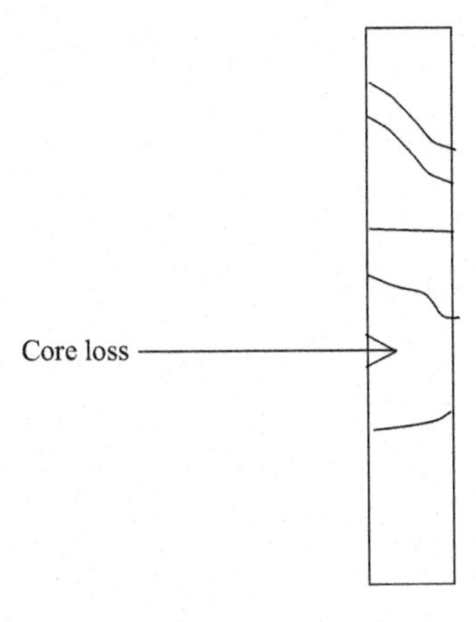

Figure 31.10 Core loss.

Core loss occurs in weak rock or in highly weathered rock.

Fractured zones: A fracture log would be able to provide fractured zones. Fracture logs of each boring can be compared to check for joints.

Drill water return information: The engineer should be able to assess the quantity of returning drill water. Drill water return could vary from 90% to 0%. This information could be very valuable in determining weak rock strata. Typically, drill water return is high in sound rock.

Water Color: Color of returning drill water can be used to identify the rock type.

- Rock Joint Parameters (Figure 31.11)
  - *Joint roughness:* Joint surface could be rough or smooth. Smooth joints could be less stable than rough joints.
  - *Joint alteration*: Alterations to the joint, such as color, filling of materials, etc.
  - *Joint filler material:* Some joints could be filled with sand while some other joints could be filled with clay. Smooth joints filled with sand may provide additional friction. However, rough joints filled with clay may reduce the friction in the joint.
  - *Joint stains:* Joint stains should be noted. Stains could be due to groundwater and various other chemicals.
- Joint Types (Figure 31.12)
  - *Extensional joints*: Joints that formed due to tension or pulling apart.
  - *Shear joints*: Joints that occurred due to shearing.

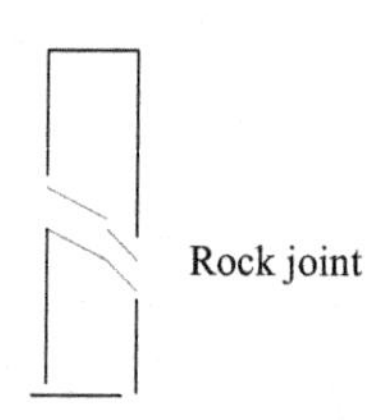

Figure 31.11 Rock joint.

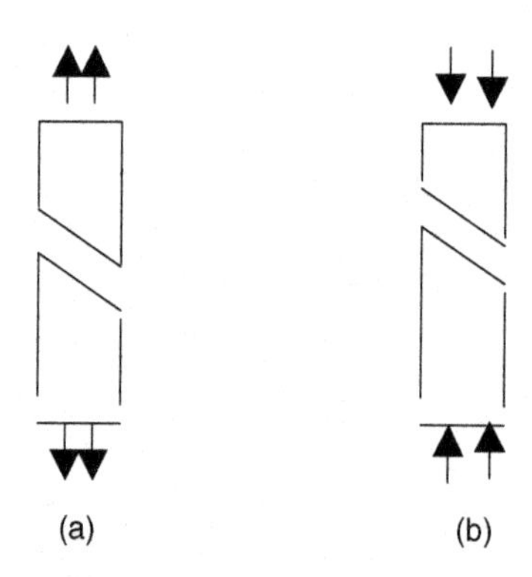

Figure 31.12 Joint types. (a) Extensional joint (tensile failure). (b) Shear joint (shearing along the failure plane).

31.5 Rock mass classification

Who is the better athlete?

Athlete A: Long jump, 23 ft.; runs 100 m in 11 s; high jump, 6.5 ft.

Athlete B: Long jump, 26 ft.; runs 100 m in 13 s; high jump, 6.4 ft.

Athlete A is very weak in long jump but very strong in 100 m. Athlete B is very good in long jump, but not very good in 100 m. Athlete A has a slight edge in high jump.

It is not easy to determine which athlete is better. Due to this reason, Olympic committee came up with a marking system for the decathlon. The best athlete is selected based on the marking system.

Similar situation exists in rock types. Let us look at an example.

Example: A geotechnical engineer has the choice to construct a tunnel in either rock type A or rock type B.

Rock Type A →	Average RQD = 60%,	Joints are smooth,	joints are filled with clay
Rock Type B →	Average RQD = 50%,	Joints are rough,	joints are filled with sand

Rock type A has a higher RQD value. On the other hand, Rock type B has rougher joints. Smooth joints in rock type A are not favorable for geotechnical engineering work. Joints in rock type A is filled with clay, while joints in rock type B are filled with sands. Based on the given information, it is not easy to select a candidate, since both rock types have good properties and bad properties. Early engineers recognized the need of a classification system to determine the better rock type. Unfortunately, more than one classification system exists. These systems are named *Rock Mass Classification Systems*.

Popular Rock Mass Classification Systems are:

- Terzaghi Rock Mass Classification System (rarely used).
- Rock Structure Rating (RSR) method.
- Rock Mass Rating system (RMR).
- Rock Tunneling Quality Index (better known as the "*Q*" system).

Recently, the "*Q*" system has gained popularity over other systems.

31.6 Q – System

The following is the equation for "*Q*" value

$$Q = \frac{\text{RQD}}{J_n} \times \frac{J_r}{J_a} \times \frac{J_w}{\text{SRF}}$$

Q, rock quality index; RQD, rock quality designation; J_n, joint set number; J_r, joint roughness number; J_a, joint alteration number; J_w, joint water reduction factor; SRF, stress reduction factor.

31.6.1 Rock quality designation (RQD)

To obtain RQD, select all the rock pieces longer than 4 in. in the rock core. Measure the total length of all the individual rock pieces greater than 4 in. . This length is given as a percentage of the total length of the core.

Total length of the core = 60 in.

Total length of all the pieces longer than 4 in. = 20 in.

RQD = 20/60 = 0.333 = 33.3%
RQD (0%–25%) = very poor
RQD (25%–50%) = poor
RQD (50%–75%) = fair
RQD (75%–90%) = good
RQD (90%–100%) = excellent

31.6.2 Joint set number (J_n)

Find the number of joint sets. When a group of joints has the same dip angle and a strike angle, that group is known as a joint set. In some cases, many joint sets exit. Assume there are 8 joints in the rock core with the following dip angles: 32, 67, 35, 65, 28, 64, 62, 30, and 31. It is clear that there are at least two joint sets. One joint set has a dip angle approximately at 30°, while the other joint set has a dip angle of approximately at 65°.

The Q system allocates the following numbers:

Zero Joints → $J_n = 1.0$
One Joint Set → $J_n = 2$
Two Joint Sets → $J_n = 4$
Three Joint Sets → $J_n = 9$
Four Joint Sets → $J_n = 15$

Higher J_n number indicates a weaker rock for construction.

31.6.3 Joint roughness number (J_r)

When subjected to stress, smoother joints may slip and failure could occur before rougher joints. Due to this reason, joint roughness plays a part in rock stability (Figure 31.13).

Slippage occurs along a smoother joint at a lower load (P).

How to obtain the joint roughness number ?

Step 1: There are three types of joint surface profiles.

- Wavy (undulating) joint surface profiles.
- Stepped joint surface profiles.
- Planar joint surface profiles.

No joint surface is either 100% planar, stepped, or wavy. Select the type, which best describes the joint surface (see Figure 31.13).

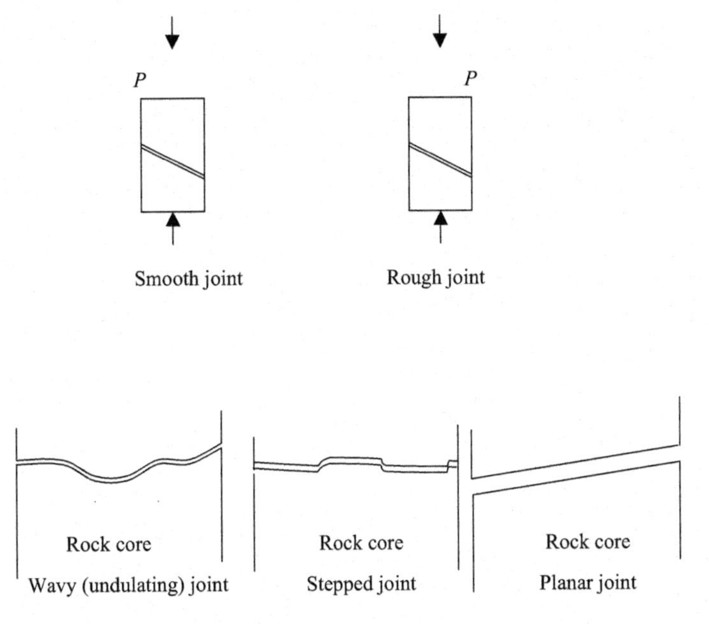

Figure 31.13 Smooth and rough joints.

Step 2: Feel the joint surface and categorize it into one of the following types.

- *Rough* – If the surface feels rough.
- *Smooth* – If the surface feels smooth.
- *Slickensided* – Slickensided surfaces are very smooth and slick. Slickensided surfaces occur when there is a shear movement along the joint. In some cases, the surface could be polished. One would notice polished (shining) patches. These polished patches indicate shear movement along the joint surface. The name slickensided was given to indicate slick surfaces.

Step 3: Use Table 31.1 to obtain J_r.

- Rock joints with stepped profiles provide the best resistance against shearing. From Table 31.1, smooth joint with a stepped profile would be better than a rough joint with a planar profile.

Table 31.1 Joint roughness coefficient (J_r)

Joint profile	Joint roughness	J_r
Stepped	Rough	4
	Smooth	3
	Slickensided	2
Undulating	Rough	3
	Smooth	2
	Slickensided	1.5
Planar	Rough	1.5
	Smooth	1.0
	Slickensided	0.5

Source: Hoek et al. (1995).

Table 31.2 **Joint alteration number (J_a)**

Description of filler material	J_a
1. Tightly healed with a nonsoftening impermeable filling seen in joints (*quartz or epidote*).	0.75
2. Unaltered joint walls. No filler material seen (surface stains only).	1.0
3. Slightly altered joint walls. Nonsoftening mineral coatings are formed. Sandy particles, clay or disintegrated rock seen in the joint.	2.0
4. Silty or sandy clay coatings, small fraction of clay in the joint.	3.0
5. Low *friction* clay in the joint (kaolinite, talc, chlorite are low friction clays).	4.0

Source: Hoek et al. (1995).

31.6.4 Joint alteration number (J_a)

Joints are altered with time. Joints could be altered due to the material filling inside them. In some cases, the filler material could cement the joint tightly. In other cases, the filler material could introduce a slippery surface creating a much more unstable joint surface. Table 31.2 can be used to obtain the J_a value.

It is easy to notice a tightly healed joint. In this case, use $J_a = 0.75$. If the joint has not undergone any alteration other than surface stains, use $J_a = 1.0$. If there are sandy particles in the joint, then use $J_a = 2.0$. If there is clay in the joint then use $J_a = 3.0$. If the clay in the joint can be considered low friction, then use $J_a = 4.0$. For this purpose, clay types existing in joints need to be identified.

31.6.5 Joint water reduction factor (J_w)

J_w (joint water reduction factor) is a measure of water in a joint. J_w cannot be obtained from boring data. A tunnel in the rock needs to be constructed to obtain the J_w value. Usually, the data from previous tunnels constructed in the same formation is used to obtain J_w. Another option is to construct a pilot tunnel ahead of the real tunnel. Table 31.3 can be used to obtain the J_w value.

31.6.5.1 Stress reduction factor (SRF)

SRF is an indication of weak zones in a rock formation. SRF cannot be obtained from boring data. A tunnel in the rock needs to be constructed to obtain SRF as in the case of J_w. Usually, data from previous tunnels constructed in the same formation is used to obtain the SRF. Another option is to construct a pilot tunnel ahead of the real tunnel.

All rock formations have weak zones. Weak zone is a region in the rock formation, which has a low RQD value. Weak zones may have weathered rock or different rock type.

The following guidelines are provided to obtain the stress reduction factor.

1. More than one weak zone occurs in the tunnel. In this case, use SRF = 10.0
2. Single weak zone of rock with clay or chemically disintegrated rock. (Excavation depth < 150 ft.) Use SRF = 5.0
3. Single weak zone of rock with clay or chemically disintegrated rock. (Excavation depth > 150 ft.) Use SRF = 2.5
4. More than one weak zone of rock without clay or chemically disintegrated rock. Use SRF = 7.5

Table 31.3 **Joint water reduction factor (J_w)**

Description	Approximate water pressure (kg/cm^2)	J_w
1. Excavation (or the tunnel) is dry. No or slight water flow into the tunnel.	<1.0	1.0
2. Water flows into the tunnel at a medium rate (water pressure 1.0–2.5 kgf/cm^2). Joint fillings are washed out occasionally due to water flow.	1.0–2.5	0.66
3. Large inflow of water into the tunnel or excavation. The rock is competent and joints are unfilled (water pressure 1.0–2.5 kgf/cm^2).	2.5–10.0	0.5
4. Large inflow of water into the tunnel or excavation. The joint filler material is washed away (water pressure 2.5–10 kgf/cm^2).	>10	0.33
5. Exceptionally high inflow of water into the tunnel or excavation (water pressure > 10 kgf/cm^2).	>10	0.1–0.2

Source: Hoek et al. (1995).

5. Single weak zone of rock without clay or chemically disintegrated rock. (Excavation depth < 150 ft.) Use SRF = 5.0
6. Single weak zone of rock without clay or chemically disintegrated rock. (Excavation depth > 150 ft.) Use SRF = 2.5.
7. Loose open joints observed. Use SRF = 5.0

Design Example 31.2

The average RQD of a rock formation was found to be 60%. Two sets of joints have been identified. Most joint surfaces are undulated (wavy) and rough. Most joints are filled with silts and sands. It has been reported that medium inflow of water has occurred during the construction of past tunnels. During earlier constructions, single weak zone containing clay was observed at a depth of 100 ft. Find the Q – value.

Solution

Step 1: $Q = \frac{\text{RQD}}{J_n} \times \frac{J_r}{J_a} \times \frac{J_w}{\text{SRF}}$

RQD = 60%

Since there are two sets of joints, $J_n = 4$.

$J_r = 3$ (from Table Table 31.1, for undulating, rough joints).

Since most joints are filled with silts and sands, $J_a = 2$ (from Table Table 31.2)

$J_w = 0.66$ (from Table 31.3).

SRF = 5

Hence $Q = (60/4) \times (3/2) \times (0.66/5) = 2.97$

Reference

Hoek, E., Kaiser, P.K., Bawden, W.F., 1995. Support of Underground Excavations in Hard Rock. A.A Balkeema Publishers, New York.

Dip angle and strike

32

32.1 Introduction

Dip angle is the angle that a set of joints extend against the horizontal plane. Each set of joints would have a dip angle and a strike. Strike is the direction that the horizontal plane and dip direction intersect. (See Figure 32.1)

Joint Plane: EBCF is the joint plane in Figure 32.1.

Dip Angle: The angle between the joint plane (EBCF) and the horizontal plane (ABCD). Dip angle is easily measured in the field.

Strike: Strike line is the horizontal line BC as shown in Figure 32.1.

Strike Direction: Right hand rule is used to obtain the strike direction.

- Open your right hand and face the palm down. Mentally lay the palm on the joint plane.
- Point four fingers (except the thumb) along the downward direction of the slope.
- The direction of the thumb indicates the strike direction.
- The clockwise angle between the strike direction and the north direction is called the strike angle. ABCD plane and north direction are in the same horizontal plane.

Dip Direction:

Dip direction is the direction of the downward slope. Dip direction is different from dip angle. Strike and dip directions are perpendicular to each other.

Notation: Dip angle and strike angle are written as follows:

35/100

The numerator in the above fraction indicates the dip angle (not the dip direction). The denominator indicates the strike direction measured clockwise from the north. The dip angle is always written in two digits, while the strike direction is always written in three digits. A joint plane with a dip angle of 40° and a strike angle of 70° is written as $40^{\circ}/70^{\circ}$.

32.2 Oriented rock coring

32.2.1 Oriented coring procedure

Step 1: A knife is installed in the core barrel to create a mark in the rock core. This mark is known as the scribe mark.

Step 2: The knife is installed in such a manner that a line drawn through the scribe mark and the center of the core would point towards the north direction (line AB). The driller would use a magnetic compass prior to coring, and locate the north direction. Then, he would be able to locate the knife (Figure 32.2).

Step 3: Draw a horizontal line going through point C, at the joint (line CD). Both the lines (line AN and line CD) are in horizontal planes.

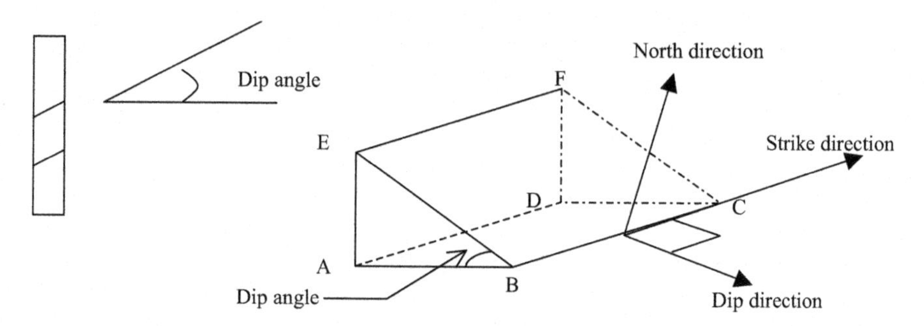

Figure 32.1 Dip angle, strike, and dip direction.

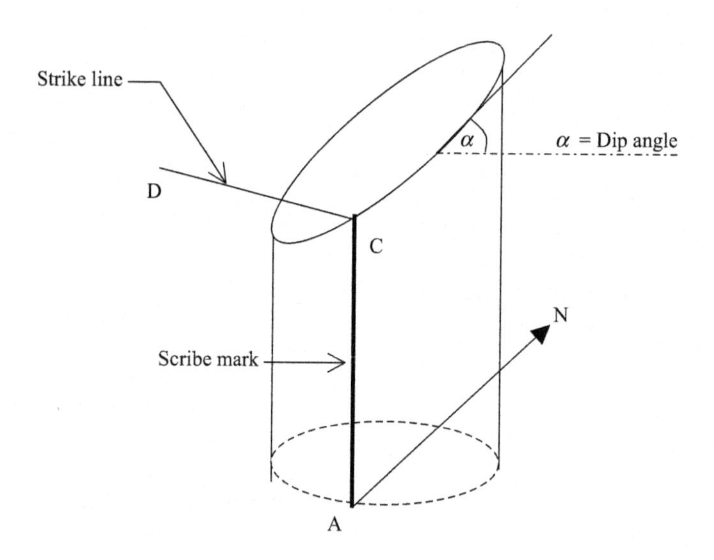

Figure 32.2 Strike line and scribe mark.

Step 4: Measure the clockwise angle between line AN (which is the north direction) and line CD. This angle is the strike angle of the joint.

32.2.2 Summary

- The driller finds the north direction using a compass. Then, the driller places the knife along the north direction. The line connecting the knife point (point A) and the center of the core will be the north-south line (line AN).
- Rock coring is conducted. A scribe mark will be created along the rock core. (The rock core does not rotate during coring).
- After the removal of the rock core from the hole, a horizontal line is drawn along the plane of the joint (line CD). The clockwise angle between line AN and line CD is the strike angle.

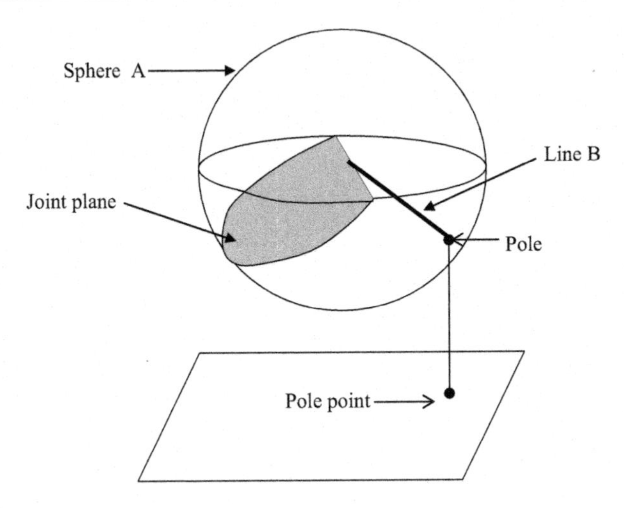

Figure 32.3 Pole and pole point.

32.3 Oriented core data

- Oriented coring would produce a *dip angle* and a *strike angle* for each joint (Figure 32.3).
- A "joint" can be represented by one point known as the pole.

32.3.1 Concept of pole

Step 1: The joint plane is drawn across the sphere A.

Step 2: A perpendicular line (line B) is drawn to the joint plane.

Step 3: The point where line B intersects the sphere is known as the pole.

Step 4: A vertical line is dropped from the pole to obtain the pole point.

Theoretically, there are two poles for any given joint – one in the lower hemisphere and the other one in the upper hemisphere. The pole in the lower hemisphere is always selected.

As you can see, it is not easy to obtain the pole point for a given joint. For that purpose, one needs a sphere and has to go through lot of trouble. There are charts available to obtain the pole point for any given joint.

Rock bolts, dowels, and cable bolts

33.1 Introduction

Nomenclature changes from area to area. Generally, rock bolts (also known as rock anchors) can be nonstressed or prestressed. Nonstressed anchors are also known as rock dowels. When there is a slight movement in the rock mass, rock dowels are tensioned and provide a resisting force.

Cable bolts are a group of wires used to create a cable. Cable bolts are used for high capacity applications.

Rock anchors are used for shallow foundations, retaining walls, bridges, and tunnels.

The basic types of rock anchors are

1. Mechanical rock anchors or rock bolts (nonstressed or prestressed)
2. Grouted (nonstressed or prestressed)
3. Resin anchors
4. Nonstressed anchors, known as rock dowels.

33.1.1 Applications

Rock anchors are widely used in retaining walls when rock is close enough for the anchors (Figures 33.1, 33.2, and 33.3).

33.2 Mechanical rock anchors

Most popular mechanical anchor system is expansion shell anchors. A hole is drilled in the rock and the rock bolt assembly is inserted. Then, the tensile force (P) is applied. When the force is applied, the cone would try to move to the right side (Figure 33.4).

The movement of the cone would expand the two *wedges* and lock the bolt into the rock.

A typical installation procedure for mechanical anchors is as follows:

Step 1: Drill a hole in the rock (Figure 33.5).
Step 2: Insert the mechanical anchor (Figure 33.6).
Step 3: Activate the mechanical wedge assembly at the end to attach the anchor to the rock. Rock anchors have a wedge assembly at the end, which expands when rotated.
Step 4: Grout the hole to avoid corrosion in the anchor.

- Initial tension applied = 70% of the total capacity of the rock bolt as recommended by E. Hoek.

(*Ref. Underground Excavations in Hard Rock,* Balkema publishers, Amsterdam.)

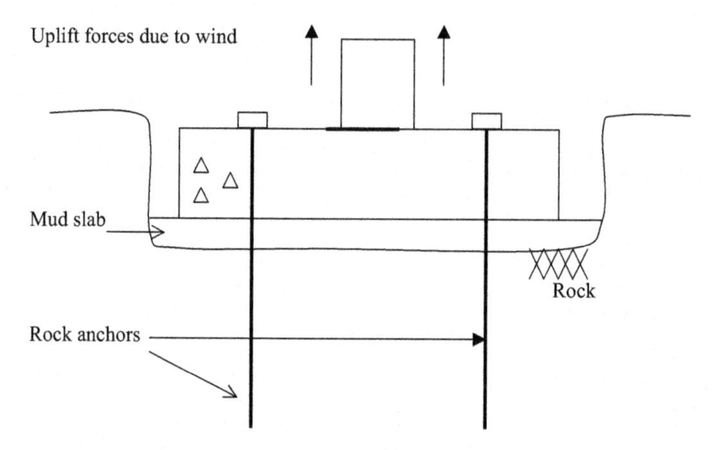

Figure 33.1 Rock anchors to resist uplift forces in a shallow foundation.

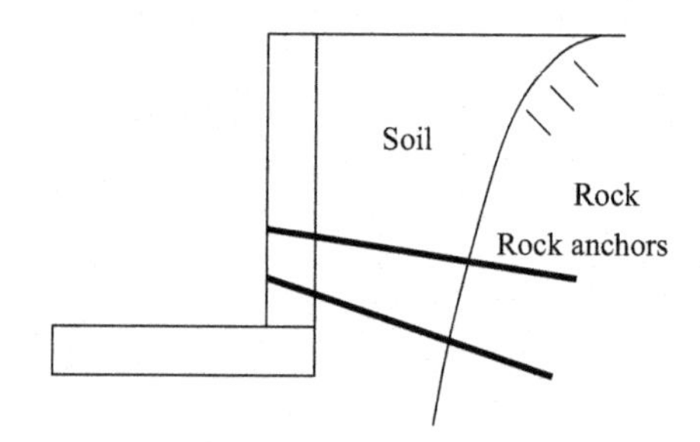

Figure 33.2 Rock anchors in a retaining wall.

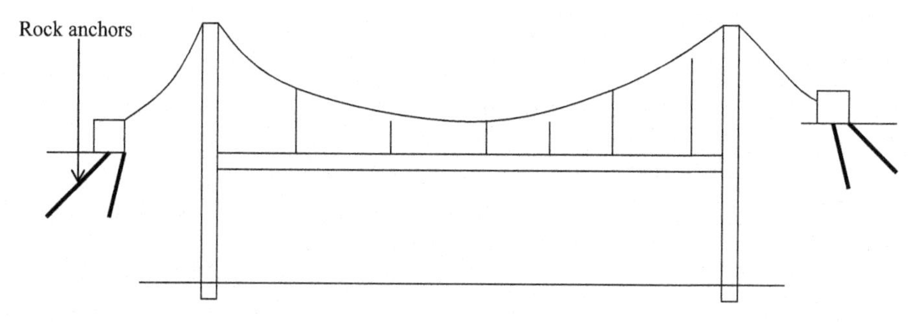

Figure 33.3 Rock anchors in a bridge.

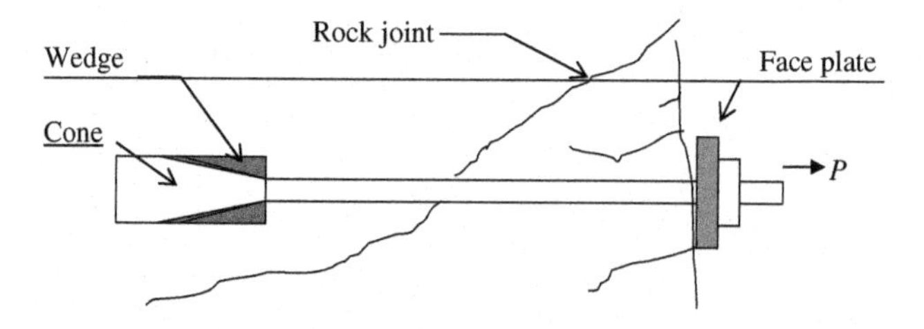

Figure 33.4 Mechanically anchored rock anchors.

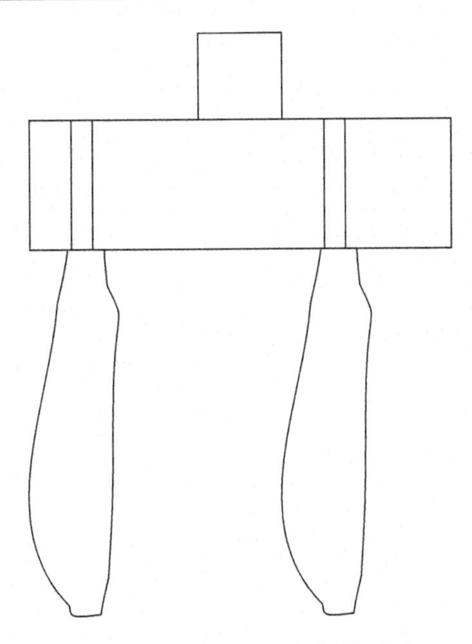

Figure 33.5 Mechanical anchor installation. Step 1: Drill a hole.

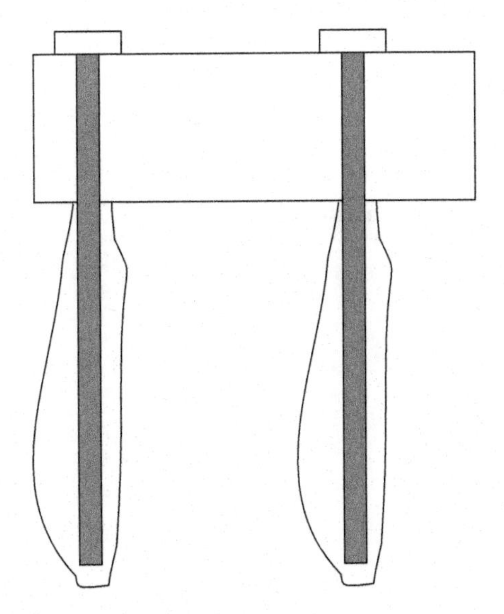

Figure 33.6 Mechanical anchor installation. Step 2: Insert anchors.

In Figure 33.7, the anchor is twisted to activate the mechanical anchor and grout the hole.

33.2.1 Mechanical anchor failure

The main reason for rock bolt failure is corrosion. Corrosion of rock bolts can be avoided by grouting the hole. Grouting of the hole is very important when there is water in the rock. Most permanent rock bolts are grouted; however, grouting may not be necessary for temporary rock bolts. It should be mentioned that groundwater in some regions could be acidic and can accelerate the corrosion process.

33.2.2 Design of mechanical anchors

The failure surface of mechanical anchors is a cone (Figure 33.8).

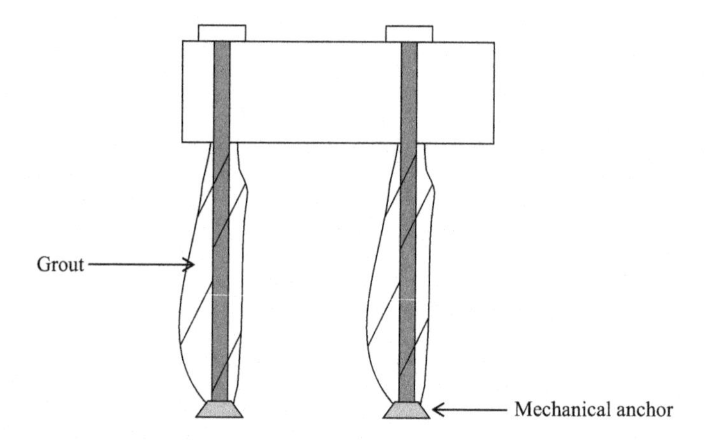

Figure 33.7 Mechanical anchor installation.

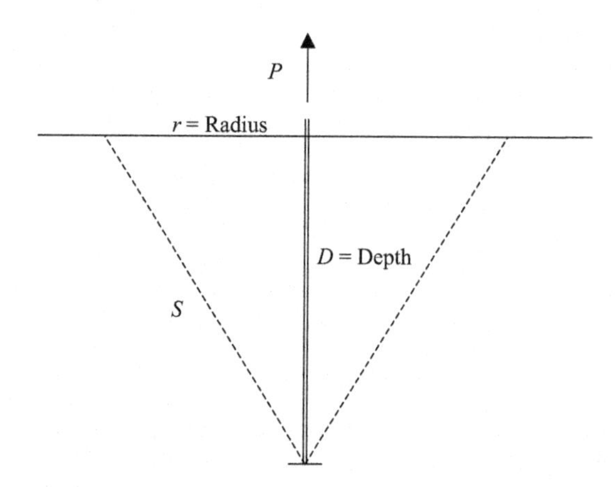

Figure 33.8 Failure cone of rock anchors.

When a mechanical anchor is subjected to failure, a failure cone is developed. The weight of the cone and rock cohesion along the cone surface acts against failure.

Resistance against failure = weight of the cone + cohesion along the surface of the cone.

The surface area and volume of cones are given by the following equations.

$$\text{Volume of a cone} = 1/3 \times (\pi \times r^2) \times D$$
$$\text{Surface area of a cone} = (\pi \times r) \times S$$

$$S = (r^2 + D^2)^{1/2}$$

Weight of the cone can be found if the density of rock is known.

Design Example 33.1

Find the failure force of the mechanical anchor shown in Figure 33.9. Following information is obtained. Assume the angle of the stress triangle to be 30°. Length of the rock anchor is 2.5 m. Ignore the weight of the rock.

Rock cohesion = 300 kPa
Rock density = 24 kN/m^3.

Solution

$$\text{Radius} = r = 2.5 \times \tan 30$$
$$r = 1.44$$

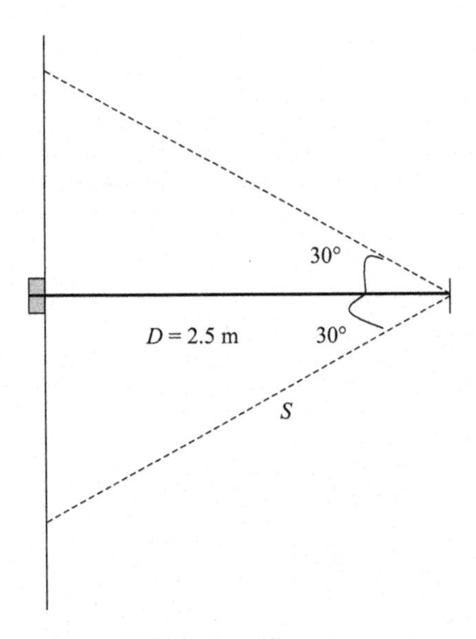

Figure 33.9 Failure mechanism of the rock anchor.

$$S^2 = r^2 + D^2 = 1.44^2 + 2.5^2$$

$$S = 2.88\ \text{m}$$

Surface area of the cone $= (\pi \times r) \times S$
Surface area of the cone $= (\pi \times 1.44) \times 2.88 = 13.01\ \text{m}^2$.

$$\begin{aligned}\text{Cohesive force} &= 13.01 \times \text{rock cohesion} \\ &= 13.01 \times 300 \\ &= 3{,}900\ \text{kN (436 tons)}\end{aligned}$$

33.2.3 Grouting methodology for mechanical rock anchors

Both standard and hollow rock bolts are anchored to the rock using a cone and wedge mechanism. The bolt tension is held by the anchor. Hence, the primary purpose of grouting is to provide corrosion protection. Due to this reason, the strength of grout is not a factor in mechanically anchored rock bolts. The other purpose of grouting is to hold the rock bolt in place so that any ground vibrations may not cause the mechanical anchor to loosen.

Mechanically anchored rock bolts could be grouted or ungrouted. Rock bolts are not grouted for most temporary applications. To make the grouting process easier, some rock bolt manufacturers have produced rock bolts with a central hole. The hole in the middle is used for the grouting process.

33.2.4 Tube method

Tubes can be inserted into the hole to grout the rock bolt. One tube is inserted to pump the grout. This tube is known as the grout injection tube. The other tube is used to remove trapped air. This tube is known as the breather tube (Figure 33.10).

The grout is injected from one tube and the entrapped air is removed from another tube. In practice, this mechanism could create problems. Often, the tubes are entangled. Two holes need to be drilled through the face plate. To avoid these problems, hollow rock bolts are used.

33.2.5 Hollow rock bolts

Hollow rock bolts were designed to make the grouting process more reliable (Figure 33.11). In hollow rock bolts, there is a hole in the center of the rock bolt. This

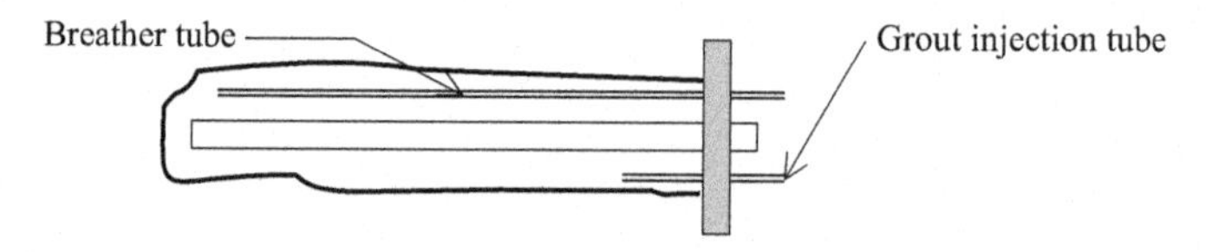

Figure 33.10 Hollow rock bolt. (Note: wedge and cone mechanism is not shown in the above figure.)

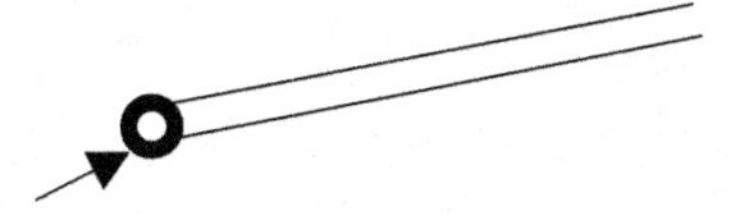

Figure 33.11 Hollow rock bolt.

hole is used for the grouting process and entangling tubes are eliminated. Hollow rock bolts may have the same cone and wedge anchor mechanism.

When the rock bolt is downward facing, the grout is inserted through the hole at the center and a short tube is inserted outside of the bolt to remove the trapped air.

When the rock bolt is upward facing, a short tube (inserted outside of the bolt) is used for the grouting process and the central hole is used as a breathing tube.

33.3 Resin anchored rock bolts

One major disadvantage of mechanically anchored rock bolts is that the mechanical anchor can loosen when there are vibrations in the rock due to blasting, train movement, traffic movement, or construction activities. Further, it is difficult to obtain a good mechanical anchor in weak and weathered rock. Shale, mudstone, and highly weathered low RQD rocks are not good candidates for mechanically anchored rock bolts. These difficulties could be avoided by using resin anchored rock bolts.

Two resin cartridges are inserted as shown in Figure 33.12. When the resin cartridges are broken (by spinning the rock bolt), the fast-setting resin would solidify first. The fast-setting resin, would anchor the rock bolt into the rock. The tensile force is applied at this point. Slow-setting resin has not solidified yet. The slow-setting portion is still in a liquid state. When the tension is applied, the rod is free to extend within the slow-setting resin area. After the tension is applied and locked in, the slow-setting resin would solidify. For temporary applications, slow-setting resin may not be necessary. Its purpose is to provide corrosion protection.

33.3.1 Disadvantages

- In fractured rock, the resin could seep into the rock and leave little resin inside the hole.

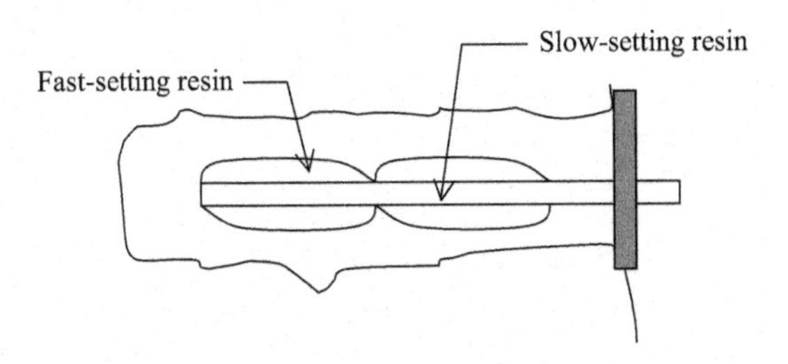

Figure 33.12 Resin anchored rock bolts.

- Some rocks contain clay seams. Resins may not be able to provide a good anchor in these types of rocks.
- Some researchers have expressed concern regarding the long-term corrosion protection ability of resins. Further groundwater chemicals may react with resins and compromise its effectiveness.

33.3.2 Advantages

- Resins work well in fractured rocks.
- Installation is very fast as compared to other methods.

33.4 Rock dowels

Unlike prestressed rock bolts, rock dowels are not tensioned. Hence, the rock dowels do not apply a positive force to the rock. Rock dowels are called into action, when there is a movement in the rock. The simplest form of rock dowel is a grouted steel bar inserted into rock. This type of rock dowel is called grouted dowels. Other types of rock dowels include split-set stabilizers and Swellex dowels.

33.4.1 Types of rock dowels

1. Cement-grouted dowels
2. Split-set stabilizers
3. Swellex dowels

33.4.1.1 Cement-grouted dowels

Figure 33.13 shows cement-grouted dowels.

Note: If the rock moves along the joint, the dowel would be tensioned.

33.4.1.2 Split-set stabilizers

These rock dowels are different from grouted dowels. What happens when a steel rod is pushed into a small hole (smaller diameter than the steel bar) drilled in the rock? The steel bar would be compressed and will be tightly entrapped inside the hole. Split set stabilizers are designed based on this principle (Figure 33.14).

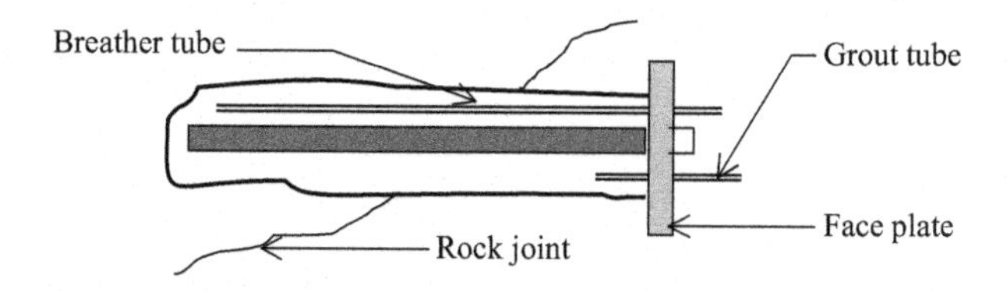

Figure 33.13 Cement grouted dowels.

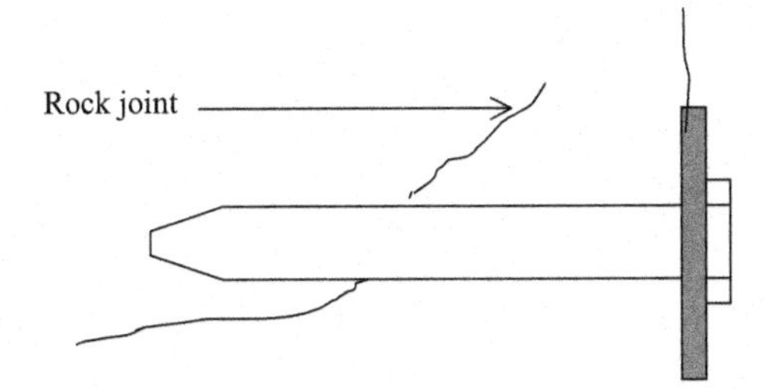

Figure 33.14 Split set stabilizers.

Note: A hole smaller than the split set dowel is drilled. Then, the dowel is pushed into the rock.

Split set stabilizers can resist a force of 10.9 tons (diameter = 33 mm). The recommended drill hole size = 31 mm (note that the hole diameter is smaller than the dowel).

33.4.1.2.1 Advantages and disadvantages

These dowels are very quick and easy to install, since there is no grouting involved. The main disadvantage is corrosion. The cost factor also may play a part in their selection.

33.4.1.3 Swellex dowels

Swellex dowels have a central hole. After installation into rock, a high-pressure water jet is sent in.

Note: a hole is drilled in the rock. Then the Swellex dowel is inserted. Next, a high-pressure water jet is sent in. The dowel would expand and hug the rock and would stay in place.

Corrosion is a major problem for Swellex dowels.

33.5 Grouted rock anchors (nonstressed)

In the case of grouted rock anchors, the bonding between rock and anchor is achieved by grout (Figure 33.15).

The anchor develops its strength through the bond strength between the rock and grout.

33.5.1 Installation procedure of grouted anchors

Step 1: Drill a hole to the desired length.
Step 2: Install the rock anchor.
Step 3: Grout the hole.
Step 4: Wait for sufficient time for the grout to harden before applying the load.

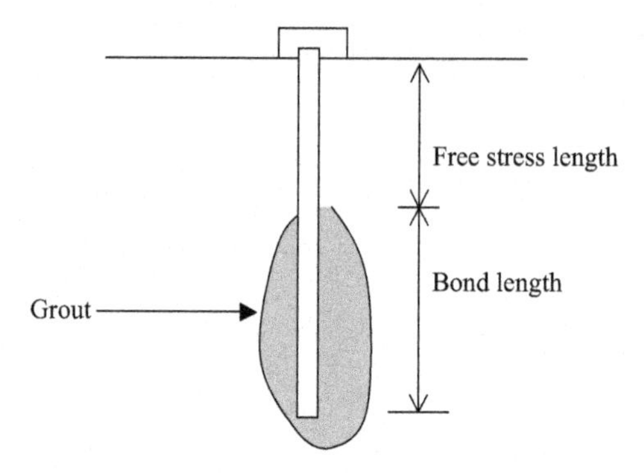

Figure 33.15 Grouted rock anchor.

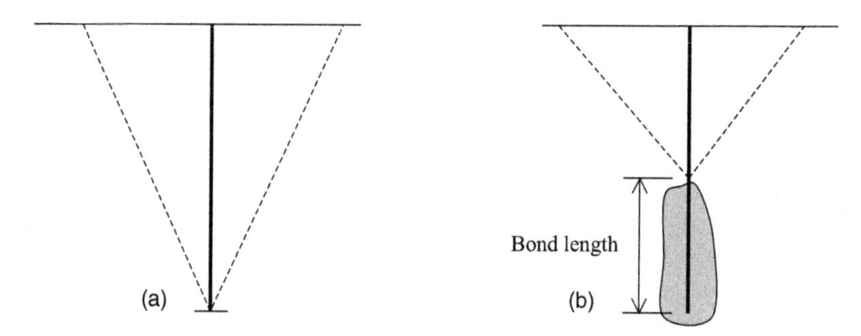

Figure 33.16 Stress cones for mechanical anchors and grouted anchors with similar length. (a) Mechanical anchors. (b) Grouted anchors.

33.5.2 Failure triangle for grouted rock anchors

Two anchors of same length are shown in Figure 33.16. The stress development cone in grouted anchors is smaller than comparable length mechanical anchors since grouted anchors need a bond length.

Table 33.1 shows the rock–grout bond strength.

33.6 Prestressed grouted rock anchors

Most rock anchors installed today are prestressed. Prestressed anchors are installed and subjected to stress.

33.6.1 Prestressing procedure

- Auger the hole.
- Place the anchor.
- Grout the desired length (Figure 33.17).

Table 33.1 Rock–grout bond strength

Rock type	Rock–grout bond strength (MPa)	Rock–grout bond strength (psi)
Granite	0.55–1.00	80–150
Dolomite/limestone	0.45–0.70	70–100
Soft limestone	0.35–0.50	50–70
Slates, strong shales	0.30–0.45	40–70
Weak shales	0.05–0.30	10–40
Sandstones	0.3–0.60	40–80
Concrete	0.45–0.90	70–130
Weak rock	0.35–0.70	50–100
Medium rock	0.70–1.05	100–150
Strong rock	1.05–1.40	150–200

Source: Wiley (1999).

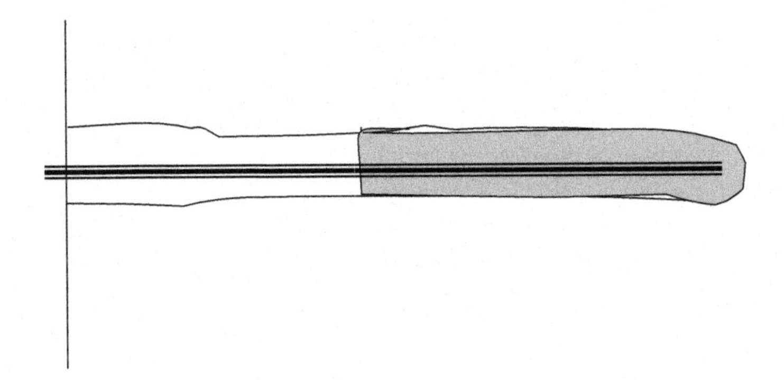

Figure 33.17 Auger a hole, place the anchor and grout.

- Wait till the grout is set.
- Stress the rock anchor to the required stress (Figure 33.18).
- Lock in the stress with a bolt (Figure 33.19).

Prestressed anchors have a number of advantages over nonstressed anchors.

33.6.2 Advantages of prestressed anchors

- Prestressed anchors will not move due to changing loads. On the other hand, nonstressed anchors move and the grout–anchor bond could be broken (Figure 33.20).
- Working load in a prestressed anchor is less than the applied prestress. In the case of nonstress anchors, the load varies. This could give rise to fatigue in steel and finally, failure could occur.

33.6.3 Anchor–grout bond load in nonstressed anchors

1. Outside load in the nonstressed anchor is 10 kN. The bond load will be equal to 10 kN (Figure 33.21).

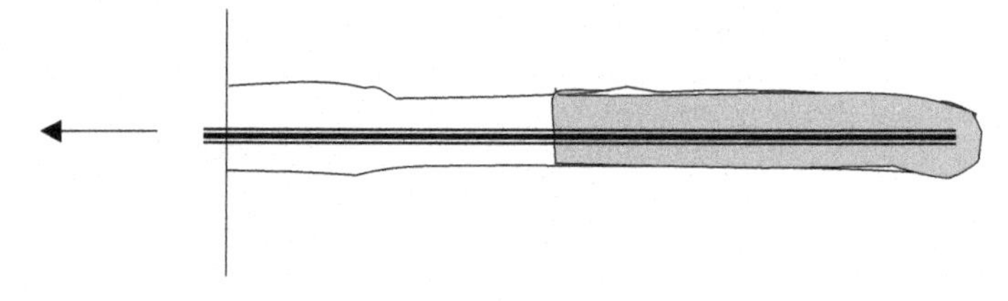

Figure 33.18 Wait until the grout is set and apply the stress using a jacking mechanism.

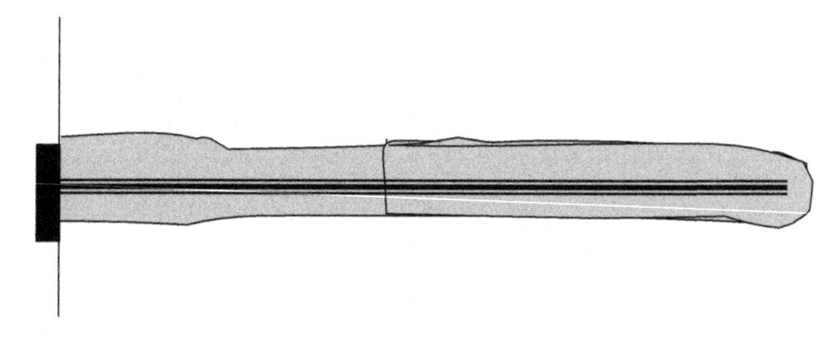

Figure 33.19 Lock in the stress and grout the rest of the anchor.

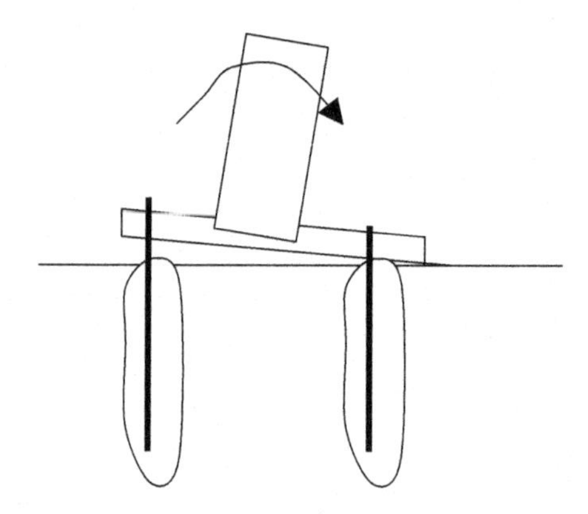

Figure 33.20 Nonstressed anchor lifted due to wind load.

2. Outside load is increased to 20 kN. The bond load will change to 20 kN. The anchor–grout bond load changes with the applied load in nonstressed anchors. This would create elongation in the anchor and damage the grout.

   Due to change in load in the anchor rod, the rod will move with the changing load.

33.6.4 Anchor–grout bond load in prestressed anchors

1. The anchor is pre-stressed to 20 kN and locked. No outside load is applied. Bond load = 20 kN (Figure 33.22).

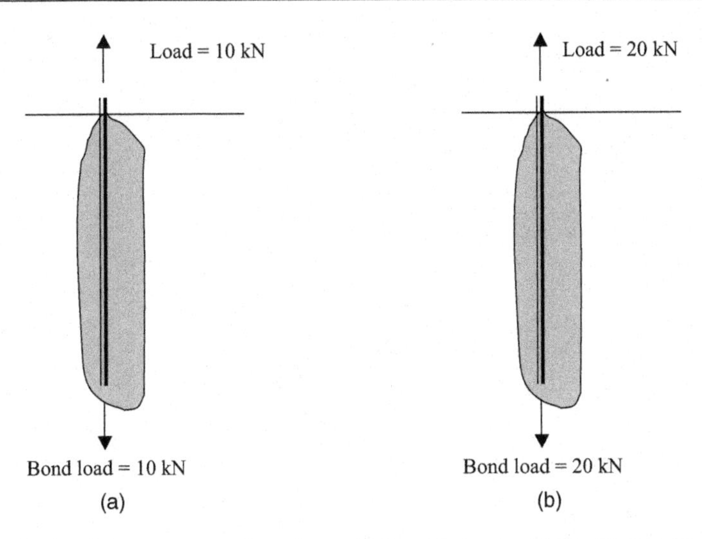

Figure 33.21 Load on nonstressed anchors. (a) Outside load = 10 kN. (b) Outside load = 20 kN.

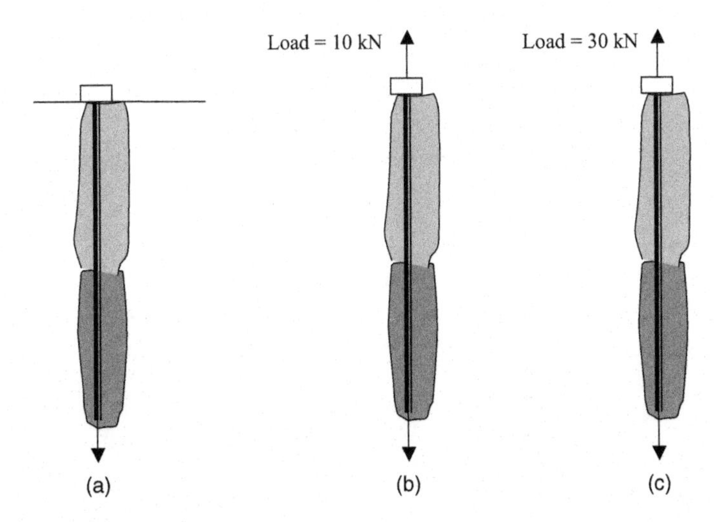

Figure 33.22 Load on prestressed anchors. (a) Prestressed load = 20 kN. (b) Bond load = 20 kN. (c) Bond load = 30 kN.

2. An outside load of 10 kN is applied. The bond load will remain at 20 kN. The rod will not move.
3. The outside load is increased to 30 kN. The bond load will now change to 30 kN and the rod will move against the grout.

There will not be any elongation in the rod until the outside load has surpassed the prestressed load.

Design Example 33.2

Find the bond length required for the rock anchor in the retaining wall. The rock grout bond strength was found to be 1.1 MPa and the diameter of the grout hole is 0.3 m (Figure 33.23).

Solution

$$\begin{aligned}\text{Lateral earth pressure at the bottom of the retaining wall} &= K_a \times \gamma \times h \\ &= 0.33 \times 18 \times 15 \\ &= 89.1\,\text{kN/m}^2.\end{aligned}$$

$$\begin{aligned}\text{Total lateral force } (F) &= \text{Area of the pressure triangle} \\ &= 15/2 \times 89.1 \\ &= 668.25\,\text{kN}\end{aligned}$$

$$\begin{aligned}\text{Overturning moment around point A} &= 668.25 \times 15/3 \\ &= 3341.25\,\text{kN. m}\end{aligned}$$

Required factor of safety = 2.5

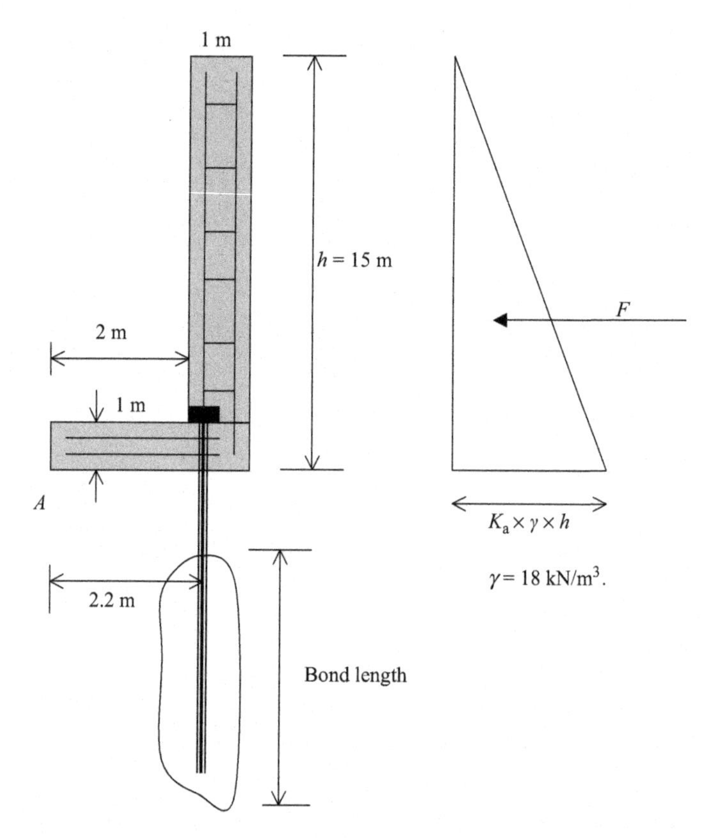

Figure 33.23 Retaining wall held by rock anchors.

$$\text{Required resisting moment} = 2.5 \times 3341.25$$
$$= 8{,}353 \text{ kN. m}$$

Resisting moment = resisting moment due to weight of the retaining wall + resisting moment due to pin pile

Resisting moment due to weight of the retaining wall is due to two parts.

Resisting moment due to weight of the retaining wall = resisting moment due to vertical portion of the wall + resisting moment due to horizontal portion of the wall

Resisting moment due to vertical portion of the wall = weight × distance to the center of gravity

$$\text{Resisting moment due to vertical portion of the wall} = (1 \times 15) \times 23.5 \times 2.5 \text{ kN. m}$$
$$= 881.25 \text{ kN. m}$$

Density of concrete = 23.5 kN/m^3.

$$\text{Resisting moment due to horizontal portion of the wall} = (2 \times 1) \times 23.5 \times 1 \text{ kN. m}$$
$$= 47 \text{ kN. m}$$

Total resistance due to weight of the retaining wall = 881.25 + 47 = 928.25 kN.m

Required resisting moment = 8353 kN.m

Resisting moment required from pin piles = 8353 − 928.25 = 7424.75 kN.m

Resisting moment from pin piles = $P \times 2.2$ kN.m

P = uplift capacity of pin piles

$P \times 2.2 = 7424.75$

$P = 3375$ kN

Bond length of pin piles = L

Pin pile diameter = 0.3 m

$P = (\pi \times \text{diameter}) \times \text{length} \times (\text{rock–grout bond strength})$

$P = 3375 = (\pi \times 0.3) \times \text{length} \times 1100$ kPa

Bond length required = 3.25 m.

Reference

Wiley, D.C., 1999. Foundations on Rock. Taylor and Francis, New York.

Soil anchors

34

Soil anchors are mainly of two types.

- Mechanical soil anchors
- Grouted soil anchors

34.1 Mechanical soil anchors

Mechanical soil anchors are installed in retaining walls, shallow foundations, and slabs that are subjected to uplift. The end of soil anchor consists of an anchor mechanism. The soil anchor is drilled into the soil (Figure 34.1). Then the anchor is pulled to activate the anchor mechanism (Figures 34.2 and 34.3).

34.2 Grouted soil anchors

The principle of grouted soil anchors is the same as the grouted rock anchors. Grouted soil anchors could be nonstressed or prestressed.

The installation procedure for nonstressed grouted soil anchors is as follows:

Step 1: Drill a hole.
Step 2: Insert the soil anchor.
Step 3: Grout the hole.

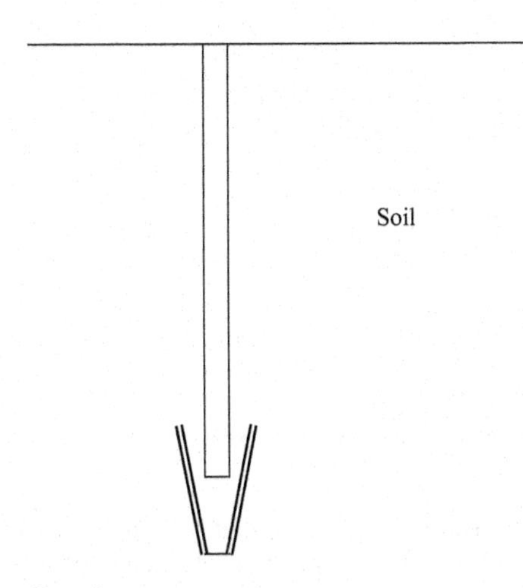

Figure 34.1 Drill the soil anchor.

Geotechnical Engineering Calculations and Rules of Thumb

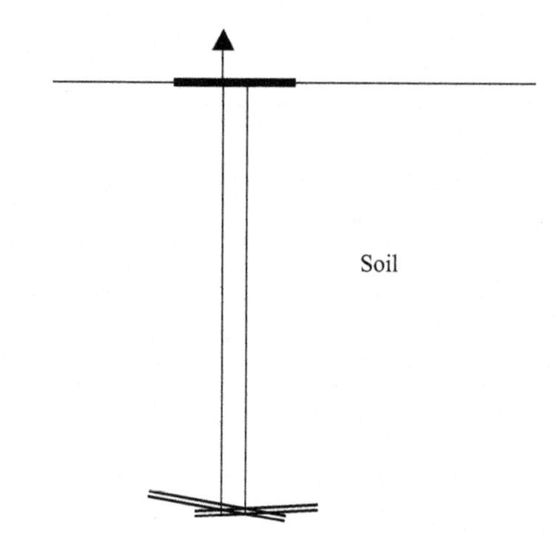

Figure 34.2 Pull the anchor to activate the anchor mechanism.

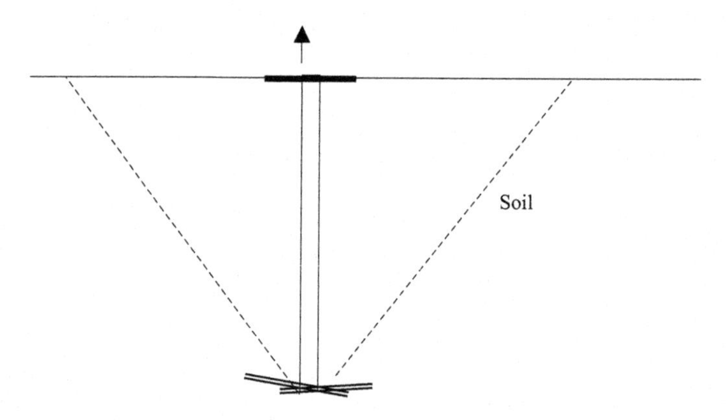

Figure 34.3 Failure cone for mechanical soil anchors.

The installation procedure for prestressed grouted soil anchors is slightly different.

Step 1: Drill a hole.
Step 2: Insert the soil anchor.
Step 3: Grout the desired length of the hole.
Step 4: Wait till the grout is set.
Step 5: Apply the desired prestress.
Step 6: Grout the rest of the hole.

Design Example 34.1

Find the grouted length required for a soil anchor installed in dense fine to medium sand (Plates 34.1 and 34.2). The soil anchor is needed to hold 6 tons and the diameter of the grouted hole is 10 in.

Plate 34.1 Soil anchor.

Plate 34.2 Soil anchors and wales.

Solution

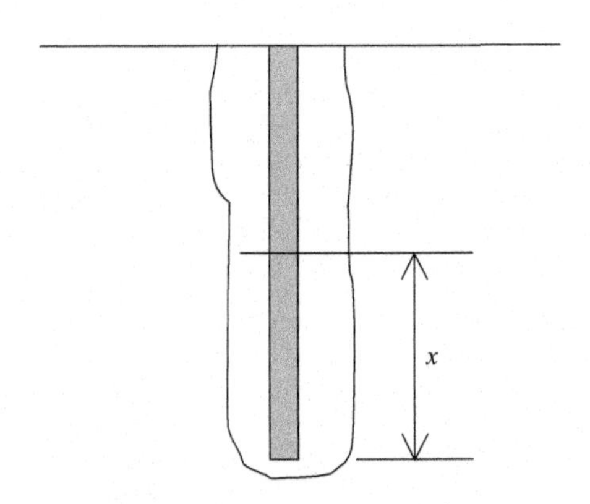

Assume that the required bond length is x.

From Table 34.1, the bond strength for dense fine to medium sand is 12–55 psi. Assume the bond strength to be 12 psi.

Total bond strength = perimeter area of the bond × bond strength

$(\pi \times D) \times \text{bond length} \times \text{bond strength} = 6 \text{ tons}$

Table 34.1 Ultimate soil–grout bond stress

Soil type	Ultimate bond stress (psi) for pressure grouted anchors
Cohesive soils (straight shaft)	
Soft silty clay	5–10
Stiff clay (medium to high plasticity)	5–10
Very stiff clay (medium to high plasticity)	10–25
Stiff clay (medium plasticity)	15–30
Very stiff clay (medium plasticity)	20–40
Cohesionless soil (straight shaft)	
Fine to medium sand (medium dense to dense)	12–55
Medium to coarse sand with gravel (medium dense)	16–95
Medium to coarse sand with gravel (dense to very dense)	35–100
Silty sand	25–50
Dense glacial till	43–75
Sandy gravel (medium dense to dense)	31–100

1 psi = 6.894 kPa.
Source: Williams form engineering corp, http://www.williamsform.com/ (modified by author).

$(\pi \times 10) \times L \times 12 = 6 \times 2000$ lbs

$L = 31.85$ in.

Assume a factor of safety of 2.0

Hence, the bond length required = 63.7 in.

Tunnel design

35.1 Introduction

Tunnel design is an interdisciplinary subject, involving geotechnical engineering, geology, geochemistry, and hydrogeology. In this chapter, the main concepts of tunnel design will be presented.

Tunnels are increasingly becoming common due to newly available equipment such as tunnel boring machines, road headers, relatively safe blasting techniques, etc. Tunnel boring machines (TBMs) are widely used today due to their ability to construct tunnels at record speeds.

The most important part of a tunnel boring machine is its cutter head. The cutter head comprises of a bunch of disc cutters that are pressed into the rock. When the machine rotates, the disc cutters would gradually cut the rock into pieces. The rock pieces that are cut would be collected in a bucket known as the muck bucket. Muck is removed from the tunnel by trains or by a conveyor belt. Gripper, as the name indicates, grips the tunnel sides so that the boring machine could be pushed hard into the rock.

35.2 Roadheaders

Roadheaders are not as powerful as tunnel boring machines and can be used for small size tunnels.

35.3 Drill and blast

Drilling and blasting was the earliest method of tunnel construction. In this techniques, holes are drilled into the rock and packed with explosives and blasted (Figure 35.1).

35.4 Tunnel design fundamentals

Tunnel excavation in solid rock may not need much roof support to hold the rock. On the other hand, weak or weathered rock may need heavy support mechanisms.

In Figure 35.2, one can notice the following significant items to consider during the design phase.

Point A to point B

The tunnel liner system has to support the soil overburden. Heavy tunnel support system is necessary in this region.

Figure 35.1 Drill and blast technique.

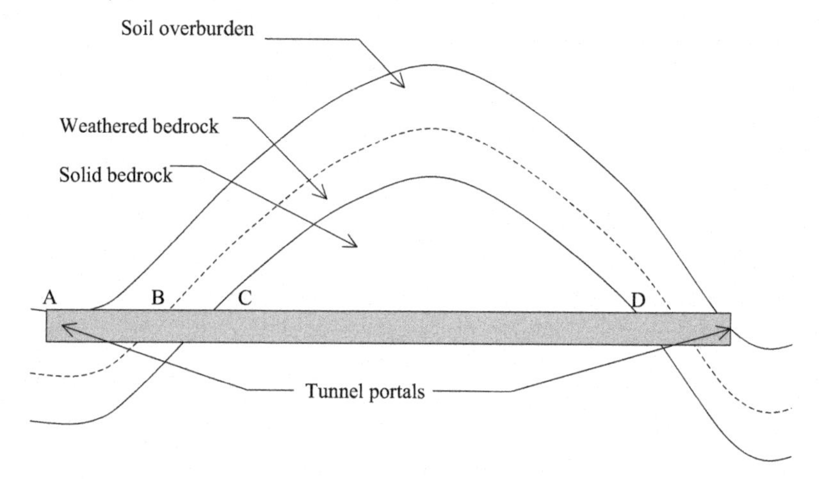

Figure 35.2 Typical tunnel through a mountain.

Point B to point C

The top few feet of the bedrock is usually weathered. This region needs more tunnel support than point C to point D since weathered bedrock exerts high loading than solid bedrock.

Point C to point D

Tunneling through solid bedrock may depend on the hardness of rock. Tunnel support system in this region would be minimal compared to solid bedrock.

35.4.1 Literature survey

1. State geological surveys – Most states have conducted geological surveys of the state. These documents would be available from the geological survey division of the state.
2. Well logs and previous borings – Well log data can be obtained from local well drillers for a nominal fee. Well logs normally contain rock stratification data, water-bearing zones, high strength rock stratums, and weak rock stratums.
3. Information on nearby tunnels – If there are tunnels nearby, the design information on these tunnels would be of immense help. The author has satisfactorily obtained design information on nearby tunnels from companies. The state department of transportation is another source for previous tunnel design work.

4. Regional university libraries contain information on various geological and Geotechnical engineering studies.

35.4.2 Subsurface investigation program for tunnels

Borings: American Association of State Highway and Transportation Officials (AASHTO) recommends borings to a depth of 50 ft. below the proposed tunnel bottom. In many cases, drilling 50 ft. below the tunnel bottom is not warranted. Most tunnel projects would have many rounds of borings. Spacing of borings should be based on information requirements.

Piezometers: Piezometers should be installed to obtain groundwater data. The piezometers should be screened through tunnel cross section. Overburden piezometers also should be installed to obtain overburden aquifer information. There are many cases where water from overburden aquifers leaking into the tunnel during construction (Figure 35.3).

Geophysical survey: Geophysical surveys are typically conducted in selected borings. Information such as elastic modulus, Poisson's ration of rock, density, porosity, and weak zones can be obtained from a geophysical survey.

Geophysical surveys are conducted by sending a seismic signal and measuring the return signal. If a large percentage of the signal returns, then the rock is hard and free of fractures. On the other hand, if most of the signal is lost, it is probably due to cracks and fissures in the rock.

Two types of Geophysical techniques are available – single hole and double hole (Figure 35.4).

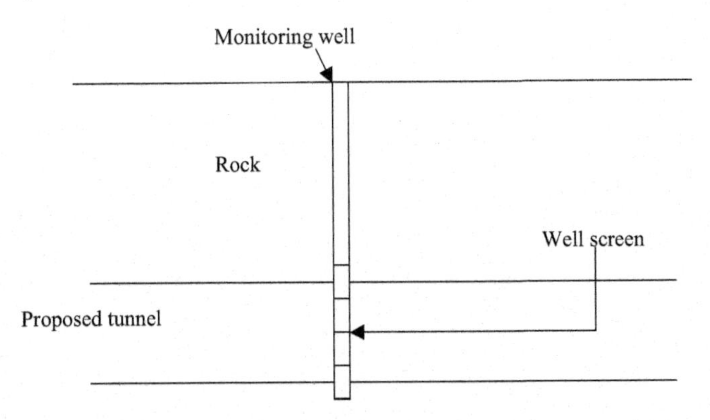

Figure 35.3 Well screens should be within the tunnel section.

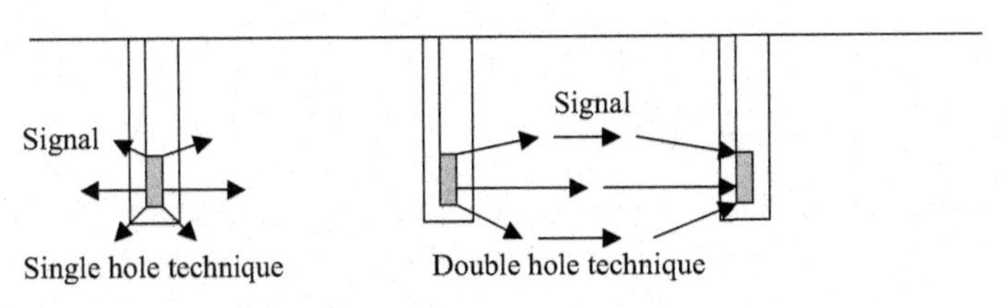

Figure 35.4 Geophysical methods.

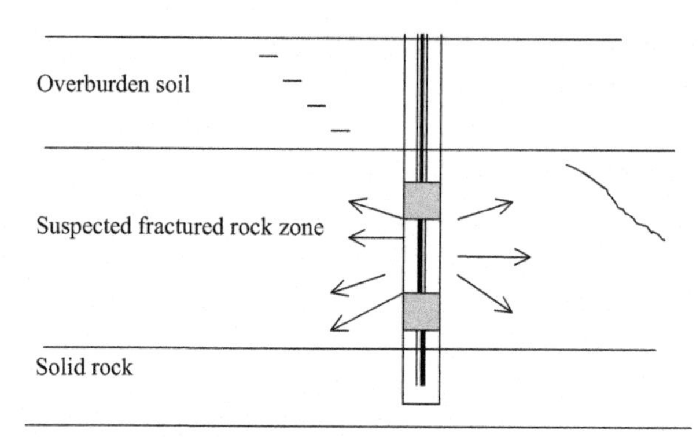

Figure 35.5 Packer inserted and pressurized.

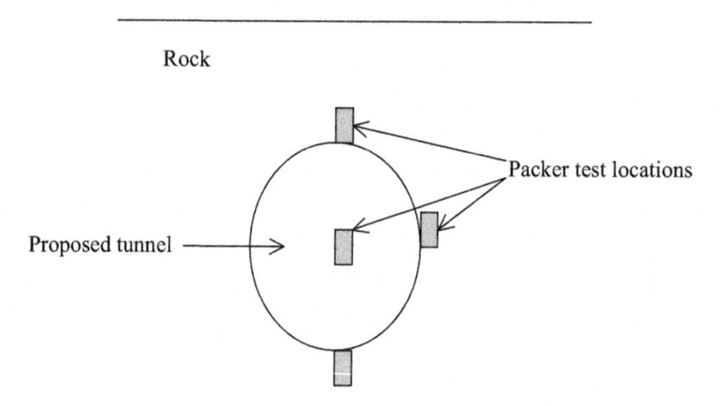

Figure 35.6 Typical Packer test locations.

Packer tests: Packer tests are conducted to measure the permeability of rock layers. Inflatable object made of rubber is inserted into the rock and pressurized. The pressure drop during the process is measured. If the rock is highly fractured, a large pressure drop will be noted (Figure 35.5).

Packer tests should be conducted at tunnel roof elevation, tunnel cross section and tunnel bottom (Figure 35.6).

Usually, during the boring program, engineers will be able to locate fractured rock zones. This can be achieved by analyzing the rock core logs and RQD values.

Groundwater plays a major role during the construction phase of the tunnel. By conducting packer tests at different locations, engineers can access the impact of groundwater.

35.4.3 Laboratory test program

Laboratory tests are conducted to investigate the strength, hardness, swelling properties, tensile strength, and other properties of rock that may influence the design process.

35.4.4 Unconfined compressive strength test

Unconfined strength test is done by placing a rock sample and crushing it to failure. This test provides the cohesion of intact rock. Unconfined compressive strength is a very important test since many correlations have been developed using unconfined compressive strength test values.

35.4.5 Mineral identification

Identification of minerals becomes very important in soft rock tunneling. Occurrence of minerals such as pyrite indicates the existence of caustic or acidic groundwater. Tunnel concrete could be subjected to chemical degradation due to acidic groundwater.

Minerals such as montmorillonite, chlorite, attapulgite, and illite indicate the rock could be subjected to swelling in the presence of moisture.

35.4.6 Petrographic analysis

Petrography is the description and systematic classification of rocks, by examination of thin sections. Petrographic analysis would include determining properties such as grain size, texture, color, fractures, and abnormalities.

35.4.7 Triaxial tests

Triaxial tests are conducted to investigate the shear strength of rock under confining pressure. Usually, triaxial tests are done on soft rocks.

35.4.8 Tensile strength test

Rocks just above the tunnel roof will be subjected to tensile forces. Hence, it is important to investigate the tensile strength of the rock stratum especially above tunnel roofs.

35.4.9 Hardness tests

Schmidt hammer test is the most popular method to investigate the hardness of rocks. Hardness of rocks is important for tunnel boring machine selection.

35.4.10 Consolidation tests

Consolidation tests are done to investigate consolidation characteristics of soft rocks. Usually, clay – shale, mudstone, argillaceous soft rocks can consolidate under pressure. Consolidation tests can be conducted to investigate the settlement properties of soft rock.

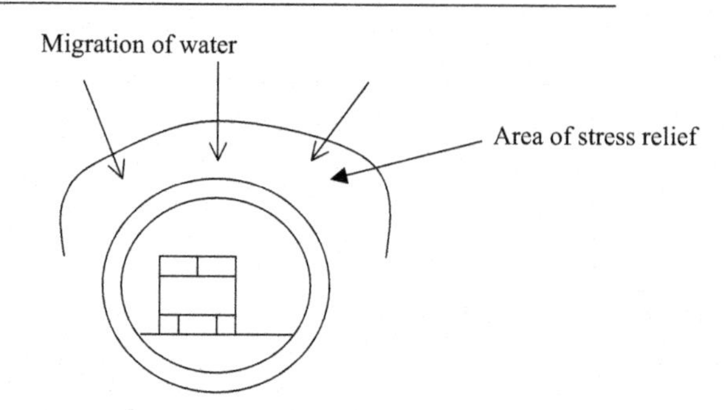

Figure 35.7 Large stresses are exerted on tunnel lining due to swelling of soft rock.

35.4.11 Swell tests

Many soft rocks swell when exposed to moisture. Swelling is very common in rocks with clay minerals. Water tends to flow from high stressed areas to low stressed areas. When a tunnel is excavated, the soil pressure in adjacent rocks would be relieved. Rock units with relatively low stress concentrations would absorb water and swell (Figure 35.7).

35.5 Tunnel support systems

All modern tunnels are protected by tunnel supporting systems. In the past, tunnels built in hard sound rock left unsupported. Tunnel support systems typically consist of steel arches and concrete segments (Figure 35.8). Concrete segments are increasingly becoming popular in the tunneling industry (Figure 35.9).

Steel ribs are also commonly used for tunnel support systems (Figure 35.10). H-beams in combination with concrete are used to provide support for the tunnel (Figure 35.11).

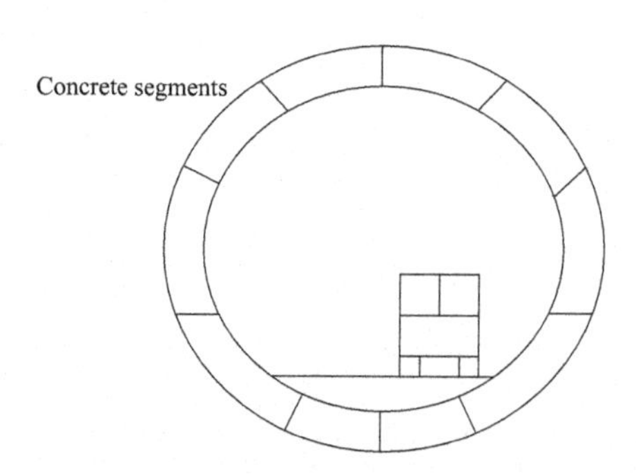

Figure 35.8 Tunnel built using concrete segments.

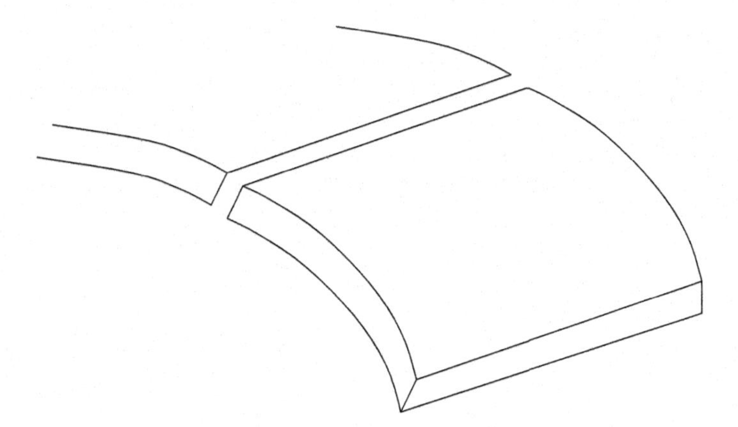

Figure 35.9 Concrete segments.

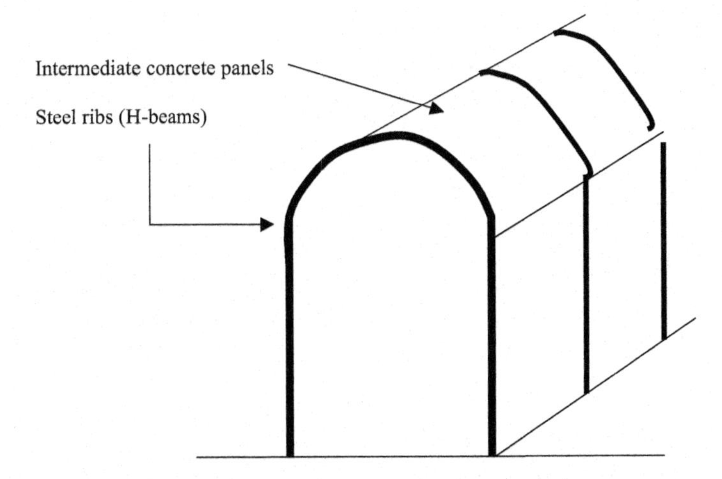

Figure 35.10 Steel rib supported tunnel.

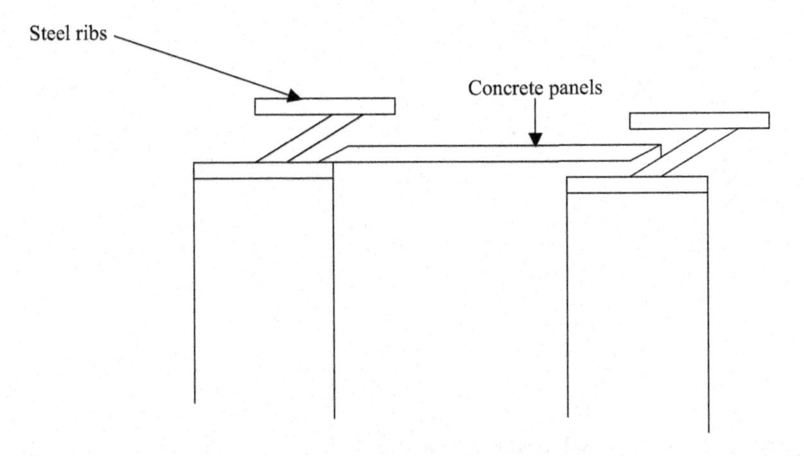

Figure 35.11 Steel supports (H-sections) and concrete panels.

35.5.1 Shotcrete

Shotcrete and rock bolts is another support mechanism that is widely used today to support tunnels.

Shotcrete is nothing but a mixture of cement, sand, and fine aggregates. Cement, sand, and fine aggregates are mixed and shot into the rock surface. Two types of techniques are available.

35.5.2 Dry mix shotcrete

Some tunnel linings are constructed using rock bolts and shotcrete (Figures 35.12 and 35.13).

35.6 Wedge analysis

During tunnel excavations, rock wedges could form due to intersection of joints (Figure 35.14). The wedge analysis needed to be performed for all excavations in hard rock.

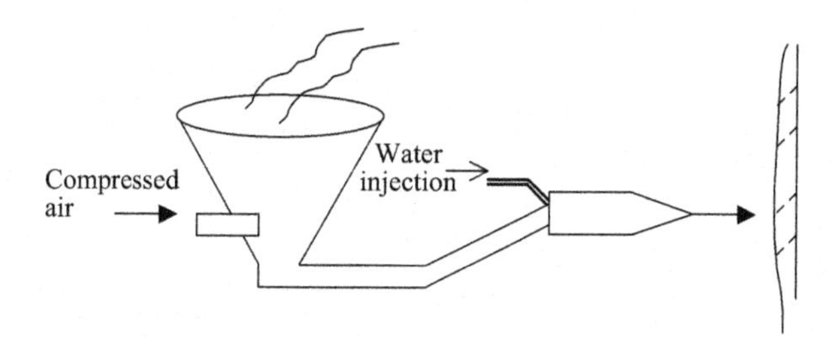

Figure 35.12 Shotcrete preparation.

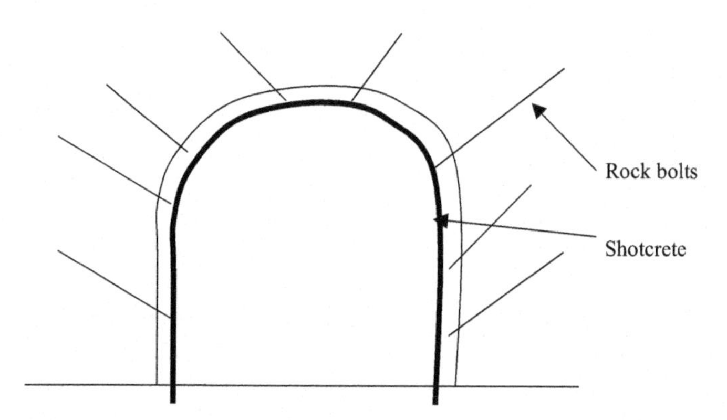

Figure 35.13 Rock bolts and shotcrete in a tunnel.

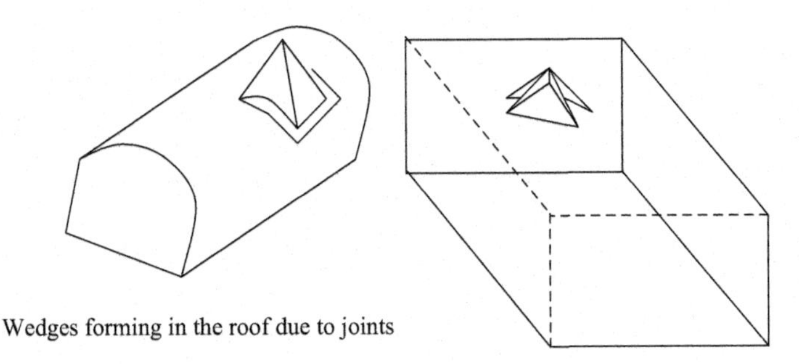

Figure 35.14 Wedges formed due to rock joints.

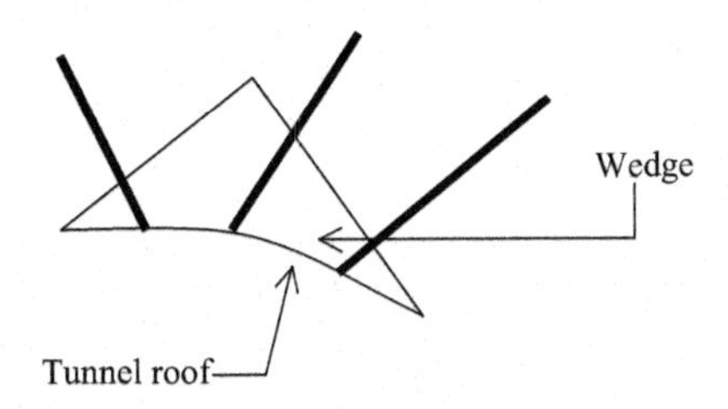

Figure 35.15 Rock bolts used to stabilize a wedge.

- Wedges are formed by the intersection of joints. When a right combination of joints is present at a given location, a wedge is formed. It is possible to find the average joint spacing between joints from the boring data.
- *Identification of potential wedges* based on rock coring data manually is a formidable task (Figure 35.15).
- Many engineers use *Computer Programs* to perform wedge analysis. As it is not realistic to perform wedge analysis manually based on coring data. UNWEDGE computer program by *Rock Engineering Group* is very popular among tunnel engineers.

Geosynthetics in geotechnical engineering

36

The following subgroups can be identified as geosynthetic products.

- *Geotextiles*: Clothes are woven out of natural fibers such as cotton and silk. Similarly, geotextiles are woven out of synthetic fibers. Synthetic fibers are manufactured using various polymers. Since, geotextiles are woven, they are relatively porous. Geotextiles are mainly used for the following purposes:
  - Separation
  - Filtration
  - Drainage
  - Reinforcement
  - Containment of contaminants
- *Geomembranes*: Geomembranes are made of thin plastic sheets. They are much more impervious than geotextiles. Due to this reason, geomembranes have found their way to landfill industry. Landfill liners, leachate collection systems and landfill covers are made of geomembranes.
- *Geosynthetic clay liners (GCLs)*: GCLs are manufactured by sandwiching a bentonite clay layer between two geotextiles or geomembranes (Figure 36.1).
- *Geogrids*: Geogrids are mainly used for reinforcing purposes (Figure 36.2).
- *Geonets*: The purpose of the geonets is to facilitate the drainage of water.
- *Geocomposites*: Geocomposites are manufactured by combining two geotextiles (or geomembranes).

  The common combinations used by the industry are:
  - *Geotextile and a geogrid*: The geotextile separates one soil from another while the geogrid provides the strength.
  - *Geonet and a geotextile*: The geonet provides drainage while the geotextile provides separation and strength.
  - *Geonet and a Geomembrane*: This combination serves the same purpose as the geonet and geotextile combination (Figure 36.3).

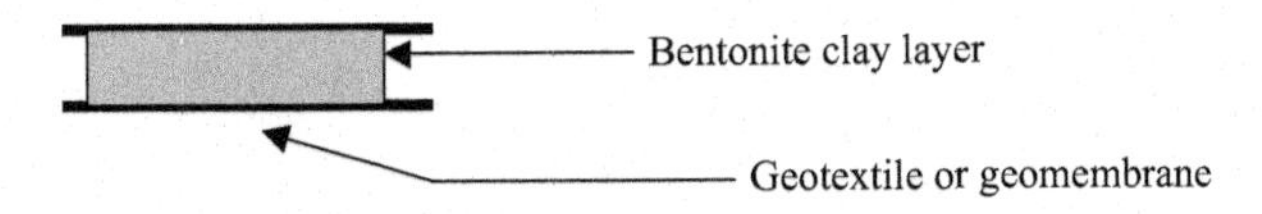

Figure 36.1 Geosynthetic clay liners (GCLs).

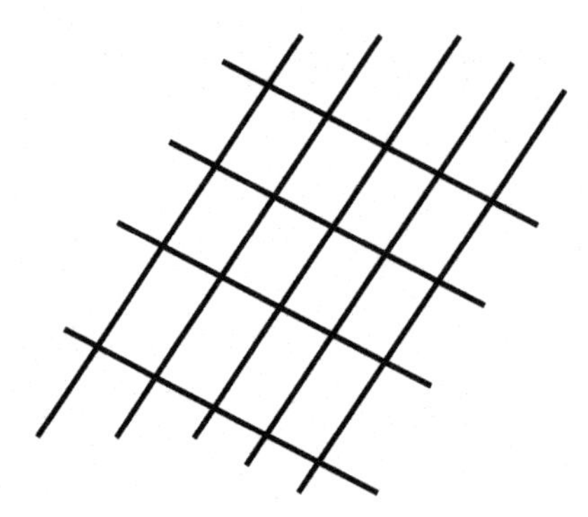

Figure 36.2 Geogrids.

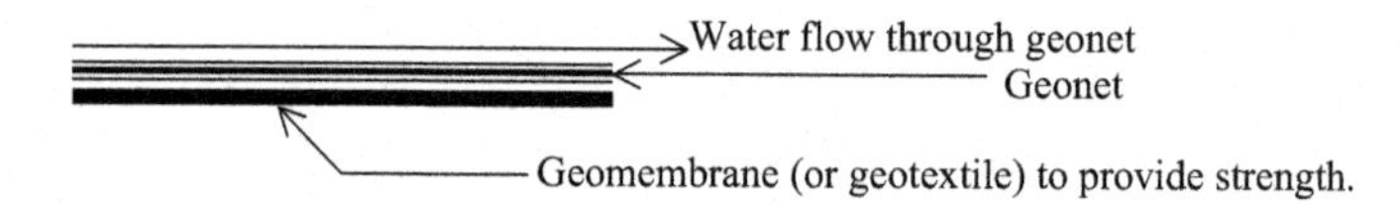

Figure 36.3 Geonet and geomembrane composite.

Slurry cutoff walls

37

Slurry cutoff walls are designed mainly for seepage control. Nowadays, slurry cutoff walls are being designed around landfills to stop contaminant migration.

A slurry cutoff wall can be constructed as shown in Figure 37.1 to stop contaminant migration. For the slurry cutoff wall to be effective, it should penetrate into an impermeable clay layer.

37.1 Slurry cutoff wall types

Two types of slurry cutoff walls are popular.

- Soil–bentonite walls (SB walls)
- Cement–bentonite walls (CB walls)

37.1.1 Soil bentonite walls (SB walls)

A trench is excavated and filled with a mixture of soil, bentonite powder, and water. Normally, the soil coming out of the trench is used. Theoretically, any type of soil can be used. More fines (smaller than sieve #200) the soil has, the permeability would be lower. If clay is in a lumpy form, the permeability will not be low, even though clay particles are finer than sieve #200 (Figure 37.2).

The slurry is pumped straight to the trench to keep the trench open. A backhoe would take the slurry out of the trench and mix with the soil, which came out of the trench. The density of a soil bentonite slurry is approximately 120 pcf.

The permeability of soil bentonite walls depends on the type of soil. If the soil contains more sandy material, then the permeability would be higher. Bulging of the trench due to high density of the soil–bentonite mixture is considered to be one of the problems of soil–bentonite walls. Cement–bentonite walls are used to avoid bulging.

37.1.2 Cement bentonite walls (CB walls)

A slurry is produced using cement and bentonite. The slurry is pumped into the trench to harden. Soil is not used. The density of a cement–bentonite slurry is approximately 70 pcf.

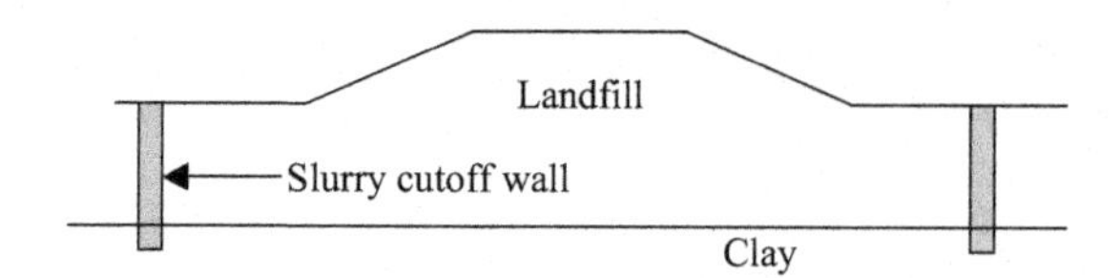

Figure 37.1 Slurry cutoff wall to stop contaminant migration.

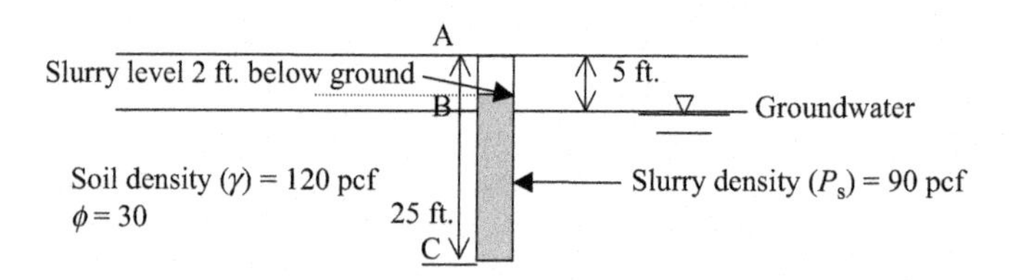

Figure 37.2 Slurry cutoff wall setup.

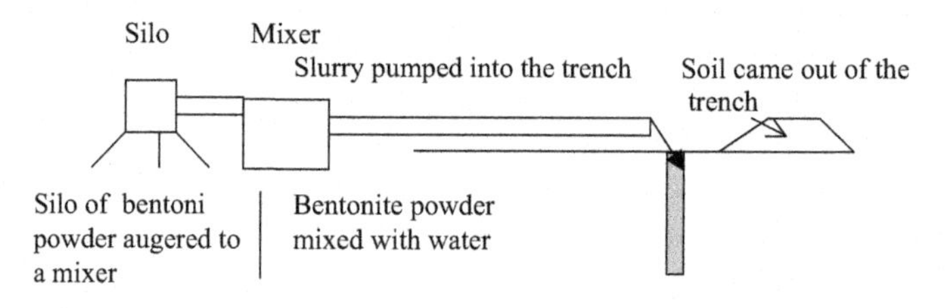

Figure 37.3 Slurry cutoff wall.

37.1.2.1 *Trench stability for slurry cutoff walls in sandy soils*

Design Example 37.1

Check the stability of a 25 ft. deep trench (Figure 37.3). GW is 5 ft. below the ground surface and the slurry level is maintained 2 ft. below the ground surface.

Solution

Step 1: Trench sidewall stability

At Point B

Vertical effective stress at B in surrounding soil = 5 × 120 = 600 psf

$$\text{Horizontal stress} = K_a \times \text{vertical effective stress}$$
$$= 0.33 \times 600 = 198\ \text{psf}$$
$$(K_a = \tan^2(45 - \phi/2) = 0.333)$$

Vertical stress in the slurry = 3 × 90 = 270 psf (slurry level is 2 ft. below the ground level)

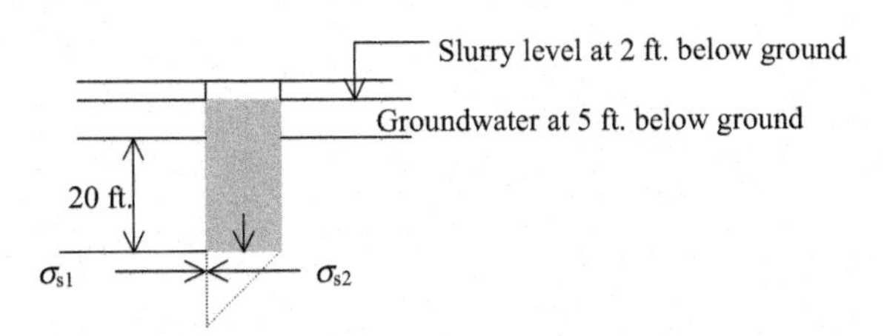

Figure 37.4 Trench bottom stability in a slurry cutoff wall.

Horizontal stress in the slurry = 270 psf ($K_a = 1$, for slurries and water. The pressure is the same in all directions).

Stress in the slurry (270) is greater than the horizontal stress due to soil (198). Hence, the trench may not cave in at point B. (FOS against caving in at point B = 270/198 = 1.4)

At Point C

$$\text{Vertical effective stress at C in surrounding soil} = 5 \times 120 + 20 \times (120 - 62.4) = 1752 \text{ psf}$$

$$\text{Horizontal stress due to soil} = K_a \times \text{vertical effective stress} = 0.33 \times 1752 = 578 \text{ psf}$$

Vertical stress in the slurry = 23 × 90 = 2070 psf
Horizontal stress in the slurry = 2070 psf
Factor of safety against trench cave in at point C = 2070/578 = 3.58

Step 2: Trench bottom stability (Figure 37.4)
The stability of the trench bottom also needs to be assured.
Vertical stress in surrounding soil at trench bottom = 1752 psf (found earlier)
Horizontal stress due to surrounding soil (σ_{s1}) = K_a × 1752 = 0.33 × 1752 = 578 psf
Stress in the slurry (P_s) = 2070 psf
The vertical stress in the soil just below the slurry = 2070 psf
Vertical effective stress in the soil at the trench bottom just below the slurry = 2070 − pore water pressure = 2070 − 20 × 62.4 = 822 psf

Note: The pore water pressure in the soil has to be reduced to obtain the vertical effective stress.
The horizontal stress due to surrounding soil (σ_{s1}) to be able to push the trench bottom up, it has to be greater than the passive pressure of the soil in the trench bottom underneath the soil.
Passive stress of the soil underneath the slurry (σ_{s2}) = K_p × vertical effective stress

$$K_p = \tan^2(45 + \phi/2) = 3.0$$
$$\sigma_{s2} = 3.0 \times 822 = 2466 \text{ psf}$$

If σ_{s1} is greater than 2466 psf, the trench bottom would be pushed up and fail.

$$\sigma_{s1} = 578 \text{ psf}$$

Factor of safety against trench bottom failure = 2466/578 = 4.26.

Earthwork

38

38.1 Excavation and embankment (Cut and Fill)

Roads are constructed over uneven ground. During the construction of a road, some locations are required to be cut and other locations have to be filled. Consider the terrain shown in the given figure. Point A to B has to be *cut* and point B to C has to be *filled.*

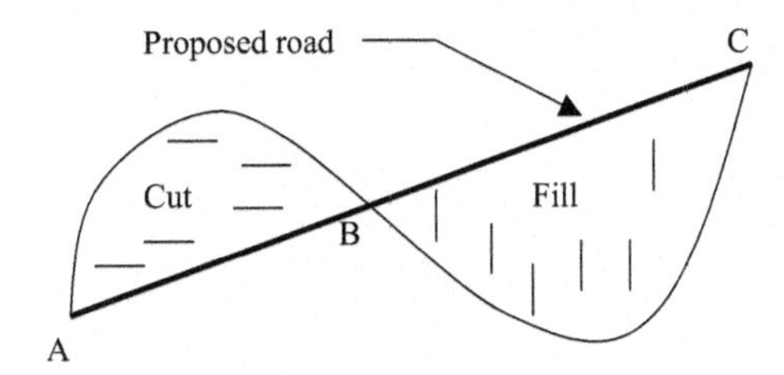

Cut and fill: In some situations, the soil removed due to a cut can be used for fill. If the soil removed is not suitable, then the suitable soil has to be imported to the site for filling purposes.

1. Borrow pit volume problems: To solve the borrow pit problems, the student needs a good knowledge of phase relationships in soil.
2. Soil phase relationships: Soil consists of solids, air, and water. Solids are soil particles. Soil matrix can be schematically represented as shown in the given soil phase diagram.

Volume		Mass
V_a	Air	$M_a = 0$ (Mass of air is taken to be zero).
V_w	Water	M_w (Mass of water)
V_s	Solid	M_s (Mass of solids)

Soil phase diagram

M_a = Mass of air = 0 (usually taken to be zero)
V_a = Volume of air (volume of air is *not* zero)

M_w = Mass of water
V_w = Volume of water

Geotechnical Engineering Calculations and Rules of Thumb

M_s = Mass of solids
V_s = Volume of solids

M = Total mass of soil = $M_s + M_w$ (mass of air ignored).
V = Total volume of soil = $V_s + V_w + V_a$
V_v = Volume of voids = $V_a + V_w$

Density of water (γ_w): Density of water can be expressed in many units.

$$\gamma_w = \frac{M_w}{V_w}$$

SI Units: $\gamma_w = 1\ \text{g/cm}^3 = 1000\ \text{g/L} = 1000\ \text{kg/m}^3 = 9.81\ \text{kN/m}^3$.
fps Units: $\gamma_w = 62.42$ pounds per cu. ft. (pcf)

Total density of soil γ_t, γ_{wet} *or* γ: Some books use γ_{wet} and some other books use γ_t or simply γ to denote the total density of soil. Total density (also known as wet density) is simply mass of soil (including water) divided by the volume of soil.

$$\gamma_{wet} = \frac{M}{V}$$

$$M = M_w + M_s \text{ and } V = V_w + V_a + V_s$$

{M = total mass of soil including water. Mass of air ignored}
{V = total volume of soil including soil, water and air. Volume of air is *not* ignored.}

Dry density of soil (γ_d): $\gamma_d = M_s/V$

M_s = mass of solid only.
V = total volume of soil including soil, water and air
$V = V_w + V_a + V_s$

Density of solids = M_s/V_s

Specific gravity (G_s): Specific gravity is defined as the density of solids divided by the density of water. Density of solids is represented by G_s or by simply G.

Specific gravity (G_s) = $M_s/(V_s\gamma_w)$

Void ratio (e): Void ratio (e) is defined as the ratio of volume of voids to volume of solids.

$$e = \frac{V_v}{V_s}$$

V_v, volume of voids (volume of water + volume of air) = $V_a + V_w$; V_s, volume of solids.

Moisture content (w): Moisture content (w) = M_w/M_s

M_w = mass of water; M_s = mass of solids

Porosity (n):

$$n = -V_v / V$$

V_v, volume of voids; V, total volume = $V_s + V_W + V_a$

What does porosity mean? If you look at the top term V_v, it basically tells us how much voids are there in the soil. The ratio between voids and total volume is given by porosity. In other words, porosity gives us an indication of pores in a soil. Soils with high porosity would have more pores than soils with low porosity. It is reasonable to assume that soils with high porosity would have a higher permeability.

Degree of saturation (S): $S = V_W/V_v$

V_W, volume of water; V_v, total of volume of voids.

When total volume of voids is filled with water, $S = 100\%$.

Degree of saturation tells us how much water is in the voids.

38.2 Some relationships to remember

Relationship 1:

$$\gamma_d = \gamma_{wet}/(1+w)$$

This relationship appears in soil compaction section as well.

It can be proven as follows.

$$\gamma_{wet} = \frac{M}{V}, \quad \text{hence} \quad V = \frac{M}{\gamma_{wet}}$$

$$\gamma_d = \frac{M_s}{V};$$

Replace V with M/γ_{wet}

$$\gamma_d = \frac{M_s}{(M/\gamma_{wet})} = \frac{M_s \times \gamma_{wet}}{M}$$

$$M = M_s + M_w \text{ (mass of air is ignored)}$$

$$\gamma_d = \frac{M_s \times \gamma_{wet}}{(M_s + \gamma_w)}$$

Divide both top and bottom by M_s.

$$\gamma_d = \gamma_{wet}/(1+w)$$

Relationship 2:

$$S \times e = G_s \times w$$

This relationship can be shown to be true as below.

$$S = \frac{V_w}{V_v};\ e = \frac{V_v}{V_s}$$

$$Se = \frac{V_w}{V_s} \tag{38.1}$$

$G_s = M_s/(V_s\gamma_w)$; $\gamma_w = M_w/V_w$; Replace γ_w in the equation.

$$G_s = M_s/(V_s) \times (V_w/M_w)$$

Hence, $G_s = (M_s \times V_w)/(V_s M_w)$;

$$w = M_w/M_s$$

$$G_s w = [(M_s \times V_w)/(V_s \times M_w)] \times (M_w/M_s)$$

By simplification;

$$G_s w = V_w/V_s \tag{38.2}$$

Equations (38.1) and (38.2) are equal. Hence $Se = G_s w$.
Relationship 3:

$$n = e/(1+e)$$

Proof:
Replace "*e*" with V_v/V_s in the above equation.

$$n = e/(1+e) = (V_v/V_s)/[1 + V_v/V_s]$$

Multiply top and bottom by V_s.

$$e/(1+e) = (V_v)/[V_s + V_v]$$
$$V_s + V_v = V$$
$$e/(1+e) = V_v/V = n\ (V_v/V \text{ is porosity})$$

Relationship 4:

$$e = n/(1-n)$$

Proof: From relationship 3;

$$n = e/(1+e)$$
$$n + ne = e$$
$$n = e - ne$$
$$n = e(1-n)$$
$$e = n/(1-n)$$

Relationship 5:

$$\gamma_d = \gamma_w G_s/[1+(w/S)\ G_s]$$

Proof:

$$\gamma_d = \frac{\gamma_w G_s}{[1+(w/S)G_s]}$$
$$\gamma_d = \frac{\gamma_w M_s/(V_s\gamma_w)}{[1+(M_w/M_s/(V_w/V_v)\times M_s/(V_s\gamma_w)]}$$
$$\gamma_d = \frac{\cancel{\gamma_w}\ M_s/(V_s\cancel{\gamma_w})}{[1+(M_w/\cancel{M_s}/(V_w/V_v)\times \cancel{M_s}/(V_s\gamma_w)]}$$
$$\gamma_d = \frac{M_s/V_s}{[1+(M_w/(V_w/V_v)\times 1/(V_s\gamma_w)]}$$
$$\gamma_d = \frac{M_s/V_s}{[1+(\cancel{\gamma_w}\ V_v)\times 1/(V_s\cancel{\gamma_w})]}$$
$$\gamma_d = \frac{M_s/V_s}{[1+(V_v)\times 1/(V_s)]}$$

Multiply top and bottom by V_s

$$\gamma_d = \frac{M_s}{[V_s+V_v]}$$
$$V_s + V_v = V$$
$$\gamma_d = \frac{M_s}{V}$$

Relationship 6:

$$\gamma_{wet} = \frac{\gamma_w G_s \times (1+w)}{[1+e]}$$

Proof:
Write down relationship 5.

$$\gamma_d = \frac{\gamma_w G_s}{[1+(w/S)G_s]}$$

Substitute for γ_d and S.

$$\gamma_d = \gamma_{wet}/(1+w) \text{ and } Se = G_s w$$

Hence, $S = G_s w/e$

$$\gamma_{wet}/(1+w) = \frac{\gamma_w G_s}{[1+(we/G_s w)G_s]}$$

$$\gamma_{wet} = \frac{\gamma_w G_s \times (1+w)}{[1+e]}$$

Relationship 7:

$$\gamma_d = \frac{\gamma_w G_s}{[1+e]}$$

Proof:

$$\gamma_d = M_s/V; \qquad G_s = M_s/V_s/\gamma_w \qquad e = V_v/V_s$$

Apply the values in the equation;

$$M_s/V = \gamma_w (M_s/V_s/\gamma_w)/(1+V_v/V_s)$$

γ_w cancels out.

$$M_s/V = (M_s/V_s)/(1+V_v/V_s)$$

M_s cancels out.

$$1/V = (1/V_s)/(1+V_v/V_s)$$

$$\begin{aligned} V_s/V &= 1/(1+V_v/V_s) \\ &= 1/[(V_s+V_v)/V_s] = V_s/[(V_s+V_v)] = V_s/V \end{aligned}$$

Left hand side and right hand side are equal.

Practice Problem 38.1

Specific gravity of a soil sample is given to be 2.65. Moisture content and degree of saturation are 0.6 and 0.7, respectively. Find the void ratio.

Solution

$$Se = Gw$$
$$0.7 \times e = 2.65 \times 0.6$$
$$e = 2.27$$

Practice Problem 38.2

Total density of a soil sample was found to be 110 pcf and moisture content to be 60%. What is the dry density of the soil sample?

Solution

$$\gamma_d = \gamma_{wet}/(1+w) = 110/(1+0.6) = 68.75 \text{ pcf}$$

Practice Problem 38.3

A soil sample was obtained from the ground and measured. The weight was found to be 1 lb and the total soil volume was found to be 0.01 ft.^{3}. The soil sample is then put in the oven and dried. The dried soil sample was weighed to be 0.7 lbs. The specific gravity of the soil is known to be 2.6.

Find the following;

1. Total density or wet density
2. Dry density
3. Porosity
4. Void ratio
5. Degree of saturation

Solution

1. Total density = M/V = 1/0.01 = 100 lbs/ft^3.
2. Dry density = $\gamma_d = \gamma/(1 + w)$
   Weight of dry soil (M_s) = 0.7 lbs

   Weight of water in the soil sample (M_w) = 1 − 0.7 = 0.3 lbs
   Water content (w) = M_w/M_s = 0.3/0.7 = 0.428
   Dry density = $\gamma_d = \gamma/(1 + w)$ = 100/(1 + 0.428) = 69.9 lbs/ft.^{3}.
3. Porosity (n) = V_v/V
   V = 0.01 ft.^{3}. This is the total volume of the soil sample
   Specific gravity (G) of the soil is given to be 2.6.

Find V_S:
$G = 2.6 = M_S/(V_S \gamma_W) = 0.7/(V_S \times 62.4)$
Since, $\gamma_W = 62.4$ lbs/ft.^{3}.
Hence, $V_S = 0.0043$ ft.^{3}.
$V = V_V + V_S$ (Total volume = volume of voids + volume of solids)
$0.01 = V_V + 0.0043$
$V_V = 0.01 - 0.0043 = 0.0057$
Porosity $(n) = V_V/V = 0.0057/0.01 = 0.57$

4. Void ratio (e) can be found using the following equation

$$e = n/(1-n)$$
$$e = 0.57/(1-0.57) = 1.326$$

5. $S \times e = G \times w$

$$S = 2.6 \times 0.428/(1.326) = 0.839$$

Practice Problem 38.4

A soil sample obtained from the ground was measured and weighed. The soil sample has a diameter of 4 in. and a height 6 in. The weight of the soil sample was measured to be 4.8 lbs. The soil sample was oven dried and weighed again. The dry weight of the soil sample was found to be 3.9 lbs. The specific gravity of the soil sample is known to be 2.65.

Find the following;

1. Total density
2. Water content
3. Dry density
4. Porosity
5. Void ratio
6. Degree of saturation

Solution

Total density:

Volume of the soil sample $(V) = \pi \times d^2/4 \times h = \pi \times (4/12)^2/4 \times (6/12) = 0.044$ ft.^{3}
Wet weight of the soil sample = 4.8 lbs
Total density (or wet density) = M / V = 4.8/0.044 = 109.1 lbs/ft.^{3}

Moisture Content (w) $= M_W/M_S$
$M_S = 3.9$ lbs; $M_W = 4.8 - 3.9$ lbs $= 0.9$ lbs
$w = 0.9/3.9 = 0.23$
$\gamma_d = \gamma/(1 + w) = 109.1/(1 + 0.23) = 88.7$ lbs/ft.^{3}.

Find V_S:

$G = 2.65 = M_s/(V_s \gamma_w) = 3.9/(V_s \times 62.4)$, Since, $\gamma_w = 62.4$ lbs/ft.^{3}.
Hence $V_s = 0.024$ ft.^{3}.
$V = V_v + V_s$ (Total volume = volume of voids + volume of solids)
$0.044 = V_v + 0.024$

$V_v = 0.02$

Porosity (n) = $V_v/V = 0.02/0.044 = 0.45$
Void ratio (e) = $n/(1 - n) = 0.45/(1 - 0.45) = 0.82$
Degree of saturation (S):

$$Se = Gw$$
$$S = 2.65 \times 0.23/0.82 = 0.74$$

Practice Problem 38.5

Degree of saturation, water content and specific gravity are 75%, 42%, and 2.68, respectively.
Find
1. Total density (γ_{wet}) 2. Void ratio (e) 3. Porosity (n)

Solution
Step 1: Use Relationship 5;

$$\gamma_d = \frac{\gamma_w G_s}{[1 + (w/S)G_s]}$$

$$\gamma_d = \frac{62.4 \times 2.68}{[1 + (0.42/0.75) \times 2.68]} = 66.9 \text{ pcf}$$

From relationship 1, $\gamma_d = \gamma_t/(1 + w)$

$$\gamma_{wet} = \gamma_d \times (1 + w) = 66.9 \times (1 + 0.42) = 95 \text{ pcf}$$

From Relationship 2,
$Se = Gw$

$$e = 2.68 \times 0.42/0.75 = 1.5$$

From Relationship 3,
$n = e/(1 + e) = 1.5/(1.5 + 1) = 0.6$

38.3 Borrow pit problems

Note: The student should master the previous chapter on soil relationships thoroughly in order to understand borrow pit problem.

Fill material for civil engineering work is obtained from borrow pits. The question is how much soil should be removed from the borrow pit for a given project.

Usually, the final product is the controlled fill or the compacted soil. Total density, optimum moisture content and dry density of the compacted soil will be available. This information can be used to obtain the mass of solids required from the borrow pit. If the soil in borrow pit is too dry, water can always be added in the site. If the water content is too high, then soil can be dried prior to use. This could take some time in the field since one has to wait for a few sunny days to get rid of water.

Water can be added or removed from soil.

What cannot be changed is the mass of *solids*. The mass of solids is the link between borrow pit soil and soil that has been transported.

Procedure

Find the mass of solids required for the compacted fill.

Excavate and transport the same mass of solids from the borrow pit.

Practice Problem 38.6

Road construction project needs compacted soil to construct a road 10 ft. wide, 500 ft. long. The project needs 2 ft. layer of soil. Soil density after compaction was found to be 112.1 pcf at optimum moisture content at 10.5%.

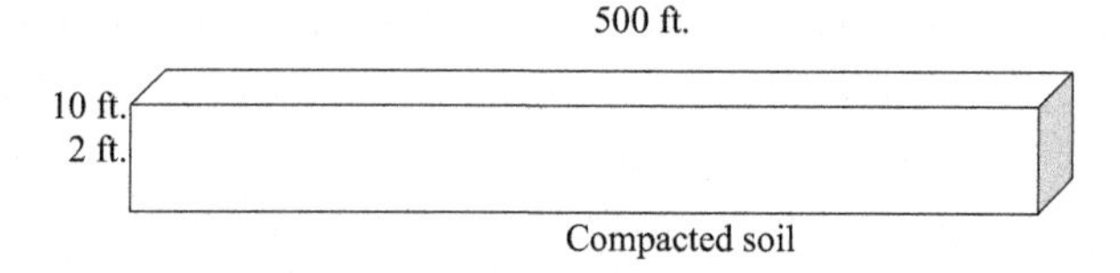

The soil in the borrow pit has following properties:
Total density of the borrow pit soil = 105 pcf
Moisture content of borrow pit soil = 8.5%.
Find the total volume of soil that need to be hauled from the borrow pit.
Total Density = $(M_W + M_S)/V$
Dry density = M_S/V
(See below for definitions of all terms).

Solution

Step 1: Find the mass of solids (M_S) required for the controlled fill:
Volume of compacted soil required = $500 \times 10 \times 2 = 10000$ ft.^{3}
Soil density after compaction (dry density) = 112.1 pcf
Moisture content required = 10.5%
Draw the phase diagram for the controlled fill:

Volume		Mass
V_a	Air	$M_a = 0$ (Usually, mass of air is taken to be zero).
V_w	Water	M_w (Mass of water)
V_s	Solid	M_s (Mass of solids)

Soil phase diagram

Definition of terms: V, Total Volume = $V_s + V_w + V_a$; M, total mass = $M_w + M_s$ (note that mass of air is taken to be zero); V_a, volume of air (volume of air is not zero); M_w, mass of water; V_w, volume of water; M_s, mass of solids; V_s, volume of solids; M, total mass of soil = $M_s + M_w$; V, total volume of soil = $V_s + V_w + V_a$; Total Density, $(M_w + M_s)/V$; Dry density = M_s/V.

Soil in the site after compaction has a dry density of 112.1 pcf and moisture content of 10.5%.

Dry density = M_s/V = 112.1 pcf
Note that total density is $(M_w + M_s)/V$
Moisture content = M_w/M_s = 10.5% = 0.105
The road needed 2 ft. layer of soil at a width of 10 ft. and length of 500 ft.
Hence, the total volume of soil = 2 × 10 × 500 = 10,000 ft.^{3}.
V = total volume = 10,000 cu. ft.^{3}.
Since M_s/V = dry density
M_s/10,000 = 112.1
M_s = 1,121,000 lbs.
M_s is the mass of solids. This mass of solids should be hauled in from the borrow pit.

Step 3: Find the mass of water in compacted soil.
Moisture content in the compacted soil = M_w/M_s = 10.5% = 0.105
M_w = 0.105 × 1,121,000 lbs = 117,705 lbs

Step 4: Find the total volume of soil that needs to be hauled from the borrow pit.
The contractor needs to obtain 1,121,000 lbs of solids from the borrow pit.
Contractor can add water to the soil in the field, if needed.
Mass of solids needed (M_s) = 1,121,000 lbs.
Density and moisture content of borrow pit soil is known.
Total density of borrow pit soil = M/V = 105 pcf
Moisture content of borrow pit soil = M_w/M_s = 8.5% = 0.085
Since M_s = 1,121,000 lbs, (M_s is the mass of solids required).
M_w/M_s = 0.085
M_w = 0.085 × 1,121,000 lbs = 95,285 lbs.

Solid mass of 1,121,000 lbs of soil in the borrow pit contains 95,285 lbs of water.
Total mass of borrow pit soil = 1,121,000 + 95,285 = 1,216,285 lbs
Total density of borrow pit soil is known to be 105 pcf.
Total density of borrow pit soil = $M/V = (M_w + M_s)/V$ = 105 pcf
Insert known values for M_s and M_w.

$M/V = (M_w + M_s)/V$ = (95,285 + 1,121,000)/V = 105 pcf
Hence, V = 11,583.7 ft.^{3}.

The contractor needs to extract 11,583.7 ft.^{3} of soil from the borrow pit.
The borrow pit soil comes with 95,285 lbs of water.
Compacted soil should have 117,705 lbs of water (See Step 3 of Practice Problem 38.6).
Hence, water needs to be added to the borrow pit soil.
Amount of water needs to be added to the borrow pit soil = 117,705 − 95,285 = 22,420 lbs.
Weight of water is usually converted to gallons. One gallon is equal to 8.34 lbs.
Amount of water needs to be added = 2,688 gallons.

Summary

Step 1: Obtain all the requirements for compacted soil.
Step 2: Find M_s or the mass of solids in the compacted soil.
This is the mass of solids that needs to be obtained from the borrow pit.

Step 3: Find the information about the borrow pit. Usually, the moisture content in the borrow pit and the total density of the borrow pit can be easily obtained.
Step 4: The contractor needs to obtain the M_s of soil from the borrow pit.
Step 5: Find the total volume of soil that needs to be removed in order to obtain the M_s mass of solids.
Step 6: Find M_w of the borrow pit (mass of water that comes along with soil).
Step 7: Find M_w (mass of water in compacted soil).
Step 8: The difference in the above two masses is the amount of water needs to be added.

Mass-haul diagrams

39

Mass-haul diagrams are used to compute the cut and fill quantity required for road construction projects (Figure 39.1).

During road projects, the road is divided into stations. Typically, stations are given as 0+00, 1+00, 2+00, etc. 1+00 means 100 ft. from the reference point and 2+00 means 200 ft. from the reference point. Similarly, 2+30 means 230 ft. from the reference point. The fill quantity required at each station varies since the ground surface tends to vary (Figure 39.2).

The fill quantity required at station 1+00 is different from the quantity required at station 2+00.

The volume of fill required is obtained by multiplying the average area of two stations by the distance between stations.

$$\text{Volume of fill required} = (A_1 + A_2)/2 \times d$$
$$d = \text{distance between stations}$$

How to obtain the areas $A1$ and $A2$?

In the real world, computer programs and planimeters are used to obtain $A1$ and $A2$ areas since they cannot be computed due to their irregular shapes. In the examination, these areas will be provided.

Practice Problem 39.1

Find the volume of fill required from station 1+00 to 1+50. Fill area at station 1+00 is found to be 60 ft^2 and the fill area at station 1+50 is found to be 80 ft^2.

Solution

$$\text{Volume of fill required} = (A_1 + A_2)/2 \times \text{distance between stations}$$
$$= (60 + 80)/2 \times 50\,\text{ft}^3.$$
$$= 3500\,\text{ft}^3 = 129.6\,\text{yd}^3.$$

39.1 Cut

During road construction, some locations may have to be cut (Figure 39.3). This happens when the proposed road is at a lower elevation than the existing ground.

In some stations, both cut and fill can occur (Figure 39.4).

Geotechnical Engineering Calculations and Rules of Thumb

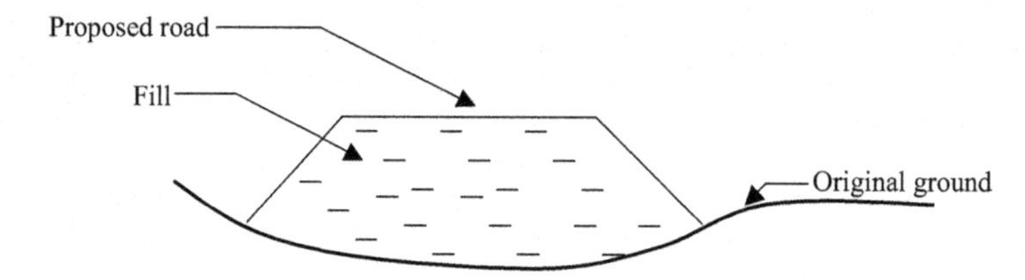

Figure 39.1 Fill quantity at a station.

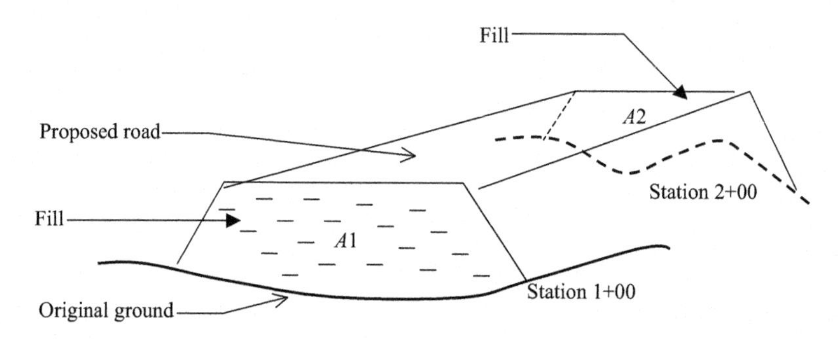

Figure 39.2 Placement of fill in road construction.

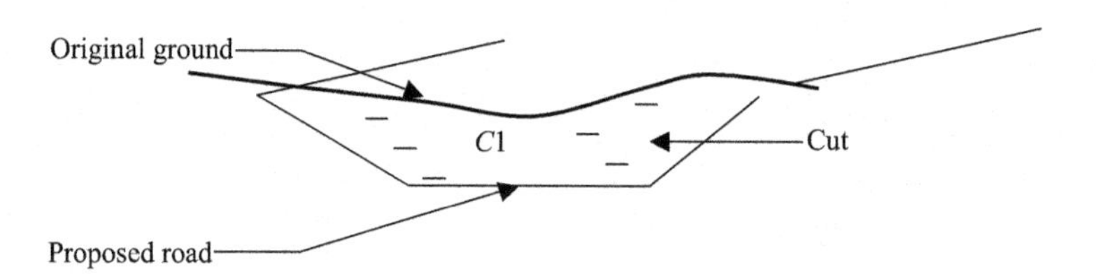

Figure 39.3 Cut in a station.

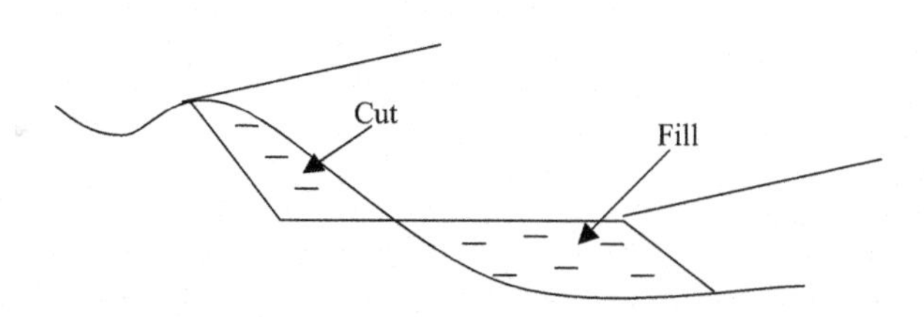

Figure 39.4 Cut and fill at the same station.

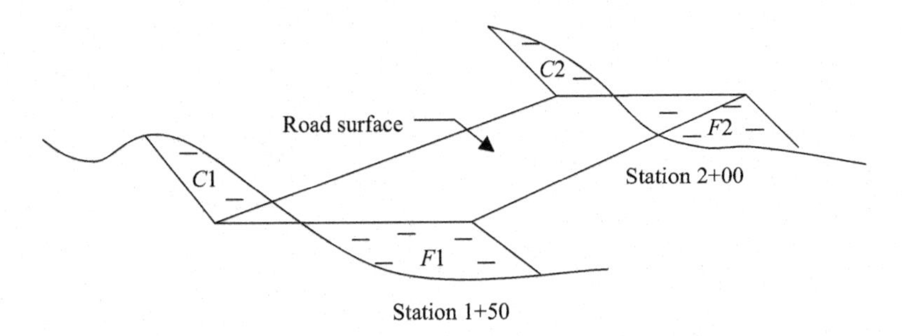

Figure 39.5 Cut and fill diagram.

When both cut and fill occur at the same station, a net value is obtained. The process is easily explained using an example.

Practice Problem 39.2

Find the net cut or fill between station 1+50 and 2+00 (Figure 39.5).

The following information has been provided.

Station 1+50, $C1 = 120\ \text{ft}^2$, $F1 = 170\ \text{ft}^2$
Station 2+00, $C2 = 200\ \text{ft}^2$, $F2 = 60\ \text{ft}^2$

Usually, this information is tabulated.

Station number	Cut area (ft^2)	Fill area (ft^2)	Cut volume	Fill volume	Net Cut (yd^3)
1+50	120	170			
2+00	200	60			

Solution

$$\begin{aligned}\text{Total cut between two stations} &= (C1 + C2)/2 \times 50 \\ &= (120 + 200)/2 \times 50 = 8000\ \text{ft}^3 = 296.3\ \text{yd}^3\end{aligned}$$

$$\begin{aligned}\text{Total fill between two stations} &= (F1 + F2)/2 \times 50 \\ &= (170 + 60)/2 \times 50 = 5750\ \text{ft}^3 = 213\ \text{yd}\end{aligned}$$

Net amount = 296.3 − 213 = 83.3 yd^3. (cut)
Cut is represented as positive while fill is represented as negative.

Station number	Cut area	Fill area	Cut volume (ft^3)	Fill volume (ft^3)	Net Cut (ft^3)
1+50	120	170			
2+00	200	60	296.3	213	83.3

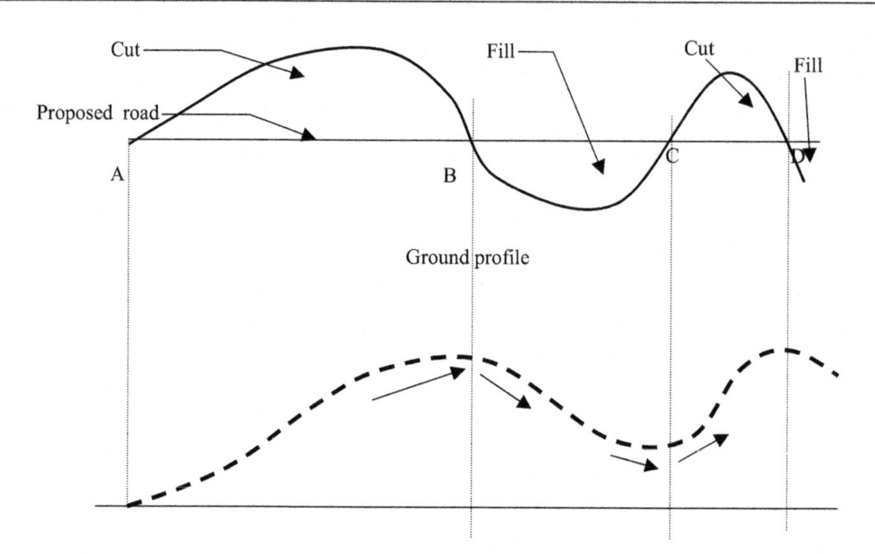

Figure 39.6 Mass-haul diagram.

39.2 Mass-haul diagrams

A mass-haul diagram is drawn using the cumulative cut values (Figure 39.6).

Figure 39.6 shows a ground profile. The mass-haul diagram for the ground profile is also shown. Maas-haul diagrams are also known as mass diagrams. Now, let us look what happens starting from point A.

Point A to B: This section is all cut. Therefore, the cumulative cut will go up from point A to B. Hence, the mass-haul diagram keeps going up.

Point B to C: This section is fill. The cumulative cut starts to reduce from point B. The mass diagram changes direction at point B and starts going down.

Point C to D: Point C to D is all cut. The mass diagram changes direction and starts going up.

Beyond Point D: Beyond point D is fill. The mass diagram changes direction and starts pointing downward.

Mass diagrams can be easily explained using an example.

Practice Problem 39.3: Mass diagram example

The cut areas and fill areas are tabulated as shown. Complete the blank columns and draw a mass diagram.

Station number	Cut area	Fill area	Cut volume (ft^3)	Fill volume (ft^3)	Net cut (ft^3)	Cumulative cut (ft^3)
0+00	175	125				
0+50	117	123				
1+00	238	250				

Station number	Cut area	Fill area	Cut volume (ft^3)	Fill volume (ft^3)	Net cut (ft^3)	Cumulative cut (ft^3)
1+50	211	240				
2+00	198	180				
2+50	140	141				
3+00	258	200				

Solution

Step 1: Complete the "Cut volume" column.

Cut volume (1st entry) = (175 + 117)/2 × 50 = 7,300
Cut volume (2nd entry) = (117 + 238)/2 × 50 = 8,875
Cut volume (3rd entry) = (238 + 211)/2 × 50 = 11,225
Cut volume (4th entry) = (211 + 198)/2 × 50 = 10,225
Cut volume (5th entry) = (198 + 140)/2 × 50 = 8,450
Cut volume (6th entry) = (140 + 258)/2 × 50 = 9,950

Station number	Cut area	Fill area	Cut volume (ft^3)	Fill volume (ft^3)	Net cut (ft^3)	Cumulative cut (ft^3)
0+00	175	125	0	0	0	
0+50	117	123	7,300			
1+00	238	250	8,875			
1+50	211	240	11,225			
2+00	198	180	10,225			
2+50	140	141	8,450			
3+00	258	200	9,950			

Step 2: Complete the "Fill volume" column.

Fill volume (1st entry) = (125 + 123)/2 × 50 = 6,200
Fill volume (2nd entry) = (123 + 250)/2 × 50 = 9,325
Fill volume (3rd entry) = (250 + 240)/2 × 50 = 12,250
Fill volume (4th entry) = (240 + 180)/2 × 50 = 10,500
Fill volume (5th entry) = (180 + 141)/2 × 50 = 8,025
Fill volume (6th entry) = (141 + 200)/2 × 50 = 8,525

Station number	Cut area	Fill area	Cut volume	Fill volume	Net cut (ft^3)	Cumulative cut
0+00	175	125	0	0	0	
0+50	117	123	7,300	6,200		
1+00	238	250	8,875	9,325		
1+50	211	240	11,225	12,250		
2+00	198	180	10,225	10,500		
2+50	140	141	8,450	8,025		
3+00	258	200	9,950	8,525		

Step 3: Complete the "Net cut" column.

Net cut (1st entry) = 7,300 − 6,200 = 1,100
Net cut (2nd entry) = 8,875 − 9,25 = −450
Net cut (3rd entry) = 11,225 − 6,200 = −1,025
Net cut (4th entry) = 7,300 − 6,200 = −275
Net cut (5th entry) = 7,300 − 6,200 = 425
Net cut (6th entry) = 7,300 − 6,200 = 1,425

Station number	Cut area	Fill area	Cut volume	Fill volume	Net cut (ft^3)	Cumulative cut
0+00	175	125	0	0	0	0
0+50	117	123	7,300	6,200	1,100	
1+00	238	250	8,875	9,325	−450	
1+50	211	240	11,225	12,250	−1,025	
2+00	198	180	10,225	10,500	−275	
2+50	140	141	8,450	8,025	425	
3+00	258	200	9,950	8,525	1,425	

Step 4: Complete the "Cumulative cut" column:

Cumulative cut is obtained by adding the present net cut to the previous net total.

Cumulative cut (1st entry) = 1,100 + 0 = 1,100
Cumulative cut (2nd entry) = 1,100 + (−450) = 650
Cumulative cut (3rd entry) = 650 + (−1,025) = −375
Cumulative cut (4th entry) = −375 + (−275) = −650
Cumulative cut (5th entry) = −650 + 425 = −225
Cumulative cut (6th entry) = −225 + 1,425 = 1,200

Station number	Cut area	Fill area	Cut volume	Fill volume	Net cut (ft^3)	Cumulative cut
0+00	175	125	0	0	0	0
0+50	117	123	7,300	6,200	1,100	1,100
1+00	238	250	8,875	9,325	−450	650
1+50	211	240	11,225	12,250	−1,025	−375
2+00	198	180	10,225	10,500	−275	−650
2+50	140	141	8,450	8,025	425	−225
3+00	258	200	9,950	8,525	1,425	1,200

Step 5: Draw the mass diagram:

Mass diagram is drawn between the "cumulative cut" and the stations.

Mass Diagram Implications

In Figure 39.7, consider point "A" shown with a vertical line. At point "A", the curve crosses the "X" axis. Point "A" lies between station 1+00 and 1+50. Let's assume point "A" to be at station 1+30.

At station 1+30, the curve crosses the "X" axis.

What does this mean?

It means that between station 0+00 and station 1+30, the total required fill is zero. In other words, all the cut material has been used for fill purposes.

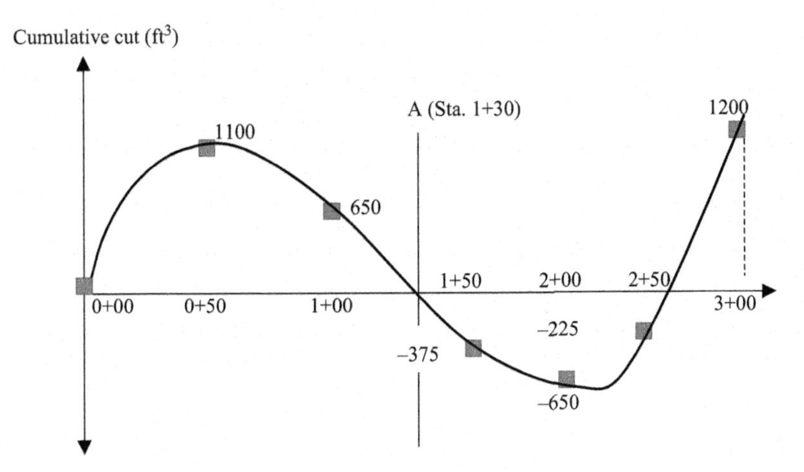

Figure 39.7 Cut and fill.

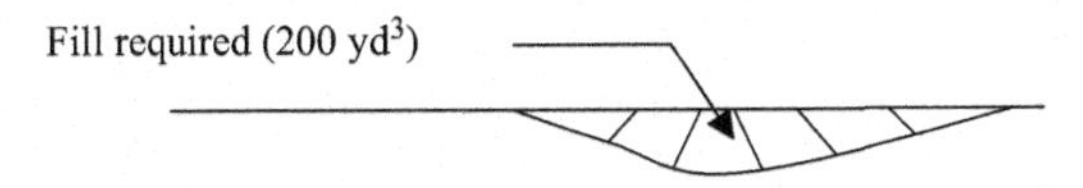

Figure 39.8 Fill required after compaction.

Balanced points

The net cut between the balanced points is zero. From stations 0+00 to 1+30, net cut is zero. Hence, stations 0+00 and 1+30 are balanced points. In other words, all the cut material in this region is utilized for fill. Similarly, stations 1+30 to 2+60 also are two balanced points.

Volume reduction due to compaction

When soil is compacted, the volume reduces. Let's assume that a contractor has found out that a portion of the road needs 200 yd^3 of fill after compaction. It is common sense for him to bring more than 200 yd^3 of fill since the volume decreases after compaction.

Volume reduction factor = additional volume of soil required/volume of soil after compaction.

Practice Problem 39.4

A contractor has to fill and compact a portion of the road. The volume of the fill needed (after compaction) is calculated to be 200 yd^3. Calculate the volume of soil that needs to be brought to the site if the volume reduction factor is 10% (Figure 39.8).

Solution

Volume reduction factor = additional volume of soil required/volume of soil after compaction

Volume reduction factor = 10%

Volume of soil after compaction = 200 yd^3

$$\text{Volume reduction factor} = 0.10 = \text{Additional volume of soil required} / 200$$

$$\text{Additional volume of soil required} = 0.10 \times 200 = 20 \text{ yd}^3$$

$$\text{Total volume of soil required} = 200 + 20 = 220 \text{ yd}^3$$

39.2.1 *Expansion*

Expansion occurs when soil is cut and removed from ground. Expansion is defined as follows:

Expansion = additional volume after cut and removal from ground/volume in ground

Soil volume increases after the soil has been cut and removed from ground.

Practice Problem 39.5

During a road project, 150 yd^3 has to be cut and removed from site. Expansion factor of soil is 7%. Each truckload can carry 10 yd^3. Find how many truckloads are required to remove the soil from site.

Solution

Expansion = additional volume after cut and removal from ground/ volume in groui

Expansion = 7%
Volume of soil in ground = 150 yd^3

0.07 = additional volume after cut and removal from ground /150

Additional volume after cut and removal from ground = $0.07 \times 150 = 10.5$ yd^3

Total volume of soil after being cut and removed from ground = 150 + 10.5 = 160.5

16 truck loads will be needed to carry 160 yd^3, since each truck can carry 10 yd^3. 1 final truck is required to carry the last remaining 0.5 yd^3. Hence, a total of 17 truckloads are required.

Practice Problem 39.6: Mass diagram example (considering expansion and volume reduction)

The cut areas and fill areas are tabulated as shown. Complete the blank columns and draw a mass diagram. The volume reduction factor is given to be 9% and the expansion factor is given to be 8%.

A	B	C	D	E	F	G	H	I
							H = E − G	
Station number	Cut area	Fill area	Cut volume (in ground)	Cut volume (after expansion)	Fill volume required	Fill volume required considering volume reduction	Net cut (ft^3)	Cumulative cut (ft^3)
0+00	175	125						
0+50	117	123						
1+00	238	250						
1+50	211	240						
2+00	198	180						

A	B	C	D	E	F	G	H	I
							H = E − G	
Station number	Cut area	Fill area	Cut volume (in ground)	Cut volume (after expansion)	Fill volume required	Fill volume required considering volume reduction	Net cut (ft³)	Cumulative cut (ft³)
2+50	140	141						
3+00	258	200						

Solution

Step 1: Complete the Cut volume (in ground) column.

Cut volume (1st entry) = (175+117)/2 × 50 = 7,300
Cut volume (2nd entry) = (117+238)/2 × 50 = 8,875
Cut volume (3rd entry) = (238+211)/2 × 50 = 11,225
Cut volume (4th entry) = (211+198)/2 × 50 = 10,225
Cut volume (5th entry) = (198+140)/2 × 50 = 8,450
Cut volume (6th entry) = (140+258)/2 × 50 = 9,950

A	B	C	D	E	F	G	H	I
							H = E − G	
Station number	Cut area	Fill area	Cut volume (in ground)	Cut volume (after expansion)	Fill volume required	Fill volume required consider-ing volume reduction	Net cut (ft³)	Cumulative cut (ft³)
0+00	175	125	0					
0+50	117	123	7,300					
1+00	238	250	8,875					
1+50	211	240	11,225					
2+00	198	180	10,225					
2+50	140	141	8,450					
3+00	258	200	9,950					

Step 2: Complete the "Cut volume after expansion" column:

As we discussed earlier, when the soil is cut and removed from the ground, it expands. The expansion factor is given to be 8%.

Cut volume in ground = 7,300
Expansion = 7,300 × 0.08 = 584
Volume of soil after expansion = 7,300 + 584 = 7,884 ft^3.
Simply, this can be done by multiplying 7,300 × (1 + 0.08) = 7,884 ft.^{3}
Similarly 8,875 × (1.08) = 9,585
11,225 × 1.08 = 12,123 ft.^{3}
10,225 × 1.08 = 11,043 ft.^{3}
8,450 × 1.08 = 9,126 ft.^{3}

$9{,}950 \times 1.08 = 10{,}746$ ft.^{3}
Input these values in column "E".

A	B	C	D	E	F	G	H	I
							H = E − G	
Station number	Cut area	Fill area	Cut volume (in ground)	Cut volume (after expansion)	Fill volume required	Fill volume required considering volume reduction	Net cut (ft^3)	Cumulative cut (ft^3)
0+00	175	125	0	0				
0+50	117	123	7,300	7,884				
1+00	238	250	8,875	9,585				
1+50	211	240	11,225	12,123				
2+00	198	180	10,225	11,043				
2+50	140	141	8,450	9,126				
3+00	258	200	9,950	10,746				

Step 3: Complete the "Fill volume" column.

Fill volume (1st entry) = $(125+123)/2 \times 50 = 6{,}200$
Fill volume (2nd entry) = $(123+250)/2 \times 50 = 9{,}325$
Fill volume (3rd entry) = $(250+240)/2 \times 50 = 12{,}250$
Fill volume (4th entry) = $(240+180)/2 \times 50 = 10{,}500$
Fill volume (5th entry) = $(180+141)/2 \times 50 = 8{,}025$
Fill volume (6th entry) = $(141+200)/2 \times 50 = 8{,}525$

A	B	C	D	E	F	G	H	I
							H = E − G	
Station number	Cut area	Fill area	Cut volume (in ground)	Cut volume (after expansion)	Fill volume required	Fill volume required considering volume reduction	Net cut (ft.^{3})	Cumulative cut (ft.^{3})
0+00	175	125	0	0	0			
0+50	117	123	7,300	7,884	6,200			
1+00	238	250	8,875	9,585	9,325			
1+50	211	240	11,225	12,123	12,250			
2+00	198	180	10,225	11,043	10,500			
2+50	140	141	8,450	9,126	8,025			
3+00	258	200	9,950	10,746	8,525			

Step 4: When fill material is compacted, the volume is reduced. Complete the "Fill volume required considering volume reduction" column:

Volume reduction factor is given to be 9%.

If required fill volume is 6,200 ft.^{3}, the contractor has to bring 6,200 × 1.09 ft.^{3} = 6,758 ft.^{3}

Similarly,

9,325 × 1.09 = 10,164 ft.^{3}

12,250 × 1.09 = 13,352 ft.^{3}

10,500 × 1.09 = 11,445 ft.^{3}

8,025 × 1.09 = 8,747 ft.^{3}

8,525 × 1.09 = 9,292 ft.^{3}

Input these values in column "G".

A	B	C	D	E	F	G	H	I
							H = E − G	
Station number	**Cut area**	**Fill area**	**Cut volume (in ground)**	**Cut volume (after expansion)**	**Fill volume required**	**Fill volume required considering volume reduction**	**Net cut (ft.^{3})**	**Cumulative cut (ft.^{3})**
0+00	175	125	0	0	0			
0+50	117	123	7,300	7,884	6,200	6,758		
1+00	238	250	8,875	9,585	9,325	10,164		
1+50	211	240	11,225	12,123	12,250	13,352		
2+00	198	180	10,225	11,043	10,500	11,445		
2+50	140	141	8,450	9,126	8,025	8,747		
3+00	258	200	9,950	10,746	8,525	9,292		

Step 5: Complete column "H".

Net cut (1st entry) = 7,884 − 6,758 = 1,126 ft.^{3}

Net cut (2nd entry) = 9,585 − 10,164 = −579 ft.^{3}

Net cut (3rd entry) = 12,123 − 13,352 = −1,229 ft.^{3}

Net cut (4th entry) = 11,043 − 11,445 = −402 ft.^{3}

Net cut (5th entry) = 9,126 − 8,747 = 379 ft.^{3}

Net cut (6th entry) = 10,746 − 9,292 = 1,454 ft.^{3}

Input these values in column "H".

A	B	C	D	E	F	G	H	I
							H = E − G	
Station number	**Cut area**	**Fill area**	**Cut volume (in ground)**	**Cut volume (after expansion)**	**Fill volume required**	**Fill volume required considering volume reduction**	**Net cut (ft.^{3})**	**Cumulative cut (ft.^{3})**
0+00	175	125	0	0	0	0	0	
0+50	117	123	7,300	7,884	6,200	6,758	1,126	

A	B	C	D	E	F	G	H	I
							H = E − G	
Station number	Cut area	Fill area	Cut volume (in ground)	Cut volume (after expansion)	Fill volume required	Fill volume required considering volume reduction	Net cut (ft.^{3})	Cumulative cut (ft.^{3})
1+00	238	250	8,875	9,585	9,325	10,164	−579	
1+50	211	240	11,225	12,123	12,250	13,352	−1,229	
2+00	198	180	10,225	11,043	10,500	11,445	−402	
2+50	140	141	8,450	9,126	8,025	8,747	379	
3+00	258	200	9,950	10,746	8,525	9,292	1,454	

Step 6: Complete the "Cumulative cut" column:

Cumulative cut is obtained by adding the present net cut to the previous net total.
Cumulative cut (1st entry) = 1,126 + 0 = 1,126
Cumulative cut (2nd entry) = 1,126 + (−579) = 547
Cumulative cut (3rd entry) = 547 + (−1,229) = −682
Cumulative cut (4th entry) = −682 + (−402) = −1,084
Cumulative cut (5th entry) = −1,084 + 379 = −705
Cumulative cut (6th entry) = −705 + 1,454 = 749
Input these values in column "I".

A	B	C	D	E	F	G	H	I
							H = E − G	
Station number	Cut area	Fill area	Cut volume (in ground)	Cut volume after expansion	Fill volume required	Fill volume required considering volume reduction	Net cut (ft.^{3})	Cumulative cut (ft.^{3})
0+00	175	125	0	0	0	0	0	0
0+50	117	123	7,300	7,884	6,200	6,758	1,126	1,126
1+00	238	250	8,875	9,585	9,325	10,164	−579	547
1+50	211	240	11,225	12,123	12,250	13,352	−1,229	−682
2+00	198	180	10,225	11,043	10,500	11,445	−402	−1,084
2+50	140	141	8,450	9,126	8,025	8,747	379	−705
3+00	258	200	9,950	10,746	8,525	9,292	1,454	749

Step 7: Draw the mass diagram.

Mass diagram is drawn between the "cumulative cut" and the "stations" (Figure 39.9).

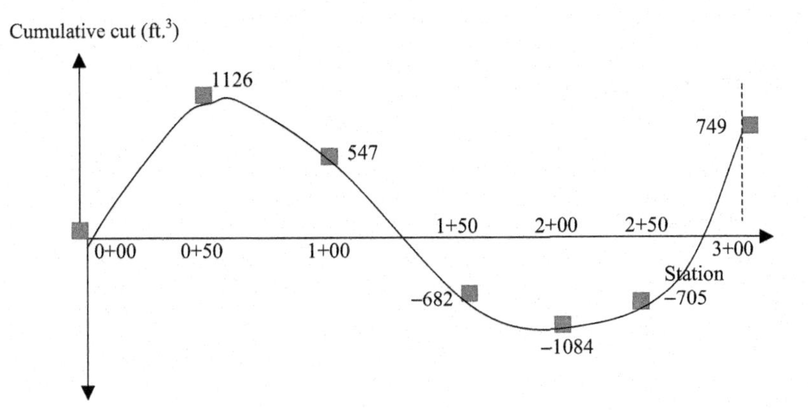

Figure 39.9

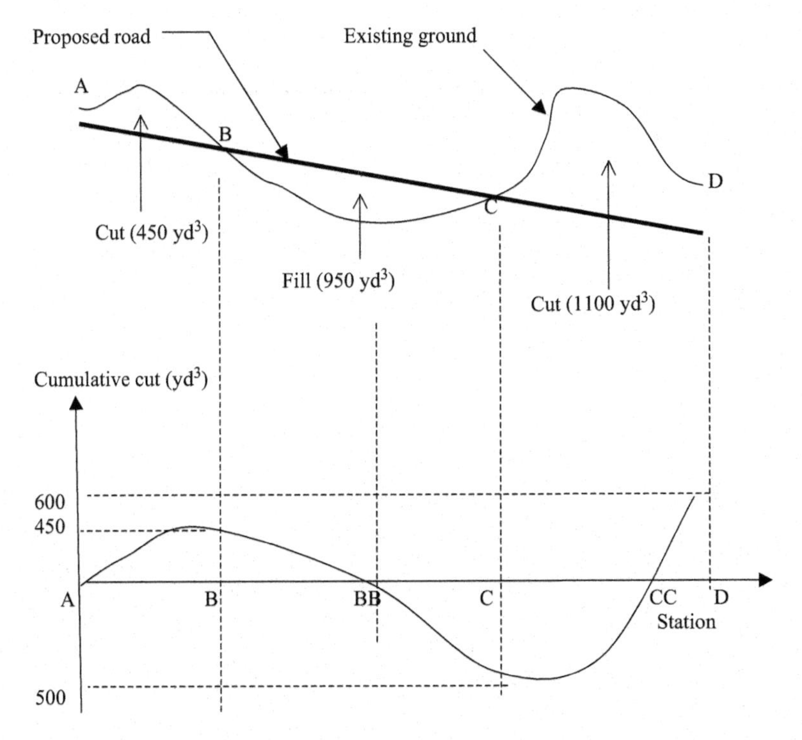

Figure 39.10

Practice Problem 39.7

Draw the mass diagram for the road construction project shown in Figure 39.10. The volume of soil cut or filled is shown in Figure 39.10.

Solution

Point A to B – The ground has to be cut. 450 yd^3 of soil will be cut from point A to B.

Point B to C – The ground has to be filled. 950 yd^3 of soil needed.

Point C to D – The ground has to be cut. 1,100 yd^3 of soil will be cut from point C to D.

Total cut = 450 + 1100 = 1550 yd^3

Total fill = 950 yd^3

Waste = 1550 − 950 = 600 yd^3

The cumulative cut volume increases from point A to B. Cumulative cut volume starts to decrease after point B, since filling has started. At point BB, the cut volume will be equal to the fill volume. There is not enough information given in this example to find point BB.

From point BB to C, the fill volume increases. Cutting starts again at point C. Hence, the cumulative cut volume starts to increase. At point CC, the cut and fill becomes equal again. There is not enough information given in this example to obtain point CC. From point CC to D, the cut volume increases. There is excess of 600 yd^3 of soil that would be left at the end of the project.

Free haul distance (FHD): Free haul distance is known as the "distance soil is moved" without additional compensation. This distance is agreed upon prior to construction.

Free haul distance can be explained using an example.

Practice Problem 39.8

The free haul distance for a cut and fill project is given to be 100 yd. Find the free haul areas of the mass diagram shown in Figure 39.11.

Point A to B = Cut volume = 600 yd^3

Point B to C = Fill volume = 600 yd^3

Point C to D = Fill volume = 350 yd^3

Point D to E = Cut volume = 500 yd^3

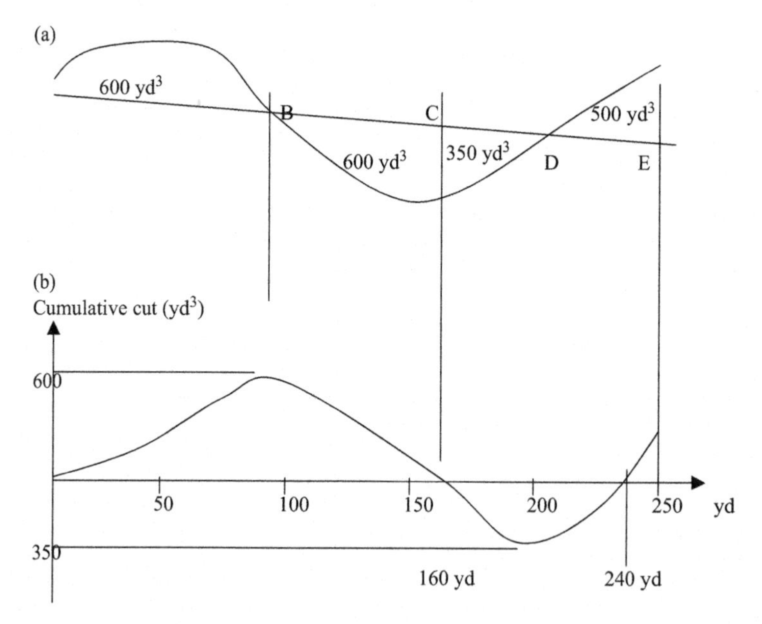

Figure 39.11 Mass diagram. (a) Ground profile. (b) Cumulative cut (yd^3).

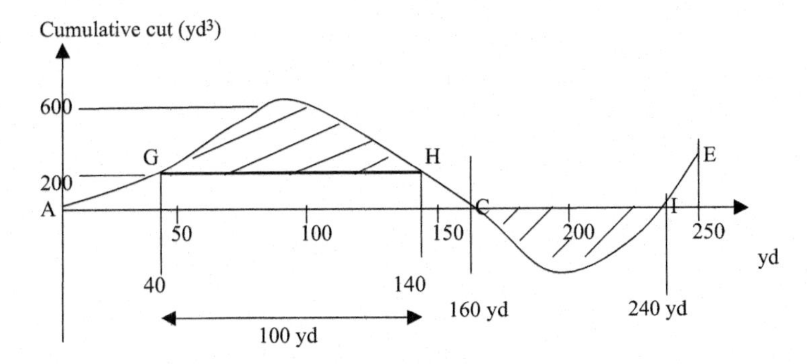

Figure 39.12

Solution

The free haul distance is given to be 100 yd. From point A to C, all cut material can be used for the fill. The distance between points A to C is 160 yd. This is greater than the free haul distance.

Draw a horizontal line (GH) at a distance of 100 yd as shown in Figure 39.12.

Find the coordinates of points G and H.

Point G = Cumulative cut = 200 yd^3, station = 40

Point H = Cumulative cut = 200 yd^3, station = 140

Between points G and H, all the cut material has been used for fill purposes. Hence, the contractor cannot charge additional funds for hauling of material in this section of the road (Point G to H is the free haul distance).

Point A to G = not within the free haul distance. Additional cost will incur for hauling material.

Point H to C = not within the free haul distance. Additional cost will incur for hauling material.

Point C to I = Distance is 80 yd (240 − 160).

This is less than the free haul distance. Hence, no additional cost will incur for points C to I.

Cut material from I to E is waste.

Overhaul

When soil is moved beyond the free haul distance, it is known as overhaul. The contractor is required to be compensated for overhaul.

Limit of profitable haul (LPH): Soil can be profitably moved from point A to point B. However, if the distance between the two points is larger than the LPH, then it is more profitable to obtain soil from an outside source.

Borrow: Volume of soil obtained from an outside source.

Waste: Volume of soil that has to be discarded.

Hauling: Soil that is cut has to be hauled to fill areas. Small distances can be hauled using dozers. Typically, distances closer to a mile are hauled using scrapers. Scrapers may not be economical for anything beyond one mile. In such situations, trucks are used. It is the judgment of the site supervisor to decide what machines are to be used for a given project (Figure 39.13).

Volume *A* has to be cut and volume *B* has to be filled. Small haul distances can be one using dozers.

Step 1: Cut the volume of soil *A*1 and fill volume *B*1. Hauling can be done using dozers since the distance is short (Figure 39.14).

*L*1 is the average transportation distance.

Cut volume *A*1 and fill volume *B*1 with that soil

*L*1 = average transportation distance = distance from center of gravity of *A*1 to the center of gravity of *B*1.

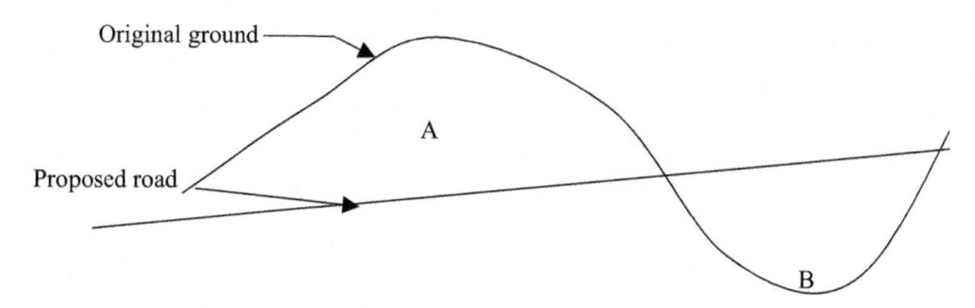

Figure 39.13 Topographic profile.

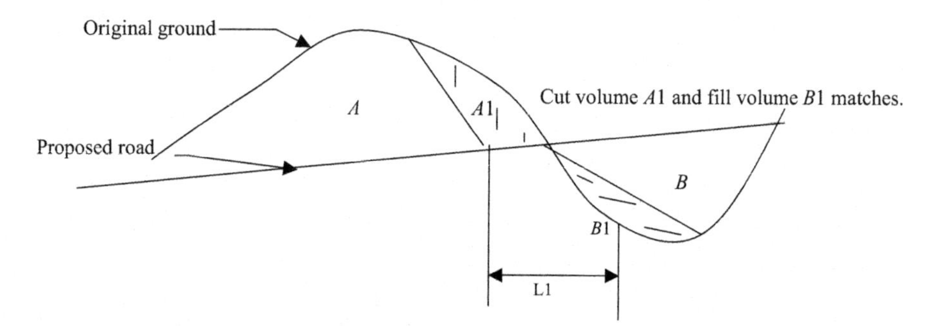

Figure 39.14 Topographic profile.

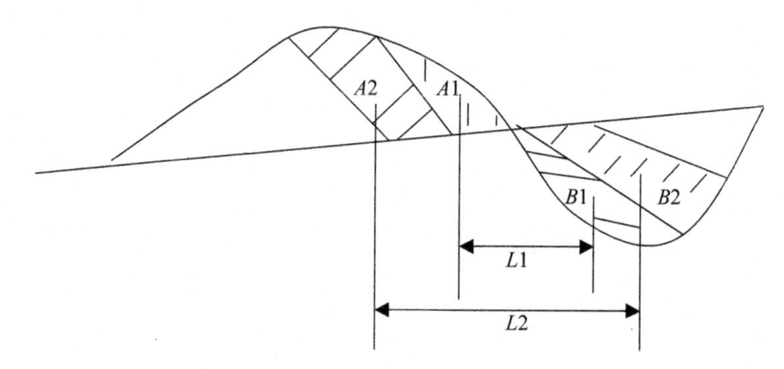

Figure 39.15 Topographic profile.

At the beginning of a cut and fill operation, soil can be moved with dozers. The distance from cut site to fill site would be small.

Step 2: Cut a volume of soil *A2* and fill volume *B2* with the same soil. (Soil at *A2* goes to *B2*) (Figure 39.15).

Cut volume *A2* and fill *B2* with that soil. *L2* is the average transportation distance from *A2* to *B2*.

One can see that the soil has to be transported longer distances $L2 > L1$.

Depending upon the site conditions, it may not be efficient to use the dozers to move soil. Hence, scrapers can be utilized.

Step 3: Cut a volume of soil at *A3* and fill volume *B3* (Figure 39.16).

Cutting occurs from X to Y. Hence, the cutting cumulative volume increases until Y. After Y, the fill begins.

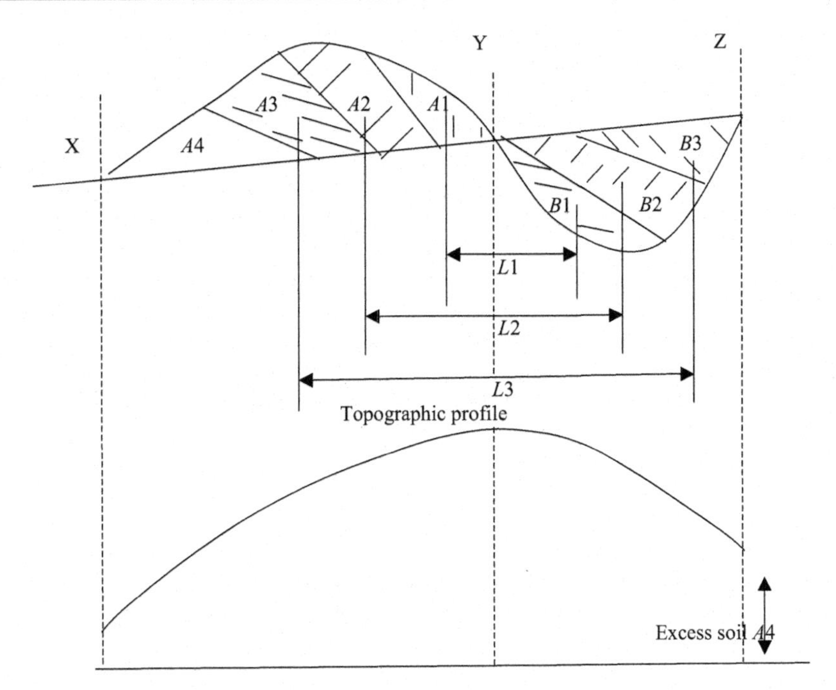

Figure 39.16 Mass diagram.

The cumulative cut volume increases from point X to Y. The fill starts at point Y. Hence, the mass diagram will start to bend down. Since there is more cut in this section, there is excess soil at the end that needs to be trucked away.

Moving soil volume *A*3 to *B*3 involves the transportation of soil for much longer distances. In such situations, trucks may be the most efficient method.

Volume of soil A4 is the surplus. This soil may be cut and transported out of the site.

Most efficient method to conduct a cut and fill operation: Cutting of soil is mostly done using dozers. Sometimes, backhoes may be needed for cutting of soil due to steep slopes or loose soil situations where dozers are unable to access the location. Moving of soil at shorter distances can be done using dozers efficiently. However, dozers become increasingly inefficient when the distance increase. In such situations, scrapers are used to scrape and move soil. One has to understand that the rental cost of a scraper may be twice as much as a dozer. Trucks are needed for much longer distances. In many situations, trucks may have to use local roads that could have significant traffic. Hence, it is advisable to move the soil early in the morning before the rush hour and then after the rush hour. The foreman of the site has to plan his work in the most efficient manner.

Contour Lines

Contour lines represent elevations.

In Figure 39.17, the contour elevation increases from 100 to 130. This represents a hill. On the other hand, Figure 39.18 represents a lake or a crater.

Cut and fill computations using contour lines

Cut and fill can be computed using contour lines. The following example shows how to compute cut and fill volumes using contour lines.

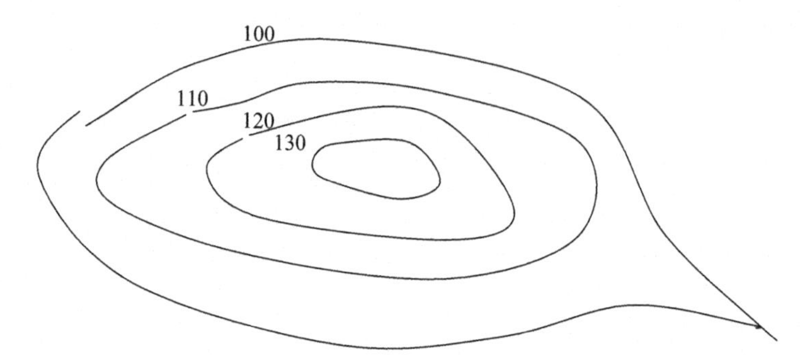

Figure 39.17 Topography.

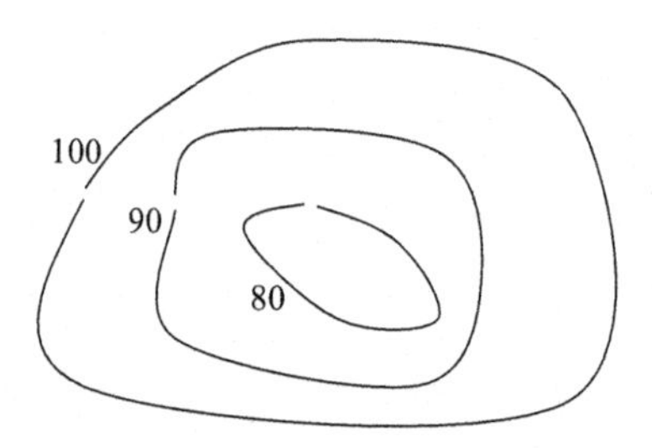

Figure 39.18 Contour lines.

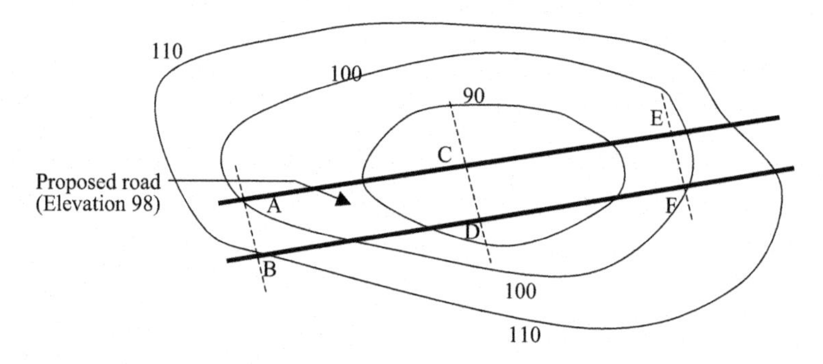

Figure 39.19 Proposed road.

Practice Problem 39.9

The width of a road is 30 ft. (Figure 39.19). BD = 500 ft. and DF = 500 ft. Elevation of the proposed road is 98 ft.

Find the following;

1. Cross sectional areas at sections AB, CD, and EF.
2. Find the cut and fill volumes from B to F.

Solution

Step 1: Find the cross sectional area at section AB.

Elevation at point A = 100 (see the contour 100)

Elevation at point B = 110

Proposed road elevation = 98 ft.

In Figure 39.20, Points P and Q are on the proposed road. Hence, elevation of both P and Q is 98.0.

$$AP = 2; \qquad BQ = 12$$

$$\text{Area of the trapezoid (APQB)} = (2+12)/2 \times 30 = 210 \text{ ft.}^2$$

Step 2: Find the cross sectional area at section CD.

Elevation at point C = 85 (Elevation at C can be *approximated* to 85 ft. Both points C and D are inside the contour 90).

Elevation at point D = 87 (Point D is closer to 90 ft. contour. Hence, it should be around 87).

Proposed road elevation = 98 ft.

In Figure 39.21, assume points R and S to be on the proposed road.

$$RC = 13 \text{ ft.}, SD = 11 \text{ ft.}$$

Area of trapezoid (RCDS) = (13 + 11)/2 × 30 = 360 ft.^{2}

Cut at station AB = 210 ft.^{2} (found in Step 1)

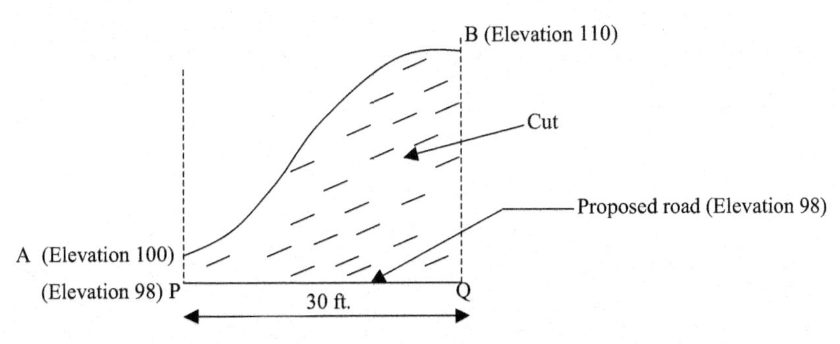

Figure 39.20 Cut portion.

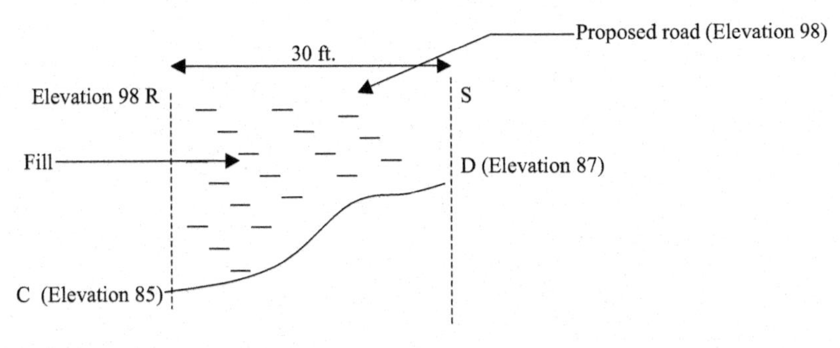

Figure 39.21 Fill portion.

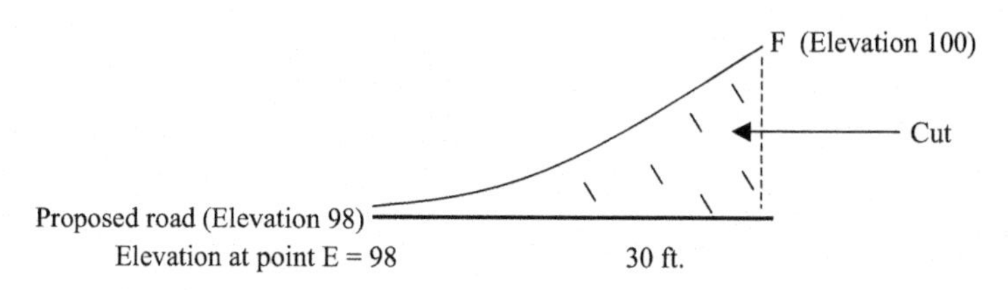

Figure 39.22 Cut portion.

Fill at station AB = 0 ft.^{2}
Cut at station CD = 0 ft.^{2}
Fill at station CD = 360 ft.^{2}
Average cut between two stations = (210 + 0)/2 × 500 ft.^{3} = 1944 yd^3
Average fill between two stations = (0 + 360)/2 × 500 ft.^{3} = 3333 yd^3

Step 3: Find the cross sectional area at section EF.

Elevation at point E = 98 (approximately)
Elevation at point F = 100
Proposed road elevation = 98 ft.
Consider Figure 39.22, there is no fill at this station.
The cut area can be calculated by considering an approximate triangle.
Cut area = 2 × 30/2 = 30 ft.^{2}
Fill area = 0 ft.^{2}
Average fill volume between stations CD and EF = (360 + 0)/2 × 500 ft.^{3} = 3,333 yd^3
Average cut volume between stations CD and EF = (0 + 30)/2 × 500 ft.^{3} = 278 yd^3.

Part 5

Pile Foundations

Pile foundations

40.1 Introduction

Piles are used to transfer the load to deeper and more stable layers of soil. Large structures were built in Egypt, Asia, and Europe prior to the advent of piles. Ancient engineers had no choice but to dig deep down to the bedrock. The second century Sri Lankan structure, known as Jethavana has a shallow foundation going 252 ft. deep down to the bedrock (Plate 40.1). Today, without piles, large structures will be economically prohibitive.

The first undisputed evidence of the use of piles occurs in the Arles circus arena, built in 148 AD, in France by Roman engineers. The wooden piles were probably driven by drop hammer mechanisms using human labor.

40.2 Pile types

- All piles can be categorized as displacement piles and nondisplacement piles. Timber piles, closed end steel pipe piles, and precast-concrete piles displace the soil when driven into the ground. These piles are categorized as displacement piles.
- Some piles displace soil during installation by a small degree. (H-piles, open end steel tubes, and hollow-concrete piles).

40.2.1 Displacement piles

40.2.1.1 Large displacement piles

Timber piles
Precast-concrete piles (reinforced and prestressed)
Closed end steel pipe piles
Jacked down solid concrete piles

40.2.1.2 Small displacement piles

Tubular-concrete piles
H-piles
Open end pipe piles
Thin-shell type
Jacked down hollow concrete cylinders

Geotechnical Engineering Calculations and Rules of Thumb

Plate 40.1 Jethavana structure with a 252 ft. deep shallow foundation.

40.2.2 Nondisplacement piles

- Steel casing withdrawn after concreting (alpha piles, delta piles, frankie piles, vibrex piles).
- Continuous flight auger drilling and concrete placement (with or without reinforcements).
- Auguring a hole and placing a thin shell and concreting.
- Drilling or auguring a hole and placement of concrete blocks inside the hole.

40.3 Timber piles

To have a 100 ft. long timber pile, one needs to cut down a tree of 150 ft. or more.

- Timber piles are cheaper than steel or concrete piles.
- Timber piles decay due to living microbes. Microbes need two ingredients to thrive. They are "*Oxygen*" and "*Moisture*". For timber piles to decay, both ingredients are needed. Below groundwater, there is ample moisture, but very little oxygen.
- Timber piles submerged in groundwater will not decay. Oxygen is needed for the fungi (wood decaying microbes) to grow. Below groundwater level, there is no significant amount of air in the soil. Due to this reason, very little decay occurs below the groundwater level.
- Moisture is the other ingredient needed for the fungi to survive. Above groundwater level, there is ample "oxygen".
- States like Nevada, Arizona, and Texas, there is very little moisture above the groundwater level. Since, microbes cannot survive without water, timber piles could last for a long period of time.

- This is not the case for the states in northern part of USA. Significant amount of moisture will be present above groundwater level due to snow and rain. Hence, both oxygen and moisture will be available for the fungi to thrive. Timber piles would decay under such conditions. Creosoting and other techniques should be used to protect the timber from decay above groundwater level.
- When the church of St. Mark was demolished due to structural defects in 1902, the wood piles driven in 900 AD were in good condition. These old piles were reused to construct a new tower in place of the old church. Venice is a city with very high groundwater level and piles had been under water for centuries.
- Engineers should consider possible future construction activities that could trigger lowering of the groundwater level (Figures 40.1 and 40.2).

40.3.1 Timber pile decay – biological agents

There are many forms of biological agents that attack timber piles. Timber is an organic substance and nature will not let any organic substance go to waste.

40.3.1.1 Fungi

Fungi belong to the plant kingdom. The main distinction of plants from animals is their ability to generate food on their own. On the other hand, all food types of animals come from plants. On that account, Fungi differ from other plants. Fungi is not capable of generating its own food since it lacks Chlorophyll, the agent that allows plants to generate organic matter using sunlight and inorganic nutrients. Due to this reason, Fungi have to rely on organic matter for its supply of food.

Fungi need organic nutrients (the pile itself), water, and atmospheric air to survive.

Identification of fungi attack: Wood piles subjected to fungi attack can be identified by a swollen, rotted, and patched surface. Unfortunately, many piles lie underground

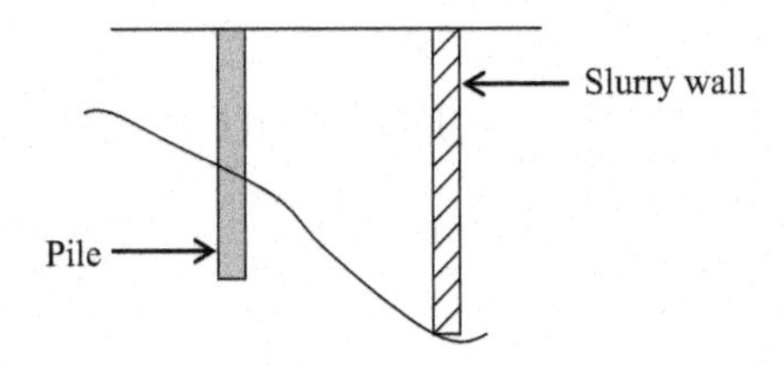

Slurry walls are constructed across landfills and around basement construction sites to lower the groundwater level.

Figure 40.1

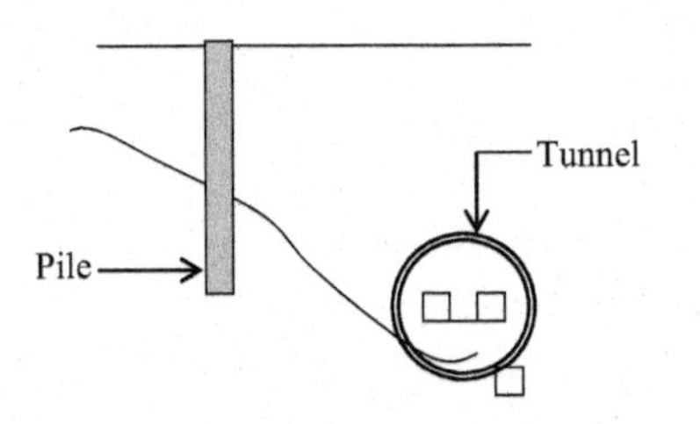

Water constantly drains into tunnels. This water is pumped out constantly. Long tunnels can pump out enough water to create a drawdown in the groundwater level.

Figure 40.2

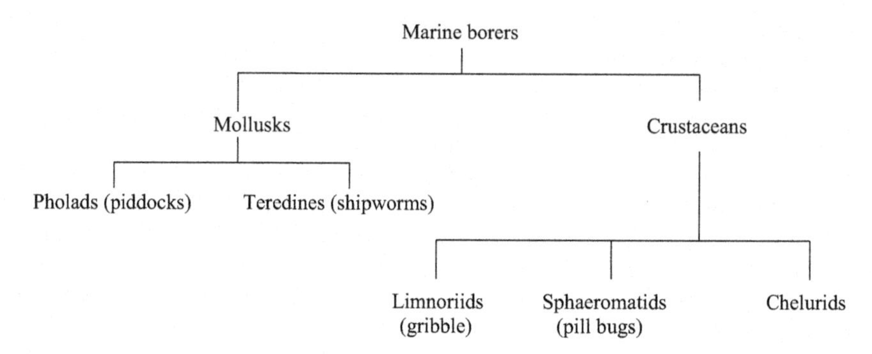

Figure 40.3

and are not visible. Due to this reason, piles that could be subjected to environmental conditions favorable to fungi growth should be treated.

40.3.1.2 Marine borers

Marine borers belong to two families – mollusks and crustaceans. Both groups live in seawater and brackish water. Water front structures usually need to be protected from them (Figure 40.3).

Pholads (*Piddocks*): These are sturdy creatures that can penetrate the toughest of wood. Pholads can hide inside their shells for prolonged periods of time.

Teredines (*Shipworms*): Teredines are commonly known as shipworms due to their worm-like appearance. They typically enter the wood at larvae stage and grow inside the wood.

Limnoriids: Limnoria belong to the crustacean group. They have 7 legs and could grow to a size of 6 mm. They are capable of digging deep into the pile and damage the pile from within, without any outward sign.

Sphaeromatids: Sphaeromatids create large size holes, approximately 0.5 in. in size and can devastate wood piles and other marine structures.

Chelurids: Chelurids are known to drive out limnoriids from their borrows and occupy them.

40.3.2 Preservation of timber piles

There are three main types of wood preservatives available.

1. Creosote
2. Oil-borne preservatives
3. Water-borne preservatives

These preservatives are usually applied in accordance with the specifications of the American Wood Preserver's Association (AWPA).

Preservatives are usually applied under pressure. Hence, the term "*pressure treated*" is used (Figure 40.4).

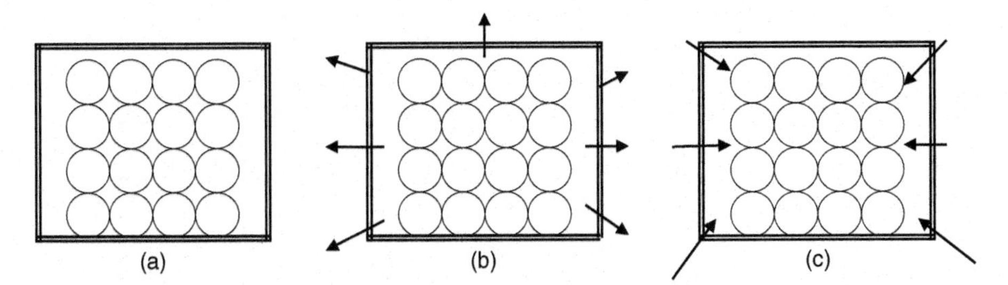

Figure 40.4 Creosoting of timber piles. (a) Timber piles are arranged inside a sealed chamber. (b) Timber piles are subjected to a vacuum. Due to the vacuum, moisture inside piles would be removed. (c) After applying the vacuum, the chamber is filled with preservatives. The preservative is subjected to high pressure until a prespecified volume of liquid is absorbed by the wood.

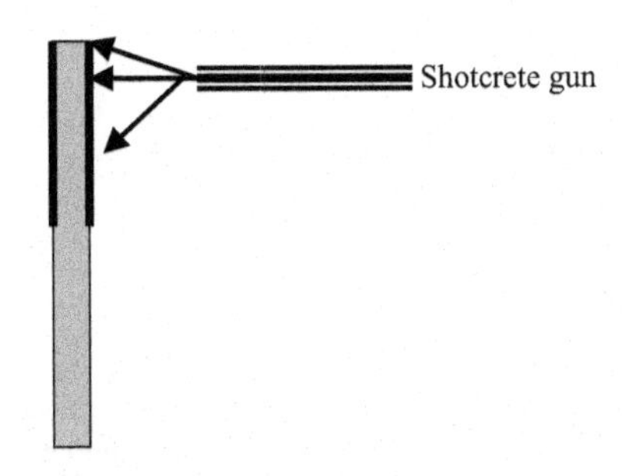

Figure 40.5 Shotcrete encasement.

40.3.3 Shotcrete encasement of timber piles

- Shotcrete is a mixture of cement, gravel, and water. High strength shotcrete is reinforced with fibers.
- Shotcrete is sprayed onto the top portion of the pile where it could possibly be above groundwater level (Figure 40.5).

40.3.4 Timber pile installation

Timber piles need to be installed with special care. Timber piles are susceptible for brooming and damage. Any sudden decrease in driving resistance should be investigated.

40.3.5 Splicing of timber piles

- Splicing of timber piles should be avoided if possible. Unlike steel or concrete piles, timber piles cannot be spliced effectively.
- The usual practice is to provide a pipe section (known as a sleeve) and bolt it to two piles (Figure 40.6).

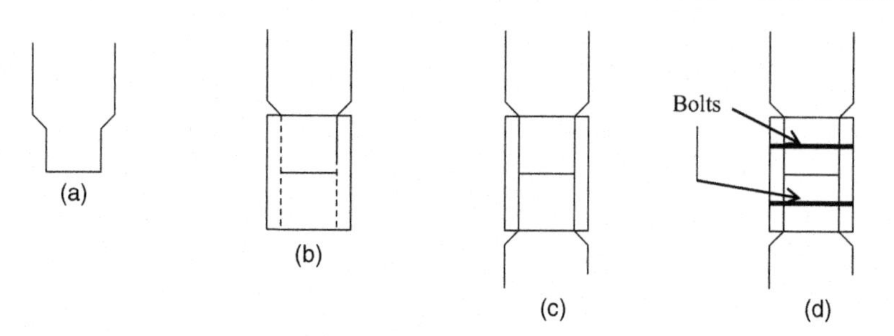

Figure 40.6 Splicing of timber piles. (a) Usually, timber piles are tapered prior to splicing. (b) The sleeve (or the pipe section) is inserted. (c) Bottom pile is inserted. (d) The pipe section is bolted to two piles.

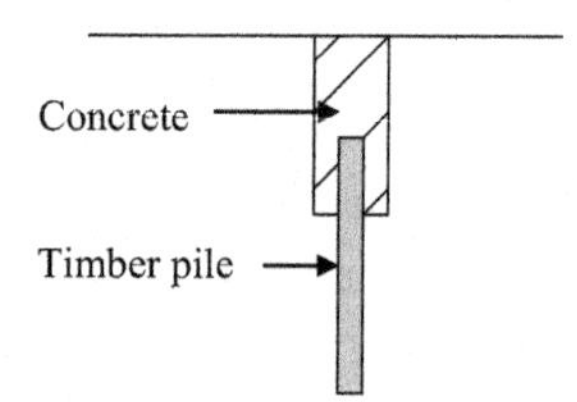

Figure 40.7 Concrete – timber composite pile.

- Sleeve joints are approximately 3 to 4 ft. in length. As one can easily see, the bending strength of the joint is much lower than the pile. Splice strength can be increased by increasing the length of the sleeve.
- Most building codes require, no splicing to be conducted on the upper 10 ft. of the pile since the pile is subjected to high bending stresses at the upper levels. If splicing is absolutely required for timber piles in the upper 10 ft. of the pile, it is recommended to construct a composite pile with upper section filled with concrete. This type of construction is much better than splicing (Figure 40.7).
- *Sleeves larger than the pile diameter*: Sleeves larger than the pile may get torn and damaged during driving and should be avoided. If this type of sleeves is to be used, the engineer should be certain that the pile sleeve would not be driven through hard stratums.
- *Uplift piles*: Timber splices are extremely vulnerable for uplift (tensile) forces and should be avoided. Other than sleeves, steel bars and straps are also used for splicing.

40.4 Steel "H" piles

- Timber piles cannot be driven through hard ground.
- Steel "H" piles are essentially end-bearing piles. Due to a limited perimeter area, "H" piles cannot generate much frictional resistance.
- Corrosion is a major problem for steel "H" piles. The corrosion is controlled by adding copper into steel.
- "H" piles are easily spliced. "H" piles are ideal for highly variable soil conditions.

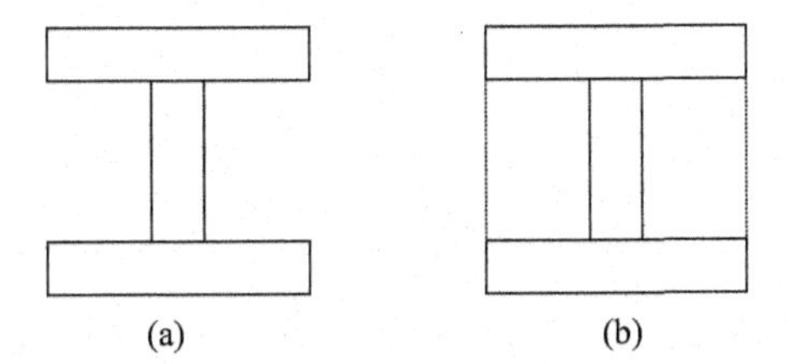

Figure 40.8 H-piles (plugging of soil). (a) Unplugged, (b) Plugged.

- "H" piles could bend under very hard ground conditions. This is known as "*dog legging*" and the pile installation supervisor needs to make sure that the piles are not out of plumb.
- H-piles can get plugged during the driving process (Figure 40.8).
- If the H-pile is plugged, the end bearing would increase due to larger area. On the other hand, the skin friction would become smaller due to lesser wall area.
- When H-piles are driven, both analyses should be done (unplugged and plugged) and the lower value for design should be used.

Unplugged: Low end bearing, high skin friction.
Plugged: Low skin friction, high end bearing.

40.4.1 Splicing of "H" piles

Figure 40.9 shows the splicing of H-piles.

- Step 1: The sleeve is inserted into the bottom part of the "H" pile as shown in Figure 40.9 and bolted to the web.
- Step 2: The top part of the pile is inserted into the sleeve and bolted.

Plate 40.2 shows a steel H-pile and Plate 40.3 shows an installed steel H-pile.

40.4.2 Guidelines for splicing (International Building Code)

The International Building Code (IBC) states that splices shall develop not less than 50% of the pile-bending capacity. If the splice is occurring in the upper 10 ft. of the pile, an eccentricity of 3 in. should be assumed for the column load. The splice should be capable of withstanding the bending moment and shear forces due to a 3-in. eccentricity.

40.5 Pipe piles

- Pipe piles are available in many sizes, and 12-inch diameter pipe piles have a range of thicknesses.
- Pipe piles can be driven either open end or closed end. When driven open end, the pipe is cleaned with a jet of water.

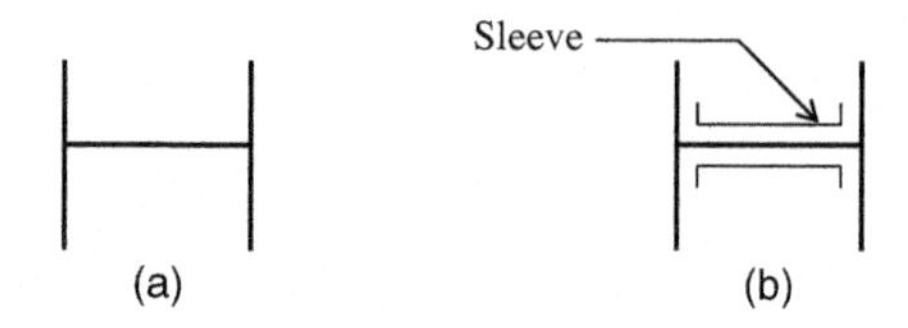

Figure 40.9 Splicing of H-piles. (a) Plan view of a steel "H" pile. (b) Plan view of a steel "H" pile, with the sleeve inserted.

Plate 40.2 Steel H-pile.

Plate 40.3 Steel H-pile installed.

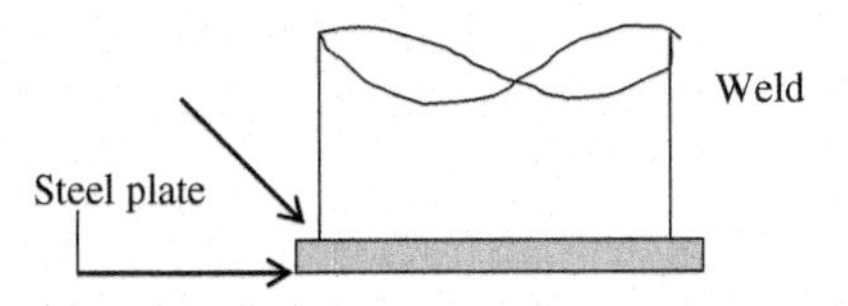

Figure 40.10 Closed end pipe pile.

40.5.1 Closed end pipe piles

- Closed end pipe piles are constructed by covering the bottom of the pile with a steel plate (Figure 40.10).
- In most cases, pipe piles are filled with concrete. In some cases, pipe piles are not filled with concrete to reduce the cost. If pipe piles were not filled with concrete, then corrosion protection layer should be applied.
- If a concrete filled pipe pile is corroded, most of the load carrying capacity of the pile would remain intact due to concrete. On the other hand, an empty pipe pile would lose a significant amount of its load carrying capacity.
- Pipe piles are a good candidate for batter piles.
- Structural capacity of pipe piles is calculated based on concrete strength and steel strength. The thickness of the steel should be reduced to account for corrosion (typically reduced by 1/16 in. to account for corrosion).

A pipe pile is covered with an end cap. The end cap is welded as shown in Plates 40.4–40.8.

Plate 40.4 Open end pipe pile.

Plate 40.5 Closed end pipe pile.

Plate 40.6 Pile hammer.

Plate 40.7 Installed pipe piles.

Plate 40.8 Concrete pipe pile.

- In the case of closed-end driving, soil heave can occur. There are occasions where open end piles also generate soil heave. This is due to plugging of the open end of the pile with soil.
- Pipe piles are cheaper than steel H-piles or concrete piles.

40.5.2 Open End Pipe Piles

- Open end pipe piles are driven and soil inside the pile is removed by a water jet (Figure 40.11).
- Open end pipe piles are easier to drive through hard soils than closed end pipe piles (Figure 40.12).

40.5.3 Ideal situations for open end pipe piles

- A soft layer of soil followed by a dense layer of soil (Figure 40.13).
- Medium dense layer of soil followed by a dense layer of soil (Figure 40.14).

40.5.4 Telescoping

Figure 40.15 shows telescoping to improve driving ability.

- Due to the smaller diameter of the telescoping pipe pile, the end bearing capacity of the pile would reduce. To accommodate the loss, the length of the telescoping pile should be increased.

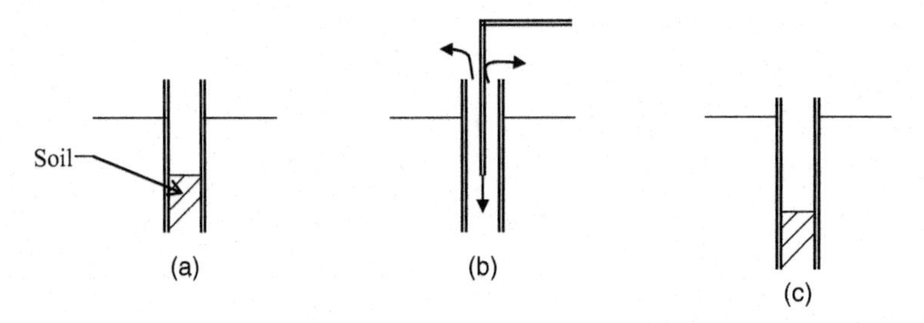

Figure 40.11 Driving of open end pipe piles. (a) Open end pipe pile is driven. (b) Soil inside the pile is washed out using a water jet. (c) The pipe is driven more and the procedure is repeated.

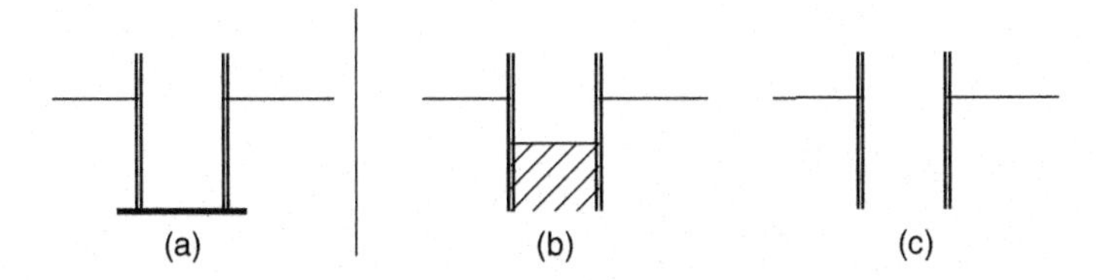

Figure 40.12 Driving of closed end pipe piles. (a) Closed end pipe pile. (b) Open end pipe pile. (c) Soil removed and ready to be driven.

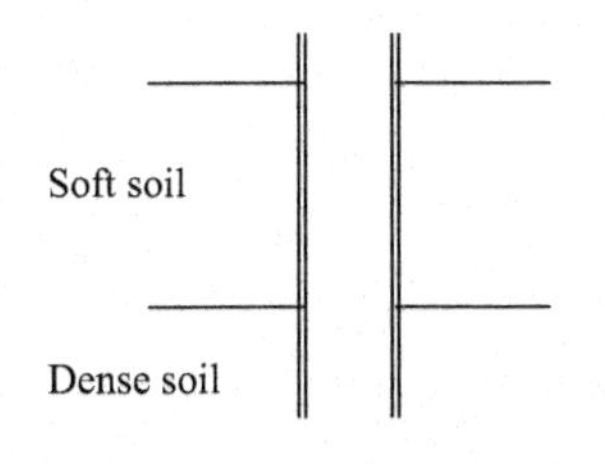

- *Site condition*: Soft layer of soil followed by a hard layer of soil.
- Open end pipe piles are ideal for this situation. After driving to the desired depth, soil inside the pipe is removed and concreted.
- Closed end pipe piles also could be considered to be ideal for this type of situation.

Figure 40.13 Pile in soft soil followed by dense soil.

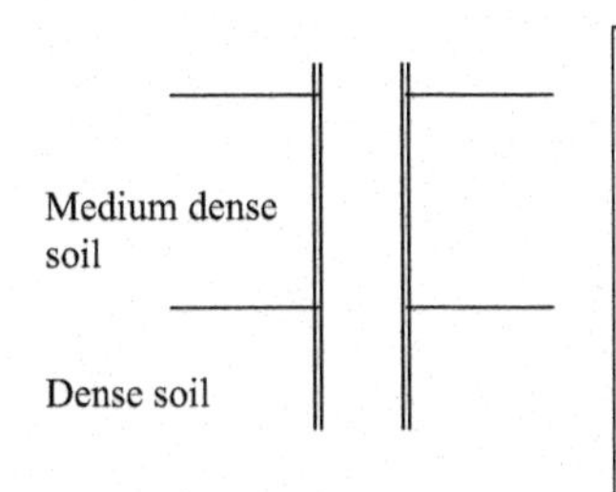

- *Site condition*: Medium dense layer of soil followed by a hard layer of soil.
- Open end pipe piles are ideal for this situation as well.
- Closed end pipe piles may *not* be a good choice, since driving closed end pipe piles through medium dense soil layer may be problematic.
- It is easier to drive open end pipe piles through a dense soil layer than closed end pipe piles.

Figure 40.14 Pile in medium dense soil underlain by dense soil.

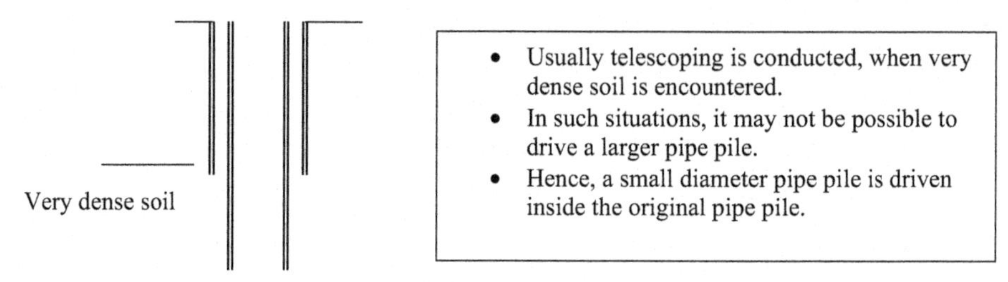

Figure 40.15 Telescoping to improve driving ability.

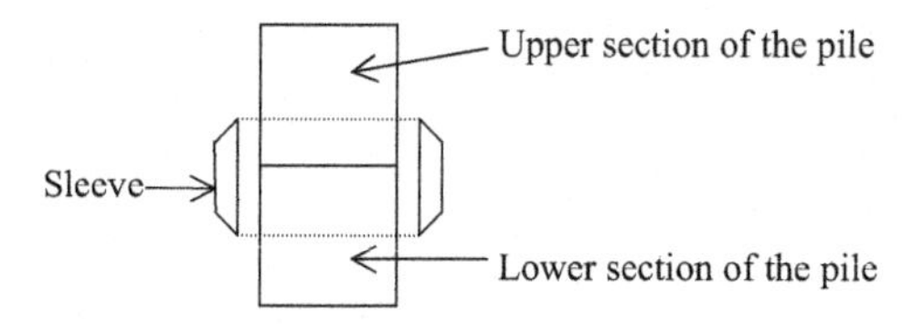

Figure 40.16 Splicing of pipe piles.

40.5.5 Splicing of pipe piles

Pipe piles are spliced by fitting a sleeve. The sleeve would fit into the bottom section of the pile as well as the top section (Figure 40.16).

40.6 Precast-concrete piles

Precast-concrete piles could be either reinforced concrete piles or prestressed concrete piles (Figure 40.17).

40.7 Reinforced concrete piles

Reinforced concrete piles are constructed by reinforcing the concrete as shown in Figure 40.18.

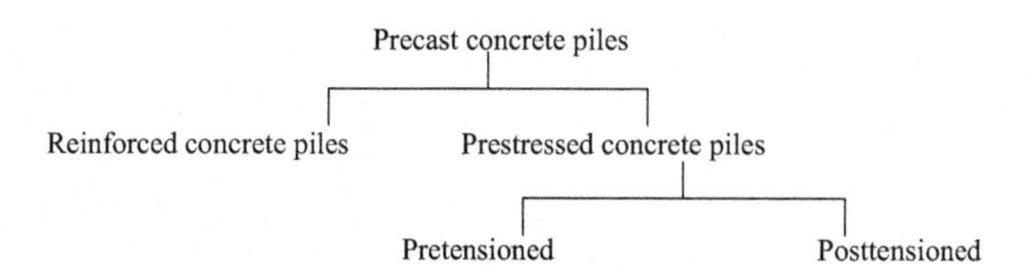

Figure 40.17 Precast-concrete piles.

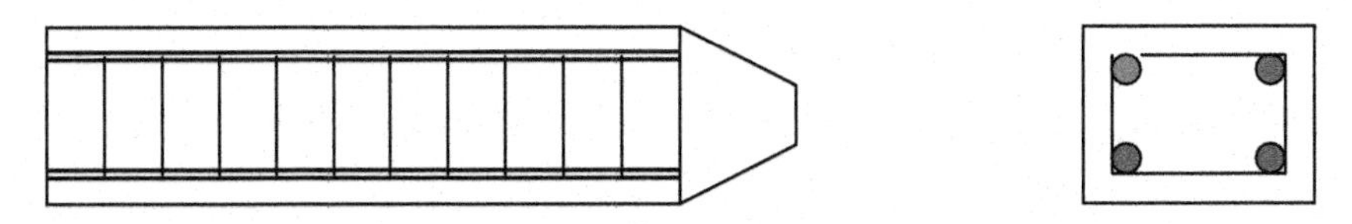

Figure 40.18 Reinforced concrete piles.

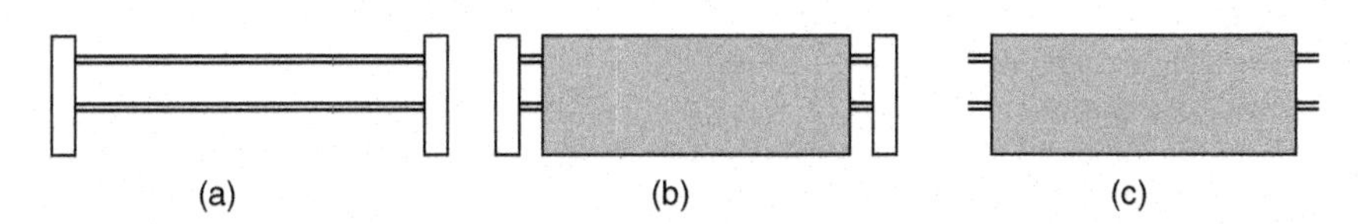

Figure 40.19 Pre-stressed concrete piles. (a) Reinforcement cables are pulled and attached to metal supports. (b) Concrete pile is cast around. (c) After the concrete is set, supports are cutoff.

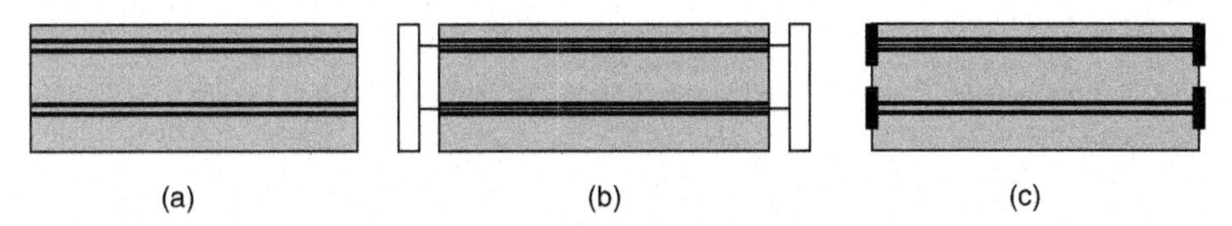

Figure 40.20 Post tensioned concrete piles. (a) Concrete pile is cast and allowed to set. Hollow tubes are inserted prior to concreting. (b) Cables are inserted through hollow tubes and pulled. (c) Supports are cutoff to allow the cables to compress the concrete.

40.8 Prestressed concrete piles

40.8.1 Pretensioning procedure

Figure 40.19 shows prestressed concrete piles.

40.8.2 Posttensioning procedure

Figure 40.20 shows posttensioned concrete piles.

IBC specifies a minimum lateral dimension (diameter or width) of 8 in. for precast-concrete piles.

40.8.3 Reinforcements for precast-concrete piles (IBC)

As per the IBC, longitudinal reinforcements should be arranged in a symmetrical manner. Lateral ties should be placed every 4 to 6 in.

40.8.4 Concrete strength (IBC)

As per IBC, 28-day concrete strength (f'_c) should be not less than 3000 psi.

40.8.5 Hollow tubular section concrete piles

- Most hollow-tubular piles are posttensioned to withstand tensile stresses.
- Hollow tubular concrete piles can be driven either closed end or open end. A cap is fitted at the end for closed end driving.
- These piles are not suitable for dense soils.
- Splicing is expensive.
- It is a difficult and expensive process to cutoff these piles. It is very important to know the depth to the bearing stratum with a reasonable accuracy.

40.9 Driven cast-in-place concrete piles

- Step 1: Steel tube is driven first.
- Step 2: Soil inside the tube is removed by water jetting.
- Step 3: Reinforcement cage is set inside the casing.
- Step 4: The empty space inside the tube is concreted while removing the casing.

40.10 Selection of pile type

Once the geotechnical engineer has decided to use piles, the next question is which piles? There are many types of piles available as discussed in Chapter 39. Timber piles are cheap but difficult to install in hard soil. Steel piles may not be good in marine environments due to corrosion.

Some possible scenarios are discussed here.

Case 1: Granular soil with boulders underlain by medium stiff clay (Figure 40.21).

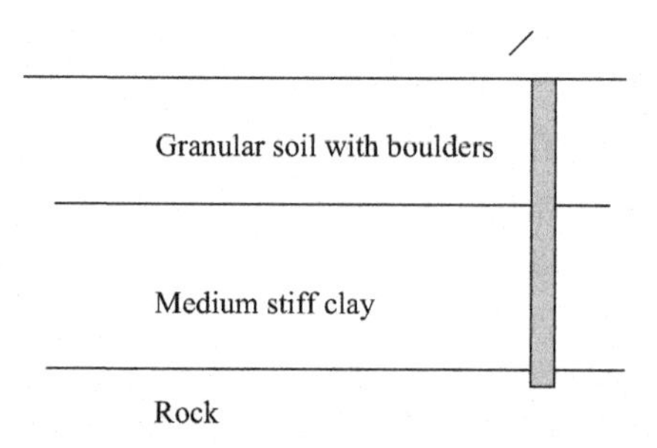

Figure 40.21 Pile in multilayer soil strata.

Timber piles are not suitable for Case 1 due to existence of boulders in the upper layers. The obvious choice is to drive piles all the way to the rock. The piles could be designed as end bearing piles. If the decision was taken to drive piles to the rock for the above configuration, H-piles will be ideal. Unlike timber piles or pipe piles, H-piles can go through boulders.

On the other hand, the rock may be too far for the piles to be driven. If the rock is too deep for piles, a number of alternatives can be envisioned.

- Drive large diameter pipe piles and place them in medium stiff clay.
- Construct a caisson and place it in the medium stiff clay.

In the first case, driving large diameter pipe piles through boulders could be problematic. It is possible to excavate and remove boulders if the boulders are mostly at shallower depths.

In the case of H-piles, one has to be extra careful not to damage the piles.

Caissons placed in the medium stiff clay is a good alternative. Settlement of caissons need to be computed.

Case 2:

Figure 40.22 shows the pressure distribution in a pile.

Piles cannot be placed in soft clay. It is possible to place piles in the medium stiff clay. In that situation, the piles need to be designed as friction piles. Pile capacity mainly come from end bearing and skin friction. End bearing piles, as the name indicates, obtain their capacity mainly from the end bearing. While friction piles obtain their capacity from skin friction.

If the piles were placed in medium stiff clay, stresses would reach the soft clay layer below. The engineer needs to make sure that settlement due to compression of soft clay is within acceptable limits.

Case 3:

Figure 40.23 shows a pile with groundwater effects considered.

Timber Piles: Timber piles could be placed in the medium stiff clay. In this situation, soft clay in the upper layers may not pose a problem for driving. Timber piles above groundwater should be protected.

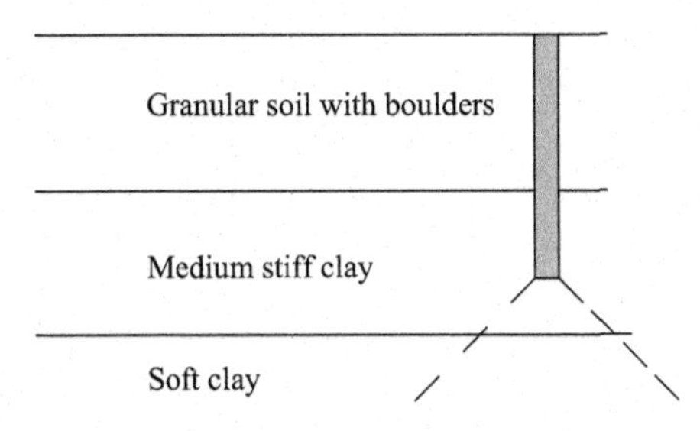

Figure 40.22 Pressure distribution in a pile.

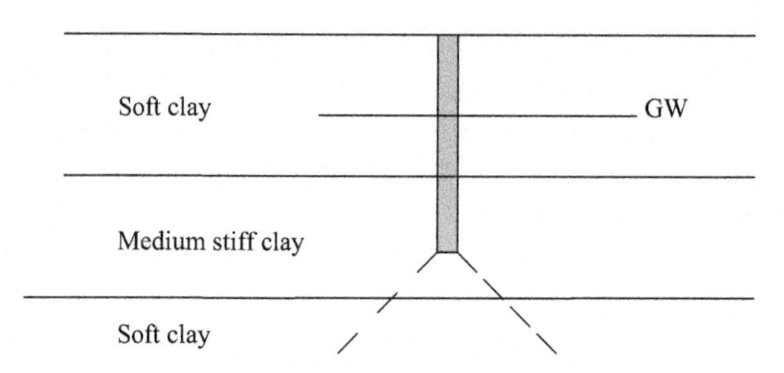

Figure 40.23 Pile with groundwater effects considered.

Pipe Piles: Pipe piles also could be used in this situation. Pipe piles cost more than timber piles. One of the main advantages of pipe piles over timber piles is that they can be driven hard. At the same time, large diameter pipe piles may be readily available and higher loads can be accommodated by few piles.

H–piles: H–piles are ideal candidates for end bearing piles. Case 3 calls for friction piles. Hence H–piles may not be suitable for the this situation.

Pile design in sandy soils

41

A modified version of Terzaghi's bearing capacity equation is widely used for pile design.

The third term or the density term in Terzaghi's bearing capacity equation is negligible in piles and hence, usually ignored. Lateral earth pressure coefficient (K) is introduced to compute the skin friction of piles.

$$P_{\text{ultimate}} = \underbrace{[\sigma'_t \times N_q \times A]}_{\text{End bearing term}} + \underbrace{[K\sigma'_v \times \tan\delta \times A_p]}_{\text{Skin friction term}}$$

P_{ultimate}, ultimate pile capacity; σ'_t, effective stress at the tip of the pile; N_q, bearing factor coefficient; A, cross-sectional area of the pile at the tip; K, lateral earth pressure coefficient (use Table 41.1 for "K" values); σ'_v, effective stress at the perimeter of the pile (σ'_v varies with the depth. Usually, σ'_v value at the mid point of the pile is obtained); $\tan\delta$, friction angle between pile and soil (see Table 41.2 for δ values); A_p, perimeter area of the pile. For round piles, $A_p = (\pi \times d) \times L$ (d = diameter, L = length of the pile).

Description of Terms:

1. *Effective Stress* (σ'): When a pile is driven, the effective stress of the existing soil around and below the pile would change. Figure 41.1 shows the effective stress prior to driving a pile. Effective stress prior to driving the pile.

$$\sigma'_a = \gamma h_1;$$
$$\sigma'_b = \gamma h_2$$

   Effective stress after driving the pile is shown in Figure 41.2. When a pile is driven, soil around the pile and below the pile would be compacted.
   Due to disturbance of soil during the pile driving process, σ'_{aa} and σ'_{bb} cannot be accurately computed. Usually, increase in effective stress due to pile driving is ignored.
2. N_q (*Bearing capacity factor*): Many researchers have provided techniques to compute bearing capacity factors. End bearing capacity is a function of friction angle, dilatancy of soil, and relative density. All these parameters are lumped into "N_q". Different methods of obtaining the N_q value will be discussed.
3. K (*Lateral earth pressure coefficient*): Prior to discussing the lateral earth pressure coefficient related to piles, it is necessary to investigate lateral earth pressure coefficients in general (Figure 41.3).
   a. K_0 – *in situ soil condition*: For the at rest condition, the horizontal effective stress is given by $K_0 \times \sigma'$. (K_0 = Lateral earth pressure coefficient at rest and σ' is the vertical effective stress)

Table 41.1 Pile type and lateral earth pressure coefficient

Pile type	K (piles under compression)	K (piles under tension – uplift piles)
Driven H-piles	0.5–1.0	0.3–0.5
Driven displacement piles (round and square)	1.0–1.5	0.6–1.0
Driven displacement tapered piles	1.5–2.0	1.0–1.3
Driven jetted piles	0.4–0.9	0.3–0.6
Bored piles (less than 24 in. diameter)	0.7	0.4

Source: NAVFAC DM 7.2 (1984).

Table 41.2 Pile type and pile skin friction angle

Pile type	δ
Steel piles	20°
Timber piles	¾ ϕ
Concrete piles	¾ ϕ

Source: NAVFAC DM 7.2 (1984).

Soil density = γ h_1 h_2
σ'_a A
σ'_b B

Figure 41.1 Effective stress at different depths.

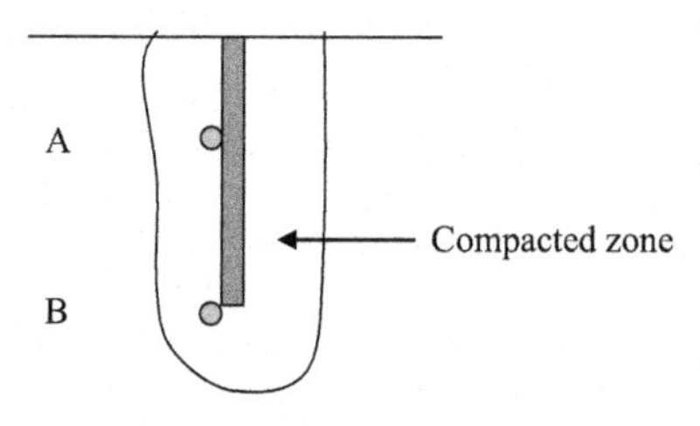

Figure 41.2 Effective stress near a pile.

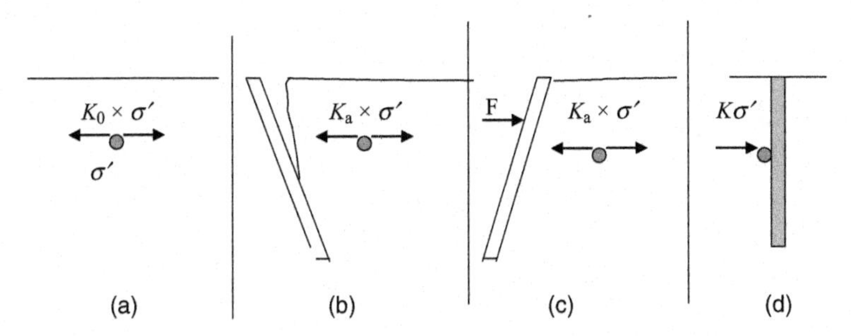

Figure 41.3 Lateral earth pressure coefficients. (a) K_0 (soil at rest). (b) K_a (active state). (c) K_p (passive state). (d) K (pile).

b. K_a – *active condition*: In this case, the soil exerts the minimum horizontal effective stress, since the soil particles have room to move (K_a = Active earth pressure coefficient). K_a is always smaller than K_0.

c. K_p – *passive condition*: In this case, the soil exerts the maximum horizontal effective stress, since the soil particles have been compressed (K_p = passive earth pressure coefficient). K_p is always greater than K_0 and K_a.

d. K – *soil near piles*: The soil near a driven pile would be compressed. In this case, the soil definitely exerts more horizontal pressure than the *in situ* horizontal effective stress (K_0). Since K_p is the maximum horizontal stress that can be achieved, K should be in between K_0 and K_p.

$$K_a < K_0 < K\,(\text{pile condition}) < K_p$$

Hence, $K = (K_0 + K_a + K_p)/3$ can be used as an approximation.
Equations for K_0, K_a, and K_p are

$$K_0 = 1 - \sin\varphi'$$
$$K_a = 1 - \tan^2(45 - \varphi'/2)$$
$$K_p = 1 + \tan^2(45 + \varphi'/2)$$

Note that $K_0 = 1 - \sin\varphi'$ is valid only for normally consolidated soils. For overconsolidated soils, the following equation should be used.

$$K_0 = (1 - \sin\varphi')\ \text{OCR}^{0.5}$$

For instance, if φ is 40° and the over consolidation ratio (OCR) is 2.5, the K_0 would be

$$K_0 = (1 - \sin\varphi')\text{OCR}^{0.5} = (1 - 0.643) \times 2.5^{1/2}$$
$$K_0 = 0.564$$

K_0 of overconsolidated clays is higher than normally consolidated clays.

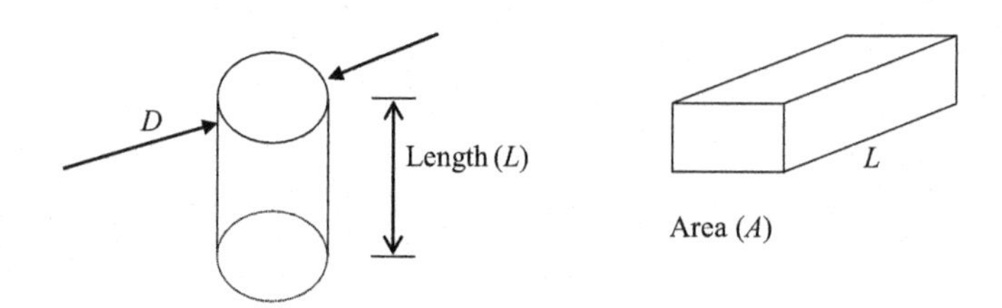

Figure 41.4 Perimeter surface area of piles.

1. *Tan δ (wall friction angle)*: The friction angle between pile material and soil "δ" decides the skin friction. Friction angle (δ) varies with the pile material and soil type. Various agencies have conducted laboratory tests and have published "δ" values for different pile materials and soils.
2. A_p (*Perimeter surface area of piles*): Figure 41.4 shows the perimeter surface area of piles.

   Perimeter surface area of a circular pile = $\pi \times D \times L$
   Perimeter surface area of a rectangular pile = $A \times L$
   Skin friction acts on the perimeter surface area of the pile.

41.1 Equations for end bearing capacity in sandy soils

There are a number of methods available to compute the end bearing capacity of piles in sandy soils.

41.1.1 API method: (American petroleum institute, 1984)

$$q = N_q \sigma'_t$$

q, end bearing capacity of the pile (units same as σ_t'); σ_t', effective stress at pile tip (maximum effective stress allowed for the computation is 240 kPa or 5.0 ksf); N_q = 8–12, for loose sand; N_q = 12–40, for medium dense sand; N_q = 40, for dense sand.

Sand consistency (loose, medium, and dense) can be obtained from Table 41.3.

41.1.2 Martin et al. (1987)

SI units: $q = C.\, N$ (MN/m^2)

fps units: $q = 20.88 \times$ C. N (ksf)

q = end bearing capacity of the pile

N = SPT value at pile tip (blows per foot)

C = 0.45 (for pure sand)

C = 0.35 (for silty sand)

Table 41.3 Depth and end bearing capacity of piles in different sandy soils

Depth		Saturated loose sand		Dry loose sand		Saturated very dense sand		Dry very dense sand	
feet (ft.)	meters (m)	tsf	MN/m^2	Tsf	MN/m^2	tsf	MN/m^2	tsf	MN/m^2
20	6.1	10	0.95	50	4.8	60	5.7	140	13.4
40	12.2	25	2.4	60	5.7	110	10.5	200	19.2
60	18.3	40	3.8	70	6.7	160	15.3	250	23.9
80	24.4	50	4.8	90	8.6	200	19.2	300	28.7
100	30.5	55	5.3	110	10.5	230	22.0	370	35.4

Source: Kulhawy (1984).

Design Example 41.1

The SPT (N) value at the pile tip is 10 blows per foot. Find the ultimate end bearing capacity of the pile assuming that pile tip is in pure sand and the diameter of the pile is 1 ft.

fps units: $q = 20.88 \times C.\ N$ (ksf)

$q = 20.88 \times 0.45 \times 10 = 94$ ksf

Total end bearing capacity = $q \times$ Area

Total end bearing capacity = $q \times \pi \times (d^2)/4 = 74$ kips = 37 tons (329 kN)

41.1.3 NAVFAC DM 7.2

$$q = \sigma'_t \times N_q$$

q, end bearing capacity of the pile (units same as σ_v'); σ_t', effective stress at pile tip.

41.1.4 Bearing capacity factor (N_q)

Table 41.4 shows that the N_q value is lower in bored piles. This is expected. During pile driving, the soil just below the pile tip would be compacted. Hence, it is reasonable to assume a higher N_q value for driven piles.

If water jetting is used, ϕ should be limited to 28°. This is due to the fact that water jets tend to loosen the soil. Hence, higher friction angle values are not warranted.

Table 41.4 Friction angle versus N_q

ϕ	26	28	30	31	32	33	34	35	36	37	38	39	40
N_q (for driven piles)	10	15	21	24	29	35	42	50	62	77	86	12	145
N_q (for bored piles)	5	8	10	12	14	17	21	25	30	38	43	60	72

Source: NAVFAC DM 7.2 (1984).

Kulhawy (1984) reported the following values for the end bearing capacity using pile load test data.

Design Example 41.2

Find the end bearing capacity of a 3 ft. (1 m) diameter caisson using the Kulhawy values (Figure 41.5). Assume that the soil is saturated at tip level of the caisson and the average SPT "*N*" value at the tip level is 10 blows/ft.

Average "*N*" values at tip level = 10

Solution

The average "*N*" values of 10 blows per foot can be considered as loose sand. Hence, the soil below the pile tip can be considered to be loose sand. The depth to the bottom of the pile is 20 ft.

Ultimate end bearing capacity for saturated loose sand at 20 ft. = 10 tsf (0.95 MN/m^2) (see Table 41.3).

Cross-sectional area of the caisson = ($\pi \times D^2/4$) = 7.07 ft.^{2} (0.785 m^2)

Ultimate end bearing capacity = Area $\times$ 10 tons = 70.7 tons (0.74 MN)

It should be pointed out that some of the values in Table 41.3 are very high. For instance, a caisson placed in very dense dry soil at 20 ft. gives 140 tsf. (see Table 41.3)

It is recommended to use other methods to check the values obtained using tables provided by Kulhawy.

41.2 Equations for skin friction in sandy soils

Numerous techniques have been proposed to compute the skin friction in piles in sandy soils. These different methodologies are briefly discussed in this chapter.

McClelland (1974) (Driven piles): McClelland suggested the following equation.

$$S = \beta \sigma_v' A_p$$

S, total skin friction; σ_v', effective stress at mid point of the pile; σ_v' changes along the length of the pile. Hence σ_v' at the mid point of the pile should be taken; A_p, perimeter surface area of the pile.

Figure 41.6 shows the skin friction acting on the perimeter of the pile.

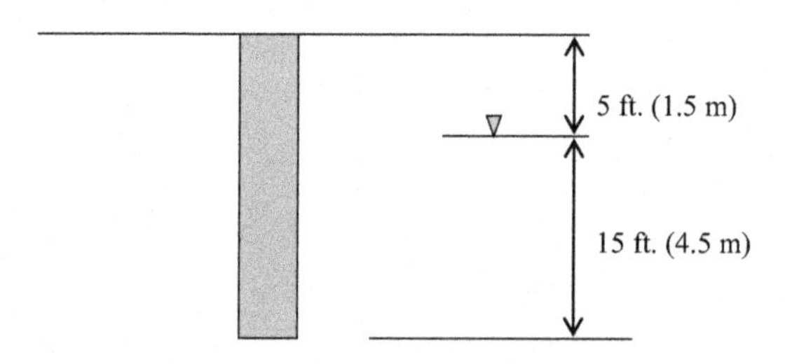

Figure 41.5 End bearing capacity in sand.

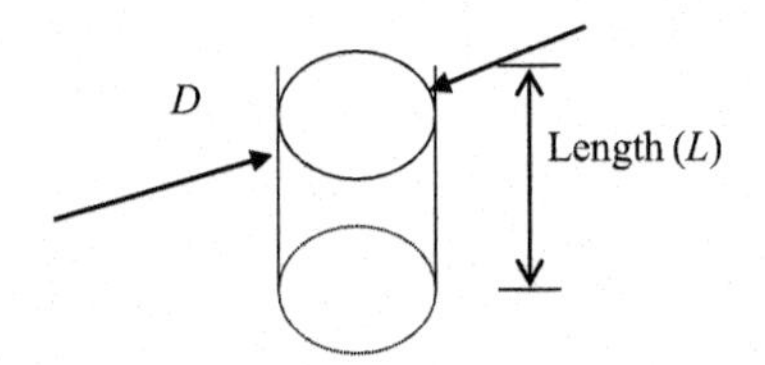

Figure 41.6 Skin friction acting on perimeter of the pile.

Perimeter surface area of a circular pile (A_p) = $\pi \times D \times L$

$\beta = 0.15$–0.35 for compression
$\beta = 0.10$–0.25 for tension (for uplift piles)

Meyerhoff (1976) (Driven piles): Meyerhoff suggested the following equation for driven piles.

$$S = \beta \sigma'_v A_p$$

S, skin friction; σ_v', effective stress at the mid point of the pile; A_p, perimeter surface area of the pile.

$\beta = 0.44$, for $\phi' = 28^\circ$
$\beta = 0.75$, for $\phi' = 35^\circ$
$\beta = 1.2$, for $\phi' = 37^\circ$

Meyerhoff (1976) (Bored piles): Meyerhoff suggested the following equation for bored piles.

$$S = \beta . \sigma'_v A_p$$

S, skin friction of the pile; σ_v', effective stress at the mid point of the pile; A_p, perimeter surface area of the pile.

$\beta = 0.10$, for $\phi = 33^\circ$
$\beta = 0.20$, for $\phi = 35^\circ$
$\beta = 0.35$, for $\phi = 37^\circ$

Kraft and Lyons (1974):

$$S = \beta \sigma'_v A_p$$

S, skin friction of the pile; σ_v', effective stress at the mid point of the pile; $\beta = C \tan (\phi - 5)$; $C = 0.7$ for compression; $C = 0.5$ for tension (uplift piles).

NAVFAC DM 7.2 (1984):

$$S = K . \sigma'_v \times \tan \delta \times A_p$$

S, skin friction of the pile; σ_v', effective stress at the mid point of the pile; K, lateral earth pressure coefficient; δ, pile skin friction angle.

41.2.1 Pile skin friction angle (δ)

Table 41.2 shows pile types and pile skin friction angle.

δ is the skin friction angle between pile material and surrounding sandy soils. Usually, smooth surfaces tend to have lesser skin friction as compared to rough surfaces.

41.2.2 Lateral earth pressure coefficient (K)

Table 41.1 shows pile types and lateral earth pressure coefficient.

Lateral earth pressure coefficient is less in uplift piles as compared to regular piles. Tapered piles tend to have the highest "K" value.

41.2.2.1 Average "K" method

Earth pressure coefficient "K" can be averaged from K_a, K_p, and K_0.

$$K = (K_0 + K_a + K_p)/3$$

Equations for K_0, K_a, and K_p are

$K_0 = 1 - \sin\varphi'$ (earth pressure coefficient at rest)

$K_a = 1 - \tan^2(45 - \varphi'/2)$ (active earth pressure coefficient)

$K_p = 1 + \tan^2(45 + \varphi'/2)$ (passive earth pressure coefficient)

Design Example 41.3: Single pile in uniform sand layer (no groundwater present)

A 10 m (32.8 ft.) long round steel pipe pile with 0.5 m (1.64 ft.) diameter is driven into a sandy soil stratum as shown in Figure 41.7. Compute the ultimate bearing capacity of the pile.

Step 1: Compute the end bearing capacity:

$$Q_{\text{ultimate}} = \underbrace{[\sigma'_t \times N_q \times A]}_{\text{End Bearing Term}} + \underbrace{[K\sigma'_p \times \tan\delta \times (\pi \times d)L]}_{\text{Skin Friction Term}}$$

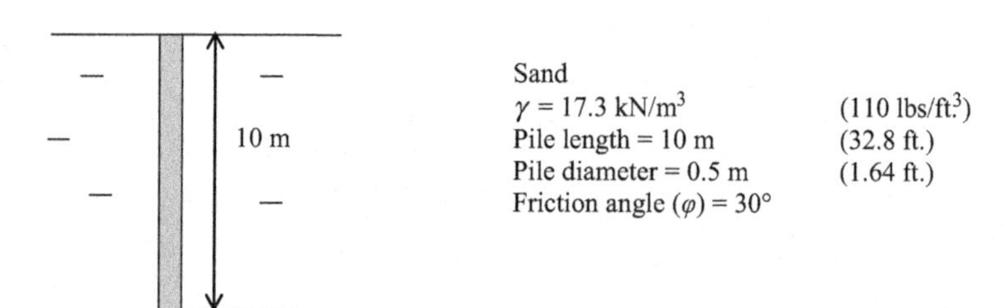

Figure 41.7 Single pile in sandy soil.

End bearing term = $\sigma'_t \times N_q \times A$ (σ'_t = effective stress at the tip of the pile)
$\sigma'_t = \gamma \times$ depth to the tip of the pile = 17.3 × 10 = 173 kN/m² *(3,610 psf)*
Find "N_q" using Table 41.4.
For a friction angle of 30, $N_q = 21$ for driven piles.

$$\text{End bearing capacity} = \sigma_t' \times N_q \times A$$
$$= 173 \times 21 \times (\pi\, d^2 / 4) = 713.3\,\text{kN}\ \textit{(160 kips)}$$

Step 2: Computation of the skin friction:
Skin friction term = $K\sigma'_p \times \tan\delta \times (\pi \times d) \times L$
Obtain the "K" value.
From Table 41.1, for driven round piles "K" value lies between 1.0 and 1.5. Hence, assume $K = 1.25$
Obtain the σ_p' (effective stress at the perimeter of the pile).
The effective stress along the perimeter of the pile varies with the depth. Hence, obtain the σ_p' value at the mid point of the pile.
The pile is 10 m long, hence use the effective stress at 5 m below the ground surface.
σ'_p (mid point) = 5 × γ = 5 × 17.3 = 86.5 kN/m². (*1.8 ksf*)
Obtain the skin friction angle (δ).
From Table 41.2, the skin friction angle for steel piles is 20°.
Find the skin friction of the pile;

$$\text{Skin friction} = K\sigma_p' \times \tan\delta \times (\pi \times d) \times L$$
$$= 1.25 \times 86.5 \times (\tan 20) \times (\pi \times 0.5) \times 10$$
$$= 618.2\,\text{kN.} \qquad \textit{(139 kips)}$$

Step 3: Compute the ultimate bearing capacity of the pile:
$Q_{ultimate}$ = ultimate bearing capacity of the pile
$Q_{ultimate}$ = end bearing capacity + skin friction
$Q_{ultimate}$ = 713.3 + 618.2 = 1331.4 kN *(300 kips)*
Assume a factor of safety of 3.0;
Hence, allowable bearing capacity of the pile = $Q_{ultimate}$/FOS
= 1331.4/3.0

Allowable pile capacity = 443.8 kN. (*99.8 kips*)
Note: 1 kN is equal to 0.225 Kips. Hence, the allowable capacity of the pile is 99.8 kips.

Design Example 41.4: Single pile in uniform sand layer (groundwater present)

A 10 m long round concrete pile with 0.5 m diameter, is driven into a sandy soil stratum as shown in Figure 41.8. Groundwater is located 3 m below the surface. Compute the ultimate bearing capacity of the pile.
Step 1: Compute the end bearing capacity:

$$Q_{\text{ultimate}} = \underbrace{[\sigma_t' \times N_q \times A]}_{\text{End Bearing Term}} + \underbrace{[K.\sigma_v' \times \tan\delta \times (\pi \times d) \times L]}_{\text{Skin Friction Term}}$$

End bearing term = $\sigma'_t \times N_q \times A$ (σ'_t = effective stress at the tip of the pile)

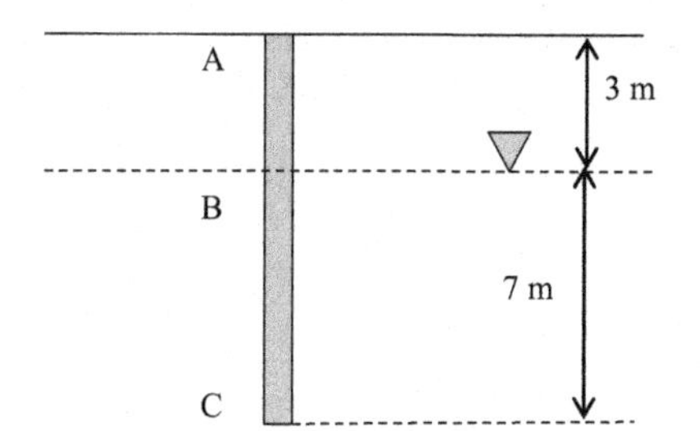

Sand
$\gamma = 17.3$ kN/m³ (110 lbs/ft.³)
Pile length = 10 m (32.8 ft.)
Pile diameter = 0.5 m (1.6 ft.)
Friction angle (φ) = 30°

Figure 41.8 Single pile in sand (groundwater considered).

$\sigma'_t = 17.3 \times 3 + (17.3 - \gamma_w) \times 7$
γ_w = density of water = 9.8 kN/m³ = 17.3 × 3 + (17.3 − 9.8) × 7 = 104.4 kN/m² (2180 psf)
Find "N_q" using Table 41.4.
For a friction angle of 30°, $N_q = 21$, for driven piles.

$$\text{End bearing capacity} = \sigma'_t \times N_q \times A$$
$$= 104.4 \times 21 \times (\pi \times 0.5^2 / 4) = 403.5 \text{ kN (96.8 kips)}$$

Step 2: Computation of the Skin Friction (A to B):
Computation of skin friction has to be done in two parts. First, find the skin friction from A to B and then find the skin friction from B to C.
Skin friction term = $K\sigma'_v \times \tan\delta \times (\pi \times d) \times L$
Obtain the "K" value.
From Table 41.1, for driven round piles "K" value lies between 1.0 and 1.5. Hence, assume $K = 1.25$
Obtain the σ_v' (effective stress at the perimeter of the pile).
The effective stress along the perimeter of the pile varies with the depth. Hence, obtain the σ_p' value at the mid point of the pile from point A to B. Length of pile section from A to B is 3 m. Hence, find the effective stress at 1.5 m below the ground surface.
Obtain the skin friction angle (δ).
From Table 41.2, the skin friction angle for steel piles is 20°.
Skin friction = $K\sigma_v' \times \tan\delta \times (\pi \times d) \times L$
σ_v' = effective stress at mid point of section A to B (1.5 m below the surface)
$\sigma_v' = 1.5 \times 17.3 = 25.9$ kN/m² (0.54 ksf)
$K = 1.25$, $\delta = 20°$,
Skin friction (A to B) = 1.25 × (25.9) × (tan20) × (π × 0.5) × 3
= 55.5 kN (12.5 kips)
Find the skin friction of the pile from B to C.
Skin friction = $K\sigma_v' \times \tan\delta \times (\pi \times d) \times L$
σ_v' = Effective stress at mid point of section B to C = 3 × 17.3 + (17.3 − 9.8) × 3.5
σ_v' = Effective stress at mid point of section B to C = 78.2 kN/m². (1,633 psf)
$K = 1.25$ and $\delta = 20°$.
Skin friction (B to C) = 1.25 × (78.2) × tan(20) × (π × 0.5) × 7
= 391 kN. (87.9 kips)
Step 3: Compute the ultimate bearing capacity of the pile:
$Q_{ultimate}$ = ultimate bearing capacity of the pile
$Q_{ultimate}$ = end bearing capacity + skin friction
$Q_{ultimate}$ = 403.5 + 55.5 + 391 = 850 kN

Assume a factor of safety of 3.0;
Hence, allowable bearing capacity of the pile = $Q_{ultimate}$/FOS
= 850.1/3.0 = 283 kN

Allowable pile capacity = 283 kN (63.6 kips)
Note: 1 kN is equal to 0.225 kips.

Design Example 41.5: Multiple sand layers with no groundwater present

A 12 m long round concrete pile with 0.5 m diameter, is driven into a sandy soil stratum as shown in Figure 41.9. Compute the ultimate bearing capacity of the pile.

Solution
Step 1: Compute the end bearing capacity.

$$Q_{ultimate} = \underbrace{[\sigma'_t \times N_q \times A]}_{\text{End Bearing Term}} + \underbrace{[K\sigma'_v \times \tan\delta \times A_p]}_{\text{Skin Friction Term}}$$

End Bearing Term = $\sigma'_t \times N_q \times A$
(σ'_t = Effective stress at the tip of the pile)

$$\sigma'_t = \gamma_1.5 + \gamma_2.7$$
$$\sigma'_t = 17.3 \times 5 + 16.9 \times 7 = 204.8\ \text{kN/m}^2. \quad (4.28\ \text{ksf})$$

Find "N_q" using Table 41.4. Use the friction angle of soil where the pile tip rests.
For a friction angle of 32, N_q = 29 for driven piles. Diameter of the pile is given to be 0.5 m.
The φ value of the bottom sand layer is used to find N_q, since the tip of the pile lies on the bottom sand layer.

$$\text{End bearing capacity} = \sigma'_t \times N_q \times A$$
$$= 204.8 \times 29 \times (\pi d^2/4) = 1166.2\ \text{kN} \quad (262.2\ \text{kips})$$

Step 2: Computation of the skin friction:
The skin friction of the pile needs to be done in two parts.

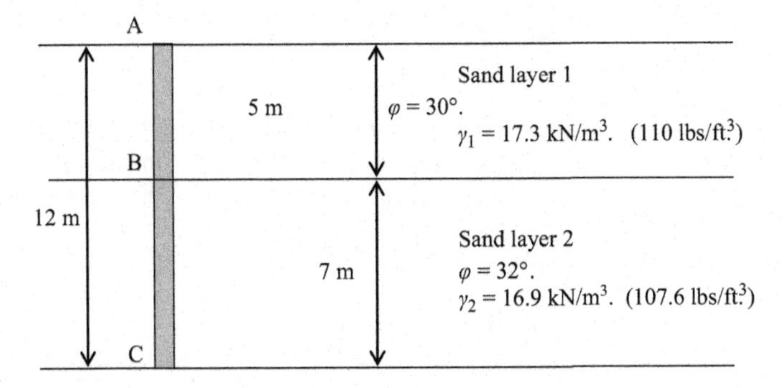

Figure 41.9 Single pile in multiple sand layers.

- Skin friction of the pile portion in sand layer 1 (A to B)
- Skin friction of the pile portion in sand layer 2 (B to C)

Compute the skin friction of the pile portion in sand layer 1: (A to B)

$$\text{Skin friction term} = K\,\sigma'_p \times \tan\delta \times A_p$$
$$A_p = (\pi \times d) \times L$$

Obtain the "K" value.
From Table 41.1, for driven round piles "K" value lies between 1.0 and 1.5.
Hence, assume $K = 1.25$
Obtain the σ_v' (average effective stress at the perimeter of the pile).
Obtain the σ_v' value at the mid point of the pile in sand layer 1.
σ'_v (mid point) = $2.5 \times (\gamma_1) = 2.5 \times 17.3$ kN/m^2.
σ'_v (mid point) = 43.3 kN/m^2. (0.9 ksf)
Obtain the skin friction angle (δ).
From Table 41.2, the skin friction angle (δ) for concrete piles is ¾ φ.
$\delta = ¾ \times 30^\circ = 22.5^\circ$. (Friction angle of layer 1 is 30°).
Skin friction in sand layer 1 = $K\sigma_v' \times \tan\delta \times (\pi \times d) \times L$
= $1.25 \times 43.3 \times (\tan 22.5) \times (\pi \times 0.5) \times 5$
= 176.1 kN. (39.6 kips)
Step 3: Find the skin friction of the pile portion in sand layer 2 (B to C)
Skin friction term = $K\sigma'_v \times \tan\delta \times (\pi \times d) \times L$
Obtain the σ_v' (effective stress at the perimeter of the pile).
Obtain the σ_v' value at the mid point of the pile in sand layer 2.
σ'_v (mid point) = $5 \times \gamma_1 + 3.5 \times \gamma_2$
σ'_v (mid point) = $5 \times 17.3 + 3.5 \times 16.9$ kN/m^2
= 145.7 kN/m^2 (3043 psf)
Obtain the skin friction angle (δ).
From Table 41.2, the skin friction angle (δ) for concrete piles is ¾ φ.
$\delta = ¾ \times 32^\circ = 24^\circ$. (Friction angle of layer 2 is 32°).
Skin friction in sand layer 2 = $K\sigma_p' \times \tan\delta \times (\pi \times d) \times L$
= $1.25 \times 145.7 \times (\tan 24) \times (\pi \times 0.5) \times 7$
= 891.6 kN (200.4 kips)
$P_{ultimate}$ = end bearing capacity + skin friction in layer 1 + skin friction in layer 2
End bearing capacity = 1166.2 kN
Skin friction in sand layer 1 = 176.1 kN
Skin friction in sand layer 2 = 891.6 kN
$P_{ultimate}$ = 2233.9 kN. (502.2 kips)
One can see that bulk of the pile capacity comes from end bearing. Next, the skin friction in layer 2 (bottom layer). Skin friction in top layer is very small. One of the reason for this is that the effective stress acting on the perimeter of the pile is very low in the top layer. This is due to the fact that effective stress is directly related to depth.
Hence, the allowable bearing capacity of the pile = $P_{ultimate}$/FOS
= 2233.9/3.0 kN
Allowable pile capacity = 744.6 kN. (167.4 kips)

Design Example 41.6: Multiple sand layers with groundwater present

A 15 m long round concrete pile with a diameter of 0.5 m is driven into a sandy soil stratum as shown in Figure 41.10. Groundwater is 3 m below the surface. Compute the ultimate bearing capacity of the pile.

Solution

Step 1: Compute the end bearing capacity:

$$Q_{\text{ultimate}} = \underbrace{[\sigma'_t \times N_q \times A]}_{\text{End Bearing Term}} + \underbrace{[K\ \sigma'_v \times \tan\delta \times A_p]}_{\text{Skin Friction Term}}$$

End bearing term = $\sigma'_t \times N_q \times A$

(σ'_t = Effective stress at the tip of the pile)

$\sigma'_t = \gamma_1 . 3 + (\gamma_1 - \gamma_w) \times 2 + (\gamma_2 - \gamma_w) \times 10$

$\sigma'_t = 17.3 \times 3 + (17.3 - 9.8) \times 2 + (16.9 - 9.8) \times 10 = 137.9\ \text{kN/m}^2$. (2880 psf)

Find "N_q" using Table 41.4.

For a friction angle of 32, $N_q = 29$ for driven piles.

The φ value of the bottom sand layer is used to find N_q, since the tip of the pile lies on the bottom sand layer.

$$\text{End bearing capacity} = \sigma'_t \times N_q \times A$$
$$= 137.9 \times 29 \times (\pi\, d^2 / 4) = 785.2\ \text{kN} \qquad (176.5\text{kips})$$

Step 2: Computation of the skin friction:

The skin friction of the pile needs to be done in three parts.

- Skin friction of the pile portion in sand layer 1 above groundwater (A to B)
- Skin friction of the pile portion in sand layer 1 below groundwater (B to C)
- Skin friction of the pile portion in sand layer 2 below groundwater (C to D)

Step 3: Compute the skin friction of the pile portion in *sand layer 1* above groundwater (A to B)

Skin friction term = $K\sigma'_v \times \tan\delta \times (\pi \times d) \times L$

Obtain the "K" value.

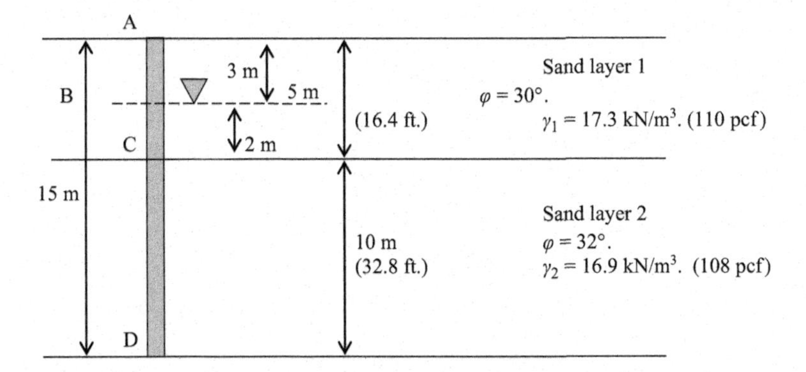

Figure 41.10 Single pile in sand (groundwater considered).

From Table 41.1, for driven round piles "K" value lies between 1.0 and 1.5.
Hence, assume $K = 1.25$

Obtain the σ_v' (effective stress at the perimeter of the pile).

Obtain the σ_v' value at the mid point of the pile in sand layer 1, above groundwater (A to B)
σ_v' (mid point) $= 1.5 \times (\gamma_1) = 1.5 \times 17.3$ kN/m^2.
σ_v' (mid point) $= 26$ kN/m^2. (543 psf)

Obtain the skin friction angle (δ).

From Table 41.2, the skin friction angle (δ) for concrete piles is ¾ φ.
$\delta = ¾ \times 30^\circ = 22.5^\circ$ (friction angle of layer 1 is 30°).
Skin friction in sand layer 1 (A to B) $= K\sigma_v' \times \tan\delta \times (\pi \times d) \times L$
$= 1.25 \times (26) \times (\tan 22.5) \times (\pi \times 0.5) \times 3$
$= 63.4$ kN. (14.3 kips)

Step 4: Find the skin friction of the pile portion in *sand layer 1* below groundwater (B to C):
Skin friction term $= K\sigma_v' \times \tan\delta \times (\pi \times d) \times L$
Obtain the σ_v' (effective stress at the perimeter of the pile).
σ_v' (mid point) $= 3 \times \gamma_1 + 1.0\,(\gamma_1 - \gamma_w)$
σ_v' (mid point) $= 3 \times 17.3 + 1.0 \times (17.3 - 9.8)$ kN/m^2.
$= 59.4$ kN/m^2. (1241 psf)
Skin friction in sand layer 1 (B to C) $= K\sigma_p' \times \tan\delta \times (\pi \times d) \times L$
$= 1.25 \times 59.4 \times (\tan 22.5) \times (\pi \times 0.5) \times 2$
$= 96.6$ kN. (21.7 kips)

Step 5: Find the skin friction in *sand layer 2* below groundwater (C to D):
Skin friction term $= K\,\sigma'_v \times \tan\delta \times (\pi \times d) \times L$
Obtain the σ_p' (effective stress at the mid point of the pile).
σ_v' (mid point) $= 3 \times \gamma_1 + 2(\gamma_1 - \gamma_w) + 5(\gamma_2 - \gamma_w)$
σ_v' (mid point) $= 3 \times 17.3 + 2.0(17.3 - 9.8) + 5(16.9 - 9.8)$ kN/m^2.
$= 102.4$ kN/m^2. (2.14 ksf)
From Table 41.2, the skin friction angle (δ) for concrete piles is ¾ φ.
$\delta = ¾ \times 32^\circ = 24^\circ$. (Friction angle of layer 2 is 32°).
Skin friction in sand layer 2 (C to D) $= K\sigma_v' \times \tan\delta \times (\pi \times d) \times L$
$= 1.25 \times 102.4 \times (\tan 24) \times (\pi \times 0.5) \times 10$
$= 895.2$ kN. (201.2 kips)

$P_{ultimate}$ = end bearing capacity + skin friction in layer 1 (above GW) + skin friction in layer 1 (below GW) + skin friction in layer 2 (below GW)

End bearing capacity	= 785.2 kN
Skin friction in layer 1 (above GW) (A to B)	= = 63.4 kN
Skin friction in layer 1 (below GW) (B to C)	= = 96.6 kN
Skin friction in layer 2 (below GW) (C to D)	= = 895.2 kN
Total	= = 1,840 kN

$P_{ultimate} = 1840.4$ kN. (413.7 kips)
In this case, bulk of the pile capacity comes from end bearing and the skin friction in the bottom layer.
Hence, allowable bearing capacity of the pile $= P_{ultimate}/\text{FOS}$
$= 1840.4/3.0$
Allowable pile capacity $= 613.5$ kN. (137.9 kips)

41.2.3 Pile design using Meyerhoff equation: correlation with SPT (N)

41.2.3.1 End bearing capacity

Meyerhoff proposed the following equation based on SPT (N) value to compute the ultimate end bearing capacity of driven piles. Meyerhoff equation was adopted by DM 7.2 as an alternative method to static analysis.

$$q_{\text{ult}} = 0.4\, C_{\text{N}} \times N \times D / B$$

q_{ult}, ultimate point resistance of driven piles (tsf); N, standard penetration resistance (blows/ft.) near pile tip.

$$C_{\text{N}} = 0.77 \log 20 / p$$

p = Effective overburden stress at pile tip (tsf)

Effective stress "p" should be more than 500 psf. It is very rare for effective overburden stress at pile tip to be less than 500 psf.

D, depth driven into granular (sandy) bearing stratum (ft.); B, width or diameter of the pile (ft); q_1, limiting point resistance (tsf), equal to 4 N for sand and 3 N for silt

If the average SPT (N) value at tip of the pile 15, the maximum point resistance for sandy soil is 4 N = 60 tsf.

41.2.3.2 Modified Meyerhoff equation

Meyerhoff developed the equation to compute the ultimate end bearing capacity of driven piles using many available load test data and obtaining the average "N" values. Pile tip resistance is a function of the friction angle. For a given SPT (N) value, different friction angles are obtained for different soils.

For a given SPT (N) value, friction angle for coarse sand is 7–8% higher compared to medium sand. At the same time, for a given SPT (N) value, friction angle is 7–8% lower in find sand as compared to medium sand. Due to this reason, the author proposes the following modified equations:

$q_{\text{ult}} = 0.45\ C_{\text{N}} \times N\, D/B$ tsf (coarse sand)
$q_{\text{ult}} = 0.4\ C_{\text{N}} \times N\, D/B$ tsf (medium sand)
$q_{\text{ult}} = 0.35\ C_{\text{N}} \times N\, D/B$ tsf (fine sand)

Design Example 41.7

Find the tip resistance of the 2 ft. (0.609 m) diameter pile shown in Figure 41.11 using Meyerhoff equation. The SPT (N) value at pile tip is 25 blows per foot.

Solution

Step 1: Ultimate point resistance for driven piles for fine sand;

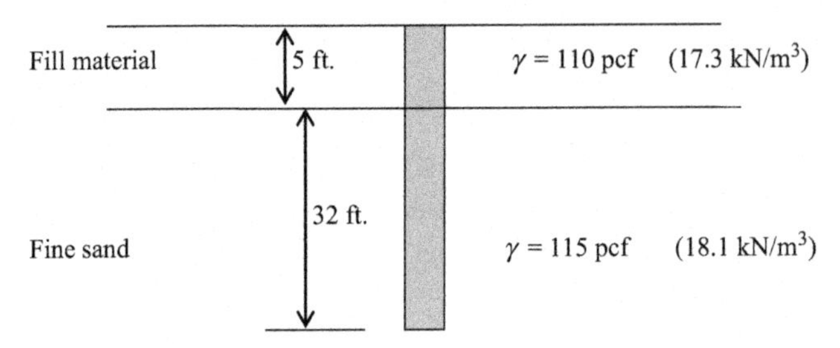

Figure 41.11 Use of Meyerhoff equation.

$q_{ult} = 0.35\ C_N \times N \times D/B$ tsf (fine sand)
$C_N = 0.77 \log 20/p$
p = effective overburden stress at pile tip (tsf)
$p = 5 \times 110 + 32 \times 115 = 4230$ psf = 2.11 tsf (0.202 MPa)
$C_N = 0.77 \log [20/2.115] = 0.751$
D = Depth driven into bearing stratum = 32 ft. (9.75 m)
Fill material is not considered to be a bearing stratum.
B = 2 ft. (Width or diameter of the pile) = 2 ft. (0.61 m)
$q_{ult} = 0.35\ C_N \times N \times D/B$ (fine sand)
$q_{ult} = 0.35 \times 0.751 \times 25 \times 32/2 = 105$ tsf (10.1 MPa)
Maximum allowable point resistance = 4 N for sandy soils
$4 \times N = 4 \times 25 = 100$ tsf
Hence, q_{ult} = 100 tsf (9.58 MPa)
Allowable point bearing capacity = 100/FOS
Assume a factor of safety of 3.0
Hence, $q_{allowable}$ = 33.3 tsf (3.2 MPa)
Total allowable point bearing capacity = $q_{allowable} \times$ Tip area = $q_{allowable} \times \pi \times (2^2)/4$
= 105 tons (934 kN)

Meyerhoff equations for skin friction:
Meyerhoff proposes the following equation for skin friction.
$f = N/50$ tsf
f = unit skin friction (tsf)
N = average SPT (N) value along the pile
Note: As per Meyerhoff, the unit skin friction "f" should not exceed 1 tsf.
The author proposes the following modified equations to account for soil gradation.
$f = N/46$ tsf, coarse sand
$f = N/50$ tsf, medium sand
$f = N/54$ tsf, fine sand

Design Example 41.8

Find the skin friction of the 2 ft. diameter pile shown in Figure 41.12 using Meyerhoff's equation. The average SPT (N) value along the shaft is 15 blows per foot.

Solution

Step 1: Ignore the skin friction in fill material.

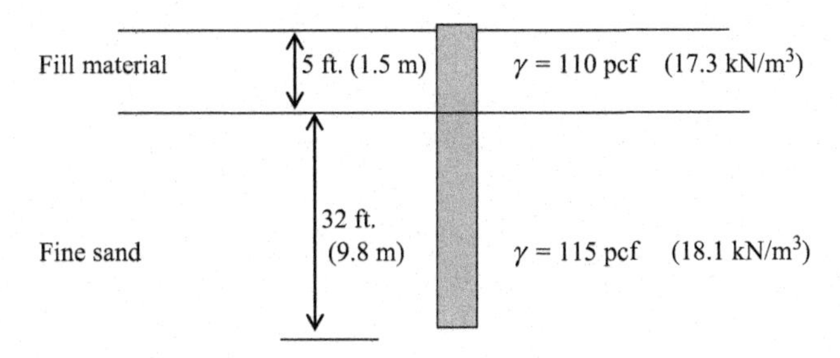

Figure 41.12 Skin friction using Meyerhoff equation.

For fine sand;
Unit skin friction; $f = N/54$ tsf
Unit skin friction; $f = 15/54$ tsf = 0.28 tsf (26.8 kN/m^2)
Total skin friction = unit skin friction × perimeter surface area
Total skin friction = $0.28 \times \pi \times D \times L$
Total skin friction = $0.28 \times \pi \times 2 \times 32 = 56$ tons
Allowable skin friction = 56/FOS
Assume a (FOS) factor of safety of 3.0
Allowable skin friction = 56/3.0 = 18.7 tons

41.3 Critical depth for skin friction (sandy soils)

Vertical effective stress (σ') increases with depth. Hence, the skin friction should increase with depth indefinitely. In reality, the skin friction *will not* increase with depth indefinitely.

It was believed in the past that skin friction would become a constant at a certain depth. This depth was named *critical depth.*

- As shown in Figure 41.13, the skin friction was assumed to increase till the critical depth and then maintain a constant value.

d_c, critical depth; S_c, skin friction at critical depth ($K\sigma'_c$. $\tan\delta$); σ'_c, effective stress at critical depth.

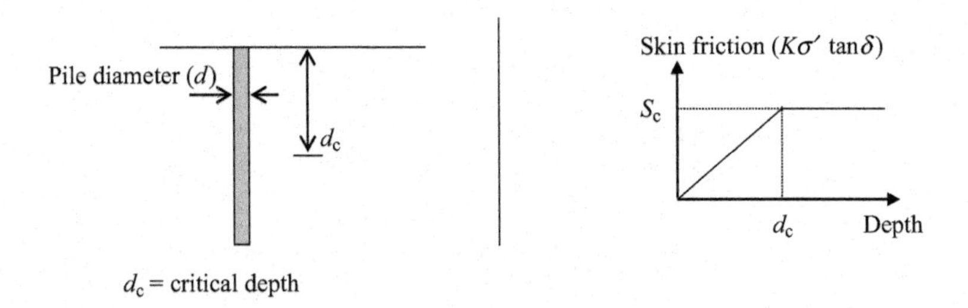

Figure 41.13 Critical depth for skin friction.

The following approximations were assumed for the critical depth.

- Critical depth for loose sand = 10*d* (*d* is the pile diameter or the width)
- Critical depth for medium dense sand = 15*d*
- Critical depth for dense sand = 20*d*

This theory does not explain the recent precise pile load test data. According to recent experiments, it is clear that skin friction will not become a constant abruptly as once believed.

41.3.1 Experimental evidence for critical depth

Figure 41.14 shows the typical variation of skin friction with depth in a pile.

- As one can see, the experimental data does not support the old theory with a constant skin friction below the critical depth.
- The skin friction tends to increase with depth. It attains its maximum value just above the tip of the pile. Skin friction would drop rapidly after that.
- Skin friction does not increase linearly with depth as once believed.
- It should be noted here that no satisfactory theory exists at the present to explain the field data.
- Due to the lack of a better theory, the critical depth theory of the past is still being used by engineers.

41.3.2 Reasons for limiting skin friction

The following reasons have been put forward to explain why skin friction does not increase with the depth indefinitely as suggested by the skin friction equation.

Unit skin friction = $K\sigma' \tan\delta$; $\sigma' = \gamma d$

1. The above "K" value is a function of the soil friction angle (ϕ'). The friction angle tends to decrease with depth. Hence "*K*" value decreases with depth (Kulhawy et al., 1983).
2. The above skin friction equation does not hold true at high stress levels due to the readjustment of sand particles.

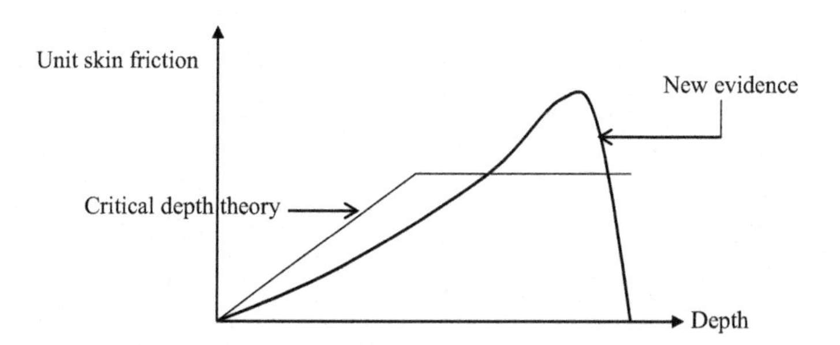

Figure 41.14 Skin friction versus depth.
Source: Randolph et al., 1994.

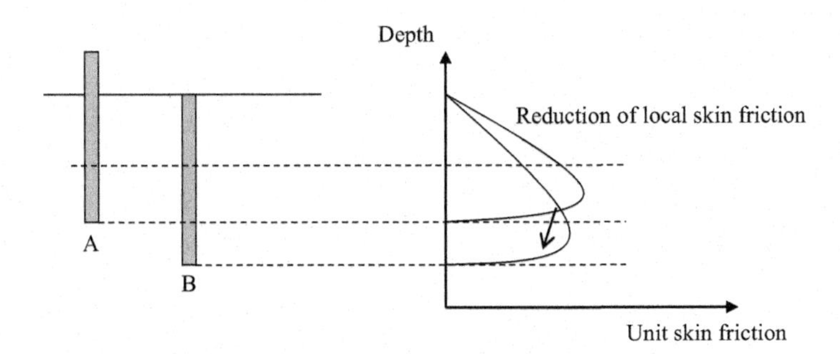

Figure 41.15 Depth versus unit skin friction.

3. There is a reduction of local shaft friction with increasing pile depth (Figure 41.15) (Randolph et al., 1994).

   Assume a pile was driven to a depth of 10 ft. and the unit skin friction was measured at a depth of 5 ft. Then assume that the pile was driven further to a depth of 15 ft. and the unit skin friction was measured at the same depth of 5 ft. It has been reported that unit skin friction at 5 ft. is less in the second case.

According to Figure 41.15, local skin friction reduces when the pile is driven further into the ground.

As per NAVFAC DM 7.2, the limiting value of skin friction and end bearing capacity is achieved after 20 diameters within the bearing zone.

41.4 Critical depth for end bearing capacity (sandy soils)

- Pile end bearing capacity in sandy soils is related to effective stress. Experimental data indicates that end bearing capacity does not increase with depth indefinitely. Due to the lack of a valid theory, engineers use the same critical depth concept adopted for skin friction for end bearing capacity as well.
- As shown in Figure 41.16, the end bearing capacity was assumed to increase till the critical depth.

  d_c, critical depth; Q_c, end bearing at critical depth; σ'_c, effective stress at critical depth.

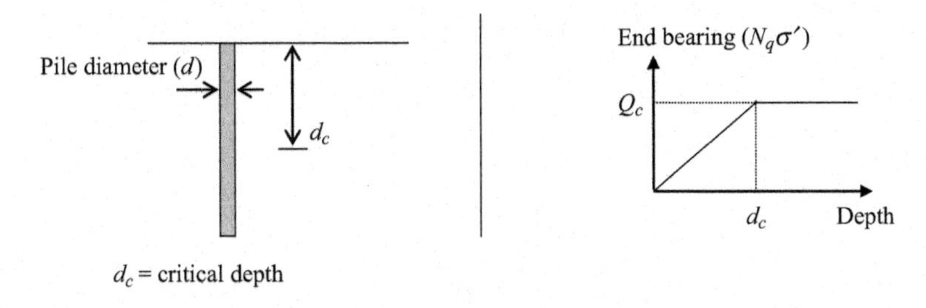

Figure 41.16 Critical depth for end bearing capacity.

The following approximations were assumed for the critical depth.

- Critical depth for loose sand = 10*d* (*d* is the pile diameter or the width)
- Critical depth for medium dense sand = 15*d*
- Critical depth for dense sand = 20*d*

Critical depth for end bearing is same as the critical depth for skin friction?

Since critical depth concept is a gross approximation that cannot be supported by experimental evidence, the question is irrelevant. It is clear that there is a connection between end bearing capacity and skin friction since same soil properties act in both cases such as effective stress, friction angle, and relative density. On the other hand, it should be noted that the two processes are vastly different in nature.

41.4.1 Critical depth example

Design Example 41.9

Find the skin friction and end bearing capacity of the pile shown in Figure 41.17. Assume that the critical depth is achieved at 20 ft. into the bearing layer (NAVFAC DM 7.2, 1984). The pile diameter is 1 ft. and other soil parameters are as shown in Figure 41.17.

Solution

The skin friction is calculated in the overburden. In this case, skin friction is calculated in the soft clay. Then, the skin friction is calculated in the bearing layer (medium sand) assuming the skin friction attain a limiting value after 20 diameters (critical depth) (Figure 41.18).

Step 1: Find the skin friction from A to B

$$\text{Skin friction in soft clay} = \alpha \times c \times \text{perimeter surface area}$$
$$= 0.4 \times 700 \times \pi \times d \times L$$
$$= 0.4 \times 700 \times \pi \times 1 \times 12 = 10{,}560 \text{ lbs } (46.9 \text{ kN})$$

Step 2: Find the skin friction from B to C:

Skin friction in sandy soils = $S = K\sigma'_v \times \tan\delta \times A_p$

S = skin friction of the pile; σ_v' = average effective stress along the pile shaft.

Average effective stress along pile shaft from B to C = $(\sigma_B + \sigma_C)/2$

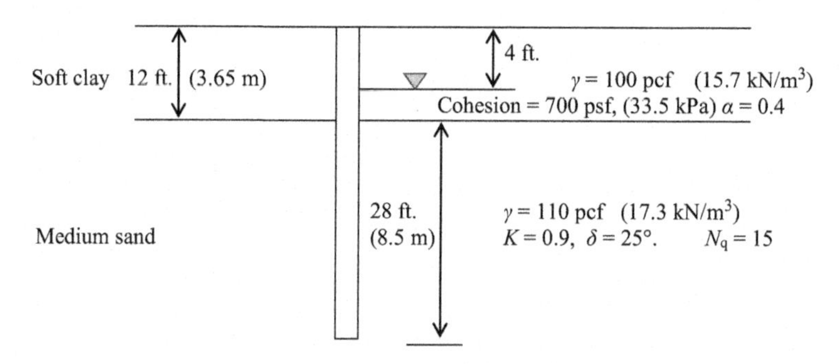

Figure 41.17 Critical depth in medium sand.

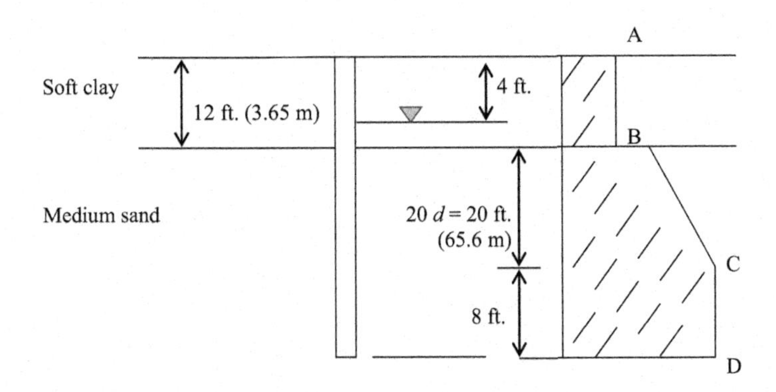

Figure 41.18 Development of skin friction.

σ_B = effective stress at B; σ_C = effective stress at C.
To obtain the average effective stress from B to C, find the effective stresses at B and C and obtain the average of those two values.

$$\sigma_B = 100 \times 4 + (100 - 62.4) \times 8 = 700.8 \text{ lbs / ft.}^2 \quad (33.6 \text{ kPa})$$
$$\sigma_C = 100 \times 4 + (100 - 62.4) \times 8 + (110 - 62.4) \times 20 = 1452.8 \text{ lbs / ft.}^2 \quad (69.5 \text{ kPa})$$

Average effective stress along pile shaft from B to C = $(\sigma_B + \sigma_C)/2$
= (700.8 + 1452.8)/2 = 1076.8 lbs/ft.^{2}.

Skin friction from B to C = $K\sigma'_v \times \tan\delta \times A_p$
= $0.9 \times 1076.8 \times \tan(25^\circ) \times (\pi \times 1 \times 20)$ = 28,407 lbs

Step 3: Find the skin friction from C to D:
Skin friction reaches a constant value at point "C", 20 diameters into the bearing layer.
Skin friction at point "C" = $K\sigma'_v \times \tan\delta \times A_p$
σ'_v at point "C" = $100 \times 4 + (100 - 62.4) \times 8 + (110 - 62.4) \times 20$ = 1452.8 lbs/ft.^{2}
Unit skin friction at point "C" = $0.9 \times 1452.8 \times \tan25^\circ$ = 609.7 lbs/ft.^{2}. (29 kPa)
Unit skin friction is constant from C to D. This is due to the fact that skin friction does not increase after the critical depth.
Skin friction from "C" to "D" = 609.7 × surface perimeter area
= $609.7 \times (\pi \times 1 \times 8)$ lbs = 15323.4 lbs (68.2 kN)

Summary
Skin friction in soft clay (A to B) = 10,560 lbs
Skin friction in sand (B to C) = 28,407 lbs
Skin friction in sand (C to D) = 15,323 lbs
Total = 54,290 lbs (241 kN)
Step 4: Compute the end bearing capacity;
End bearing capacity also reaches a constant value below the critical depth.
End bearing capacity = $q \times N_q \times A$
q, effective stress at pile tip; N_q, bearing capacity factor (given to be 15); A, cross sectional area of the pile.
If the pile tip is below the critical depth, "q" should be taken at critical depth. In this example, the pile tip is below the critical depth, which is 20 diameters into the bearing layer. Hence "q" is equal to the effective stress at critical depth (point C).

Effective stress at point "C" = $100 \times 4 + (100 - 62.4) \times 8 + (110 - 62.4) \times 20 = 1452.8$ lbs/ft.^{2}
End bearing capacity = $q \times N_q \times A = 1,452.8 \times 15 \times (\pi \times d^2/4)$ lbs = 17,115 lbs
Total ultimate capacity of the pile = Total skin friction + end bearing
= 54,290 + 17,115 = 71,405 lbs = 35.7 tons (317.6 kN).

References

American Petroleum Institute, 1984. Recommended Practice for Planning, Designing and Constructing Fixed Offshore Platforms, fifteenth ed. API RP2A.

Kraft, L.M., Lyons, C.G., State of the Art: Ultimate Axial Capacity of Grouted Piles. In: Proc. Sixth Annual Offshore Technology Conference, Houstan paper OTC 2081, 487–503.

Kulhawy, F.H., 1984. Limiting Tip and Side Resistance, Analysis and Design of Pile Foundations. ASCE, NY.

Kulhawy, F.H., et al.,1983. Transmission Line Structure Foundations for Uplift-Compression Loading. Electric Power Research Institute, Palo Alto, Report EL. 2870.

Martin, et al.,1987. Concrete pile design in tidewater. J Geotechnical Eng. 113 (6), 6–15.

McClelland, B., 1974. Design of deep penetration piles for ocean structures. J. Geotechnical Eng. GT7, 705–747.

Meyerhoff, G.G., 1976. Bearing capacity and settlement of pile foundations. J. Geotechnical Eng. GT3, 195–228.

NAVFAC DM 7.2 – Foundation and Earth Structures, Department of the Navy, 1984.

Randolph, M.F., Dolwin, J., Beck, R., 1994. Design of driven piles in sand. Geotechnique 44 (3), 427–448.

Pile design in clay soils

42

Figure 42.1 shows skin friction in clay soil.

There are two types of forces acting on piles.

1. End bearing acts on the bottom of the pile.
2. Skin friction acts on the sides of the pile.

To compute the total load that can be applied to a pile, one needs to compute the end bearing and the skin friction acting on sides of the pile.

The modified Terzaghi's bearing capacity equation is used to find the pile capacity

$$P_u = Q_u + S_u$$
$$P_u = 9cA_c + \alpha c A_p$$

P_u, ultimate pile capacity; Q_u, ultimate end bearing capacity; S_u, ultimate skin friction; c, cohesion of the soil; A_c, cross sectional area of the pile; A_p, perimeter surface area of the pile, α, adhesion factor between pile and soil.

As per the modified Terzaghi's bearing capacity equation, clay soils with a higher cohesion would have a higher end bearing capacity. End bearing capacity of piles in clayey soils is usually taken to be 9c.

Skin friction: Assume that a block is placed on a clay surface. Now, if a force is applied to move the block, the adhesion between the block and the clay would resist the movement. If the cohesion of clay is "c", the force due to adhesion would be "αc" (α is the adhesion coefficient) (Figure 42.2).

α, adhesion coefficient depending on pile material and clay type; c, cohesion.

Highly plastic clays would have a higher adhesion coefficient. Typically, it is assumed that adhesion is not dependent on the weight of the block. This is not strictly true as explained later.

Now let's look at a pile outer surface (Figure 42.3).

Ultimate skin friction = $S_u = \alpha \times c\, A_p$

Usually, it is assumed that ultimate skin friction is independent of the effective stress and depth. In reality, skin friction is dependent on effective stress and cohesion of soil.

The skin friction acts on the perimeter surface of the pile.

For a circular pile, surface area of the pile is given by πDL

$(\pi \times D)$ = circumference of the pile; L = length of the pile.

Perimeter surface area of a circular pile $(A_p) = \pi \times D \times L$

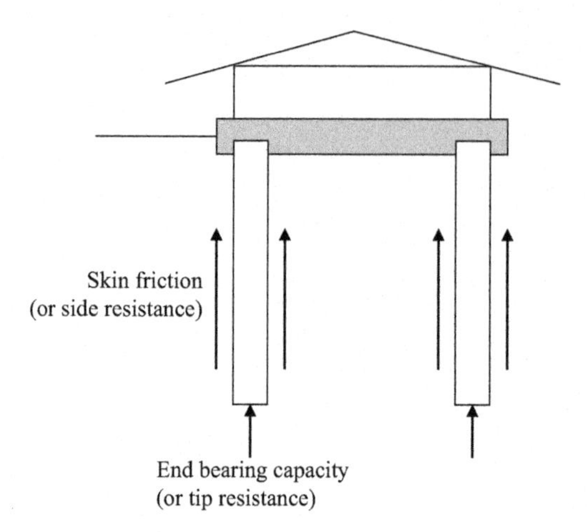

Figure 42.1 Skin friction in clay soil.

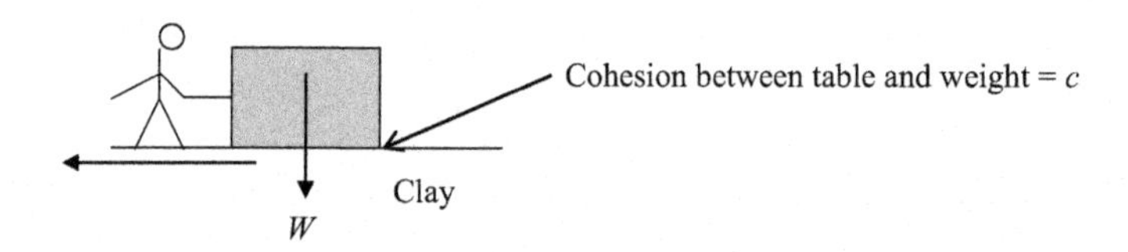

Figure 42.2 Cohesive forces.

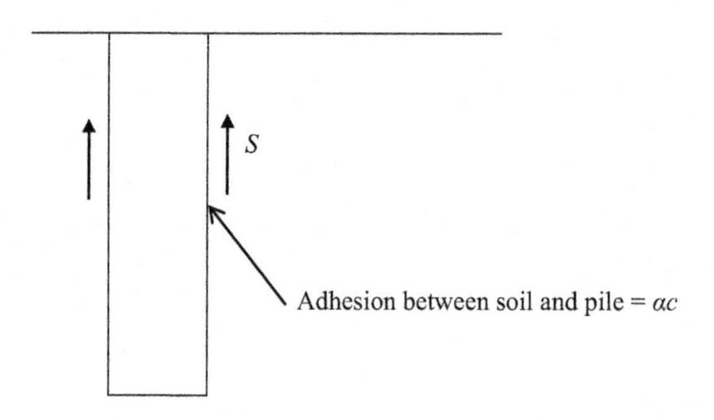

Figure 42.3 Adhesion.

42.1 End bearing capacity in clay soils (different methods)

42.1.1 Driven piles

Skempton (1959): The equation proposed by Skempton is widely being used to find the end bearing capacity in clay soils.

$$q = 9C_u$$

q, end bearing capacity; C_u, cohesion of soil at the tip of the pile.

Martin et al. (1987):

$$q = C \times N \text{ MN/m}^2$$

$C = 0.20$; N = SPT value at pile tip.

42.1.2 Bored piles

Shioi and Fukui (1982):

$$q = C \times N \text{ MN/m}^2$$

$C = 0.15$; N = SPT value at pile tip.

NAVFAC DM 7.2 (1984):

$$q = 9C_u$$

q, end bearing capacity; C_u, cohesion of soil at the tip of the pile.

Equations based on undrained shear strength (cohesion) $\rightarrow f_{ult} = \alpha \times S_u$

f_{ult}, ultimate skin friction; α, skin friction coefficient; S_u, undrained shear strength or cohesion; S_u, $Q_u/2$ (Q_u = unconfined compressive strength).

These type of equations ignore effective stress effects (Figure 42.4).

42.1.3 Driven Piles

American Petroleum Institute (1984): API provides the following equation to find the skin friction in clay soils.

$$f = \alpha C_u$$

f, unit skin friction; C_u = cohesion

$\alpha = 1.0$, for clays with $C_u < 25$ kN/m^2 (522 psf)

$\alpha = 0.5$, for clays with $C_u > 70$ kN/m^2 (1460 psf)

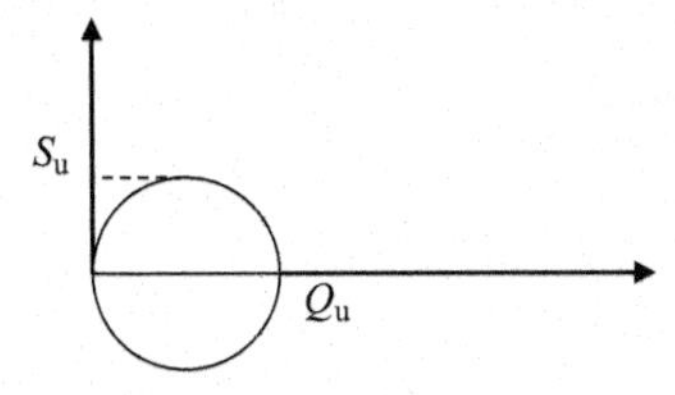

Figure 42.4 Mohr's circle diagram.

Table 42.1 Pile type, cohesion, and adhesion

Pile type	Soil consistency	Cohesion range (kN/m²)	α
Timber and concrete piles	Very soft	0–12	0–1.0
	Soft	12–24	1.0–0.96
	Medium stiff	24–48	0.96–0.75
	Stiff	48–96	0.75–0.48
	Very stiff	96–192	0.48–0.33
Steel piles	Very soft	0–12	0.0–1.0
	Soft	12–24	1.0–0.92
	Medium stiff	24–48	0.92–0.70
	Stiff	48–96	0.70–0.36
	Very stiff	96–192	0.36–0.19

Source: NAVFAC DM 7.2 (1984).

Interpolate for the "α" value for cohesion values between 25 kN/m^2 and 70 kN/m^2

As per the API method, the skin friction is solely dependent on cohesion. Effective stress changes with depth and API method disregards the effective stress effects in soil.

NAVFAC DM 7.2 (1984):

$$f = \alpha C_u A_p$$

S, skin friction; C_u, cohesion; A_p, perimeter surface area of the pile.

Table 42.1 shows pile type, cohesion, and adhesion.

As in the API method, effective stress effects are neglected in DM 7.2 method.

42.1.4 Bored piles

Fleming et al. (1985):

$$f = \alpha C_u$$

f = unit skin friction; C_u = cohesion.

$\alpha = 0.7$, for clays with $C_u < 25$ kN/m^2 (522 psf)

$\alpha = 0.35$, for clays with $C_u > 70$ kN/m^2 (1460 psf)

Note: "α" value for bored piles is chosen to be 0.7 times the value for driven piles.

Equations based on vertical effective stress $\rightarrow f_{ult} = \beta \times \sigma'$

f_{ult}, ultimate skin friction; β, skin friction coefficient based on effective stress; σ', effective stress.

These type of equations ignore cohesion effects.

Burland (1973):

$$f = \beta \sigma'_v$$

f, unit skin friction; σ_v', effective stress; $\beta = (1 - \sin\phi')\tan\phi'\,(\mathrm{OCR})^{0.5}$

OCR = over consolidation ratio of clay

The Burland method does not consider the cohesion of soil. One can argue that OCR is indirectly related to cohesion.

Equation based on both cohesion and effective stress:

A new method was proposed by Kolk and Van der Velde (1996) that considers both cohesion and effective stress to compute the skin friction of piles in clay soils.

Kolk and Van der Velde (1996): *Kolk and Van der Velde* method, considers both cohesion and effective stress.

$$f_{\mathrm{ult}} = \alpha \times S_{\mathrm{u}}$$

f_{ult} = ultimate skin friction; α = skin friction coefficient obtained using the correlations provided by Kolk and Van der Velde. α is a parameter based on both cohesion and effective stress; S_u, undrained shear strength (cohesion).

"α" in the Kolk and Van der Velde equation is based on the ratio of undrained shear strength and effective stress. Large database of pile skin friction results were analyzed and correlated to obtain α values.

According to Table 42.2; "α" decreases when "S_u" increases and "α" increases when σ' (effective stress) increases.

Design Example 42.1

Find the skin friction of the 1 ft. diameter pile shown in Figure 42.5 using the Kolk and Van der Velde method.

Solution

The Kolk and Van der Velde method depends on both effective stress and cohesion. Effective stress varies with the depth. Hence, obtain the average effective stress along the pile length. In this case, obtain the effective stress at midpoint of the pile.

Effective stress at the midpoint of the pile = 110 × 7.5 psf = 825 psf (39.5 kPa)

Cohesion = S_u = 1000 psf (47.88 kPa)

S_u/σ' = 1000/825 = 1.21

From Table 42.2, provided by Kolk and Van der Velde, α = 0.5, for S_u/σ' = 1.2.

Hence, use α = 0.5

Ultimate unit skin friction = $\alpha \times S_u$ = 0.5 × 1000 = 500 psf (23.9 kPa)

Ultimate skin friction of the pile = 500 × ($\pi \times d \times L$) = 500 × π × 1 × 15 lbs = 23,562 lbs (104.8 kN)

Table 42.2 Skin friction factor

S_u/σ'	0.2	0.3	0.4	0.5	0.6	0.7	0.8	0.9	1.0	1.1	1.2	1.3	1.4
α	0.95	0.77	0.7	0.65	0.62	0.60	0.56	0.55	0.53	0.52	0.50	0.49	0.48
S_u/σ'	1.5	1.6	1.7	1.8	1.9	2.0	2.1	2.2	2.3	2.4	2.5	3.0	4.0
α	0.47	0.42	0.41	0.41	0.42	0.41	0.41	0.40	0.40	0.40	0.4	0.39	0.39

Source: Kolk and Van der Velde (1996).

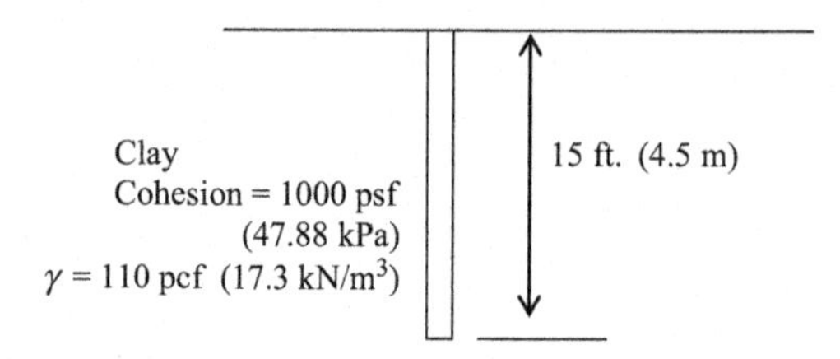

Figure 42.5 Skin friction in clay.

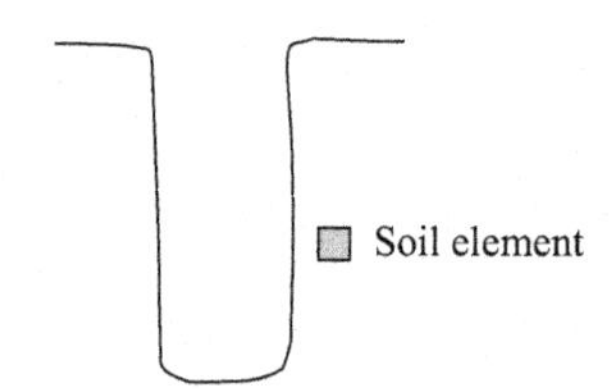

Figure 42.6 Drilling in soil.

Skin friction using undrained shear strength of clay:

As mentioned in previous chapters, attempts to correlate skin friction with undrained shear strength were not very successful. Better correlation was found with skin friction and (S_u/σ') ratio. "S_u" being the undrained shear strength and σ' being the effective stress.

What would happen to a soil element when a hole is drilled (Figure 42.6)?

The following changes occur in a soil element near the wall after the drilling process:

- The soil element shown would be subjected to a stress relief. Undrained shear strength would reduce due to the stress relief.
- Reduction of undrained shear strength would reduce the skin friction as well.

Computation of skin friction in bored piles:

- The same procedure used for driven piles could be used for bored piles, but with a lesser undrained shear strength. The question is how much reduction should be applied to the undrained shear strength for bored piles?
- Undrained shear strength may reduce as much as 50% due to the stress relief in bored piles. On the other hand, measured undrained shear strength is already reduced due to the stress relief occurring when the sample was removed from the ground. By the time, the soil sample reaches the laboratory, soil sample has undergone stress relief and the measured value already indicates the stress reduction. Considering these two aspects, reduction of 30% in the undrained shear strength would be realistic for bored piles.

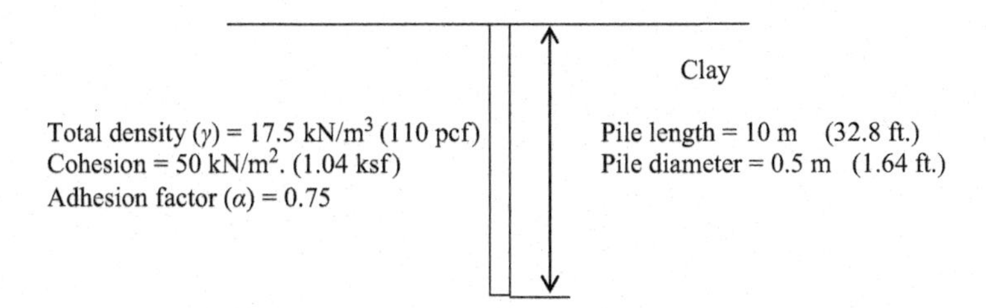

Figure 42.7 Single pile in clay soil.

Design Example 42.2: Single pile in a uniform clay layer

Find the capacity of the pile shown in Figure 42.7. The length of the pile is 10 m and the diameter of the pile is 0.5 m. Cohesion of the soil is found to be 800 psf. Note the groundwater level is at 2 m below the surface.

Solution

Step 1: Find the end bearing capacity.

End bearing capacity in clay soils = $9cA$.

c = cohesion = 50 kN/m^2,

$N \times c = 9$,

$A = \pi \times D^2/4 = \pi \times 0.5^2/4$ m^2 = 0.196 m^2

Ultimate end bearing capacity = 9 × 50 × 0.196 = 88.2 kN. (19.8 kips)

Step 2: Find the skin friction.

Ultimate skin friction = $S_u = \alpha \times c \times A_p$

Ultimate skin friction = 0.75 × 50 × A_p

A_p = perimeter surface area of the pile = $\pi \times D \times L = \pi \times 0.5 \times 10$ m^2.

A_p = 15.7 m^2

Ultimate skin friction = 0.75 × 50 × 15.7 = 588.8 kN. (132 kips)

Step 3: Find the ultimate capacity of the pile

Ultimate pile capacity = ultimate end bearing capacity + ultimate skin friction

Ultimate pile capacity = 88.2 + 588.8 = 677 kN. (152 kips)

Allowable pile capacity = ultimate pile capacity/FOS

Assume a factor of safety of 3.0.

Allowable pile capacity = 677/3.0 = 225.7 kN. (50.6 kips)

Note: The skin friction was very much higher than the end bearing in this situation.

Design Example 42.3: Single pile in a uniform clay layer with groundwater present

Find the allowable capacity of the pile shown in Figure 42.8. Pile diameter is given to be 1 m and the cohesion of the clay layer is 35 kPa. Groundwater is 2 m below the surface. Find the allowable capacity of the pile.

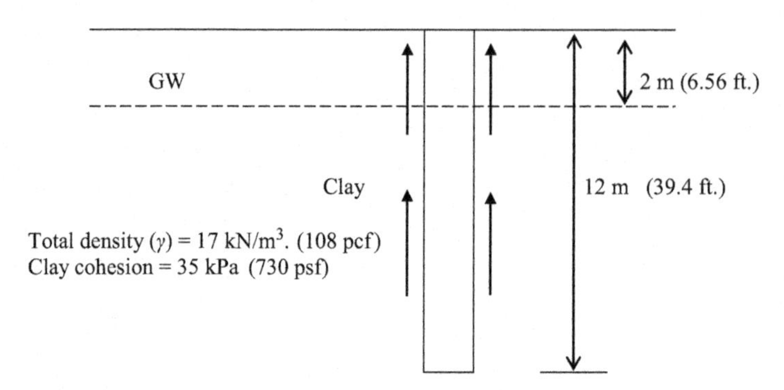

Figure 42.8 Skin friction (groundwater considered).

Solution

Unlike sands, groundwater does not affect the skin friction in clayey soils.

Step 1: Find the end bearing capacity.

End bearing capacity in clay soils = $9 \times c \times A$.

c = cohesion = 35 kN/m^2,

$N_c = 9$,

$A = \pi \times D^2/4 = \pi \times 1^2/4\ \text{m}^2 = 0.785\ \text{m}^2$

Ultimate end bearing capacity = $9 \times 35 \times 0.785 = 247.3$ kN. (55.5 kips)

Step 2: Find the skin friction.

Ultimate skin friction = $S_u = \alpha \times c \times A_p$

Adhesion factor α, is not given. Use the method given by American Petroleum Institute (API).

$\alpha = 1.0$, for clays with cohesion = < 25 kN/m^2.

$\alpha = 0.5$, for clays with cohesion = > 70 kN/m^2

Since, the cohesion is 35 kN/m^2 interpolate to obtain the adhesion factor (α).

25	1.0
35	X
70	0.5

$$(X - 1.0)/(35 - 25) = (X - 0.5)/(35 - 70)$$
$$(X - 1.0)/10 = (X - 0.5)/-35$$
$$-35X + 35 = 10X - 5$$
$$45X = 40$$
$$X = 0.89$$

Hence, α at 35 kN/m^2 is 0.89.

Ultimate skin friction = $0.89 \times 35 \times A_p$

A_p = perimeter surface area of the pile = $\pi \times D \times L = \pi \times 1.0 \times 12\ \text{m}^2$.

$A_p = 37.7\ \text{m}^2$

Ultimate skin friction = $0.89 \times 35 \times 37.7$ kN.

Ultimate skin friction = 1174.4 kN (264 kips)

Step 3: Find the ultimate capacity of the pile

Ultimate pile capacity = ultimate end bearing capacity + ultimate skin friction

Ultimate pile capacity = 247.3 + 1174.4 = 1421.7 kN.

Allowable pile capacity = ultimate pile capacity/FOS
Assume a factor of safety of 3.0.
Allowable pile capacity = 1421.7/3.0 = 473.9 kN. (106.6 kips)

Design Example 42.4: Computation of the skin friction using the Kolk and Van der Velde method

A 3 m sand layer is underlain by a clay layer with cohesion of 25 kN/m^2. Find the skin friction of the pile within the clay layer. Use the Kolk and Van der Velde method. Density of both sand and clay are 17 kN/m^3. Diameter of the pile is 0.5 m (Figure 42.9).

Solution

Find the skin friction at the top of the clay layer and bottom of the clay layer. Obtain the average of the two values.

Step 1:

At point A: $\sigma' = 3 \times 17 = 51$ kN/m^2; (1065 $lbs/ft.^2$)
$S_u = 25$ kN/m^2;
$S_u/\sigma' = 25/51 = 0.50$
From the Kolk and Van der Velde table; $\alpha = 0.65$
At point B: $\sigma' = 6 \times 17 = 102$ kN/m^2.
$S_u = 25$ kN/m^2;
$S_u/\sigma' = 25/102 = 0.25$
From the Kolk and Van der Velde table; $\alpha = 0.86$

Step 2:

Ultimate skin friction at point A:
$f_{ult} = \alpha \times S_u = 0.65 \times 25 = 16$ kN/m^2
Ultimate skin friction at point B:
$f_{ult} = \alpha \times S_u = 0.86 \times 25 = 21$ kN/m^2
Assume the average of two points to obtain the total skin friction.
Average = (16 + 21)/2 = 18.5 kN/m^2.
Total skin friction = 18.5 × perimeter × length = 18.5 × ($\pi \times$ 0.5) × 3 = 87 kN (19.6 kips)

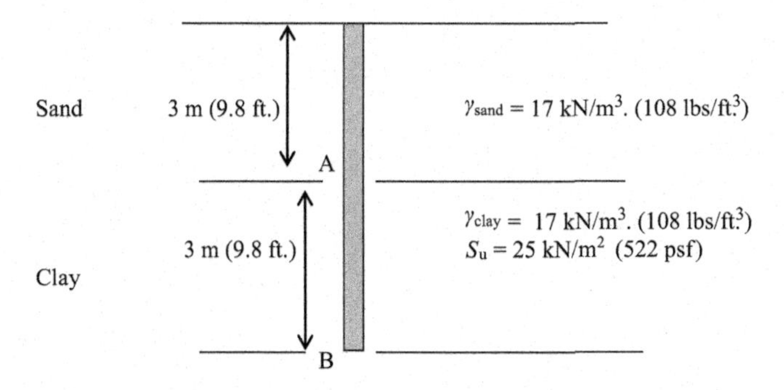

Figure 42.9 Pile in multiple layers.

Table 42.3 Summary of equations

	Sand	Clay
Pile end bearing capacity	$N_q \times \sigma_v \times A$	$9cA$
Pile unit skin friction	$K \times \sigma_v \times \tan\delta \times A_p$	$\alpha c A_p$

Summary of equations:

Table 42.3 shows a summary of equations.

A, bottom cross-sectional area of the pile; A_p, perimeter surface area of the pile; σ_v, vertical effective stress; N_q, Terzaghi's bearing capacity factor; c, cohesion of the soil; K, lateral earth pressure coefficient; α, adhesion factor; δ, soil and pile friction angle.

42.2 Case study – foundation design options

D'Appolonia and Lamb (1971) describe the construction of several buildings at MIT. The soil conditions of the site are given in Figure 42.10. Different foundation options were considered for the building.

General soil conditions:

Note: All clays start as normally consolidated clays. Over years consolidated clays had been subjected to higher pressures in the past than existing *in situ* pressures. This happens mainly due to glacier movement, fill placement, and groundwater change. On the other hand, normally consolidated soils are presently experiencing the largest pressure it had ever experienced. Due to this reason, normally consolidated soils tend to settle more than over consolidated clays.

Foundation option 1: Shallow footing placed on compacted backfill (Figure 42.11):

Construction procedure:

- Organic silt was excavated to the sand and gravel layer.
- Compacted backfill was placed and footing was constructed.
- This method was used for light loads.

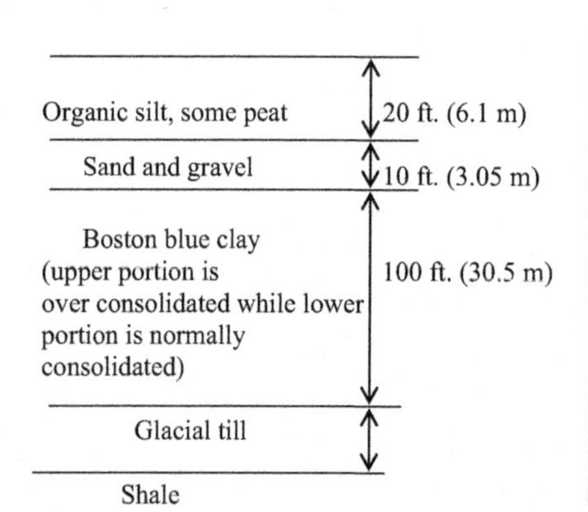

- Organic silt was compressible and can not be used for shallow foundations.
- Sand and gravel was medium dense and can be used for shallow foundations.
- Upper portion of Boston blue clay was over consolidated. Lower portion was normally consolidated.
- Normally consolidated clays settle appreciably more than over consolidated clays.
- Glacial till can be used for end bearing piles.

Figure 42.10 General soil conditions.

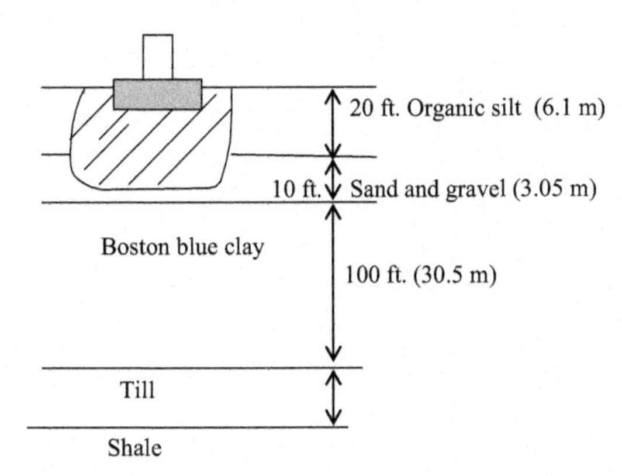

Figure 42.11 Shallow footing option.

- It has been reported that Boston blue clay would settle by more than 4 in. when subjected to a stress of 400 psf. (Aldrich, 1952). Hence, engineers had to design the footing, so that the stress on Boston blue clay was less than 400 psf.

Foundation option 2: Timber piles ending on sand and gravel layer (Figure 42.12)
Construction procedure:

- Foundations were placed on timber piles ending in sand and gravel layer.
- Engineers had to make sure that underlying clay layer was not stressed excessively due to piles.

This option was used for light loads.
Foundation option 3: Timber piles ending in Boston blue clay layer (Figure 42.13)
Construction procedure:

- Pile foundations were designed on timber piles ending in Boston blue clay.
- This method was found to be a mistake, since huge settlements occurred due to consolidation of the clay layer.

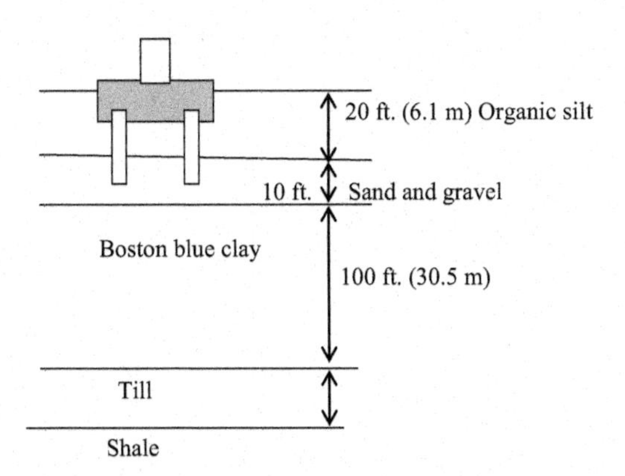

Figure 42.12 Short pile option.

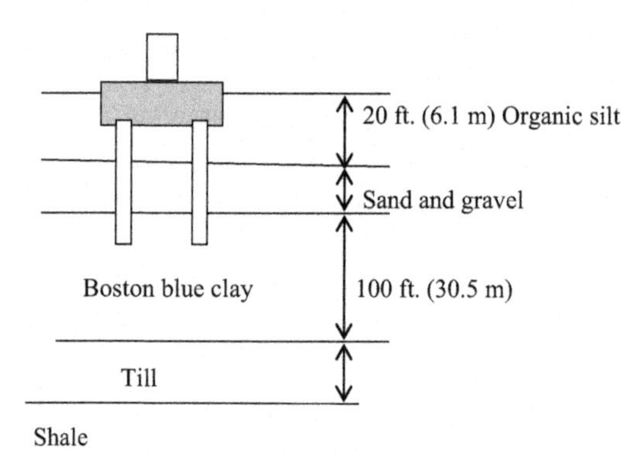

Figure 42.13 Medium length piles.

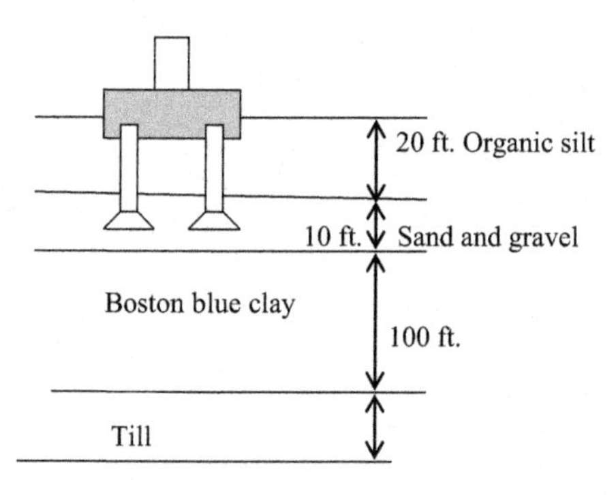

Figure 42.14 Short-belled caissons.

Foundation option 4: Belled piers ending in sand and gravel (Figure 42.14).

Foundations were placed on belled piers ending in sand and gravel layer. It is not easy to construct belled piers in sandy soils. This option had major construction difficulties.

Foundation option 5: Deep piles ending in till or shale (Figure 42.15).

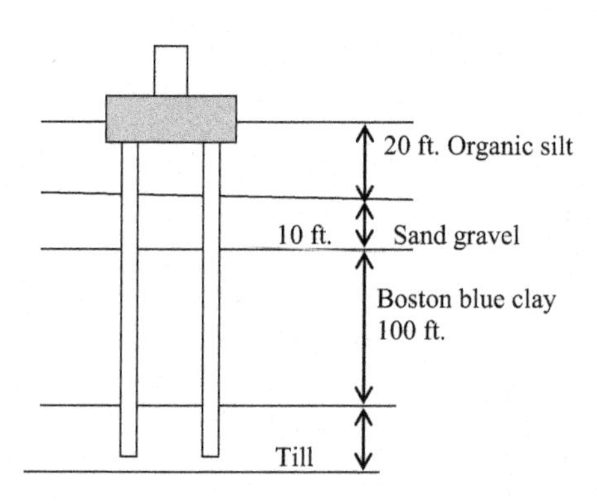

Figure 42.15 Deep piles.

- Foundations were placed on deep concrete pipe piles ending in till.
- These foundations were used for buildings with 20–30 stories.
- Their performance was found to be excellent. Settlement readings in all buildings were less than 1 in.
- Closed end concrete filled pipe piles were used. These piles were selected over H-piles due to their lower cost.
- During pile driving, adjacent buildings underwent slight upheaval. After completion of driving, buildings started to settle. (Great settlement in adjacent buildings within 50 ft. of piles was noticed).
- Measured excess pore water pressures exceeded 40 ft. of water column, 15 ft. away from the pile. Excess pore pressures dropped significantly after 10–40 days.
- Due to upheaval of buildings, some piles were preaugured down to 15 ft., prior to driving. Preauguring reduced the generation of excess pore water pressures. In some cases, excess pore water pressures were still unacceptably high.
- Piles in a group were not driven at the same time. After, one pile was driven, sufficient time was allowed for the pore water pressures to dissipate prior to driving the next pile.
- Another solution was to drive the pipe piles open end and then cleanout the piles. This was found to be costly and time consuming.

Foundation option 6: Floating foundations placed on sand and gravel (rafts) (Figure 42.16).

- Mat or floating foundations were placed on sand and gravel.
- Settlement of floating foundations were larger than deep piles (Option 5).
- Settlement of floating foundations varied from 1.0 in. to 1.5 in. Performance of rafts was inferior to deep piles. Average settlements in raft foundations were slightly higher than deep footings.
- There was a concern that excavations for rafts could create settlements in adjacent buildings. This found out to be a false alarm since adjacent buildings did not undergo any major settlement due to braced excavations constructed for raft foundations.
- When the cost of driving deep piles was equal to the cost of raft foundations, it is desirable to construct raft foundations, since rafts would give a basement.
- Basement may not be available in the piling option, unless it is specially constructed with additional funds (Table 42.4).

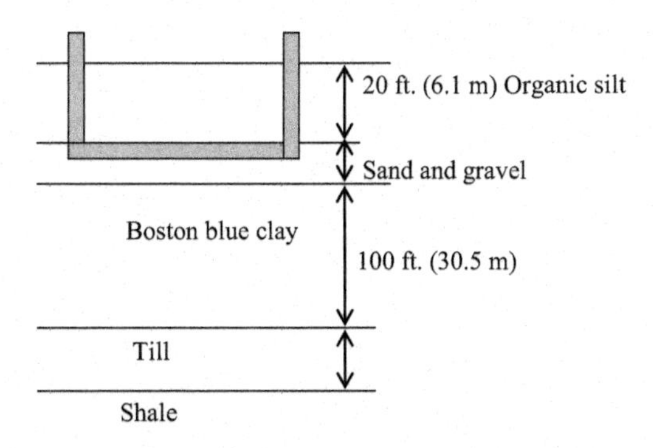

Figure 42.16 Floating foundation.

Table 42.4 **Settlement of buildings**

Foundation type	Bearing stratum	Number of buildings	Average number of stories	Settlement
Timber piles	Boston blue clay	27	1–6	1–16 in.
Belled caissons	Sand and gravel	22	1–8	1–3 in.
Raft foundations	Sand and gravel	7	6–20	1–1.5 in.
Deep piles	Shale or till	5	6–30	0.5

References

Aldrich, H.P., 1952. Importance of the Net Load to Settlement of Buildings in Boston. Contributions of Soil Mechanics, Boston Society of Civil engineers, New York.

American Petroleum Institute, 1984. Recommended Practice for Planning, Designing and Constructing Fixed Offshore Platforms", fifteenth ed. API RP2A.

Burland, J.B., 1973. Shaft friction of piles in clay – a simple fundamental approach. Ground Eng. 6 (3), 30–42.

D'Appolonia, D.J., Lamb, W.T., 1971. Performance of four foundations on end bearing piles. J. Soil Mech. Found. Eng.

Fleming, et al., 1985. Piling Engineering. Surrey University Press, New York.

Kolk, H.J., Van der Velde, A., 1996. A reliable method to determine friction capacity of piles driven into clays. Houston, TX. Proceedings of the Offshore Technological Conference, vol. 2.

Martin, et al., 1987. Concrete pile design in tidewater. J. Geotech. Eng. 113 (6).

NAVFAC DM 7.2., 1984. Foundation and Earth Structures, US Department of Navy.

Shioi, Y., Fukui, J., 1982. Application of "N" Value to Design of Foundations in Japan. Proc. ESOPT2, Amsterdam 1, 159–164.

Skempton, A.W., 1959. Cast-in-situ bored piles in London clay. Geotechnique, 153–173.

Pile installation and verification

43

After installing piles, they need to be inspected and verified. Typically, an inspection firm would inspect the following items:

1. Straightness of the pile
2. Damage to the pile
3. Plumbness

43.1 Straightness of the pile

During installation, the pile can bend. This is known as dog legging (Figure 43.1).

A flashlight attached to a rope is lowered to the pile to find out whether the pile is bent (Figure 43.2).

If the flashlight is not visible from the top, it means that the pile is bent.

Also some inspectors drop a 4–5 ft. long rebar attached to a rope into the pile. The length of the rope is the same as the length of the pile. If the rebar goes all the way to the bottom of the pile, then the pile has no bend. If the rebar gets stuck in the pile, then the pile is bent.

In a straight pile, rebar will go all the way to the bottom of the pile as shown in Figure 43.3. This can be verified by measuring the length of the rope. In a bent pile, the rebar get stuck in the middle (Figure 43.4). More rope would be left on the ground level.

43.2 Damage to the pile

Other than the bends, piles also could be damaged. If a pile is damaged, water would go into the pile (Figure 43.5).

If water is detected inside the pile, the pile is damaged. Make sure that rain water had not gone into the pile. Typically, the top of the piles are capped to avoid rain water from entering the pile.

43.3 Plumbness of piles

Plumbness of a pile can be assessed using a plumb meter (Figure 43.6). The plumb meter is attached to the side of the pile. It would give the angle to the vertical.

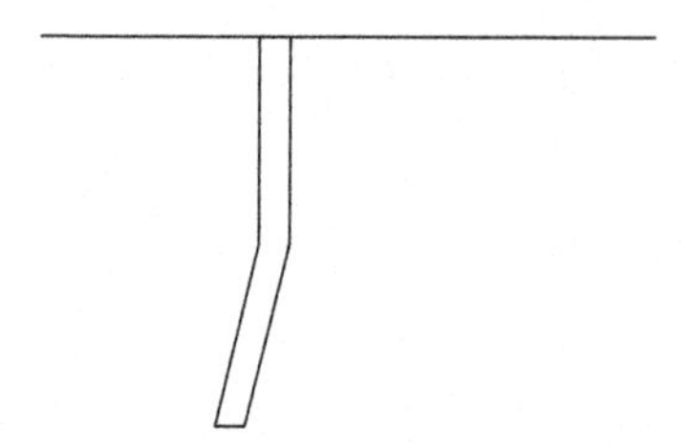

Figure 43.1 Dog legging or bending of piles.

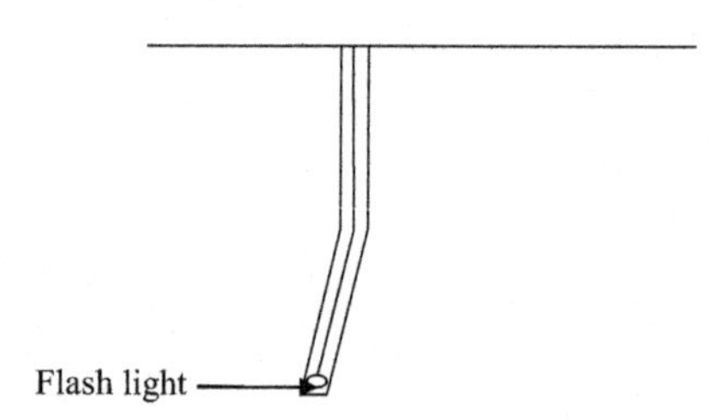

Figure 43.2 Flashlight is lowered to the bottom of the pile.

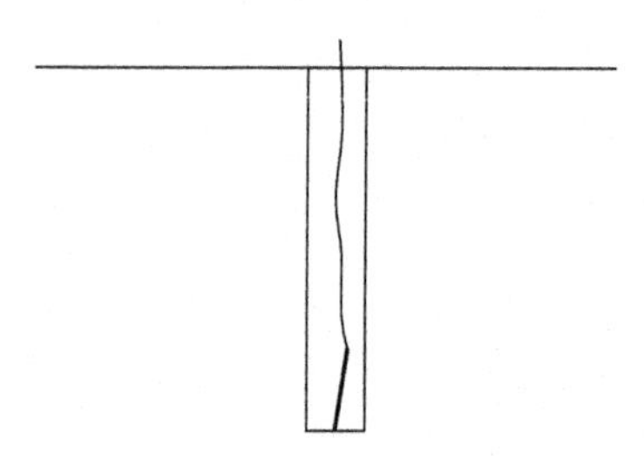

Figure 43.3 In a straight pile, rebar goes all the way to the bottom.

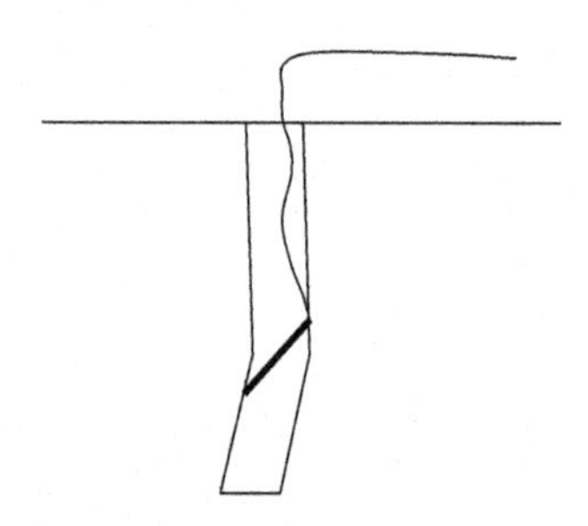

Figure 43.4 In a bent pile, rebar gets stuck in the bent.

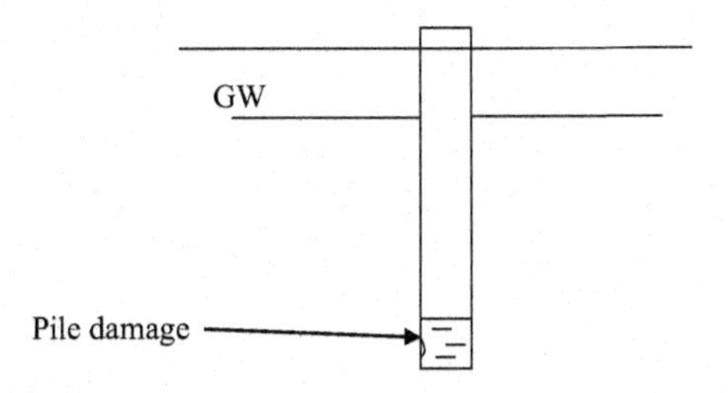

Figure 43.5 Damaged pile with water at the bottom.

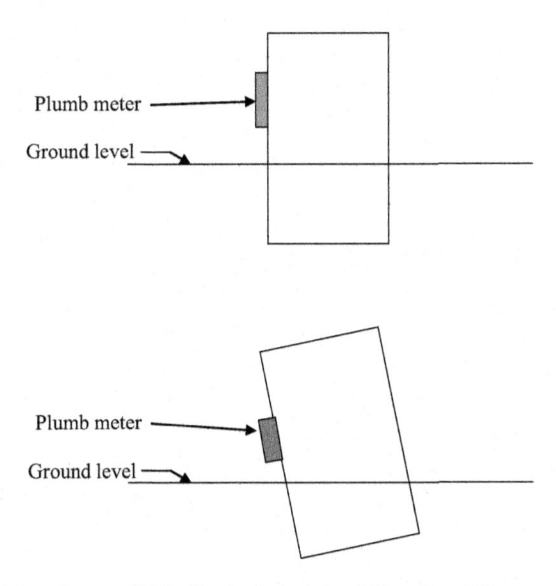

Figure 43.6 Plumb meter would indicate the angle of the pile with regard to the vertical.

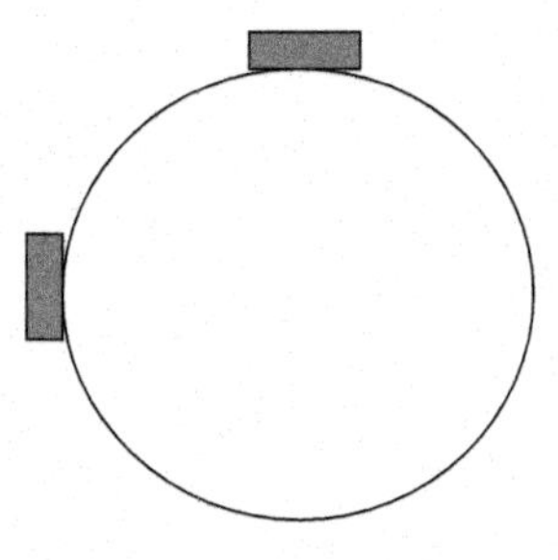

Figure 43.7 Plan view of a pile.

Plumb meter should be attached to the pile on one side and then it should be attached to another side to assess the plumbness in the opposite direction.

Attach the plumb meter to two sides as shown in Figure 43.7. This way plumbness on either direction can be found out.

Design of pin piles – semiempirical approach

44.1 Theory

- A few decades ago, no engineer would have recommended any pile less than 9 in. in diameter. Today, some piles could be as small as 4 in. in diameter.
- These small diameter piles are known as "*Pin Piles*". Other names such as *Mini piles*, *Micro piles*, *GEWI piles*, *Pali radice*, *Root piles*, and *Needle piles* are also used to describe the small-diameter piles.

44.1.1 Construction of a pin pile

Drill a hole using a roller bit or an auger (Figure 44.1a). A casing is installed to stop the soil from dropping into the hole. If the hole is steady, a casing may not be necessary. Tremie grout the hole (Figure 44.1b). Insert the reinforcement bar or bars. In some cases more than one bar may be necessary (Figure 44.1c). Lift the casing and pressure grout the hole (Figure 44.1d). Fully remove the casing (Figure 44.1e). Some Engineers prefer to leave part of the casing. This would increase the strength and the cost.

44.2 Concepts to Consider

- *Drilling the hole* (Figure 44.1a): Auguring could be a bad idea for soft clays, since augers tend to disturb the soil more than roller bits. This would decrease the bond between the soil and grout and would lower the skin friction. Drilling should be conducted with water. Drilling mud should not be used. Since casing is utilized to stop soil from falling in. Hence, drilling mud is not necessary. Drilling mud would travel into the pores of the surrounding soil. When drilling mud occupies the pores, grout may not spread into the soil pores. This would reduce the bond between grout and soil.
- *Tremie grouting* (Figure 44.1b): After the hole is drilled, tremie grout is placed. Tremie grout is a cement–water mix (typically with a water content ratio of 0.45–0.5 by weight).
- *Placement of reinforcement bars* (Figure 44.1c): High strength reinforcement bars (or number of bars wrapped together) or steel pipes can be used at the core. High strength reinforcement bars specially designed for pin piles are available from *Dywidag Systems International* and *William Anchors* (leaders in the industry). These bars can be spliced easily, so that any length can be accommodated.

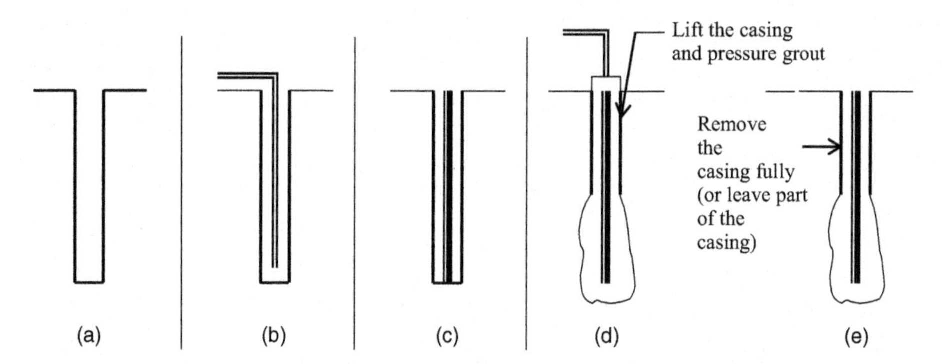

Figure 44.1 Construction of a pin pile. (a) Drill a hole and install a casing. (b) Tremie grout the hole. (c) Insert the reinforcement bar or cage. (d) Lift the casing and pressure grout the hole. (e) Casing removal.

- *Lift the casing and pressure grout* (Figure 44.1d): During the next step, the casing is lifted and pressure grouted. The pressure should be adequate to force the grout into the surrounding soil to provide a good soil–grout bond. At the same time, pressure should not be large enough to fracture the surrounding soil. Failure of surrounding soil would drastically reduce the bond between grout and soil. This would decrease the skin friction. Typically, grout pressure ranges from 0.5 MPa (10 ksf) to 1.0 MPa (1 MPa = 20.9 Ksf).
- *Ground heave:* There are many instances where ground heave occurs during pressure grouting. This aspect needs to be considered during the design phase. If there are nearby buildings, action should be taken to avoid any grout flow into these buildings.
- *Remove the casing* (Figure 44.1e): At the end of pressure grouting, the casing is completely removed. Some engineers prefer to leave a part of the casing intact. Obviously this would increase the cost. The casing would increase the rigidity and the strength of the pin pile. Casing would increase the *lateral resistance* of the pile significantly. Further it is guaranteed, that there is a pin pile with a diameter not less than the diameter of the casing. Diameter of a pin pile could reduce due to soil encroachment into the hole. In some occasions, it is possible for the grout to spread into the surrounding soil and create an irregular pile.

The grout could spread in an irregular manner as shown in Figure 44.2. This could happen when the surrounding soil is loose. The main disadvantage of leaving the casing in the hole is the additional cost.

Figure 44.2 Irregular spread of grout.

44.2.1 *Design of pin piles in sandy soils*

Design Example 44.1

Compute the design capacity of the pin pile shown in Figure 44.3. The diameter is 6 in. The surrounding soil has an average SPT (N) value of 15.

Solution

Step 1: Ultimate unit skin friction in gravity grouted pin piles is given by (Suzuki et al., 1972)

$$\tau = 21 \times (0.007N + 0.12)\ \text{Ksf}$$

Note: The above equation has been developed based on empirical data.

N = SPT value; τ = unit skin friction in Ksf.

For N value of 15

$$\tau = 21 \times (0.007 \times 15 + 0.12)\ \text{Ksf}$$
$$\tau = 4.725\ \text{Ksf}$$

The diameter of the pile = 6 in. and the length of the grouted section is 13 ft.

Step 2:

$$\text{Skin friction below the casing} = \pi \times (6/12) \times 13 \times 4.725\ \text{Kips}$$
$$= 96.5\ \text{Kips} = 48.2\ \text{tons}$$

Note: Most engineers ignore the skin friction along the casing. End bearing capacity of pin piles is not significant in most cases due to low cross sectional area. If the pin piles are seated on a hard soil or rock stratum, the end bearing needs to be accounted.

Assume a factor of safety of 2.5

Allowable capacity of the pin pile = 48.2/(2.5) = 19 tons

Notes:

1. The Suzuki's equation was developed for pin piles grouted using gravity. For pressure grouted pin piles, the ultimate skin friction can be increased by 20–40%.
2. Littlejohn (1970) proposed another equation.

   τ = 0.21N Ksf (where "N" is the SPT value.)

   For N = 15, τ = 0.21 × 15 = 3.15 Ksf

   The Suzuki's equation yielded τ = 4.725 Ksf

For this case, Littlejohn's equation provides a conservative value for skin friction. If Littlejohn's equation was used for the above pin pile:

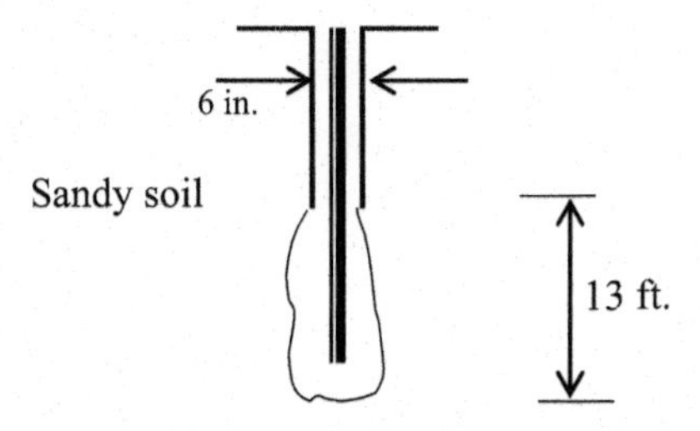

Figure 44.3 Pin pile in sandy soil.

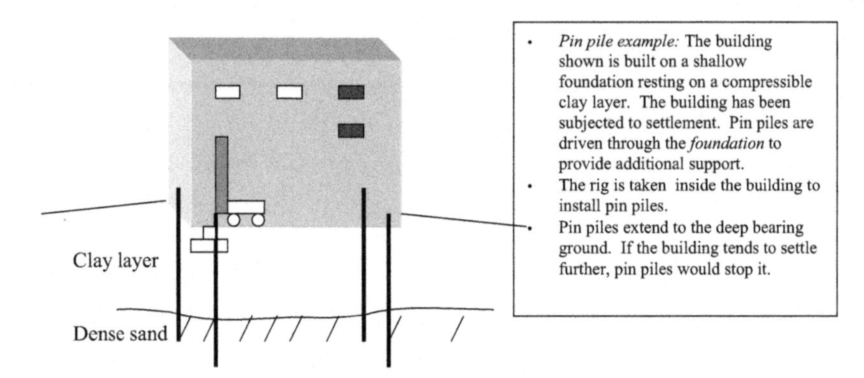

Figure 44.4 Pin piles located in dense sand.

Ultimate skin friction = $\pi \times (6/12) \times 13 \times 3.15 = 64.3$ Kips

Allowable skin friction = $64.3/2.5 = 25$ Kips = 12.5 tons

Suzuki's equation yielded 19 tons for the above pin pile (Figure 44.4).

References

Littlejohn, G.S., 1970. Soil Anchors. Ground Engineering Conference. Institution of Civil Engineers, London, pp. 33–44.

Suzuki et al., 1972. Developments Nouveaux danles Foundations de Plyons pour Lignes de Transport THT du Japon. Conf. Int. de Grand Reseaux Electriques a haute tension, paper 21-01, 13 pp.

Neutral plane concept and negative skin friction

45.1 Introduction

- During the pile driving process, the soil around the pile would be stressed.
- Due to the high stress generated in surrounding soil during pile driving, water is dissipated from the soil as shown in Figure 45.1.
- After pile driving is completed, water would start to come back to the previously stressed soil region around the pile (Figure 45.2).
- Due to migration of water, soil around the pile would consolidate and settle.
- It should be mentioned here that settlement of soil surrounding the pile is very small, yet large enough to exert a downward force on the pile. This downward force is known as "residual compression".
- Due to the downward force exerted on the pile, the pile would start to move downward.
- Eventually the pile would come to equilibrium and stop.
- At equilibrium, *upper* layers of soil would exert a downward force on the pile while *lower* layers of soil would exert an upward force on the pile.
- The direction of the force reverses at neutral plane.

45.1.1 Soil and pile movement: above the neutral plane

- Both soil and the pile move downward above the neutral plane. But the *soil* would move slightly more in downward direction than the *pile*. Hence relative to the pile, soil moves downward.
- This downward movement of soil relative to the pile, would exert a downward drag on the pile.

45.1.2 Soil and pile movement: below the neutral plane

- Both soil and the pile move downward below the neutral plane as in the previous case.
- But this time, pile would move slightly more in downward direction relative to the soil. Hence relative to the pile, soil moves upward.
- This upward movement of soil relative to the pile, would exert an upward force on the pile.

45.1.3 Soil and pile movement: at neutral plane

- Both soil and the pile move downward at the neutral plane as in previous cases.
- But at neutral plane both soil and pile move downward by the same margin.
- Hence relative movement between soil and pile is zero.
- At neutral plane, no force is exerted on the pile.

Geotechnical Engineering Calculations and Rules of Thumb

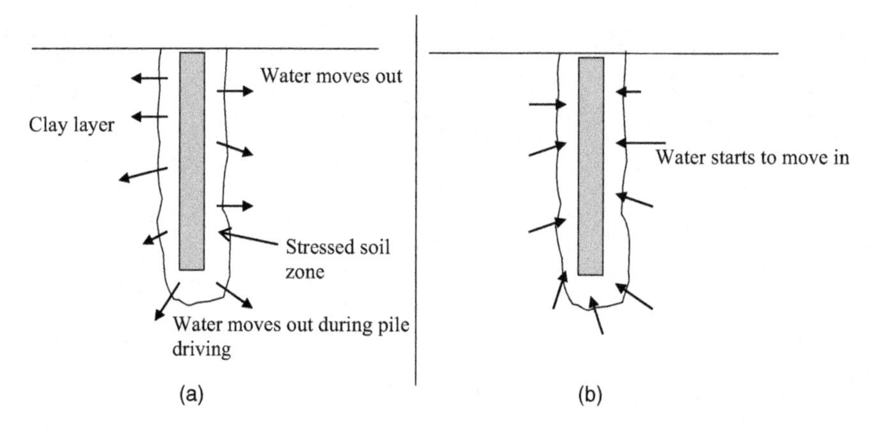

Figure 45.1 Water movement near a pile. (a) During pile driving. (b) After driving is completed.

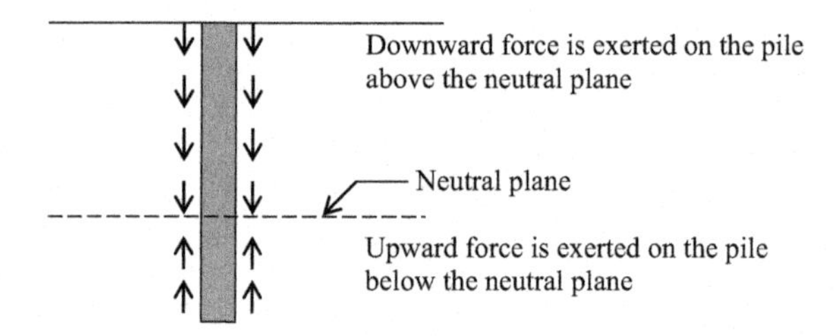

Figure 45.2 Neutral plane.

45.1.4 Location of the neutral plane

- Exact location of the neutral plane cannot be estimated without elaborate techniques involving complicated mathematics.
- For floating piles, neutral plane is taken at 2/3 of the pile length (Figure 45.3).
- In the case of full end bearing piles, (piles located on solid bedrock) neutral plane lies at the bedrock surface. For an upward force to be generated on the pile, the pile has to move downward relative to the surrounding soil.
- Due to the solid bedrock, the pile is incapable of moving downward. Hence the neutral plane would lie at the bedrock surface.

45.2 Negative skin friction

Negative skin friction is the process where skin friction acts downward due to soil. When skin friction is acting downwards it would decrease the pile capacity.

Negative skin friction (also known as downdrag), could be a major problem in some sites.

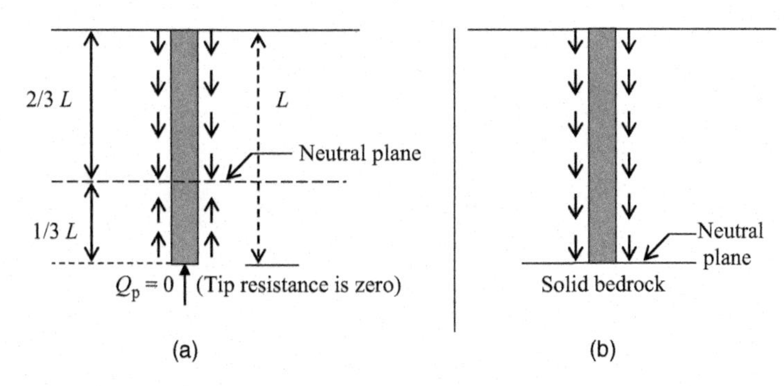

Figure 45.3 Neutral plane location. (a) Floating pile (zero end bearing). (b) Full end bearing pile.

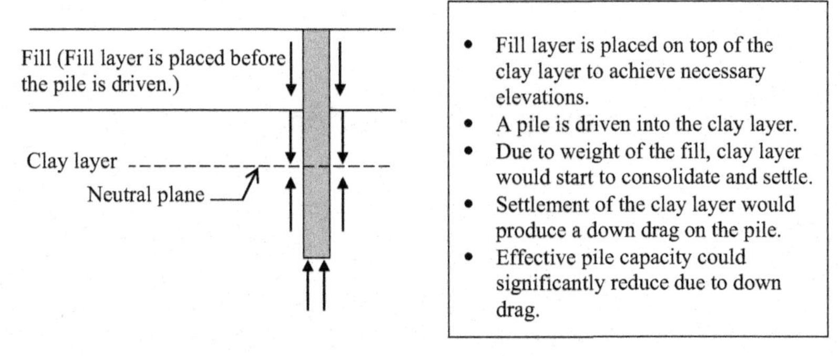

Figure 45.4 Negative skin friction.

Causes of negative skin friction are as follows:

- *Placement of fill*: Fill needs to be placed to gain required elevation for the floor slab. Additional fill would consolidate clay soils underneath. Settling clays would drag the pile down.
- *Change in groundwater level*: When groundwater level in a site goes down, buoyancy force would diminish. Hence the effective stress in clay would increase causing the clay layer to settle. Settling clay layer would generate a negative skin friction on the pile (Figure 45.4).
- In most situations, fill has to be placed to achieve necessary elevations. Weight of the fill would consolidate the clay layer.
- Generally the piles are driven after placing the fill layer.
- The settlement of the clay layer (due to the weight of the fill) would induce a downdrag force on the pile.
- If the piles were driven after full consolidation of the clay layer has occurred, no downdrag would occur.
- Unfortunately full consolidation may not be completed for years and many developers may not wait that long to drive piles.

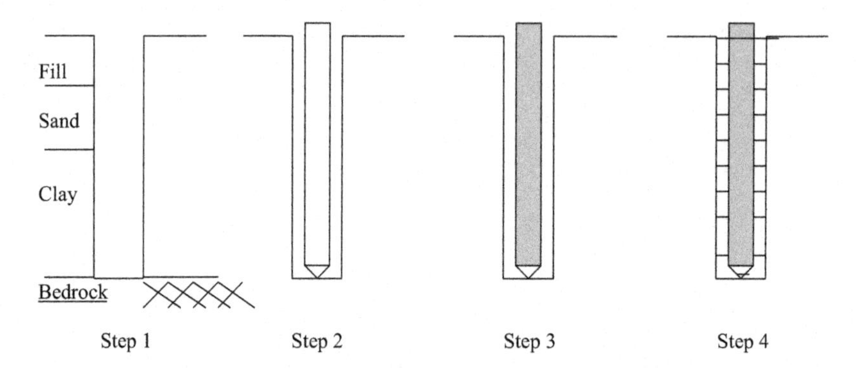

Figure 45.5 Bitumen coated pile installation. Step 1: A hole is bored. Step 2: The bitumen coated pipe pile is placed. Step 3: The pipe pile is concreted. Step 4: The annular space is grouted.

45.3 Bitumen coated pile installation

- Negative skin friction can be effectively reduced by providing a bitumen coat around the pile (Figure 45.5).

45.3.1 How bitumen coating would work against downdrag

Figure 45.6 shows bitumen coated pile mechanism.

45.3.1.1 Typical situation

Figure 45.7 shows bitumen coated pile mechanism.

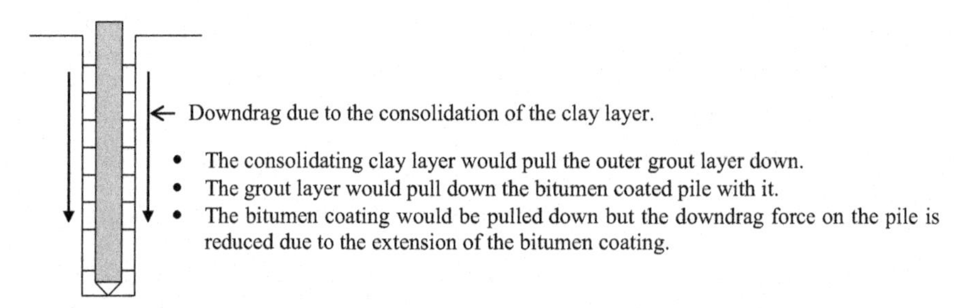

Figure 45.6 Bitumen coated pile mechanism.

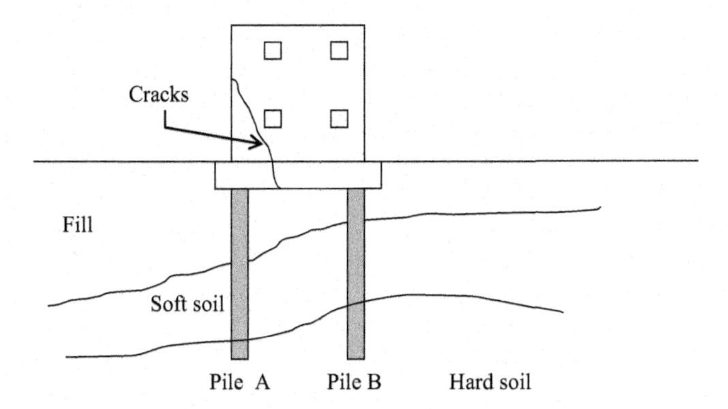

Figure 45.7 Negative skin friction due to fill. New building construction.

Note: due to larger fill load acting on left half of the building, negative skin friction acting on pile A would be greater than pile B. Hence pile A, would undergo larger settlement than pile B. Tension cracks could form as shown due to differential settlement.

Design of caissons

46

The term "caisson" is normally used to identify large bored concrete piles. Caissons are constructed by drilling a hole in the ground and filling it with concrete. A reinforcement cage is placed prior to concreting. The diameter of caissons could be as high as 15 ft.

Following is a list of other commonly used names to identify caissons.

1. Drilled shafts
2. Drilled caissons
3. Bored concrete piers
4. Drilled piers

A bell can be constructed at the bottom to increase the capacity. Caissons with a bell at the bottom are known as belled piers, belled caissons, or under-reamed piers.

46.1 Brief history of caissons

It was clear to many engineers that large diameter piles were required to transfer heavy loads to deep bearing stratums. The obvious problem was the inability of piling rigs to drive these large diameter piles.

The concept of excavating a hole and filling with concrete was the next progression. The end bearing of the caisson was improved by constructing a bell at the bottom.

Early excavations were constructed using steel rings and timber lagging. This was an idea borrowed from the tunneling industry (Figure 46.1).

46.2 Machine digging

With time, hand digging was replaced with machine digging. The following two types of machines are in use:

- Auger types
- Bucket types

Auguring is the most popular method to excavate for caissons.

46.3 Caisson design in clay soil

The equations for caissons in clay soils are similar to that of piles.

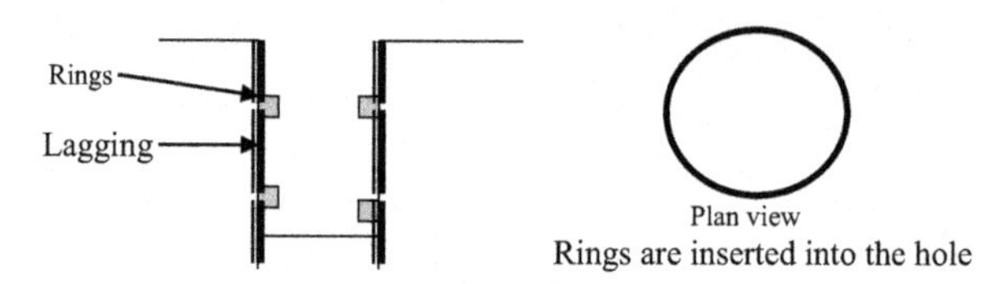

Figure 46.1 Ring structures.

46.3.1 Different methods

Ultimate caisson capacity = ultimate end bearing capacity + ultimate skin friction

$$P_u = Q_u + S_u$$

P_u, ultimate capacity of the caisson; Q_u, ultimate end bearing capacity; S_u, ultimate skin friction.

End bearing capacity (Q_u) = $9cA$

c, cohesion; A, cross-sectional area

Skin friction = $\alpha c A_p$

"α" is obtained from the Table 46.1.

c, cohesion; A_p, perimeter surface area.

Notes:

1. *Filter cake*: When drilling mud is used, a layer of mud gets attached to the soil. This layer of mud is known as the filter cake. In some soils, this filter cake gets washed away during

Table 46.1 "α" value and limiting skin friction

Caisson construction method	α (for soil to concrete)	Limiting unit skin friction (f)
		kPa
Uncased caissons		
1. Dry or using lightweight drilling slurry	0.5	90
2. Using drilling mud where filter cake removal is uncertain (see Note 1)	0.3	40
3. Belled piers on about same soil as on shaft sides		
a. By dry method	0.3	40
b. Using drilling mud where filter cake removal is uncertain	0.15	25
4. Straight or belled piers resting on much firmer soil than around shaft	0.0	0
5. Cased caissons (See Note 2)	0.1–0.25	

concreting. In some situations, it is not clear whether the filter cake will be removed during concreting. If the filter cake does not get removed during concreting, the bond between concrete and soil will be inferior. Hence, a low "α" value is expected.
2. *Cased caissons*: In the case of cased caissons, the skin friction would be between soil and steel casing. "α" is significantly low for steel–soil bond as compared to concrete–soil bond.

Factor of safety: Due to the uncertainties of soil parameters, such as cohesion and friction angle, the factor of safety should be included. The suggested factor of safety values in literature ranges from 1.5 to 3.0. Factor of safety of a caisson affects the economy of a caisson significantly. Assume that the ultimate capacity was found to be 3000 kN and a factor of safety of 3.0 was used. In this case, the allowable caisson capacity would be 1000 kN. On the other hand, a factor of safety of 2.0 would give an allowable caisson capacity of 1500 kN, or a 50% increase in the design capacity. A lower factor of safety value would result in a much more economical design. Yet on the other hand, failure of a caisson can never be tolerated. In the case of a pile group, the failure of one pile probably may not have a significant effect on the group. Caissons stand alone and are typically not used as a group.

When selecting the factor of safety, the following procedure is suggested.

- Make sure that enough borings are conducted so that soil conditions are well known in the vicinity of the caisson.
- Sufficient unconfined strength tests should be conducted on all different clay layers to obtain an accurate value of for cohesion.
- Sufficient borings with SPT values should be conducted to obtain the friction angle of sandy soils.

If the subsurface investigation program is extensive, lower factor of safety of 2.0 could be justified. If not, a factor of safety of 3.0 may be the safest bet. On the other hand, as a middle ground, a factor of safety of 3.0 for skin friction and a factor of safety of 2.0 for end bearing can be used.

Weight of the caisson: Unlike in piles, the weight of the caisson is significant and needs to be considered for the design capacity.

$$P_{\text{allowable}} = Q_{\text{u}}/\text{FOS} + S_{\text{u}}/\text{FOS} - W + W_{\text{s}}$$

Q_{u}, ultimate end bearing capacity; S_{u}, ultimate skin friction; $P_{\text{allowable}}$, allowable column load; W, weight of the caisson; W_{s}, weight of soil removed.

Logically, the weight of the soil removed during construction of the caisson should be reduced from the weight of the caisson (Figure 46.2).

Ignore skin friction on top and bottom of the shaft: Reese and O'Neill suggest to ignore the skin friction at top 5 ft. (1.5 m) of the caisson in clay since the load tests suggested that the skin friction in this region is negligible. For the skin friction to mobilize, there needs to be a slight relative movement between caisson and the surrounding soil. The top 5 ft. of the caisson does not have enough relative movement between the caisson and the surrounding soil. It has been reported that the clay at the top 5 ft. can desiccate and crack and result in negligible skin friction. At the same time, the bottom of the shaft will undergo very little relative displacement in relation to the

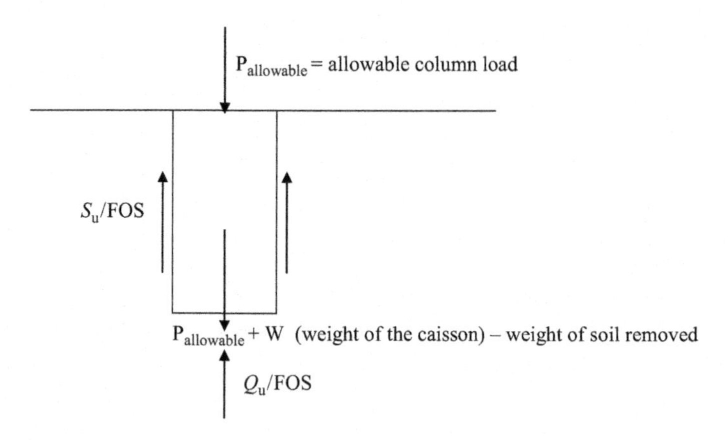

Figure 46.2 Forces on caissons.

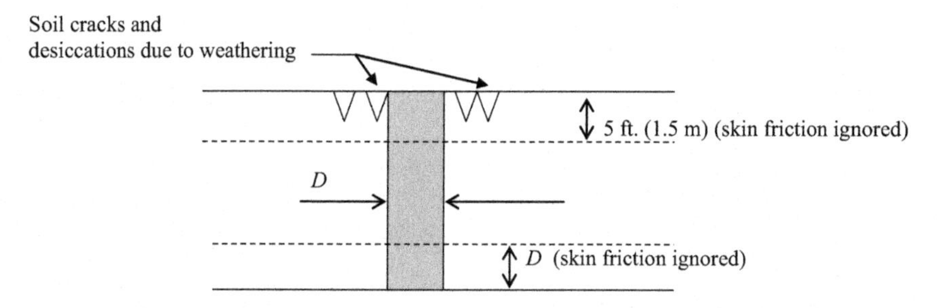

Figure 46.3 Desiccations on top of the caisson.

soil as well. Hence, Reese and O'Neill suggest to ignore the skin friction at the bottom distance equal to the diameter of the caisson as well.

Figure 46.3 shows the desiccations on top of the caisson.

D = diameter of the shaft

Ignore the skin friction in

- Top 5 ft. (1.5 m)
- The bottom distance equal to the diameter of the shaft (D).

American Association of State Highway and Transportation Officials (AASHTO method): AASHTO proposes the following method for caisson design in clay soils.

$$P_u = Q_u + S_u$$

P_u, ultimate capacity of the caisson; Q_u, ultimate end bearing capacity; S_u, ultimate skin friction.

End bearing capacity (Q_u) = $9cA$

c, cohesion; A, cross-sectional area.

Skin friction = $\alpha c A_p$

Table 46.2 Adhesion factor and maximum allowable skin friction

Location along drilled shaft	Value of α	Maximum allowable unit skin friction
From ground surface to depth along drilled shaft of 5 ft.	0	0
Bottom 1 diameter of the drilled shaft	0	0
All other portions along the sides of the drilled shaft	0.55	5.5 ksf (253 kPa)

Source: AASHTO (American Association of State Highway and Transportation Officials).

"α" adhesion factor is obtained from the Table 46.2.
c, cohesion; A_p, perimeter surface area.
It should be mentioned here that the AASHTO method has only one value for "α".

Design Example 46.1: Caisson design in single clay layer

Find the allowable capacity of a 2 m diameter caisson placed at 10 m below the surface (Figure 46.4). The soil was found to be clay with a cohesion of 50 kPa. The caisson was constructed using dry method. The density of soil = 17 kN/m^3. Groundwater is at 2 m below the surface.

Solution

Step 1: Find the end bearing capacity.

$$P_u = Q_u + S_u$$

End bearing capacity (Q_u) = $9cA$

$$\text{End bearing capacity } (Q_u) = 9 \times 50 \times (\pi d^2)/4$$
$$= 9 \times 50 \times (\pi 2^2)/4 = 1413.8 \text{ kN (318 kips)}$$

Step 2: Find the skin friction.
The cohesion of soil and the adhesion factors may be different above groundwater level in the unsaturated zone. Unless the cohesion of the soil is significantly different above groundwater, such differences are ignored.

Skin friction = $\alpha c A_p$

$\alpha = 0.5$ (From Table 46.1. The caisson was constructed using the dry method)

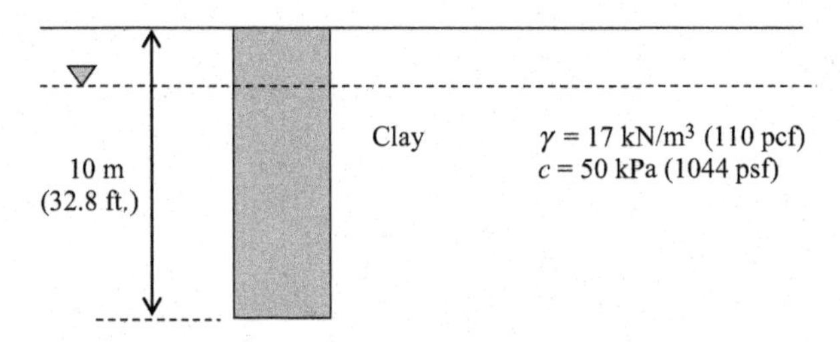

Figure 46.4 Caisson in clay.

$A_p = \pi \times d \times L = \pi \times 2 \times L$
Total length of the shaft = 10 m
Ignore the top 1.5 m of the shaft and bottom "D" of the shaft for skin friction.
Diameter of the shaft is 2 m.
Effective length of the shaft = 10 − 1.5 − 2 = 6.5 m (21.3 ft.)
$A_p = \pi \times d \times L = \pi \times 2 \times L = 40.8\ m^2$.
Cohesion, $c = 50$ kPa

$$\text{Skin friction} = 0.5 \times 50 \times 40.8\ \text{kN}$$
$$= 1020\ \text{kN} \quad (229\ \text{kips})$$

Step 3: Find the allowable caisson capacity:
$P_{allowable} = S_u/FOS + Q_u/FOS$ − weight of the caisson + weight of soil removed
Weight of the caisson (W) = volume of the caisson × density of concrete
Assume the density of concrete to be 23.5 kN/m³.

$$\text{Weight of the caisson } (W) = (\pi \times D^2/4 \times L) \times 23.5\ \text{kN}.$$
$$= (\pi \times 2^2/4 \times 10) \times 23.5\ \text{kN}.$$

Weight of the caisson (W) = 738.3 kN (166 kips)
Assume a factor of safety of 3.0 for skin friction and 2.0 for end bearing.
$P_{allowable} = S_u/FOS + Q_u/FOS$ − weight of the caisson + weight of soil removed
$P_{allowable} = 1413.8/3.0 + 1020/2.0 - 738.3 = 243$ kN (55 kips)
In this example, the weight of soil removed has been ignored.

Note: the $P_{allowable}$ obtained in this case is fairly low. Piles may be more suitable for this situation. If there is stiffer soil available at a lower depth, the caisson can be placed at a much stronger soil. Another option is to consider a bell.

Design Example 46.2: Caisson design in multiple clay layers

Find the allowable capacity of a 1.5 m diameter caisson placed at 15 m below the surface (Figure 46.5). The top layer was found to be clay with a cohesion of 60 kPa and the bottom layer was found to have a cohesion of 75 kPa. The top clay layer has a thickness of 5 m. The caisson was constructed using drilling mud where it is not certain that the filter cake will be removed during

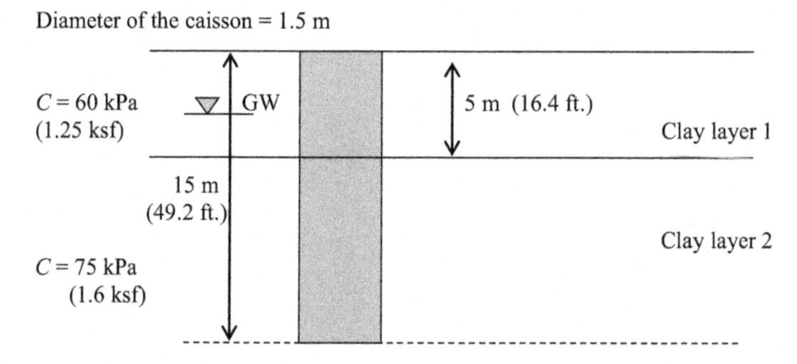

Figure 46.5 Caisson in multiple clay layers.

concreting. The density of soil = 17 kN/m^3 for both layers and groundwater is at 2 m below the surface.

Solution

Step 1: Find the end bearing capacity.

$$P_u = Q_u + S_u$$

End bearing capacity (Q_u) = 9cA

$$\text{Endbearing capacity } (Q_u) = 9 \times 75 \times (\pi \times d^2)/4$$
$$= 9 \times 75 \times (\pi \times 1.5^2)/4 = 1192.8\,\text{kN} \quad (268\,\text{kips})$$

Step 2: Find the skin friction in clay layer 1:

Skin friction = $\alpha c A_p$

α = 0.3 (From Table 46.1. Assume that the filter cake will be not be removed).

$A_p = \pi \times d \times L = \pi \times 1.5 \times L$

Ignore the top 1.5 m of the shaft and bottom "D" of the shaft for skin friction.

Effective length of the shaft in clay layer 1 = 5 − 1.5 = 3.5 m

$A_p = \pi \times d \times L = \pi \times 1.5 \times 3.5 = 16.5$ m^2.

Cohesion, c = 60 kPa

$$\text{Skin friction} = 0.3 \times 60 \times 16.5\,\text{kN}$$
$$= 297\,\text{kN} \quad (67\,\text{kips})$$

Step 3: Find the skin friction in clay layer 2:

Skin friction = $\alpha \times c \times A_p$

α = 0.3 (From Table 46.1. Assume that the filter cake will be not be removed).

$A_p = \pi \times d \times L = \pi \times 1.5 \times L$

Ignore the bottom "D" of the shaft.

Effective length of the shaft in clay layer 2 = 10 − diameter = 10 − 1.5 = 8.5 m

$A_p = \pi \times d \times L = \pi \times 1.5 \times 8.5 = 40.1$ m^2.

Cohesion, c = 75 kPa

$$\text{Skin friction} = 0.3 \times 75 \times 40.1\,\text{kN}$$
$$= 902.2\,\text{kN} \ (203\,\text{kips})$$

Total skin friction = 297 + 902.2 = 1199.2 kN

Step 4: Find the allowable caisson capacity:

$P_{allowable} = S_u$/FOS + Q_u/FOS − weight of the caisson + weight of soil removed

Weight of the caisson (W) = volume of the caisson × density of concrete

Assume density of concrete to be 23.5 kN/m^3.

$$\text{Weight of the caisson } (W) = (\pi \times D^2/4 \times L) \times 23.5\,\text{kN}.$$
$$= (\pi \times 1.52/4 \times 15) \times 23.5\,\text{kN}.$$

Weight of the caisson (W) = 622.9 kN

Assume a factor of safety of 3.0 for skin friction and 2.0 for end bearing.

$P_{allowable} = S_u$/FOS + Q_u/FOS − weight of the caisson

$P_{allowable}$ = 1,199.2/3.0 + 1,192.8/2.0 − 622.9 = 373.2 kN (84 kips)

Weight of soil removed is ignored in this example.

46.4 Meyerhoff's equation for caissons

End bearing capacity

Meyerhoff proposed the following equation based on SPT (N) value to compute the ultimate end bearing capacity of caissons. Meyerhoff's equation was adopted by DM 7.2 as an alternative method to static analysis.

$$q_{\text{ult}} = 0.13C_{\text{N}} \times N \times D / B$$

q_{ult}, ultimate point resistance of caissons (tsf); N, standard penetration resistance (blows/ft.) near pile tip; $C_N = 0.77 \log 20/p$; p, effective overburden stress at pile tip (tsf).

Note: "p" should be more than 500 psf. It is very rare for effective overburden stress at pile tip to be less than 500 psf.

D, depth driven into granular (sandy) bearing stratum (ft.); B, width or diameter of the pile (ft.); q_1, limiting point resistance (tsf), equal to 4 N for sand and 3 N for silt.

46.4.1 Modified Meyerhoff equation

Meyerhoff developed the above equation using many available load test data and obtaining average "N" values. Pile tip resistance is a function of the friction angle. For a given SPT (N) value, different friction angles are obtained for different soils.

For a given SPT (N) value, friction angle for coarse sand is 7–8% higher as compared to medium sand. At the same time for a given SPT (N) value, friction angle is 7–8% lower in fine sand as compared to medium sand. Hence, the author proposes the following modified equations:

$$q_{\text{ult}} = 0.15C_{\text{N}} \times N\, D / B \,\text{tsf (coarse sand)}$$
$$q_{\text{ult}} = 0.13C_{\text{N}} \times N\, D / B \,\text{tsf (medium sand)}$$
$$q_{\text{ult}} = 0.12C_{\text{N}} \times N\, D / B \,\text{tsf (fine sand)}$$

Design Example 46.3

Find the tip resistance of the 4 ft. diameter caisson shown in Figure 46.6 using the modified Meyerhoff's equation. SPT (N) value at caisson tip is 15 blows per foot.

Solution

Pile capacity comes from tip resistance and skin friction. In this example, only the tip resistance is calculated.

Step 1:

Ultimate point resistance for driven piles for fine sand = $q_{\text{tip}} = 0.12\ C_N \times N \times D/B$ tsf (fine sand)

$C_N = 0.77 \log 20/p$

p = Effective overburden stress at pile tip (tsf)

$p = 5 \times 110 + 22 \times 115 = 3080$ psf = 1.54 tsf (147 kPa)

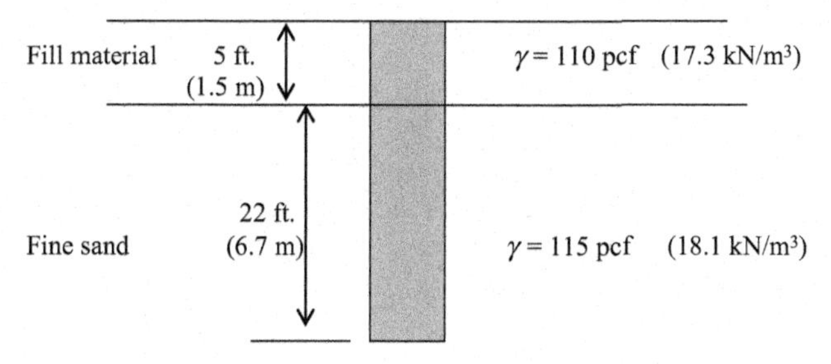

Figure 46.6 Caisson in fine sand.

$C_N = 0.77 \log [20/1.54] = 0.86$
D = Depth into bearing stratum = 22 ft. (6.7 m)
Fill material is not considered to be a bearing stratum.
B = 4 ft. (width or diameter of the pile)
$q_{tip} = 0.12\ C_N \times N \times D/B$ (fine sand)
$q_{tip} = 0.12 \times 0.86 \times 15 \times 22/4 = 8.5$ tsf (814 kPa)
Maximum allowable point resistance = $4 \times N$ tsf for sandy soils
$4 \times N = 4 \times 15 = 60$ tsf
Hence, $q_{tip} = 8.5$ tsf
Allowable point bearing capacity = 8.5/FOS
Assume a factor of safety of 3.0
Hence, the total allowable point bearing capacity,
$q_{tip \times allowable} = 2.84$ tsf (272 kPa)

$$q_{tip \times allowable} \times \text{tip area} = q_{tip \times allowable} \times \pi \times (4^2)/4$$
$= 36$ tons (320 kN)

Only the tip resistance was computed in this example.

Meyerhoff's equation for skin friction:

Meyerhoff proposes the following equation for skin friction for caissons.

$$f = N/100 \text{ tsf}$$

f, unit skin friction (tsf); N, average SPT (N) value along the pile.
Note: as per Meyerhoff, unit skin friction "f" should not exceed 1 tsf.
The author has modified Meyerhoff equations to account for soil gradation.

$f = N/92$ tsf (coarse sand)
$f = N/100$ tsf (medium sand)
$f = N/108$ tsf (fine sand)

Design Example 46.4

Find the skin friction of a 4 ft. diameter caisson shown in Figure 46.7 using Meyerhoff's equation. The average SPT (N) value along the shaft is 15 blows per foot. Ignore the skin friction in fill material.

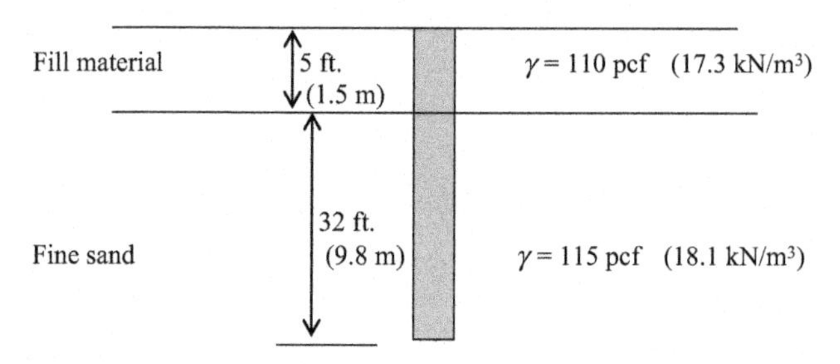

Figure 46.7 Caisson design using Meyerhoff method.

Solution

Only the skin friction is calculated in this example.

Step 1: For fine sand.

Unit skin friction; $f = N/108$ tsf

Unit skin friction; $f = 15/108$ tsf = 0.14 tsf

Total skin friction = unit skin friction × perimeter surface area

Total skin friction = $0.14 \times \pi \times D \times L$

Total skin friction = $0.14 \times \pi \times 4 \times 32 = 56$ tons (498 kN)

Allowable skin friction = 56/FOS

Assume a factor of safety (FOS) of 3.0

Allowable skin friction = 56/3.0 = 18.7 tons (166 kN)

AASHTO method for end bearing capacity: AASHTO adopts the method proposed by Reese and O'Neill.

The ultimate end bearing capacity in caissons, placed in sandy soils is given by:

Q_u, $q_t A$; $q_t = 1.20 \times N$ ksf ($0 < N < 75$); N, standard penetration test value (blows/ft.)

Metric units, $q_t = 57.5 \times N$ kPa ($0 < N < 75$)

A = Cross sectional area at the bottom of the shaft

For all "N" values above 75,

q_t = 4310 kPa (in metric units)

q_t = 90 ksf (in psf units)

Design Example 46.5: Caisson design in sandy soil

Find the ultimate capacity of a 2 m diameter caisson placed 10 m below the surface (Figure 46.8). The soil was found to be medium dense. The density of soil = 17 kN/m^3 and the friction angle of the soil is 30°.

Solution

$$P_u = Q_u + S_u$$

End bearing capacity (Q_u) = $(q_p/\alpha) \times A$

Skin friction (S_u) = $K \times p_v \times \tan\delta. (A_p)$

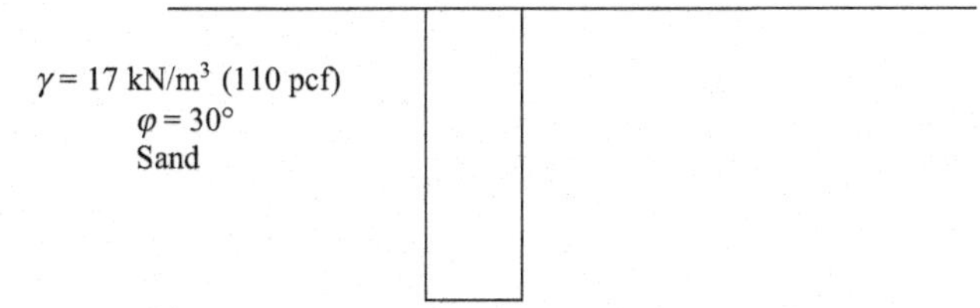

Figure 46.8 Caisson in sandy soil.

Step 1: Find the end bearing capacity.
End bearing capacity (Q_u) = (q_p/α) × A_p
q_p = 1600 kPa from the Table 46.1 for medium dense sand.
α = 2.0 B for SI units = 2.0 × 2 = 4
A, cross-sectional area of the caisson = ($\pi D^2/4$); L = ($\pi 2^2/4$) = 3.14 m^2.
End bearing capacity (Q_u) = (1600/4) × 3.14 kN = 1256.6 kN (282 kips)
Step 2: Find the skin friction.
Skin friction = Kp_v × tanδ × (A_p)
K = 0.6 (From Table 46.1 provided by Reese. The depth to base is 10 m); p_v, effective stress at the mid point of the caisson; p_v = 17 × 5.0 = 85 kN/m^2;
$\delta = \varphi$ for sandy soils; Hence, δ = 30°.
A_p = (πd) × L = (π2) × 10 = 62.8 m^2.
Ultimate skin friction = $Kp_v\tan\delta(A_p)$
Ultimate skin friction (S_u) = 0.6 × 85 × (tan30) × 62.8 = 1849 kN (416 kips)
Step 3: Find the allowable caisson capacity.
$P_{allowable}$ = S_u + Q_u/FOS − weight of the caisson + weight of soil removed
Weight of the caisson (W) = volume of the caisson × density of concrete
Assume the density of concrete to be 23.5 kN/m^3.

$$\text{Weight of the caisson } (W) = (\pi \times D^2/4L) \times 23.5 \text{ kN.}$$
$$= (\pi \times 2^2/4L) \times 23.5 \text{ kN.}$$

Weight of the caisson (W) = 738.3 kN
Assume a factor of safety of 3.0
$P_{allowable}$ = S_u + Q_u/FOS − weight of the caisson + weight of soil removed
$P_{allowable}$ = 1849 + 1256.6/3.0 − 738.3 = 1530 kN (344 kips)
Weight of soil removed is ignored in this example.

Design Example 46.6: Caisson design in sandy soil – multiple sand layers

Find the ultimate capacity of a 2 m diameter caisson placed 12 m below the surface (Figure 46.9). Soil strata are as shown in the figure. Groundwater is 2 m below the surface. Find the allowable capacity of the caisson.

Solution

$$P_u = Q_u + S_u$$

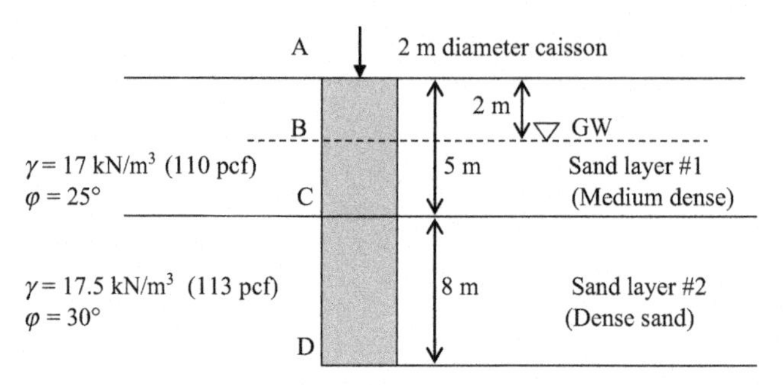

Figure 46.9 Caisson in multiple sand layers.

End bearing capacity $(Q_u) = (q_p/\alpha) \times A$
Skin friction $(S_u) = Kp_v \tan\delta(A_p)$

Step 1: Find the end bearing capacity.
End bearing capacity $(Q_u) = (q_p/\alpha) \times A$
q_p = 4000 kPa from Table 46.1 for dense sand.
$\alpha = 2.0$ B, for SI units $= 2.0 \times 2 = 4$
A = cross-sectional area of the caisson $= (\pi D^2/4)\ L = (\pi 2^2/4) = 3.14\ \text{m}^2$.
End bearing capacity $(Q_u) = (4000/4) \times 3.14$ kN = 3140 kN (706 kips)

Step 2: Find the skin friction in sand layer #1 (above groundwater) (A to B):
Skin friction $= Kp_v \tan\delta(A_p)$
$K = 0.5$ (the depth to base is 13 m, hence, from Table 46.1, $K = 0.5$); p_v, effective stress at the mid point of the sand layer 1 from A to B.
$p_v = 17 \times 1 = 17\ \text{kN/m}^2$.
$\delta = \varphi$ for sandy soils; Hence, $\delta = 25^\circ$ for sand layer 1.
$A_p = (\pi d) \times L = (\pi 2) \times 2 = 12.6\ \text{m}^2$.
Ultimate skin friction $= Kp_v \tan\delta(A_p)$
Ultimate skin friction (S_u) (A to B) $= 0.5 \times 17 \times (\tan 25) \times 12.6 = 49.9$ kN

Step 3: Find the skin friction in sand layer #1 (below groundwater) (B to C):
Skin friction $= Kp_v \tan\delta(A_p)$
$K = 0.5$ (the depth to base is 13 m, hence, from Table 46.1, $K = 0.5$); p_v = effective stress at the mid point of the sand layer 1 from B to C.
$p_v = \gamma \times 2 + (\gamma - \gamma_W) \times 1.5 = 17\ \text{kN/m}^2$.
$p_v = 17 \times 2 + (17 - 9.8) \times 1.5 = 44.8\ \text{kN/m}^2$.
$\delta = \varphi$ for sandy soils; Hence, $\delta = 25^\circ$ for sand layer 1.
$A_p = (\pi d) \times L = (\pi 2) \times 2 = 12.6\ \text{m}^2$.
Ultimate skin friction $= Kp_v \tan\delta(A_p)$
Ultimate skin friction (S_u) (B to C) $= 0.5 \times 44.8 \times (\tan 25) \times 12.6 = 131.6$ kN (29.7 kips)

Step 4: Find the skin friction in sand layer #2 (below groundwater) (C to D):
Skin friction $= K.p_v \tan\delta(A_p)$
$K = 0.5$ (the depth to base is 13 m, hence, from Table 46.1, $K = 0.5$); p_v = effective stress at the midpoint of the sand layer 2 from C to D.
$p_v = \gamma_1 \times 2 + (\gamma_1 - \gamma_W) \times 3 + (\gamma_2 - \gamma_W) \times 4 = 17\ \text{kN/m}^2$.
$p_v = 17 \times 2 + (17 - 9.8) \times 3 + (17.5 - 9.8) \times 4 = 86.4\ \text{kN/m}^2$. (1804 psf)
$\delta = \varphi$ for sandy soils; Hence, $\delta = 30^\circ$ for sand layer 2.
$A_p = (\pi d) \times L = (\pi 2) \times 8 = 50.3\ \text{m}^2$.
Ultimate skin friction $= Kp_v \tan\delta(A_p)$

Ultimate skin friction (S_u) (C to D) = 0.5 × 86.4 × (tan30) × 50.3 = 1254 kN
Total skin friction:
Skin friction from A to B = 49.9
Skin friction from B to C = 131.6
Skin friction from C to D = 1254
Total skin friction from A to D = 49.9 + 131.6 + 1254 = 1436 kN (322 kips)

Step 3: Find the allowable caisson capacity:

$P_{allowable} = S_u + Q_u/FOS$ − weight of the caisson + weight of the soil removed
Weight of the caisson (W) = volume of the caisson × density of concrete
Assume density of concrete to be 23.5 kN/m^3.

$$\text{Weight of the caisson } (W) = (\pi \times D^2/4 \times L) \times 23.5 \text{ kN.}$$
$$= (\pi \times 2^2/4 \times 13) \times 23.5 \text{ kN.}$$

Weight of the caisson (W) = 959.8 kN (216 kips)
Assume a factor of safety of 3.0 for skin friction and 2.0 for end bearing.
$P_{allowable} = S_u/FOS + Q_u/FOS$ − weight of the caisson
$P_{allowable}$ = 1436/3.0 + 3140/2.0 − 959.8 = 1088 kN (244 kips)
Weight of soil removed was ignored in this example.

46.5 Belled caisson design

Belled caissons are used to increase the end bearing capacity. Unfortunately, one loses the skin friction in the bell area since experiments have shown that the skin friction in the bell is negligible. In addition Reese and O'Neill and ASHTO suggest to exclude the skin friction in a length equal to the shaft diameter above the bell (see Figure 46.10).

Belled caissons are usually placed in clay soils. It is almost impossible to create a bell in sandy soils. On rare occasions, belled caissons are constructed in sandy soils with special equipment.

Belled caissons are used to increase the end bearing capacity. Unfortunately, one loses the skin friction in the bell area and one diameter above the bell.

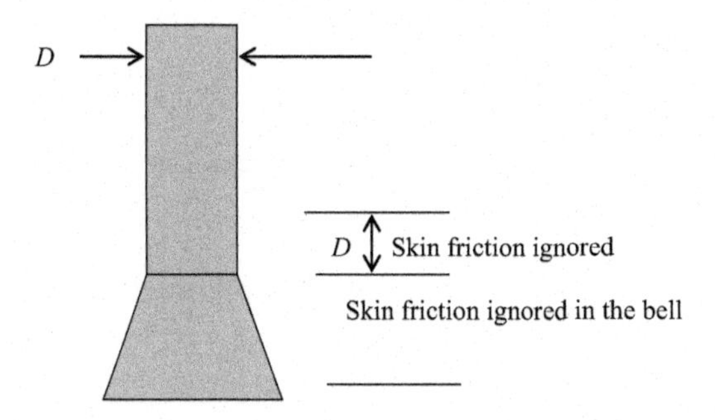

Figure 46.10 Belled caisson.

In Figure 46.10,

D = diameter of the shaft.

Skin friction in the bell and one diameter of the shaft (D) above the bell is ignored.

The ultimate capacity of belled caissons are given by the following equation.

Ultimate capacity of the belled caisson = ultimate end bearing capacity + ultimate skin friction

$$P_u = Q_u + S_u$$

Q_u = 9*c*bottom cross-sectional area of the bell; *c*, cohesion.

Area of the bell = $\pi \times d_b^2/4$ (d_b = bottom diameter of the bell)

$S_u = \alpha \times c \times$ perimeter surface area of the shaft (ignore the bell)

$S_u = \alpha \times c \times (\pi \times d \times L)$

d, diameter of the shaft; *L*, length of the shaft portion; $\alpha = 0.55$ for all soil conditions unless found by experiments (AASHTO).

Design Example 46.7

Find the allowable capacity of the belled caisson shown in Figure 46.11. The diameter of the bottom of the bell is 4 m and the height of the bell is 2 m. Diameter of the shaft is 1.8 m and the height of the shaft is 10 m. Cohesion of the clay layer is 100 kN/m². Adhesion factor (α) was found to be 0.55. Ignore the skin friction in the bell and one diameter of the shaft above the bell.

Solution

Step 1: Ultimate caisson capacity (P_u) = $Q_u + S_u - W$

$$\begin{aligned} Q_u &= \text{Ultimate end bearing capacity} \\ &= 9 \times c \times (\text{area of the bottom of the bell}) \\ &= 9 \times 100 \times (\pi \times 4^2/4) = 11{,}310\ \text{kN}\ (2{,}543\ \text{kips}) \end{aligned}$$

S_u = ultimate skin friction

W = weight of the caisson

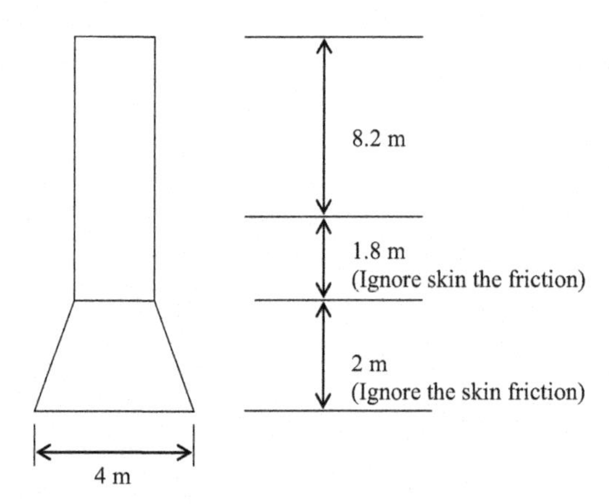

Figure 46.11 Skin friction in belled caisson.

Step 2: Find the ultimate skin friction.

Ultimate skin friction (S_u) = $\alpha \times c \times (\pi \times d \times L)$

Ultimate skin friction (S_u) = $0.55 \times 100 \times (\pi \times 1.8 \times 8.2) = 2550$ kN (573 kips)

Height of the shaft is 10 m and the length equal to one diameter of the shaft above the bell is ignored. Hence, the effective length of the shaft is 8.2 m.

Step 3: Find the weight of the caisson.

Assume the density of concrete to be 23 kN/m^3.

$$\text{Weight of the shaft} = (\pi \times d^2/4) \times 10 \times 23$$
$$= (\pi \times 1.8^2/4) \times 10 \times 23\,\text{kN} = 585.3\,\text{kN} \qquad (131\,\text{kips})$$

Find the weight of the bell.

Average diameter of the bell (d_a) = (1.8 + 4)/2 = 2.9 m

Use the average diameter of the bell (d_a) to find the volume of the bell.

The volume of the bell = $\pi \times d_a^2/4 \times h$

h = height of the bell.

$\pi \times 2.9^2/4 \times 2 = 13.21$ m^3.

Weight of the bell = volume × density of concrete = 13.21 × 23 = 303.8 kN (68.3 kips)

Step 4:

Ultimate caisson capacity (P_u) = $Q_u + S_u$ − weight of the caisson + weight of soil removed

Allowable caisson capacity = 11,310/FOS + 2318/FOS − 585.3 − 303.8

Assume a factor of safety of 2.0 for end bearing and 3.0 for skin friction. Since the weight of the caisson is known fairly accurately no safety factor is needed. Weight of soil removed is ignored in this example.

Allowable caisson capacity = 11,310/2.0 + 2550/3.0 − 585.3 − 303.8

Allowable caisson capacity = 5,615 kN (1,262 kips)

Design Example 46.8

Find the allowable capacity of the shaft in Design Example 46.7, assuming a bell was not constructed (Figure 46.12). Assume that the skin friction is mobilized in the full length of the shaft.

Solution

Step 1: Ultimate end bearing capacity (Q_u) = $9 \times c \times (\pi \times 1.8^2/4)$

Since, there is no bell, new diameter at the bottom is 1.8 m.

Ultimate end bearing capacity (Q_u) = $9 \times 100 \times (\pi \times 1.8^2/4) = 2{,}290$ kN (515 kips)

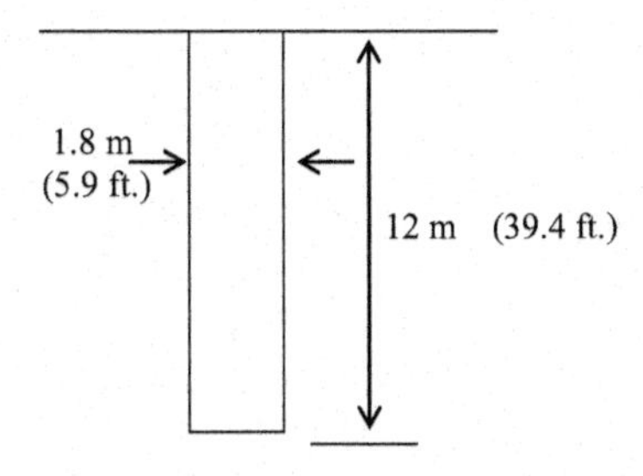

Figure 46.12 Caisson without a bell.

Step 2: Ultimate skin friction (S_u) = $\alpha \times c \times (\pi \times 1.8) \times 12$

New height of the shaft is 12 m since a bell is not constructed.

Ultimate skin friction (S_u) = $0.55 \times 100 \times (\pi \times 1.8) \times 12 = 3{,}732$ kN (839 kips)

Note: assumption is made that the skin friction of the shaft is mobilized along the full length of the shaft for 12 m.

Step 3: Find the weight of the shaft;

W = weight of the shaft = $(\pi \times d^2/4) \times L \times$ density of concrete

W = weight of the shaft = $(\pi \times 1.8^2/4) \times 12 \times 23 = 702.3$ kN (158 kips)

Step 4: Find the allowable caisson capacity.

Allowable caisson capacity = Q_u/FOS + S_u/FOS − weight of caisson + weight of soil removed

Assume a factor of safety of 2.0 for end bearing and 3.0 for skin friction. Since the weight of the caisson is known fairly accurately no safety factor is needed.

$$\begin{aligned}\text{Allowable caisson capacity} &= 2290/2.0 + 3732/3.0 - 702.3\ \text{kN}\\ &= 1686.7\ \text{kN}\end{aligned}$$

Note: weight of soil removed is ignored in this example.

In Design Example 46.7, we found the allowable capacity with the bell to be 5,615 kN, significantly higher than the straight shaft.

Design Example 46.9

Find the allowable capacity of the belled caisson shown in Figure 46.13. The diameter of the bottom of the bell is 4 m and the height of the bell is 2 m. Diameter of the shaft is 1.8 m and the height of the shaft is 11.8 m. Cohesion of the clay layer is 100 kN/m^2. Adhesion factor (α) was found to be 0.5. Ignore the skin friction in the bell and one diameter above the bell.

Solution

Step 1: Ultimate caisson capacity (P_u) = Q_u + S_u − weight of caisson + weight of soil removed

$$\begin{aligned}Q_u &= \text{ultimate end bearing capacity}\\ &= 9 \times c \times (\text{area of the bottom of the bell})\\ &= 9 \times 100 \times (\pi \times 4^2/4) = 11{,}310\ \text{kN (2,542 kips)}\end{aligned}$$

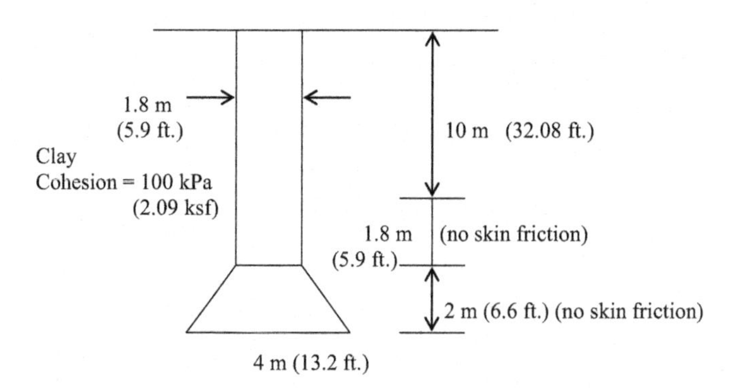

Figure 46.13 Belled caisson in clay.

S_u = ultimate skin friction

Step 2: Find the ultimate skin friction.

Ultimate skin friction (S_u) = $\alpha \times c \times (\pi \times d \times L)$

Ultimate skin friction (S_u) = $0.5 \times 100 \times (\pi \times 1.8 \times 10) = 2827$ kN (636 kips)

Skin friction in 1.8 m (length equal to diameter of the shaft) is ignored.

Step 3: Find the weight of the caisson.

Assume the density of concrete to be 23 kN/m^3.

$$\text{Weight of the shaft} = (\pi \times d^2 / 4) \times 11.8 \times 23$$
$$= (\pi \times 1.8^2 / 4) \times 11.8 \times 23\,\text{kN} = 690.6\,\text{kN} \quad (155\,\text{kips})$$

Find the weight of the bell.

Average diameter of the bell (d_a) = (1.8 + 4)/2 = 2.9 m (9.5 ft.)

Use the average diameter of the bell (d_a) to find the volume of the bell.

The volume of the bell = $\pi \times d_a^2/4 \times h$

h = height of the bell.

$\pi \times 2.9^2/4 \times 2 = 13.21$ m^3.

Weight of the bell = volume × density of concrete = 13.21 × 23 = 303.8 kN

Step 4: Ultimate caisson capacity (P_u) = Q_u + S_u − weight of caisson + weight of soil removed

Allowable caisson capacity = 11,310/FOS + 2827/FOS − 690.6 − 303.8

Note: weight of soil removed ignored in this example.

Assume a factor of safety of 2.0 for end bearing and 3.0 for skin friction. Since the weight of the caisson is known fairly accurately no safety factor is needed.

Allowable caisson capacity = 11,310/2.0 + 2827/3.0 − 690.6 − 303.8

Allowable caisson capacity = 5,602 kN (1,259 kips)

46.6 Caisson design in rock

46.6.1 Caissons under compression

Figure 46.14 shows caisson in rock.

Note: most of the load is taken by skin friction in rock. In most cases, end bearing is less than 5% of the total load. In other words, 95% or more load will be carried by skin friction.

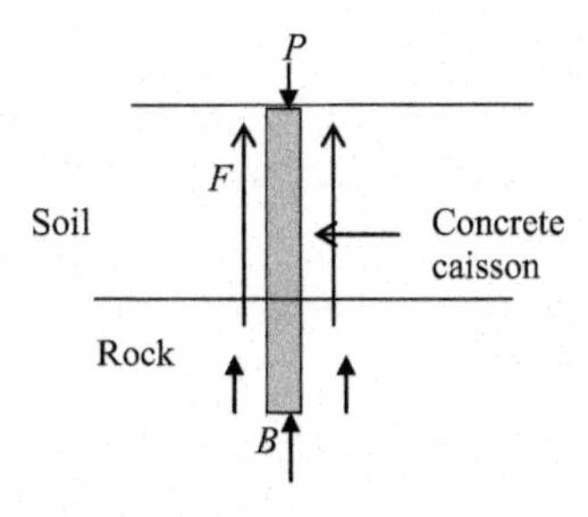

Figure 46.14 Caisson in rock.

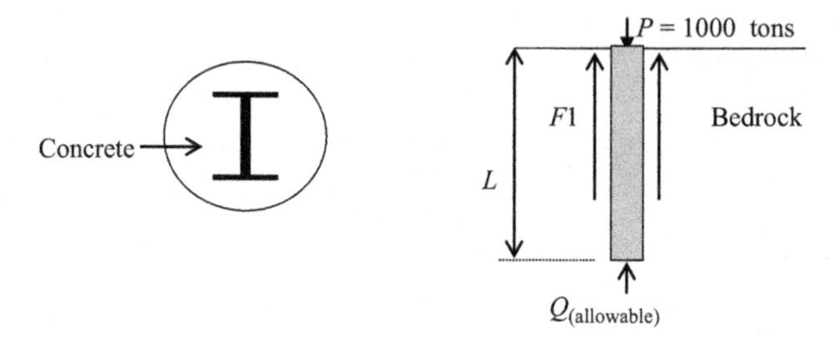

Figure 46.15 Concrete caisson with a W-section.

Design Example 46.10

Design a concrete caisson with a W-section (steel) at the core to carry a load of 1000 tons (Figure 46.15). Assume the skin friction to be 150 psi and end bearing to be 200 psi.

The following parameters are given:
Ultimate steel compressive strength = 36,000 psi
Ultimate concrete compressive strength = 3,000 psi

Simplified design procedure

Simplified design procedure is first explained. In this procedure, the composite nature of the section is ignored.

Step 1: Structural design of the caisson

Assume a diameter of 30 in. for the concrete caisson. Since, E (elastic modulus) of steel is much higher than concrete, a major portion of the load is taken by steel. Assume that 90% of the load is carried by steel.

Load carried by steel = 0.9 × 1000 tons = 900 tons
Allowable steel compressive strength = 0.5 × 36,000 = 18,000 psi
(NYC Building Code, Table 11.3 recommends a factor of safety of 0.5. Check your local building code for the factor of safety value).

$$\text{Steel area required} = \frac{900 \times 2{,}000}{18{,}000} = 100\ \text{in.}^2$$

Check the manual of steel construction for an appropriate W-section.

Use $W14 \times 342$. This section has an area of 101 in.^{2} The dimensions of this section are given in Figure 46.16.

Step 2: Check whether this section fits inside a 30 in. hole.

Distance along a diagonal (Pythagoras theorem) = $(16.36^2 + 17.54^2)^{1/2} = 23.98$ in.

This value is smaller than 30 in. Hence, the section can easily fit inside a 30 in. hole.

Step 3: Compute the load carried by concrete

- Concrete area = area of the hole − area of steel
  = 706.8 − 101 = 605.8
- Allowable concrete compressive strength = 0.25 × ultimate compressive strength
  = 0.25 × 3000 = 750 psi

  (NYC Building Code, Table 11.3 recommends a factor of safety of 0.25. Engineers should refer to local building codes for the relevant factor of safety values).

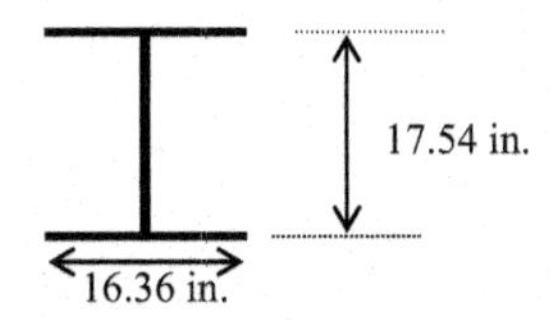

Figure 46.16 I-beam.

- Load carried by concrete = concrete area × 750 psi = 605.8 × 750 lbs = 227.6 tons
- Load carried by steel = 900 tons (computed earlier)
- Total capacity of the caisson = 900 + 227.6 = 1127.6 tons > 1000 tons

Note: the designer can start with a smaller steel section to optimize the above value.
Step 4: Compute the required length (L) of the caisson

- The skin friction is developed along the perimeter of the caisson.
- Total perimeter of the caisson = $\pi \times$ (diameter) × (length) = $\pi \times D \times L$
- Total skin friction = $\pi \times 30 \times L \times$ unit skin friction of rock; (in this case 150 psi)
- Total skin friction = $\pi \times 30 \times L \times 150$ lbs ("L" should be in inches)
- Design the caisson so that 95% of the load is carried by skin friction.

$$0.95 \times 1000 \text{ tons} = 950 \text{ tons}$$

- Hence, the total load carried through skin friction = 950 tons
  Total skin friction = $\pi \times 30 \times L \times 150$ = 950 tons = 950 × 2000 lbs
  (L) Length of the caisson required = 134 in. = 11.1 ft.

Design Example 46.11

The following parameters are given (Figure 46.17).
Caisson diameter = 4 ft.
Compressive strength of steel = 36,000 psi
$E_r/E_c = 0.5$

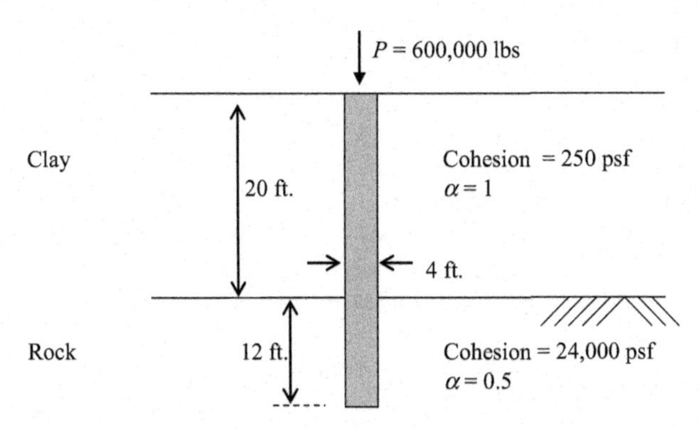

Figure 46.17 Caisson in rock (example).

(E_r = elastic modulus of rock; E_c = elastic modulus of concrete)
Cohesion of the bedrock = 24,000 psf
Adhesion coefficient (α) for rock = 0.5
Adhesion coefficient (α) for clay = 1.0

Solution

Step 1: Compute the ultimate end bearing capacity:

q_u = ultimate end bearing strength of the bedrock = $N_c \times$ cohesion
$N_c = 9$
Hence, $q_u = 9 \times$ cohesion = 216,000 psf
Ultimate end bearing capacity (Q_u) = area $\times q_u$
$Q_u = (\pi \times 4^2/4) \times 216{,}000 = 1{,}357$ tons
Allowable end bearing capacity ($Q_{allowable}$) = 1,357/3 = 452 tons

Step 2: Compute the ultimate skin friction:

Ultimate unit skin friction per unit area within the rock mass (f) = $\alpha \times C = \alpha \times 24{,}000$
Adhesion coefficient (α) = 0.5
Hence, ultimate unit skin friction (f) = $0.5 \times 24{,}000 = 12{,}000$ psf
Assuming an FOS of 3.0, the allowable unit skin friction per unit area within the rock mass [$f_{allowable}$] = 12,000/3.0 = 4,000 psf = 2 tsf

Step 3: Find the skin friction within the soil layer.

The skin friction generated within the soil layer can be calculated as in a pile.
Soil skin friction (f_{soil}) = αC
α = adhesion factor
(f_{soil}) = $1.0 \times 250 = 250$ psf
Skin friction mobilized along the pile shaft within the clay layer = (f_{soil}) $\times$ perimeter
$= 250 \times (\pi \times d) \times 20 = 62{,}800$ lbs
Allowable skin friction = 62,800/3.0 = 20,900 lbs = 10 tons
(Factor of safety of 3.0 is assumed).
Load transferred to the rock (F) = (P − 20,900) lbs = 600,000 − 20,900 = 579,100 lbs
Note: it is assumed that allowable skin friction within soil is fully mobilized.

Step 4: Load transferred to the rock:

The load transferred to the rock is divided between the total skin friction (F_{skin}) and the end bearing at the bottom of the caisson [Q_{base}].
(f_{skin}) = unit skin friction mobilized within the rock mass
(F_{skin}) = total skin friction = $f_{skin} \times$ perimeter area within the rock mass
(q_{base}) = end bearing stress mobilized at the base of the caisson.
(Q_{base}) = $q_{base} \times$ area of the caisson at the base
Step 5: Find the end bearing (Q_{base}) and skin friction (F_{skin}) within the rock mass
End bearing ratio (n) is defined as the ratio between the end bearing load of the rock mass (Q base) and the total resistive force mobilized ($Q_{base} + F_{skin}$) within the rock mass.
End bearing ratio (n) = $Q_{base}/[Q_{base} + F_{skin}]$
"n" is obtained from Table 46.3.
Q_{base}, end bearing load generated at the base; F_{skin}, total skin friction generated within the rock mass.
L, length of the caisson within the rock mass; a, radius of the caisson = 2 ft.

$$E_r / E_c = \frac{\text{Elastic modulus of rock}}{\text{Elastic modulus of concrete}} = 0.5 \text{ (given)}$$

Total load transferred to the rock mass = 579,100 lbs = $Q_{base} + F_{skin}$ (see Step 3)

Table 46.3 **End bearing ratio (n)**

$E_r/E_c = 0.5$		$E_r/E_c = 1.0$		$E_r/E_c = 2.0$		$E_r/E_c = 4.0$	
L/a	n	L/a	n	L/a	n	L/a	n
1	0.5	1	0.48	1	0.45	1	0.44
2	0.28	2	0.23	2	0.20	2	0.16
3	0.17	3	0.14	3	0.12	3	0.08
4	0.12	4	0.08	4	0.06	4	0.03

Source: Osterberg and Gill (1973).

Assume a length (L) of 8 ft. (Since, a = 2 ft., $L/a = 4$)
From Table 46.3 for L/a of 4 and E_r/E_c of 0.5, end bearing ratio (n) = 0.12

$$n = 0.12 = Q_{base} / (Q_{base} + F_{skin})$$
$$0.12 = Q_{base} / 579{,}100$$

Hence, $Q_{base} = 579{,}100 \times 0.12 = 69{,}492$ lbs = 35 tons
$Q_{allowable}$ = 452 tons (see Step 1)
$Q_{allowable}$ is greater than the end bearing load (Q_{base}) generated at the base.
F_{skin} = load transferred to the rock − end bearing load
$F_{skin} = 579{,}100 - 69{,}492 = 509{,}608$ lbs = 255 tons
F_{skin} should be less than $F_{allowable}$.
$F_{allowable} = f_{allowable} \times$ perimeter of the caisson within the rock mass.
$f_{allowable}$ = 2 tsf. (see Step 2).
Since a length (L) of 8 ft. was assumed within the rock mass
$F_{allowable} = 2 \text{ tsf} \times (\pi \times 4) \times 8 = 201$ tons
Skin friction generated = 255 tons (as calculated earlier)
$F_{allowable}$ is less than the skin friction generated. Hence, the pile diameter or length of the pile should be increased.

Reference

Osterberg, J.O., Gill, S.A., 1973. Load Transfer Mechanism for Piers Socketed in Hard Soils or Rock. In: Proceedings of the Canadian Rock Mechanics Symposium, 235–261, Montreal.

Design of pile groups

47

Typically, piles are installed in a group and provided with a pile cap. The column will be placed on the pile cap so that the column load is equally distributed among the individual piles in the group (Figure 47.1).

Capacity of a pile group is obtained by using an efficiency factor.

Pile group capacity = efficiency of the pile group × single pile capacity × number of piles

If the pile group contains 16 piles and capacity of a single pile is 30 tons and the group efficiency is found to be 0.9, the group capacity is 432 tons.

Pile group capacity = $0.9 \times 30 \times 16 = 432$ tons

It is clear that high pile group efficiency is desirable. The question is how to improve the group efficiency? The pile group efficiency is dependent on the spacing between piles. When the piles in the group are closer together, the pile group efficiency decreases. When the piles are placed far apart, the efficiency increases. When the piles are placed far apart, the size of the pile cap has to be larger increasing the cost of the pile cap.

47.1 Soil disturbance during driving

What happens in a pile group?

When piles are driven, the soil surrounding the pile will be disturbed (Figure 47.2). Disturbed soil has lesser strength than undisturbed soil. Some of the piles in the group are installed in partially disturbed soil causing them to have a lesser capacity than others. Typically, piles in the center are driven first.

Soil disturbance caused by one pile, impacts the capacity of the adjacent piles. The group efficiency can be improved by placing the piles at a larger spacing. In clay soils, the shear strength will be reduced due to disturbance.

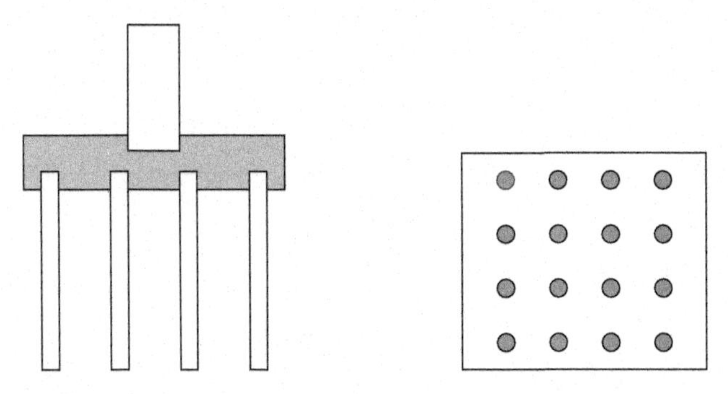

Figure 47.1 Pile group.

Geotechnical Engineering Calculations and Rules of Thumb

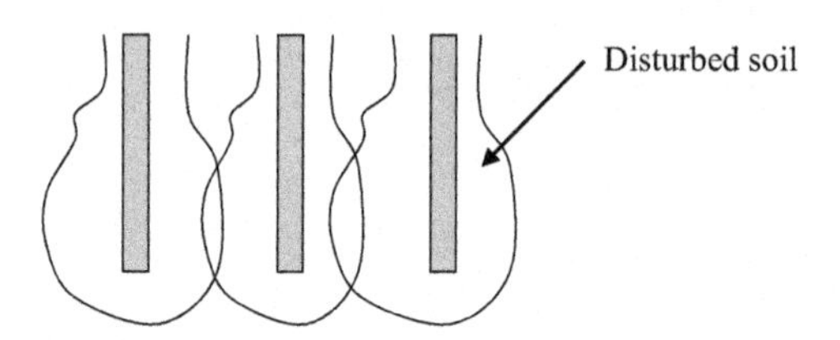

Figure 47.2 Disturbed soil due to driving.

47.2 Soil compaction in sandy soil

When driving piles in sandy soils, the surrounding soil will be compacted. Compacted soil tends to increase the skin friction of piles. Pile group placed in sandy soils may have a larger than one group efficiency. Soil compaction due to pile driving will be minimal in clay soils.

47.3 Pile bending

When driving piles in a group, some piles could be bent due to soil movement. This effect is more pronounced in clayey soils.

Assume that pile A is driven first and pile B is driven next. The soil movement caused by pile "B" can bend pile "A" as shown in Figure 47.3. This in return would create a lower group capacity.

47.4 End bearing piles

Piles, which mainly rely on end bearing capacity may not be affected by other piles in the group (Figure 47.4).

Piles ending in very strong bearing stratum or in rock: When piles are not dependent on the skin friction, a group efficiency of 1.0 can be used.

Various guidelines for computing group capacity are discussed.

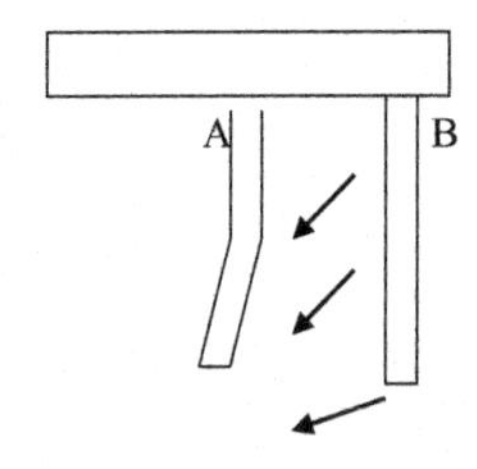

Figure 47.3 Pile bending due to hard driving.

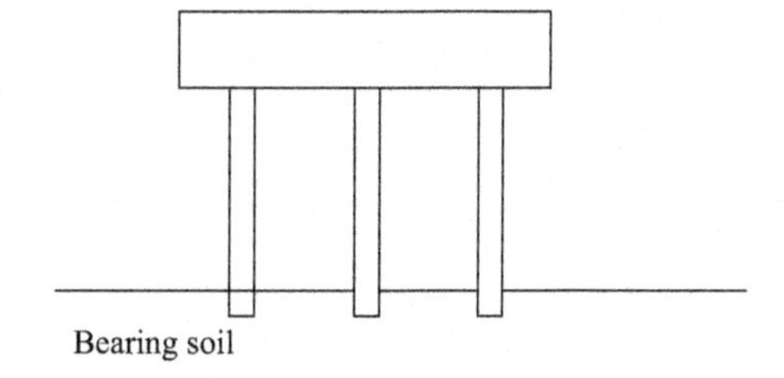

Figure 47.4 End bearing piles.

47.4.1 AASHTO (1992) guidelines

American Association of State Highway and Transportation (AASHTO) provides the following guidelines.

AASHTO considers three situations.

1. Pile group in cohesive soils (clays and clayey silts).
2. Pile group in noncohesive soils (sands and silts).
3. Pile group in strong soil overlying weaker soil.

Author has constructed Tables 47.1 and 47.2 using AASHTO guidelines for pile group efficiency in cohesive soils.

Design Example 47.1: Pile group capacity

A pile group is constructed in clayey soils as shown in Figure 47.5. Single pile capacity was computed to be 30 tons and each pile is 12 in. in diameter. The center to center distance of piles is 48 in. Find the capacity of the pile group using the AASHTO method.

Table 47.1 **Pile group efficiency for clayey soils**

Pile spacing (center to center)	Group efficiency
3*D*	0.67
4*D*	0.78
5*D*	0.89
6*D* or more	1.00
D = diameter of piles	

Source: AASHTO (1992).

Table 47.2 **Pile group efficiency for sandy soils**

Pile spacing (center to center)	Group efficiency
3*D*	0.67
4*D*	0.74
5*D*	0.80
6*D*	0.87
7*D*	0.93
8*D* or more	1.00
D = diameter of piles	

Source: AASHTO (1992).

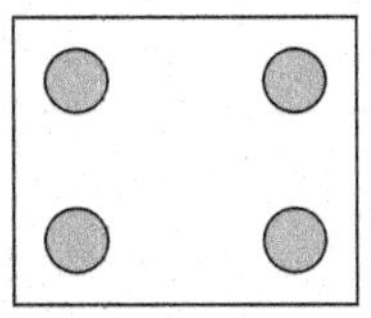

Figure 47.5 Pile group with four piles.

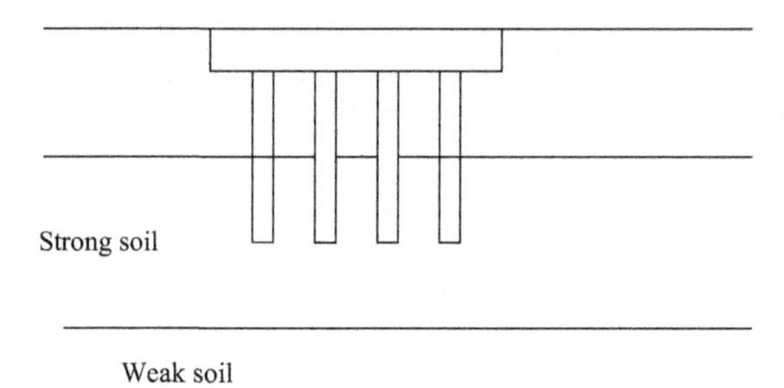

Figure 47.6 Pile group in strong soil overlying weak soil.

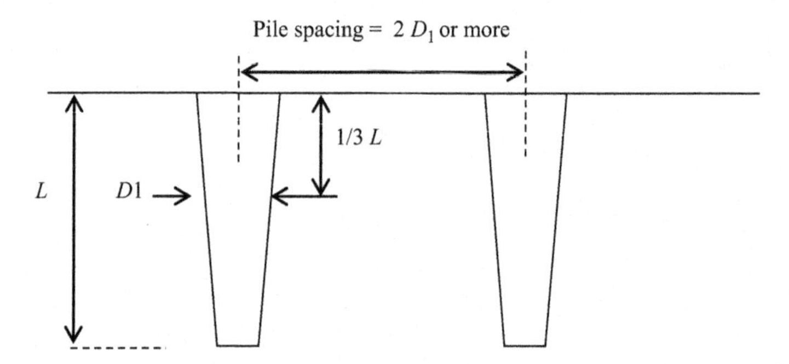

Figure 47.7 Pile spacing.

Solution

Pile group capacity = efficiency of the pile group × single pile capacity × number of piles

Center to center distance between piles = 48 in.

Since the diameter of piles is 12 in., the center to center distance is $4D$.

Efficiency of the pile group = 0.78 (AASHTO Table 47.1)

Pile group capacity = $0.78 \times 30 \times 4 = 93.6$ tons

Pile group capacity when strong soil overlying weaker soil (AASHTO):

Usually piles are ended in strong soils. In some cases, there could be a weaker soil stratum underneath the strong soil strata. In such situations, the settlement due to weaker soil underneath has to be computed (Figure 47.6).

Pile spacing (center to center distance): International Building Code Guidelines:

The minimum distance should not be less than 24 in. in any case (Figure 47.7)..

For circular piles: Minimum distance (center to center) = twice the average diameter of the butt.

Rectangular piles: Minimum distance (center to center) = ¾ times the diagonal for rectangular piles

Tapered piles: Minimum distance (center to center) = twice the diameter at 1/3 of the distance of the pile measured from the top of pile.

Reference

AASHTO, 1992. American Association of State Highway and Transportation.

Subject Index

S

V

W

Y

Printed in the United States
By Bookmasters